Pollutants and Their Ecotoxicological Significance

Pollutants and Their Ecotoxicological Significance

Edited by

H. W. Nürnberg
Institut für Angewandte Physikalische Chemie der Kernforschungsanlage Jülich, FRG

A Wiley–Interscience Publication

JOHN WILEY & SONS

Chichester · New York · Brisbane · Toronto · Singapore

Library of Congress Cataloging in Publication Data:
Main entry under title:

Pollutants and their ecotoxicological significance.

Chiefly English, some French.
"A Wiley–Interscience publication."
1. Pollutants—Environmental aspects. 2. Pollutants—Toxicology. I. Nürnberg, H. W.
QH545.A1P65 1984 628.5. 84–7540
ISBN 0 471 90509 7 (U.S.)

British Library Cataloguing in Publication Data:
Pollutants and their ecotoxicological significance.
1. Pollution
I. Nürnberg, H.W.
363.7′3 TD174

ISBN 0 471 90509 7

Printed in Great Britain by Galliard (Printers) Ltd, Great Yarmouth

Contents

Section II: Aquatic Environment

Section III: Terrestrial Environment and Man

Section IV: Regulatory and General Aspects

List of Contributors

R. AMAVIS	*Commission of the European Communities, Rue de la Loi 200, 1049 Brussels, Belgium*
H. P. BERTRAM	*Institut für Pharmakologie und Toxikologie, Westfälische Wilhelms Universität, Domagstrasse 12, 44 Münster, FRG*
C. J. BRANDT	*VGB-Technische Vereinigung der Grosskraftwerksbetrieber, Klinkestrasse 27–31, 4300 Essen 1, FRG*
R. BREDER	*Institut für Angewandte Physikalische Chemie der Kernforschungsanlage Jülich, Postfach 19 13, D-5170 Jülich 1, FRG.*
K. DELBEKE	*Vrije Universiteit Brussel, Laboratorium voor Ekologie en Systematiek, Pleinlaan 2, 1050 Brussels, Belgium*
R. G. DERWENT	*Environmental and Medical Sciences Division, AERE Harwell, Oxfordshire OX11 0RA, UK*
J. C. DUINKER	*Institut für Meereskunde an der Universität Kiel, Düsternbrooker Weg 20, D-2300 Kiel 1, FRG*
D. H. EHHALT	*Institut für Atmosphärische Chemie der Kernforschungsanlage Jülich GmbH, Postfach 19 13, D-5170 Jülich 1, FRG.*
I. ELSKENS	*Vrije Universiteit Brussel, Faculteit der Wetenschappen, Laboratorium voor Ekologie en system, Pleinlaan 2, 1050 Brussels, Belgium*
J. FAUCHERRE	*Université Paris VII, Laboratoire de Chimie Minérale, 2, Place Jussieu, Tour 45—5 étage, Paris cédex 05, France*
S. W. FOWLER	*International Atomic Energy Agency, International Laboratory of Marine Radioactivity, Musée Océanographique, Principality of Monaco*
F. FÜHR	*Institut für Radioagronomie der Kernforschungsanlage Jülich, Postfach 19 13, D-5170 Jülich 1, FRG.*

M. GELDMACHER-v. MALLINCKRODT — *Institut für Rechtsmedizin, der Universität Erlangen-Nürnberg, Universitätsstrasse 22, 852 Erlangen, FRG*

H-W. GEORGII — *Institut für Meteorologie und Geophysik der Johann Wolfgang Goethe-Universität, Feldbergstrasse 47, D-6000 Frankfurt-am-Main, FRG*

L. GOEYENS — *Vrije Universiteit Brussel, Faculteit der Wetenschappen, Laboratorium voor Ekologie en system, Pleinlaan 2, 1050 Brussels, Belgium*

G. GRIMMER — *Biochemisches Institut für Umweltkarcinogene, Siegerlandstrasse 19, D-2070 Ahrensburg, FRG*

C. JOIRIS — *Vrije Universiteit Brussel, Laboratorium voor Ekologie en Systematiek, Pleinlaan 2, 1050 Brussels, Belgium*

A. JUNOD — *Institut Suisse de météorologie, Station aérologique Payerne, 1530 Payerne, Switzerland*

W. KLEIN — *Gesellschaft für Strahlen- und Umweltforschung GmbH, Ingolstädter Landstrasse 1, D-8042 Neuherberg, FRG*

D. KLOCKOW — *Universität Dortmund, Abteilung Chemie, Otto-Hahn-Strasse, D-4600 Dortmund, FRG*

F. KORTE — *Gesellschaft für Strahlen- und Umweltforschung GmbH, Ingolstädter Landstrasse 1, D-8042 Neuherberg, FRG*

P. LEHMANN — *Institut Suisse de météorologie, Station aérologique Payerne, 1530 Payerne, Switzerland*

R. MARGALEF — *Universidad de Barcelona, Facultad de Biologia, Departmento de Ecologia, Barcelona, Spain*

P. MÜLLER — *Lehrstuhl für Biogeographie, Fachbereich 6: Geographie, Universität de Saarlands, D-6600 Saarbrücken, FRG*

H. W. NÜRNBERG — *Institut für Angewandte Physikalische Chemie der Kernforschungsanlage Jülich, Postfach 19 13, D-5170 Jülich 1, FRG.*

A. C. POSTHUMUS — *Research Institute for Plant Protection, Binnehaven 12, POB 42, Wageningen, the Netherlands*

B. PRINZ — *Landesanstalt für Immissionsschutz, Wallneyer Strasse 6, 43 Essen 1, FRG*

O. RAVERA — *Joint Research Centre, Department of Physical and Natural Sciences, 21020 Ispra (Varese), Italy*

P. RAVUSSIN — *Chaire de la Mecanique de la Turbulence, Ecole Polytechnique Federale de Lausanne, avenue de cour 33, CH-1007 Lausanne, Switzerland*

C. RONNEAU — *Laboratorie de Chimie Inorganique et Nucléaire, Chemin du Cyclotron 2, B-1348 Louvain-la-Neuve, Belgium*

U. SCHMIDT — *Institut für Atmosphärische Chemie der Kernforschungsanlage Jülich, Postfach 19 13, D-5170 Jülich I, FRG.*

F. SCHMIDT-BLEEK — *Chemicals Division, Environment Directorate, OECD, 2 Rue André Pascal, F-75 775 Paris cédex 16, France*

P. SCHMIDT DI FRIEDBERG — *Casa Postale 4100, Larog donegani 1.2, 20124 Milan, Italy*

J. SMEETS — *Commission of the European Communities, Rue de la Loi 200, 1049 Brussels, Belgium*

H. P. STIEF-TAUCH — *Commission of the European Communities, Rue de la Loi 200, 1049 Brussels, Belgium*

M. STOEPPLER — *Institut für Angewandte Physikalische Chemie der Kernforschungsanlage Jülich, Postfach 19 13, D-5170 Jülich I, FRG.*

Preface

Many thousands of chemicals have been and are released from a number of different anthropogenic sources into the environment. Among them is a large variety of organic substances predominantly strange to the environment and of purely man-made origin. Furthermore there are substantial anthropogenic emissions of certain hazardous inorganic substances, such as SO_2, NO_x and toxic heavy metals, chemicals which also occur naturally in the environment as a consequence of the action of natural phenomena, e.g. volcanism, bush fires, weathering. The introduction of chemicals not occurring naturally in the environment and the enhancement of levels of naturally existing organic and inorganic substances generally constitutes pollution of the affected ecosystems. In a number of cases pollution caused by human activities nevertheless remains bearable and tolerable for the respective aquatic or terrestrial ecosystem or for man as long as it does not exceed critical levels.

Certain organic substances, e.g. halocarbons, polyaromatic hydrocarbons, chlorinated pesticides, as well as inorganic chemicals, e.g. SO_2, NO_x, and hazardous heavy metals (cadmium, lead, mercury, chromium, nickel, arsenic and at higher doses a number of other metals) have already been shown to exert significant ecotoxicity partially, even at rather low levels. Research in environmental chemistry and ecotoxicology has been focused on these groups of organic and inorganic chemicals having either immediate ecotoxic effects or being the precursors of ecotoxic substances formed from them in the respective ecosystems.

The ecological chemistry of these ecotoxic substances is complex and their fate, transfer and pathways in the atmosphere, the various aquatic and terrestrial ecosystems and their numerous divergent components (water, suspended matter, sediments, plankton, aquatic organisms, plants, soil, animals) depend on a number of chemical, physical, hydrological, geological, biological and ecological parameters, specific for the type of ecosystem involved. This is well reflected by the treatment of these topics and the aforementioned categories of organic and inorganic chemical pollutants in the corresponding chapters of this book.

A frequently occurring difficulty to be mastered in ecochemical studies of these pollutants is to be emphasized in this context. Many hazardous substances exist in the ecosystems at trace levels as a consequence of their distribution after emission. Therefore, their ecochemistry is basically often demanding applied trace chemistry. This means that a number of special methodological problems and requirements must be scrupulously met if accurate and reliable results are to be obtained, which is a mandatory prerequisite for relevant judgments and conclusions on the ecotoxicity of the studied pollutant under the given ecological conditions.

The ecotoxic significance of chemical pollutants can be more specific as well as rather general. For instance under the given circumstances some will be more toxic for plants or microorganisms, while others—even environmental chemicals—will also have impacts on higher animals and man, causing not only toxic effects but also induced genetic damage.

In view of the high significance and interdisciplinary character of the subject it was the concept of this book to assemble a number of experts to deal with the various important aspects of the related ecochemistry and ecotoxicology of the known focal organic and inorganic pollutants in the atmosphere, hydrosphere, terrestrial environment and in man. This is supplemented by contributions on regulatory aspects and some general reflections relevant to the subject. The preparation of such a multi-authored text of course requires time. Yet the aim of this book was not to compile a report on the most recent results but to prepare a rather comprehensive treatment of the subject exemplified by instructive examples and based on the experience, expertise and significant findings of leading experts in the relevant disciplines of environmental science.

It is to be hoped that this book, which provides an overview of the methodology and scientific working philosophy for relevant studies on the ecochemical behaviour and ecotoxicological significance of recently investigated focal pollutants and on already well-established findings and results, will provide a stimulus for urgently needed future research on the pollution of the environment, an area which constitutes one of the major present problems for society and challenges to all branches of natural sciences.

The editor thanks all authors for their contributions and the publishers for their helpful and efficient cooperation. Special thanks go to his collaborator Dr R. Breder for preparing the subject index of the book and for his valuable support in the final proof-reading.

Hans Wolfgang Nürnberg
Jülich, October 1984

SECTION I: ATMOSPHERE

Pollutants and Their Ecotoxicological Significance
Edited by H. W. Nürnberg

CHAPTER 1

Atmospheric Pollution by SO_2 and Oxidants

H-W. GEORGII

SUMMARY

The following chapter reviews the sources, distribution and sinks of two important pollutants, namely SO_2 and photochemically produced oxidants. It also deals with the oxidation products of SO_2, the sulphate-aerosols and briefly with the main precursor of photo-oxidants, the nitrogen oxides. It can be shown that SO_2 as well as oxidants are transported over distances of several hundred kilometres, so that unpolluted areas are influenced by these pollutants. Under certain atmospheric situations, SO_2, sulphate aerosols and oxidants can be spread over wide regions of Western Europe. Therefore, international co-operation in this field is of great importance.

EMISSIONS

Anthropogeneous sources of atmospheric pollutants are mainly industry, domestic combustion of fossil fuel and automobile traffic. These man-made sources produce sulphur dioxide, carbon monoxide, hydrocarbons and particulate matter, besides other minor constituents. In this context we do not discuss carbon dioxide which is also produced by all combustion processes. Table 1.1 gives a survey of the most important emissions and the differentiation according to the different sources for the Federal Republic of Germany for the year 1976 according to a survey compiled by the Federal Environmental Agency.[1]

In this context we have to consider that automobile exhaust gases with a fraction of 50 per cent CO emissions, 25 per cent NO_x emissions and 19 per cent HC emissions are emitted directly above street level. Therefore, their proportion at a level of 1–2 m above the street is higher than their total contribution to air

Table 1.1 Anthropogeneous emissions in the Federal Republic of Germany (10^6 t yr^{-1})

	CO	SO_2	NO_x	Particulates
Automobile exhaust	7.0	0.1	0.4	0.1
Industry	—	2.9	1.0	0.6
Domestic combustion	1.5	0.6	—	0.1

pollution. During day-time the fraction of automobile exhaust gases in the streets of Cologne, related to total pollution, is estimated as follows: 94–99 per cent CO, 76–90 per cent HC and 83–93 per cent of NO_x.[2] According to a study by the Federal Environmental Agency it is expected that the emissions of particulate matter will decrease, the SO_2 emission will remain constant and the NO_x emission will further increase with the main source, the traffic in cities.

TRANSFORMATION IN THE ATMOSPHERE

Trace-constituents emitted from different sources have a different fate which is determined by turbulent diffusion and convective processes in the atmosphere, but also by sinks like incorporation into clouds and precipitation and by homogeneous and heterogeneous oxidation processes which lead to a transformation of the primary emission within the atmosphere. The rate of transformation, the number and distribution of sinks and the presence and concentration of reaction partners determine the tropospheric residence time of pollutants. These residence times are important for the spatial distribution of pollutants in a regional or global scale. In some cases of a global distribution of a pollutant an accumulation with a world-wide increase of the concentration is to be observed—the global increase of CO_2 or of chlorfluormethanes is mentioned in this connection. In the following chapters we shall mainly deal with two pollutants, the sulphur dioxide and its oxidation products and the products of the reaction of nitrogen oxides and reactive hydrocarbons under the influence of solar radiation, the so-called 'oxidants', also well known under the heading 'photochemical smog'.

SULPHUR DIOXIDE AND ITS OXIDATION PRODUCTS

The total sulphur emitted from anthropogeneous sources into the atmosphere has been increased during the last 30 years. Cullis and Hirschler[3] give a global increase of the SO_2 emissions from 72×10^6 t yr^{-1} in 1940 to 149×10^6 t yr^{-1} in 1965 and 187×10^6 t yr^{-1} in 1974. The man-made production of SO_2 is at present increasing at a rate of 2.2 per cent per year. Eighty per cent of the SO_2 emission in Europe is generated from combustion processes and the emission inventory of the OECD shows that, in Europe in 1970, about 50×10^6 t SO_2 y^{-1} are emitted into the atmosphere.[4] A comparison of SO_2 emissions in the northern and southern hemisphere show that 94 per cent of SO_2 are produced in the northern hemisphere. Only the emissions due to the smelting of non-ferrous ores are at all comparable in both hemispheres.

The dominant fraction of SO_2 is of anthropogeneous nature. This is the case for about 90 per cent of atmospheric SO_2. The fraction originating from H_2S or from organic sulphur compounds like COS or dimethylsulphide is regionally negligible. With respect to the immision concentration of SO_2 on the basis of continuous records gained in industrial and densely populated areas, we generally find a decrease of SO_2 concentration since the middle of the sixties. This is shown in Figure 1.1 for the city of Frankfurt-on-Main. The decrease of the SO_2 concentration is about 50 per cent of the value of the year 1963/64. It should be stressed that the immission concentration of NO_x has steadily increased during the same period according to measurements at the same measuring site.[5] This result has been confirmed in several industrial areas or large cities in Germany, but also in England, where in spite of increasing SO_2 emissions, the immission concentration dropped.[6] This evidence is mainly found near the large cities and the industrial regions. In contrast to that, in the so-called unpolluted regions, the SO_2 background concentration did not drop but remained constant or increased slightly. This impression can be gained when the measurements at the 'background-stations' of the German Environmental Agency are evaluated.[7] We can therefore summarize that the decreasing SO_2 concentration in cities is compensated by a constant or slightly increasing SO_2 concentration in rural

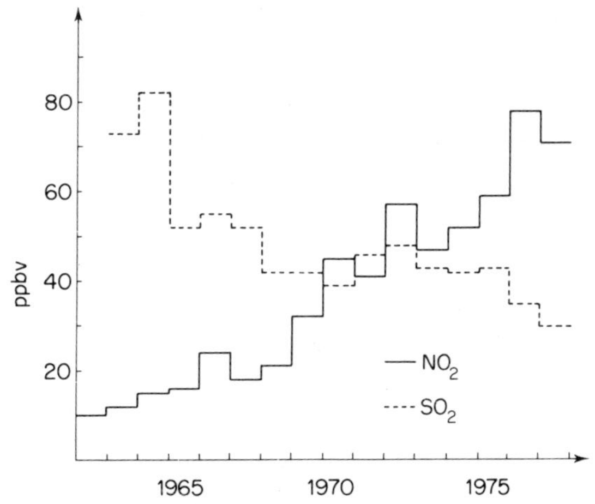

Fig. 1.1 Temporal trend of NO_2- and SO_2-concentration in Frankfurt-on-Main (according to Pilotstation of Federal Environmental Agency)

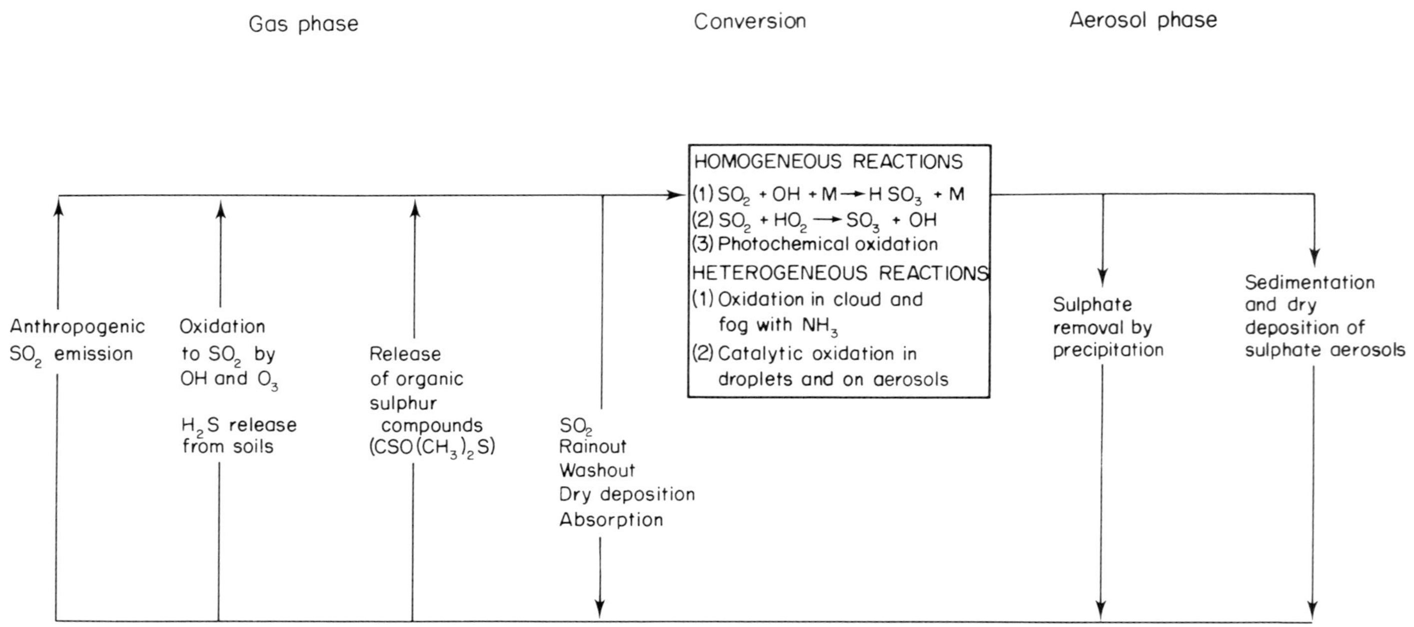

Fig. 1.2 Cycle of excess sulphur

areas. The decrease of the SO_2 concentration in polluted regions is generally explained by the construction of tall chimneys, but also by the burning of oil instead of coal and by the use of light oil and gas in domestic heating.

About 20–50 per cent of the atmospheric SO_2 is being transformed to sulphate particles within 1 or 2 days by homogeneous and heterogeneous oxidation processes. In this review, we have not the space to go into detail of the different transformation mechanisms. They are summarized in Figure 1.2 together with the main sources and sinks. Recent research on the homogeneous oxidation of SO_2 has been summarized by Calvert *et al.*,[8] while a state of art of heterogeneous SO_2 oxidation has been described by Beilke and Gravenhorst.[9] The oxidation product of SO_2 is a sulphate aerosol, the main cation being H^+ and NH_4^+. The size distribution of that aerosol shows that the main fraction consists of particles below 1 μ radius (Georgii and Vitze[10]). The tropospheric residence time of these sulphate particles is therefore approximately 1 week, while the SO_2 residence time is only 2–5 days. The longer residence time of the sulphate aerosols has the consequence that aerosols produced from the precursor SO_2 are noticeable within a radius of 2000 km around the source areas. This explains the more homogeneous distribution of sulphate aerosols with little spatial variations and the extension of aerosols into regions far from immediate source areas. The fact that aerosols are a secondary pollutant means that their concentrations tend to be more dependent on total emissions and less on the height of emission than is the case for SO_2.[11] Summarizing the results of sulphate analysis at several measuring sites in the United Kingdom it can be seen that the concentration of sulphate-containing aerosols increased during the last 20 years by a factor of about 2. The authors of that article stress that the increasing trend is still stronger when only the atmospheric sulphate concentration during the summer months is considered.[6] Homogeneous oxidation of SO_2 under the influence of solar radiation leads to the formation of small-sized particles which are thought to be responsible for the increasing trend of sulphate aerosols in air in Britain. It is supposed that heterogeneous oxidation in cloud droplets or catalytic oxidation on the surface of aerosol particles are of minor importance for the aerosol load of the atmosphere. This can also be concluded from the work by Penkett[12] or Runca-Köberich.[13]

Sources at the ground, the vertical wind and temperature profile and the depth of the mixing layer determine the vertical distribution of sulphur compounds in the atmosphere. In the case of intensive vertical mixing we should expect that production near the ground and dispersion in the atmosphere are equal leading to a constant mixing ratio with altitude. In the case of SO_2 and sulphates we have to consider sinks in the atmosphere leading to a decrease of the SO_2 concentration with altitude. This is also proved by aircraft ascents over Western Europe which show a decrease of the SO_2 concentration from an average of $20\,\mu g\,m^{-3}$ near the ground during winter and $10\,\mu g\,m^{-3}$ during summer, reaching $5\,\mu g\,m^{-3}$ in 2–3 km altitude and $1\,\mu g\,m^{-3}$ in 5 km altitude. The difference between winter and summer is only observed in the lower troposphere (Georgii[14]). Individual ascents show frequently a two-layered mass distribution of SO_2 with high concentrations within the mixing layer in the lower troposphere which is separated from the middle and upper troposphere by a temperature inversion. Above this inversion the SO_2 concentration drops very suddenly to background level. With respect to the vertical sulphate distribution it is an interesting fact that the sulphate concentration is more constant with altitude than the SO_2 concentration with the consequence that the $SO_4^{=}/SO_2$ ratio increases with increasing height (Meszaros[15]). This is caused partly by the gradual conversion of SO_2 to sulphate particles as well as by the difference of deposition velocity of gases and particles. Meszaros[16] has estimated the total burden of sulphur compounds in the atmosphere. Values for SO_2—S and $SO_4^{=}$—S are shown in Table 2.2. Estimates of the burden of all reduced sulphur amount to approximately 0.1×10^6 t S.

Table 1.2 Global atmospheric sulphur burden

	SO_2—S	$SO_4^{=}$—S
Oceans	0.17×10^6 t	0.3×10^6 t
Continents	0.28×10^6 t	0.5×10^6 t

Transport of SO_2 from the source areas downwind can be observed over a distance of several hundred kilometres. At a distance of 200 km an effect of a large industrial area on the SO_2 distribution can still be observed. In contrast to the SO_2 distribution, the sulphate concentration was only weakly influenced by SO_2 sources. Sulphur dioxide emissions from tall stacks lead to an increase of the SO_2 concentration in the free atmosphere at distances up to 100 km (Bingemer *et al.*[17]).

PHOTOCHEMICAL OXIDANTS

Not long ago photochemical pollution was mainly restricted to cities and industrial areas in subtropical latitudes like Australia, South Africa, Mexico or California. These photochemical oxidants are produced under the influence of the UV fraction of solar radiation with wave lengths between 290 μm and 430 μm from a mixture of reactive hydrocarbons and nitrogen oxides. Owing to the increasing immission concentration of nitrogen oxides, these oxidants have also become a problem of summer pollution in temperate latitudes. In Europe, oxidant concentrations above average are found in areas several hundred kilometres downwind of industrial areas and large cities.[18]

For several years an extensive study on the formation and distribution of oxidants was performed in the Bonn–Köln area. Measuring stations at the ground as well as aircraft ascents furnished a three-dimensional picture of the occurrence and concentration of oxidants and on their formation downwind of the source areas of the precursors.[19]

High concentrations of ozone, that means half-hour values above 75 ppb are mainly observed on calm summer days with intensive sun radiation. The correlation between ozone concentration and intensity of sun radiation was clearly established, the photochemical activity follows the diurnal trend of incoming sun radiation. We can summarize the following results:

1 The presence of NO_2 and solar radiation is necessary for the formation of photochemical smog.
2 Nitrous oxide is a precursor of NO_2 and reacts rapidly with ozone. Ozone formation depends on the rapid oxidation of freshly emitted NO to NO_2. The ozone formation is therefore retarded and takes at some distance from the primary emissions.
3 The reaction of hydrocarbons with oxygen leads to the formation of oxidants.

Aircraft ascents showed an increase of the oxidant concentration horizontally at a distance of 50 km from the source areas with 90 ppb and at 100 km with 110 ppb. In the vertical the maximum concentration was found in an altitude of 700 m above the ground. Upwind of the source area the maximum ozone concentration was only 60 ppb, providing the anthropogeneous origin of the ozone found downwind.

On some days a maximum of 160 ppb ozone in 1200 m altitude above ground was observed about 100 km downwind of the source area of the precursors. Above 1200 m a decrease of the ozone concentration could be clearly established, indicating the anthropogeneous origin of the maximum. At 3 km altitude the ozone concentration was dropped to the normal background of 45–50 ppb. During summer 1975, a flight from Munich to Rotterdam revealed a large scale increase of the oxidant concentration over wide parts of Western Europe. Unpolluted regions downwind at more than 100 km of industrial centres were affected by photo-oxidants. These showed residence times of more than 1 day.[20] The general tendency of increasing precursor concentrations like NO_x opens the possibility of a more frequent occurrence of photo-smog episodes during the summer months in Europe in the future.

REFERENCES

[1] Jost, D. (1978). In *Natur- und Umweltschutz in der Bundesrepublik Deutschland* (Ed. G. Olschowy), Paul Parey Verlag, p. 671.
[2] Emissionskataster Köln, TÜV-Rheinland 1972.
[3] Cullis, C.F., and Hirschler, M.M. (1979). Proceedings of the International Symposium on Sulphur Emissions and the Environment, London, p. 1.

[4] Semb, A. (1978). *Atm. Environment*, **12**, 455.
[5] Umweltbundesamt (1977). *Materialien zum Umweltschutzbericht 1977*, Erich Schmidt Verlag, pp. 90 and 95.
[6] Salmon, L. *et al.* (1978). *Science of the Total Environment*, **9**, 161.
[7] Köhler, A. (1974). Kommission zur Erforschung der Luftverunreinigung, DFG Mitt. XV.
[8] Calvert, J.G. *et al.* (1978). *Atm. Environment*, **12**, 197.
[9] Beilke, S., and Gravenhorst, G. (1978). *Atm. Environment*, **12**, 231.
[10] Georgii, H-W., and Vitze, W. (1971). *Idöjaras*, 294.
[11] Keddie, A.W.C. (1979). Proceedings of the International Symposium on Sulphur Emissions and the Environment, London, p. 45.
[12] Penkett, S.A. (1978). *VDI Berichte*, 314, 41.
[13] Runca-Köberich, D. (1979). *J. Diss. Univ. Frankfurt.*
[14] Georgii, H-W. (1978). *Atm. Environment*, **12**, 681.
[15] Meszaros, E. (1979). Proceedings of the International Symposium on Sulphur Emissions and the Environment, London, p. 36.
[16] Meszaros, E. (1978). *Atm. Environment*, **12**, 699.
[17] Bingemer, H., and Georgii, H-W. (1978). Proceedings, Interregional Transport of Air Pollutants, Umweltbundesamt, p. 119.
[18] Brasser, L.J. *et al.* (1977). *VDI-Berichte*, 270, 25.
[19] Georgii, H-W. *et al.* (1977). *VDI-Berichte*, 270, 19.
[20] Fricke, W., and Rudolf, W. (1977). *Staub*, **37**, 341.

Pollutants and Their Ecotoxicological Significance
Edited by H. W. Nürnberg

CHAPTER 2

Critical Review of Sampling Techniques and Determination Methods for Atmospheric Strong Acids

D. KLOCKOW, R. NIESSNER, B. JABLONSKI, H. DENZINGER AND G. RÖNICKE

SUMMARY

A trend, which can be observed on a world-wide scale, is the increasing 'acidification' of the atmosphere. This shift of the natural equilibrium between acidic and alkaline substances has been indicated mainly by long-term investigations of precipitation during the past decades. The results of these investigations clearly indicate, that strong acids, such as sulphuric acid and (or) nitric acid must contribute significantly to 'atmospheric acidity'.

The methods at present available for the determination of sulphuric and nitric acid in ambient air are briefly reviewed. The results obtained with most of these methods are uncertain because of an insufficient knowledge of the influence of the applied sampling procedures on the composition of the material under study. Therefore, special emphasis is placed on the discussion of sampling artefacts and on new techniques, which offer the possibility to avoid such effects.

The occurrence of strong acids in the atmosphere gained substantial attention since an increasing acidification of precipitation could be documented in 1967 by the results of a European atmospheric chemistry network.[1] Later on it was shown, that 'acid rain' is a regional environmental problem not only in parts of Europe but also in the eastern United States.[2] The data obtained by long-term pH measurements of rain water (see, for example, ref. 3) and by analysis of the samples for some major anions and cations, indicates that the increasing acidity of precipitation must be related to the emission and conversion of sulphur- and (or) nitrogen-containing pollutants.[3-6] We have analysed rain-water samples collected at Mount Schauinsland (Black Forest) in the south-western part of Germany for thirteen bulk anions and cations, including protons of strong acids. We were able to show that mainly sulphuric acid is responsible for the low pH values observed (down to 3.5) by using frequency distributions of different anion-to-cation concentration ratios or surface loads of the major ions as a function of time.[7]

Because the deposition of acidic material is related to its concentrations in air, the knowledge of these concentrations is of substantial interest. Therefore, a considerable number of methods for the determination of sulphuric acid (H_2SO_4) and nitric acid (HNO_3) in air-borne matter has been developed, which can be attributed to a few physical or chemical principles applied for obtaining specific signals (Tables 2.1 and 2.2). It has been shown, however, that there exist many difficulties in speciation of atmospheric strong acids because these must be detected besides the simultaneously present gaseous precursors (SO_2, NO_x) and

Table 2.1 Analytical principles applied to the determination of H_2SO_4 in aerosols

Processes essential for signal formation	Sampling	References
I *Thermal treatment*		
(a) Volatilization	Filter (separated by microdiffusion)	8–10
	Prefilter/impr. filter	11
(b) Evaporation, ionization (mass spectrometer)	Impaction	12
	Filter	13
(c) Evaporation, reduction to SO_2 (determination of SO_2)	Filter; Impactor	14
	Filter; combined with *in-situ* formation of amine salt	15
(d) Evaporation, reduction to S_2^* (flame emission)	Filter for separation of SO_2 and H_2SO_4	16
	Continuous, cyclic; heated diffusion tube for separation of H_2SO_4 from sulphates	17–21
II *Protolytic reactions*		
(a) Reaction with amines	Filter; sampling of amine salt	15
	Filter; fuming with amine	22, 23
	Impregnated filter	24
(b) Solvent extraction	Filter	25–28
(c) Esterification (GC)	Filter	29
III *IR absorption*	Impactor	30, 31
IV *Variation of droplet size by changing relative humidity*	*In-situ* nephelometry	32
	In-situ electrical mobility analysis	33

the corresponding salts (e.g. $(NH_4)_2SO_4$, NH_4NO_3). This makes necessary the combination of the detection system with an appropriate sampling and sample pretreatment procedure, except for long path IR spectroscopy.

As can be seen from Tables 2.1 and 2.2 most of the techniques described use filters (H_2SO_4) or a combination of prefilter and collection filter (H_2SO_4, HNO_3) for extraction of the acids from ambient air. Several authors have shown (see literature cited in refs 48 and 51) that without special precautions the use of filters in collection of reactive atmospheric trace constituents leads to sampling artifacts which are difficult to control. These artifacts may result from irreversible interactions of reactive compounds with the filter material employed or from gas–particle and particle–particle reactions on filter surfaces. We have experimentally identified artifact formation in filter sampling of H_2SO_4 and acidic sulphates.[48] In order to avoid such unwanted effects, one can try to selectively mask the compound to be determined (e.g. H_2SO_4) prior to[15] or during[24,49] filtration.

A most promising approach to the determination of H_2SO_4 in aerosols containing neutral and acidic sulphates is based on the different thermal behaviour of the sulphate species and avoids air filtration.[17–21] Thermal pretreatment of the aerosol under study at suitable temperatures results in an evaporation of H_2SO_4 droplets and in a practically complete loss of the acid at the wall of a diffusion separator (denuder). However, $(NH_4)HSO_4$, $(NH_4)_2SO_4$ and other sulphate particles pass through the denuder tube unaffected. The combination of thermal decomposition and diffusion separation with a flame photometric sulphur detector has led to different types of H_2SO_4 monitors.[17–21]

Table 2.2 Analytical principles applied to the determination of air-borne nitric acid[34]

Processes essential for signal formation	Sampling	References
I *Reduction*		
(a) To NO (chemilum.)	Continuous, cyclic; nylon trap for separation of HNO_3 from NO_x	35, 36
(b) To NO_2^- (photom.)	Prefilter/impr. filter (NaCl)	37, 38
(c) To NH_4^+ (photom.)	Prefilter/nylon filter or impr. filter	39, 40
II *Protolytic reaction*	Absorber solution (coulometer), nylon filter for separation of HNO_3 from other acidic compounds	41, 42
III *Nitration of aromatic compounds* (GC)	Prefilter/nylon or cotton fibres	43, 44
IV *IR absorption*	*In-situ* long path spectroscopy	45–47

A set-up for the determination of gaseous HNO_3 by using a denuder has also been proposed.[50]

With the diffusion separation techniques so far described results are obtained, which are based on the difference of two measurements (salts plus acid, and salts without acid). Such an approach may lead to erroneous information about atmospheric strong acid concentration especially at low mixing ratios. Therefore we have modified the thermo-denuder conception for H_2SO_4 in such a way that the acid vapour, formed during thermal treatment of the aerosol in a heated denuder tube and trapped at the wall of this tube, is quantitatively determined.[51] By this means a direct signal for H_2SO_4 is obtained, and it has been shown that the same principle can be applied to the determination of HNO_3 also. Further extension of this work has led to the development of a train of several denuder tubes, operated at different temperatures, by which it should be possible to study the system NH_3, HNO_3, H_2SO_4 and ammonium salts in ambient air.[52]

REFERENCES

1 Odén, S. (1976). In *Proceedings of the First International Symposium on Acid Precipitation and the Forest Ecosystem*, USDA Forest Service General Technical Report NE-23, Upper Darby, Pennsylvania, USA, pp. 1–36.

2 Likens, G.E., and Borman, F. H. (1974). *Science*, **184**, 1176–1179.

3 Kayser, K., Jessel, U., Köhler, A., and Rönicke, G. (1974). *Mitteilung IX der Kommission zur Erforschung der Luftverunreinigung*, Deutsche Forschungsgemeinschaft, Bonn-Bad Godesberg.

4 Marsh, A.R.W. (1978). *Atmos. Environ.*, **12**, 401–406.

5 Granat, L. (1978). *Atmos Environ.*, **12**, 413–424.

6 Slanina, J., Möls, J.J., Baard, J.H., van der Sloot, H.A., van Raaphorst, J.G., and Asman, W. (1979). *Intern. J. Env. Anal. Chem.*, **7**, 161–176.

7 Klockow, D., Denzinger, H., and Rönicke, G. (1978). *Sauerstoffhaltige Schwefelverbindungen*, VDI-Berichte Nr. 314, VDI-Verlag, Düsseldorf, pp. 21–26.

8 Dubois, L., Baker, C.J., Teichman, T., Zdrojewski, A., and Monkman, J.L. (1969). *Mikrochim. Acta*, 269–279.

9 Dubois, L., Zdrojewski, A., Teichman, T., and Monkman, J.L. (1972). *Intern. J. Environ. Anal. Chem.*, **1**, 259–266.

10 Maddalone, R.F., Thomas, R.L., and West, P.W. (1976). *Env. Sci. Technol.*, **10**, 162–168.

11 Mason, D.W., and Miller, H.C. (1979). In *Ion Chromatographic Analysis of Environmental Pollutants*, vol. 2, (Eds E. Sawicki and J.D. Mulik), Ann Arbor Science Publ., pp. 193–201.

12 Schulten, H-R., and Schurath, U. (1975). *J. Phys. Chem.*, **79**, 51–57.

13 Harvey, T.M., and Schuetzle, D. (1977). Society of Automotive Engineers (SAE), Report Nr. 770063, Detroit, Michigan.

14 Scaringelli, F.P., and Rehme, K.A. (1969). *Anal. Chem.*, **41**, 707–713.

15 Snyder, R.E., Reed, T.J., and McKissick, A.M. (1977). US Environmental Protection Agency, EPA-600/2-77-010.

16 Crider, W.L. (1965). *Anal. Chem.*, **37**, 1770–1773.

17 Huntzicker, J.J., Hoffman, R.S., and Ling, Chaur-Sun (1978). *Atmos. Environ.*, **12**, 83–88.

[18] Cobourn, W.G., Husar, R.B., and Djukič-Husar, J. (1978). *Atmos. Environ.*, **12**, 89–98.
[19] Kittelson, D.B., McKenzie, R., Vermeersch, M., Dorman, F., Pui, D., Linne, M., Liu, B., and Whitby, K. (1978). *Atmos. Environ.* **12**, 105–111.
[20] Tanner, R.L., D'Ottavio, T., Garber, R., and Newman, L. (1980). *Atmos. Environ.*, **14**, 121–127.
[21] Cobourn, W.-G., Djukič-Husar, J., and Husar, R.B. (1980). *J. Geophys. Res.*, **85**, 4487–4494.
[22] Huygen, C. (1975). *Atmos. Environ.*, **9**, 315–319.
[23] Dzubay, T.G., Snyder, G.K., Reutter, D.J., and Stevens, R.K. (1979). *Atmos. Environ.*, **13**, 1209–1212.
[24] Dasgupta, P.K., Lundquist, G.L., and West, P.W. (1979). *Atmos. Environ.*, **13**, 767–774.
[25] Barton, S.G., and McAdie, W.G. (1970). *Proceedings of the Second International Clean Air Congress*, Washington, D.C., pp. 279–382.
[26] Leahy, D., Siegel, R., Klotz, P., and Newman, L. (1975). *Atmos. Environ.*, **9**, 219–229.
[27] Tanner, R.L., Cederwall, R., Garber, R., Leahy, D., Marlow, W., Meyers, R., Phillips, M., and Newman, L. (1977). *Atmos. Environ.*, **11**, 955–966.
[28] Appel, B.R., Wall, S.M., Haik, M., Kothny, E.L., and Tokiwa, Y. (1980). *Atmos. Environ.*, **14**, 559–563.
[29] Penzhorn, R.D., and Filby, W.G. (1976). *Staub-Reinh. Luft*, **36**, 205–207.
[30] Cunningham, P.T., and Johnson, S.A. (1976). *Science*, **191**, 77–79.
[31] Kellner, R., and Novakov, T. (1978), University of California, Lawrence Berkeley Laboratory, Annual Report 1977–78, LBL-8696, pp. 20–27.
[32] Charlson, R.J., Vanderpol, A.H., Covert, D.S., Waggoner, A.P., and Ahlquist, N.C. (1974). *Atmos. Environ.*, **8**, 1257–1267.
[33] Liu, B.Y.H., Pui, D.Y.H., Whitby, K.T., Kittelson, D.B., Kousaka, Y., and McKenzie, R.L. (1978). *Atmos. Environ.*, **12**, 99–104.
[34] Stevens, R.K. (Ed.) (1979). *Current Methods to Measure Atmospheric Nitric Acid and Nitrate Artifacts*, US Environmental Protection Agency, EPA-600/2-79-051.
[35] Joseph, D.W., and Spicer, C.W. (1978). *Anal. Chem.*, **50**, 1400–1403.
[36] Kelly, T.J., and Stedman, D.H. (1979). In *Current Methods to Measure Atmospheric Nitric Acid and Nitrate Artifacts*, US Environmental Protection Agency, EPA-600/2-79-051., pp. 37–43.
[37] Okita, T., Morimoto, S., Izawa, M., and Konno, S. (1976). *Atmos. Environ.*, **10**, 1085–1089.
[38] Forrest, J., Tanner, R.L., Spandau, D., D'Ottavio, T., and Newman, L. (1980). *Atmos. Environ.*, **14**, 137–144.
[39] Lazrus, A.L., Gandrud, B.W., and Greenberg, J.P. (1979). In *Current Methods to Measure Atmospheric Nitric Acid and Nitrate Artifacts*, US Environmental Protection Agency, EPA-600/2-79-051, pp. 45–50.
[40] Huebert, B.J. (1980). *Geophys. Res. Letters*, **7**, 325–328.
[41] Miller, D.F., and Spicer, C.W. (1975). *J. Air Poll. Control. Ass.*, **25**, 940–942.
[42] Spicer, C.W., Ward, G.F., and Gay, B.W., Jr. (1978). *Anal. Letters*, **A11**, 85–95.
[43] Hare, R.J., Wininger, M.T., and Ross, W.D. (1979). In *Current Methods to Measure Atmospheric Nitric Acid and Nitrate Artifacts*, US Environmental Protection Agency, EPA-600/2-79-051, pp. 63–65.
[44] Tesch, J.W., and Sievers, E. (1979). In *Current Methods to Measure Atmospheric Nitric Acid and Nitrate Artifacts*, US Environmental Protection Agency, EPA-600/2-79-051, pp. 67–71.

[45] Murcray, D.G., Kyle, T.G., Murcray, F.H., and Williams, W.J. (1969). *J. Opt. Soc. Am.*, **59**, 1131–1134.

[46] Tuazon, E.C., Winer, A.M., Graham, R.A., and Pitts, J.N., Jr. (1979). In *Current Methods to Measure Atmospheric Nitric Acid and Nitrate Artifacts*, US Environmental Protection Agency, EPA-600/2-79-051, pp. 9–26.

[47] Doyle, J.J., Tuazon, E.C., Graham, R.A., Mischke, T.M., Winer, A.M., and Pitts, J.N., Jr. (1979). *Env. Sci. Technol.*, **13**, 1416–1419.

[48] Klockow, D., Jablonski, B., and Nießner, R. (1979). *Atmos. Environ.*, **13**, 1665–1676.

[49] Klockow, D., Nießner, R., and Jablonski, B. (1980). *Anal. Letters*, **13**, 1397–1408.

[50] Shaw, R. W., Dzubay, T.G., and Stevens, R.K. (1979). In *Current Methods to Measure Atmospheric Nitric Acid and Nitrate Artifacts*, US Environmental Protection Agency, EPA-600/2-79-051, pp. 79–84.

[51] Nießner, R., and Klockow, D. (1980). *Intern. J. Environ. Anal. Chem.*, **8**, 163–175.

[52] Slanina, J., v. Lamoen-Doornenbal, L., Lingerak, W.A., Meilof, W., Klockow, D., and Nießner, R. (1981). *Intern. J. Environ. Anal. Chem.*, **9**, 59–70.

Pollutants and Their Ecotoxicological Significance
Edited by H. W. Nürnberg

CHAPTER 3

Energy Production and Air Pollution with Particular Emphasis to the Medium Term within the European Community

H. P. STIEF-TAUCH

SUMMARY

Energy production and use, particularly the combustion of fossil fuels, is at the origin of most of the air pollution encountered within EEC as well as in other industrialized countries. Therefore policies destined to combat such pollution have to consider what contribution to overall emissions is possibly coming from combustion process, what kind of fuel is used, where such use takes place and how emissions are made. A number of important questions need to be answered before choosing rational policy solutions, e.g. what is the current, and what will be the future pattern of energy production and use? Where is this energy to be used and in which equipment? Which pollutants are likely to be emitted, in which quantities and under what conditions? What may be the impact of these emissions on man and environment once they appear as ground-level concentrations?

Only a thorough analysis of the energy situation in terms of supply and demand can give more or less precise answers to these questions on which policies can then be built which aim at achieving the air quality required at regional or overall scale with respect to specific pollutants. A short review of the energy situation is given together with some general remarks concerning the role of fossil fuel use and its implications.

The analysis of the energy situation within EEC and forecasts at medium term (1990) constitute the basis for a detailed assessment of emissions of some major pollutants like SO_2, particulates and CO_2. The current and future pattern of fossil fuel use within EEC is of particular importance in this context. Emission estimates based on these patterns are inherently uncertain, as any prediction, moreover such estimates must necessarily rely on assumed averaged conditions, the real measurements of emissions made being still too scarce today.

The role of emissions in the resulting pollutants concentration in ambient air is highlighted, while indicating some of the actions taken in this field by the EEC Commission. The problems connected particularly with the emissions of SO_2 and particulate matter are analysed in detail; international research activities and international agreements on long-range, trans-boundary, air pollution are discussed.

In view of the increasing world energy use concern is growing on the increasing emissions of CO_2 which may eventually induce climatic changes at regional or global scale. It becomes apparent from these examples of pollution problems that a new international dimension must be added to national or EEC policies dealing with air pollution.

INTRODUCTION

On opening the newspaper there is hardly a day when the reader is not confronted with more or less alarming headlines on an imminent energy crisis, on the dangers of nuclear power plants or on the necessity of using nuclear energy, on the prediction of environmental threats and consequent damages, on the economic consequences of less energy and more environmental protection, etc. Some 12 years ago such wide and sometimes passionate public interest in the issues of energy and environment would have been quite unthinkable. So what has brought about this change in public attitude?

The economic growth experienced in most of the industrially developed countries in the sixties and up to the early seventies was associated with rapid industrial and urban development. One result of this activity was increased wealth. One other result was a growing pressure on the environment threatening and deteriorating what is today generally called the 'quality of life'.

Governments were consequently faced with the challenge of how to continue reaping the benefits of economic growth while avoiding unacceptable or undesirable deterioration of the environment. To this end a new dimension aimed at the protection of the environment was introduced into the legislation and into the assessment of new projects and undertakings.

Of all human activities, the production and use of energy, in its various forms, has probably the most important impact on the environment or parts of it, particularly on the pollution of the air we breathe, on the water we drink and on the landscape we live in. This fact was recognized as early as the thirteenth century in England where the noxious fumes from open burning of coal gave rise to concern for health, and measures were taken to limit the use of coal in some places or during certain periods.[1]

An overall analysis of the sources of air pollution has to be based upon a close look on the energy scene, on a careful evaluation of the situation and its predictable evolution. Therefore, the first part of this chapter tries to give, after some general statements, a thorough analysis of the energy scene as foreseeable for the whole EEC up to 1990, together with an outlook to 2000.*

PRINCIPAL FORMS OF ENVIRONMENTAL IMPACT FROM ENERGY SUPPLY AND USE

Energy is generally not available in a simple form but requires rather complex systems for its production, made up of a number of components which are necessary to bring a basic energy resource from its natural state to the place in which it is used. For most energy systems, this involves extraction, processing, often conversion into a more useful form, transportation between the various operations, transmission to the ultimate user, its final use, and, eventually, disposal of residues.

Of necessity, all forms of energy production and use have an impact on the environment. The extent of that impact will depend firstly on the *quantity* of energy required and the *form* in which it is supplied and consumed, and secondly on the *measures* taken to reduce the impact.

There are mainly *three principal forms of impact*: disturbance at the local level, effluent in the atmosphere and possible releases of radioactivity. There are, of course, many others: thermal pollution from discharge of cooling water, noise, secondary impacts associated with increased water needs, possible climatic disturbances from CO_2, particulate emissions, unavoidable and accidental effects of offshore oil developments, etc.

It is important to stress that these different forms of impact cannot be directly compared. The impact can result in alterations of, or damage to the environment.

PRIMARY ENERGY DEMAND NOW AND IN THE FUTURE

Review of the Situation and Forecasts at World Scale

On a global scale the most important fraction of energy requirements today is met by the combustion of fossil fuels: coal, oil and natural gas. Together they provide about 95 per cent of the present supply and even by the year 2000 they most probably will still meet about 75 per cent of the needs. The share of hydropower in the total energy balance will decrease, while the part of solar energy will increase in the near future. Nuclear power is likely to play an increasing part in electricity generation, at least in the near future, and within the next decades nuclear energy may progressively replace the combustion of fossil fuels for many other energy uses.

Based on scenario studies on possible world energy demand patterns, carried out and published in 1978 by the World Energy Conference,[2] Table 3.1 indicates the range of possible energy demand up to 2000. Depending on the assumption

* Editor's note: The topic is treated on the basis of the information sources available until 1980.

Table 3.1 Range of world energy demand by fuel type to the year 2000 (Ep in 10^9 toe)

Fuel	1972		1975		1985		1990		2000	
	Ep	%	Ep	%	Ep	%	Ep	%	Ep	%
Fossil (coal, wood, oil, natural gas)	5.7	94	5.85	94	6.8–7.7	86	7.3–8.8	82	8.7–11.7	74
Other (nuclear, hydro, solar, etc.)	0.35	6	0.4	6	1.1–1.25	14	1.6–1.8	18	3.1–4.0	26
Total	6.05	100	6.25	100	7.9–8.95	100	8.9–10.6	100	11.8–15.7	100
Average annual percentage growth rates[a]										
World population (10^9)	3.75		3.97		4.88		5.42		6.7	
Specific energy usage (toe per inhabitant)	1.61		1.57		1.62–1.83		1.64–1.95		1.76–2.35	

Source: World Energy Conference, 1978.

[a] Primary energy demand, 2.4–3.4 per cent; economic growth, 3.1–4.2 per cent.

of growth rates of world primary energy demand (2.4–3.4 per cent per year) the use of fossil fuels is likely to double within the next 20 years, whereas total primary energy demand may go up by a factor of 2.5 compared with the situation today. Relating these figures to the world population of about 3.75 billion people in 1972 and, at an estimated constant growth rate of 2.1 per cent per year, to about 6.7 billion people living in 2000, the specific energy consumption of 1.6 toe per capita (toe = tons of oil equivalent) in 1972 remains fairly constant up to 2000, when the projected lower growth rate of primary energy demand (2.4 per cent) is considered. This tends to suggest that the lower range of growth rates considered by the World Energy Conference is only sufficient to sustain the actual average standard of living (in terms of goods and services produced) unless a substantial overall improvement in the efficiency of energy use occurs within the next 20 years.

Statistical evidence exists, at least for the past, for a direct correlation between economic growth measured in terms of gross national product and total growth in energy demand. For virtually all nations, developing as well as developed, the larger the per capita gross national product the larger the commercial energy consumption.[3,4] The transition from an agrarian to an industrial economy has required a large increase in energy consumption, both for powering industry and for raising productivity in agriculture. At the same time, the development of tight-woven transportation systems has also required a substantial share of total energy needs. As long as nations continue their efforts towards more welfare, it is reasonable to expect that the per capita energy consumption will also rise, although the rate of increase may vary. Taking into account the projected upper growth rate of primary energy demand in Table 3.1 (3.4 per cent per year), this assumption could become true if a sensible improvement of the standard of living is to be achieved for the average of the world population in the year 2000. These basic trends highlight the global prospects in terms of probable overall energy demand and per capita demand, while hiding completely the extremely large regional differences in energy use, in terms of total or of per capita figures. In 1975 North America alone consumed 31 per cent of the total world's energy production of 8.3 toe per capita, which is five times the world's per capita average: the EEC took 14.5 per cent or 3.47 toe per capita (Table 3.2).

Table 3.2 Primary energy demand and population (1975)

	Energy			Population 10^9	toe per capita
	10^9 tce[a]	10^9 toe	%		
World	8.9	6.25	100	3.970	1.57
North America	2.8	1.95	31	0.235	8.31
EEC (9 member states)	1.3	0.9	14.5	0.259	3.47

[a] tce = tons of coal equivalent.

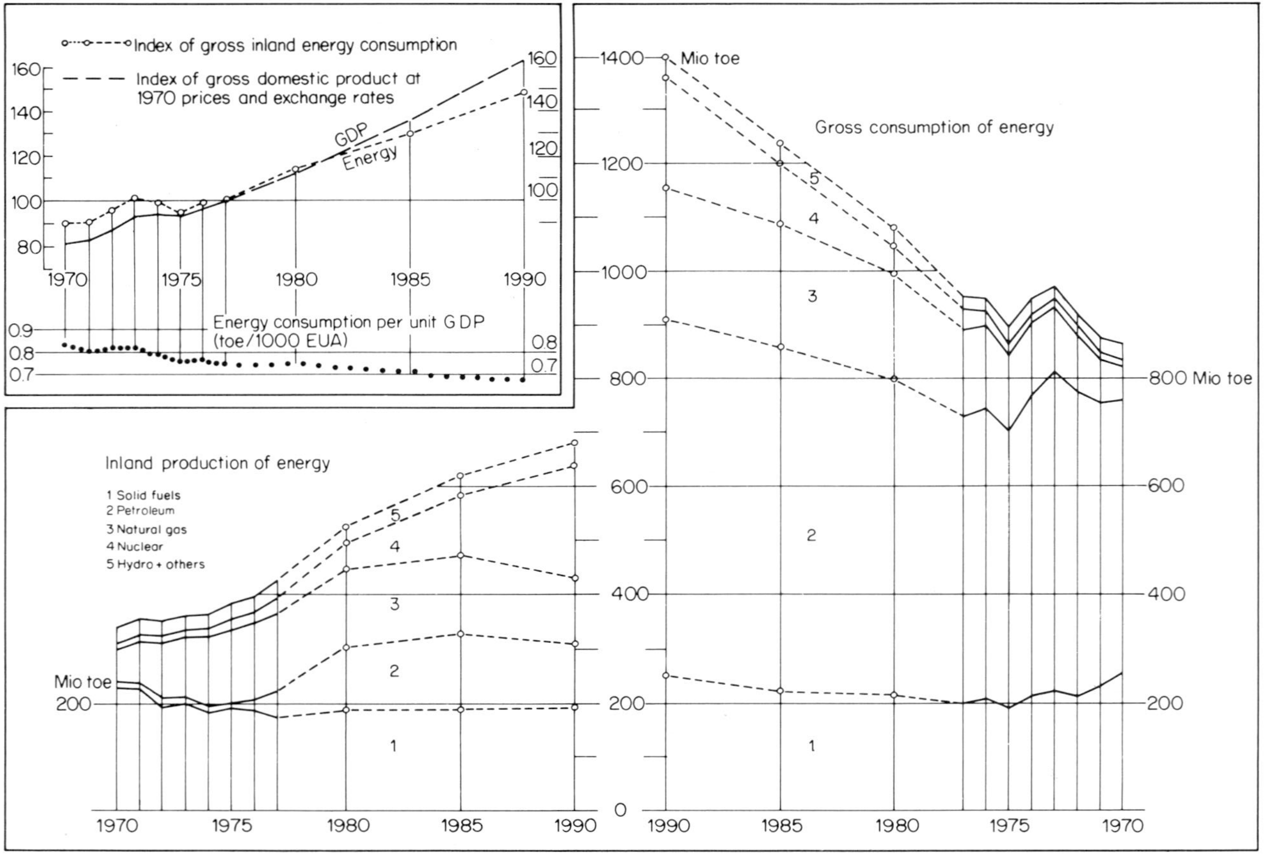

Fig. 3.1 EEC gross consumption of energy 1970–1990. *Source:* Sum of member states programmes

Table 3.3 Evolution of primary energy usage in the EEC 1970–1979 (without bunkers)

	1970	1973	1974	1975	1976	1977	1978	1979
10^6 toe	832	936	917	864	908	926	937	965[a]

[a] Forecast SEPC-3

Review of the Energy Scene and Forecasts at EEC Scale

The relation between growth rate of primary energy demand to growth rate of gross domestic product, sometimes called 'energy coefficient', remained close to 1 because energy was inexpensive. Between 1959 and 1973 consumption rose, without interruption by an average, and relatively constant, 4.5 per cent per annum but this trend ceased abruptly in 1974 and was followed by 4 years of stagnation. Within the EEC, gross internal consumption of primary energy (without bunkers) has shown the trend given in Table 3.3.

This has been an exceptional phenomenon, and one which previous experience since the Second World War made it absolutely impossible to predict. There has, in fact, been a decrease in energy consumption for two consecutive years, with recovery in 1978 only (Figures 3.1 and 3.2). A breakdown of total EEC primary energy demand into the different energy vectors is given in Table 3.4.

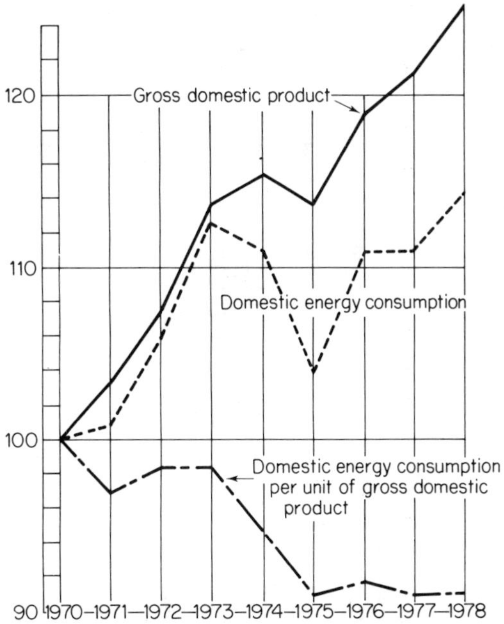

Fig. 3.2 Gross domestic product and domestic energy consumption in EEC (1970 = 100)

Table 3.4 The EEC's pattern of primary energy demand including non-energetic usage and bunkers (mtoe = 10^6 toe)

Source	1970		1973		1975		1976		1980		1985		1990		2000	
	mtoe	%	mtoe	%	mtoe	%	mtoe	%	mtoe	%	mtoe	%	mtoe	%	mtoe	%
Solid fuels	264	80.5	222.1	22.8	194.0	21.6	207	21.7	217	20.1	224	18.1	248	17.7	260–255	19.3–15.4
Oil	499	57.6	592.7	60.9	507.9	56.7	542.5	56.9	585	54.2	631–641	51.0–51.8	652–667	46.4–47.4	430–650	32–39.5
Natural gas	62	7.2	117.9	12.1	142.3	15.9	155.9	16.4	193	17.9	231–221	18.7–17.9	253–238	18.0–17.0	310–335	23–20.3
Nuclear	10.5	1.2	14.1	1.4	19.5	2.2	21.5	2.3	48	4.5	113	9.1	208	14.8	205–245	15–14.8
Hydro and others	30.8	3.5	27.1	2.8	32.7	3.6	26	2.7	36	3.3	38	3.1	43	3.1	145–165	10.7–10
Total	866.3	100	973.9	100	'896.4	100	952.9	100	1079	100	1237	100	1404	100	1350–1650	100

The forecast for 1990 is around 1400 mtoe. Recent studies[5–8] on different primary energy demand and supply scenarios likely to be encountered in the EEC (of nine member states) in the year 2000 indicate a range of some 1350–1650 mtoe, while anticipating and putting strong emphasis on energy conservation measures.

These recent estimates contrast remarkably with those made before the 1973 oil price increase, and which forecast for the year 2000 a level in energy demand of 3 to 3.7 times the 1970 level. In 1978 forecasts, the probable level of energy demand in 2000 was estimated to be around 1.6–1.9 times the 1970 level, and in the light of the latest events on the international energy scene this estimation has probably already to be revised as well. In fact, the straightforward predictions made in the past by assuming a 'surprise-free' evolution no longer hold which makes medium-term planning very hazardous. Nevertheless, it is indispensable to have at least some indications on the probable range of total energy demand as well as on the share of different energy sources likely to satisfy this demand. In particular the latter task will be subject to large uncertainties because of fluctuations in the availability of fuels.

A breakdown of total primary energy demand into fossil energy and other energy is given in Table 3.5. The proportion of fossil fuels (solid, liquid, gaseous ones) constituting the total primary energy demand increased to about 96 per cent in 1973. After the oil price rise of 1973 the tendency has been reversed towards a stronger use of non-fossil fuels, mainly by producing more primary electricity from nuclear plants and, in the longer term, by using more renewable energy (geothermal, solar, wind etc.), although the latter is not likely to cover more than 5 per cent of the total energy demand in 2000.

Table 3.5 Share of fossil fuels in EEC primary energy demand

	1970 %	1973 %	1975 %	1976 %	1980 %	1985 %	1990 %	2000 %
Solid fuels, Oil, Natural gas	95.3	95.8	93.9	95.0	91.2	87.8	82.1	75
Nuclear, Hydro and others	4.7	4.2	6.1	5.0	8.8	12.2	17.9	25

SEPC-3

After this somewhat lengthy but necessary analysis of the energy scene, to which I shall refer back later, I will now give a view on the air pollution scene as conditioned by energy production and use within the EEC. Some indications will be given on the actions undertaken by the EEC Commission against air pollution and lines of development will be shown.

AIR POLLUTION FROM ENERGY PRODUCTION AND USE

It is generally acknowledged that combustion of fossil fuels is the most important source of air pollution. Among combustion products it is mainly sulphur dioxide (SO_2), nitrogen oxides (NO_x) and particulate matter that are considered as particularly detrimental to the environment and possibly hazardous to human health. Recently concern is growing on the increasing release of CO_2 from the combustion processes. I will come to that particular problem later.

Other minor pollutants are released as well, e.g. trace elements of heavy metals (Hg, Pb, Cd), halides Cl, F, hydrocarbons and other compounds. That is why in industrialized countries during the last few years special attention has been given to these pollutants, to their origin, propagation and to possibilities of abatement.

Overall estimates of air pollution for the important pollutants identify three principal sources of pollution; (a) fuel combustion in stationary sources (like boilers, furnaces, incinerators, domestic heating, etc.); (b) other industrial installations (process equipment, ore roasting, chemical processing, paper production, etc.); and (c) combustion in mobile sources (traffic).

Stationary combustion installations are in fact estimated to create between 70 and 90 per cent of total SO_2 and particulate emissions. Moreover, they contribute about 50 per cent of total NO_x emissions, a further 40 per cent being produced by traffic. Traffic is also the main source of carbon monoxide (CO) pollution (more than 90 per cent) and a very important source of hydrocarbon (HC) emissions (about 50 per cent). The radioactive gaseous effluents from nuclear power stations and reprocessing plants must be considered in this context as well, although at present they do not make any significant contribution to the total level of radioactivity encountered in the atmosphere. Energy consumption for traffic purposes also produces pollutants which are potentially toxic if they concentrate in the atmosphere above certain levels. In 1973, for instance, about 36 000 t of lead were emitted from automobile exhausts within the member states of the EEC. This figure is based on a total gasoline consumption of approximately 70 million tons, with an average lead content of about $0.55\ g\ l^{-1}$ and 70 per cent of total lead contained being emitted in the atmosphere.

PROBLEMS IN ESTIMATING EMISSIONS

If it is already difficult to establish a reliable estimate of the primary energy demand and energy use, the correct assessment of air pollution caused by this energy use is even more complicated. The diversity of equipment used, its technological perfection and the varying operating modes resulting in more or less important emissions do not permit a precise calculation of the quantity of pollutants released on an overall basis. At best one can use simplifying assumptions for one particular type of process or equipment operated under average conditions with certain types of fuel, and thus determine a probable range of emissions or of emission factors (the latter linking emission of a specific pollutant to, for example, unit of energy produced or used, or unit of product yielded). An increasingly applied, although expensive, method for producing more reliable estimates for a limited region is the establishment of emission inventories where, in a systematic and rather detailed way, the different emitting sources are investigated and the missions measured or calculated.

It is under these premises that all figures indicating levels of emissions are to be understood and which also explains why the literature often contains quite different values for emissions from apparently identical devices or processes. Obviously such differences may become even larger when considering the overall situation.

This general warning on the relative reliability of emission figures should not mean that it is not practicable to produce such estimates but that they should not be taken at face value, rather as an indication of the magnitude of the emissions encountered.

Looking at the phenomenon of air pollution not only in absolute figures of emissions but also in terms of effects likely to be produced by the former, one has to consider as well *where* the impact of such pollution occurs. Depending on the pattern of energy supply and use the impact from energy production, transport and use may have local, regional or global implications; it may result in damage and will often go hand-in-hand with phenomena other than air pollution.

Atmospheric pollution caused principally by fossil-fuel use has both a local and regional, and possibly global, impact. The effects on health, vegetation and materials of SO_2, NO_x and particulates are well documented. It is now generally accepted that SO_2, NO_x and small particles may be transported over large distances. Recent OECD investigations show the EEC to be net exporters of sulphur to other countries.

The accidental release of highly radioactive material during the burning of nuclear fuel or during the operation of the fuel cycle, including its final disposal, would cause serious local or regional damage of either a short- or long-term nature. Low-level radioactive release also occurs in effluent emitted from nuclear plants, and from waste disposal (as well as from coal-fired plants). Also associated with nuclear power development is the possible threat of nuclear proliferation and terrorist activity.

The principal form of local damage that is difficult to ameliorate is the effect on amenity values caused by the installations for the extraction and processing of fossil fuels and for generation and transmissions of electricity. Disposal of solid wastes can make a significant impact locally, and land set aside for the disposal of nuclear wastes must also be classed as a serious disamenity.

EMISSIONS AND AMBIENT AIR CONCENTRATIONS

The knowledge of the magnitude of emissions constitutes only one side of the coin. In order to judge what effects air pollutants may have on man and the environment the concentrations occurring in ambient air must obviously be known as well. Therefore, a lot of skill has been invested in developing and setting up appropriate monitoring methods and equipment capable of telling us what these concentrations are and how they change. Similarly, epidemiological studies have been carried out to establish relationships between the dose of air pollutants impacting on human beings and the effects observed from that impact. These relationships called 'criteria' have served as a basis for setting up goals for air quality, in the form of legally compulsory air-quality standards or of recommendations destined to protect human health.

The very theme of the Sixth Session of the European Environmental Summer School was endeavouring to make a contribution to the still more complicated task of establishing criteria for the protection of ecosystems and parts of it against pollutants. Hopefully, the results may enable public authorities to build on its conclusions when they are confronted with the task of legislating in order to protect not only public health but the environment as well.

I will not enter into considerations on concentrations measured within the EEC member states for specific air pollutants. I only note that in many regions of the EEC measured concentrations of specific pollutants are still well above the limits recommended by the WHO for health protection.

The two EEC action programmes on the environment put much emphasis on that particular aspect of air pollution. Moreover, an exchange of information between the surveillance and monitoring networks based on data relating to atmospheric pollution caused by certain sulphur compounds and suspended particulates was established by decision of the Council of Ministers in June, 1975. For the years 1976–78 the results of some 400 stations throughout the EEC are being collected, the reports will be edited in due time. This project is a pilot exercise destined to help the Commission and the member states' authorities towards a better understanding of what was measured, how the results compare, and how the situation in certain highly polluted areas develops. It has revealed already the need for a much stronger harmonization of the measurement methods currently in use if valid conclusions should be drawn.

The comparability of air pollution results measured by different methods becomes particularly crucial when legally compulsory air-quality standards are to be respected at EEC level. Obviously, any significant deviation from the standards, induced by specific, nationally used measurement equipment other than the specified reference method, may bring about differences in the efforts which member states have to make to bring down air-pollution levels to the required values.

I will not discuss fully methods and means for abating air pollution. A lot of discussion is going on concerning the best way of bringing down the concentrations of the main air pollutants like SO_2, particulate matter, NO_x and others, mainly within highly polluted areas. Several principle ways are possible: e.g. to decrease emissions at the source (by emission control, by using less energy, by using less polluting energy forms); to disperse the sources of emissions (by an appropriate siting policy) and to dilute emissions (by spreading it out over larger areas thus decreasing the impact near the source, while making use of the regenerative capacity of the atmosphere). There is no patented solution fitting for all cases; apparently, varying combinations of all means may render the optimum result in terms of cost and effectiveness, depending on local and regional circumstances.

Without arguing on exact levels of air pollutant concentrations likely to be found in some EEC regions now and in the future, a look on the magnitude of future emissions from energy use will give a valuable indication of the possible evolution of the former, bearing in mind that energy use constitutes the biggest source of air pollution. Although imprecise by nature, such estimates of future emissions are helpful for taking appropriate legal action in time.

EMISSION OF THE MOST IMPORTANT POLLUTANTS AND SOME INTERNATIONAL PROBLEMS RAISED

Sulphur Dioxide

Sulphur dioxide and particulate matter have been recognized as major pollutants in urban areas for more than two decades. Their control was often joined because of early suspicions that their effects were augmented when they appeared together.

Sulphur compounds affect human health, plants, trees, rivers, lakes and building materials. Emissions of sulphur dioxide increased in all industrialized countries during the 1960s in parallel with the burning of more coal and oil. The upward trend was reversed in the EEC as well as in a number of other countries after 1973. A recent OECD study shows that for OECD as a whole, SO_2 emissions from fuel combustion were at about the same level in 1974 as in 1968, although energy consumption had continued to grow.[9]

These results can be attributed to: (a) lower economic growth plus increased energy conservation and, therefore, a lower rate of growth in energy use; (b) greater use of natural gas in some countries and increased use of low-sulphur-content crude oil, as well as rapid growth in the use of scrubbers; and (c) other techniques for the removal of sulphur. Differences in economic structure, in industrial activities and in emission controls account, however, for substantial variations in air pollution and its abatement in different countries.

A detailed and consistent study[10] was made in 1977 for the EEC Commission with the aim of assessing the present and future SO_2 emissions likely to occur within the nine member states. The study assessed the primary energy consumption, country by country, in terms of coal, oil, gas and electricity for the years 1972, 1975, 1980, 1985. Within this framework particularly the sulphur-containing energy consumption was considered in the three end-use sectors—

Table 3.6 SO_2 emissions from fossil fuel combustion within EEC ($10^6\,t\,SO_2\,y^{-1}$)

Source	1972	1975	1980	1985	1990	2000
Coal	5.845	5.610	6.230	6.475	7.200	7.500
Fuel oil	9.805	8.540	10.950	10.800–10.950	11.110–11.360	6.800–11.150
Gas oil	1.835	1.600	1.080	1.085–1.105	1.125–1.150	0.700–1.140
Total	17.485	15.750	18.260	18.360–18.530	19.435–19.710	15.000–19.790

SEPC-3

power generation, industry, domestic transport; for each sector the split was made between solid fuel, fuel-oil and gas-oil. Variations in the average sulphur content of the fuels used due to legal requirements already in force in some member states or to be introduced soon are taken into account.

The updated emission estimates incorporating the latest forecasts on EEC primary energy consumption pattern up to 1990 are indicated in Table 3.6 together with an outlook to 2000. Sulphur dioxide emissions from fuel oil are, and will continue to be, the biggest source, increasing by some 30 per cent over the 1975 level if no further abatement measures are taken soon. Emissions from coal show a similar increase, whereas, after the introduction of Community legislation on the maximum sulphur content of gas-oils in 1976, the SO_2 emissions from that fuel decrease sharply.

The emission picture is not uniform at a country-by-country basis; particularly in the Netherlands SO_2 emissions are likely to increase sharply in the near future because the natural-gas reserves are running out and fuel oil or coal will have to compensate for the missing energy.

Particulate Matter

Considerable efforts have been made in the last decade to control emissions of particulate matter, not only because of the annoyance they cause but because of their effect on health when combined with SO_2. Due to the complex nature of particulates, which may be emitted from combustion processes, and to the difficulties in sampling and measuring the true concentrations of particulate matter occurring in ambient air, estimates on particulate-matter emissions are affected by large uncertainties.

The OECD[11] has recently assembled trends (see Table 3.7) in particulate matter emissions for a number of countries covering the period 1965–75, among which are several EEC member states. According to this report,

> trends in emissions of such pollutants over the period 1970–75 show a greater decrease than for SO_2 in a number of OECD countries. Emissions have been cut by about half in Germany and France. Smoke has been reduced to the same extent in the United Kingdom and emissions of all forms of particles have

been reduced by more than 30 per cent in the United States. Decreases in the exposure of people to particulate matter have been recorded in several countries and particularly in more severely polluted areas.

In the United Kingdom the hours during which sunshine penetrated to the ground in central London and Manchester in the winter have been rising. Improvements in air quality in the United Kingdom are due mainly to domestic smoke control.

In many other countries decreases in emissions and in ambient concentrations are the result of increased use of industrial emission-control equipment. Such equipment can effectively collect coarse particles, over 2–3 μ in diameter. It does not, however, eliminate finer particulates. The result is a decrease in the total volume of particulate emissions, but an increase in the emission of fine particulates.[12] This fine dust can be deposited in the human body causing bronchitis, impaired breathing and increased susceptibility to asthma and the common cold. It can also affect visibility and perhaps temperature.

Unlike inland water pollution, which remains confined to the river basin concerned, air pollution is unbound by nature; the pollutants, once emitted, will be carried to all directions by the wind and spread everywhere. In travelling they often react physically and chemically in the atmosphere and may form new secondary pollutants. Indeed the complex transformations which the major air pollutants (SO_2, particulates, NO_x, HC, CO_2, O_3 etc.) undergo in the atmosphere are still poorly understood and large research efforts are made to gain a better knowledge on their behaviour. At this point I would like to recall the activities of the EEC Commission which has promulgated an international coordination of research, both within and outside the EEC, in the frame of the COST 61a programme (COST = *CO*opération *S*cientifique et *T*echnique). This

Table 3.7 Emissions from selected countries as compiled by OECD for 1965 and 1975 (base = 100 in 1970)

Country	Particulate matter		NO_x		Hydrocarbons	
	1965	1975	1965	1975	1965	1975
France	81.6	49.4	65.3	132.0	76.0	93.4
Germany	202.8	47.6	82.5	114.0	79.4	98.4
Italy	70.0	—	63.8	112.9	69.6	79.5
Netherlands	136.0	73.3	82.8	105.6	101.7	100.9
United Kingdom	159.7	54.2	—	93.3	—	102.9
Sweden	—	150.0	67.0	170.0	—	252.0
Canada	—	100.6	—	153.0	—	88.6
United States	—	63.7	—	108.8	—	88.2
Japan	—	—	48.0	107.9	—	—

programme, which is just being reconducted for another 4-year period, aims at research into the physico-chemical behaviour of SO_2 in the atmosphere.

It has been generally recognized that the physical and chemical reactions of SO_2 and particulates in the atmosphere give rise to the formation of aerosols or small particles of solid or liquid matter dispersed throughout the air. The resulting secondary pollutants may reduce visibility, cause acid rainfalls and affect health.

During the last few years, growing concern was raised on a phenomenon known as 'long-range transport of atmospheric pollutants'. In countries with high precipitation and poorly buffered soils, particularly in the northern parts of Scandinavia, the rainfalls, as well as the inland waters, have an increasing acidity, lowering the pH down to 3.5 in some cases and damaging the biological chain in the water.

Studies by OECD on long-range transport of air pollutants in Europe show that sulphur is carried across frontiers.[13] Where both SO_2 and sulphate aerosols were transported over distances up to a few thousand kilometres they contributed to the acidity of rain and other precipitation. A first estimate of the relative importance of sulphur deposition of national and foreign origin was made by OECD using data collected in 1974 (Table 3.8). Further investigations in 1975 and 1976 showed the importance of changes in the weather, particularly the wind, to the contributions made by countries of origin to the long-range transport of pollutants.

At international level this concern on the long-range transport of air-pollution has condensated in action. In the frame of UN–ECE Geneva during the past months, an international convention was negotiated for the ECE region as a whole (including USA and Canada as well) on long-range transboundary air pollution, starting with SO_2. This convention will be signed at a high-level meeting which will be convened in November this year.

The phenomenon of long-range transport of air pollution is an example for the regional, if not global, impact of such pollution. Another case although still less-well understood, is the issue of CO_2 pollution.

Carbon Dioxide

Measurements reveal that atmospheric CO_2 levels have been increasing for at least a century, rising from 290 parts per million (ppm) in the 1860–90 period to 332 ppm in 1976. Over the last 20 years, the increase has been 5 per cent.

Most experts consider the major cause of the CO_2 increase to be the combustion of fossil fuels. Recently, the increasing destruction of the world's tropical forests, which otherwise store large quantities of CO_2, has also been identified as an important contributor to the problem. There is little dispute among investigators that atmospheric CO_2 will continue to increase, based on projections of energy demand and supply into the next century, and the absence

Table 3.8 Estimated deposition of sulphur,[a] Europe, 1974

Receivers	Emitters														
	Austria	Belgium	Denmark	Finland	France	Germany	Italy	Netherlands	Norway	Sweden	Switzerland	United Kingdom and Ireland	Other areas[c]	Unattributed	Total[a] received from all areas	Annual emission 1973
Austria	60	6	0	0	20	40	30	2	0	0	5	20	67	30	300	221
Belgium	0	100	0	0	30	20	0	5	0	0	0	30	6	10	200	499
Denmark	0	1	60	0	3	6	0	1	0	2	0	10	11	10	100	312
Finland	0	2	8	100	4	10	0	2	2	30	0	10	137	70	400	274
France	2	40	1	0	600	50	30	10	0	0	6	100	57	150	1000	1616
Germany	8	60	7	0	100	700	7	40	0	2	7	100	120	90	1300	1964
Netherlands	0	10	1	0	10	10	0	60	0	0	0	30	6	10	150	391
Norway	0	4	8	1	9	10	0	4	30	9	0	60	22	100	250	91
Sweden	0	7	30	10	10	30	0	6	6	100	0	40	108	100	500	415
Switzerland	1	2	0	0	20	7	6	1	0	0	30	10	6	20	100	76
United Kingdom	0	8	2	0	20	10	0	4	0	0	0	800	14	100	1000	2883[b]
Other areas[c]	60	60	80	40	200	400	900	40	9	50	10	600	700	1000	11000	—
Total emitted to above areas[a]	100	300	200	150	1000	1300	1000	200	40	200	60	1800	8200	1900	17000	na

NOTES:

a) Unit: 10^3 tons of sulphur. These data are based on 1973 emission data. The 1974 meteorological conditions are used to calculate the 1974 sulphur deposition. Number rounded to one significant figure and accurate to within ±50 per cent. The sums are calculated from unrounded figures and thereafter rounded separately.

b) Including 80 000 tons of sulphur from Ireland.

c) Other areas refer to some other European countries not included in the list of either the emitting countries or the receiving countries.

of alternative fuels or strong sociopolitical motivations for significantly reducing reliance on fossil fuels in the foreseeable future.

Concern about CO_2 build-up relates to a predicted gradual increase in average global temperature with secondary effects on precipitation patterns, ocean currents, degree of cloudiness and overall world climate and weather. The warming is associated with a 'greenhouse' effect created when higher concentrations of atmospheric CO_2 trap more and more of the heat that would otherwise radiate away from the earth into outer space.

There is, at present, no firm evidence that global temperature is actually being affected by CO_2 increase. Most investigators believe that there is, in fact, a CO_2 effect, but that it is being obscured by natural climatic variations. They further believe that, by the end of this century, the effect of increased fossil fuel combustion will begin to be detectable in the form of significant increases in global atmospheric temperature, leading to discernible shifts in worldwide climate.

Major uncertainties exist, however, because of the large gaps in our knowledge of how CO_2 moves through, and is stored by, the environment. The oceans, for example, might increase their uptake of carbon dioxide and thus mask the impact on the atmosphere; or, rising temperatures over the oceans may induce cloudiness, which would have an offsetting, cooling effect. If CO_2 increases from the use of fossil fuel do occur as predicted, it is still unclear what the actual change in global temperature will be. Predictive models suggest that increases of from 1.5 to 3 °C are possible by the year 2025 if atmospheric CO_2 concentrations double.

The near-term, major concern relates to the effect on world food production in the face of shifting temperature and rainfall patterns. Any global temperature increase is predicted to be distributed non-uniformly. At higher latitudes, agricultural productivity and general livability are likely to be improved, particularly where the length of the growing season is now marginal or submarginal. Certain countries and regions might benefit from CO_2-induced changes of a modest nature. It is currently unknown who the actual 'losers' and any 'winners' would be over the long term.

For national leaders, the issue involves significant risk in the face of a great deal of uncertainty. Given the world reliance on fossil fuels, the fact that a future CO_2-induced climate change cannot be proven, and the expectation that drastic curtailment of fossil-fuel usage is neither economically, technically or politically feasible, options appear to be limited. Reducing fossil-fuel use enough to affect CO_2 build up is extraordinarily difficult. An international scenario[14] based on halving annual growth rate of fossil-fuel energy consumption to 1.6 per cent would require, for example, a freeze on the expansion of Japanese and Western European fossil-fuel use, a 50 per cent per capita reduction in United States use, and only 1.7 per cent per capita annual growth rate of energy use in the Third World. If such a solution were initiated in 2000, CO_2 doubling, according to model calculations, would be delayed roughly 20 years.

The consequences of this situation are obvious: the unresolved scientific problems must be investigated as quickly as possible to know if mankind is likely, within a relatively short time, to touch natural limits of the use of energy, resulting in an unacceptable impact on the environment.

Nitrogen Oxides, Hydrocarbons and Photochemical Oxidants

Nitrogen oxides (NO_x) are produced upon burning of fossil fuels particularly when the flame temperature is high. The main sources are stationary combustion installations, like power plants, incinerating and steam-raising plants, domestic heating installations and mobile sources like automobiles with spark ignition or diesel engine and aircraft gas turbines.

Nitrogen oxides, at high concentrations, are toxic to man and most animal species. In the atmosphere the nitrogen oxide (NO) emitted is rapidly oxidized to nitrogen dioxide (NO_2), a toxic gas that decomposes in sunlight and starts a series of complex reactions that can, in the presence of hydrocarbons (HC), result in 'photochemical oxidants' such as ozone (O_3). Such reactions have been found to occur mainly near to large industrialized and densely populated areas, also within the EEC.

Emissions of NO_x have increased in many countries in the last years and are likely to continue that way. The increases were and are due to the growth in number and size of stationary sources, to the increased use of natural gas in domestic heating and, last but not least, to an ever rising number of motor vehicles (their number may rise within EEC from 70 million in 1975 to some 110–125 million units in 2000).

In 1975 some 5 million t of NO_x were emitted within EEC, this figure may go up to 6 million in 1985 and to 7 million t in 2000 if no new control technologies are being introduced on top of already existing ones (e.g. tighter EEC limits for motor vehicle NO_x emissions).

Hydrocarbons are the best known for their role in the formation of photochemical oxidants. They are among the most difficult air pollutants to monitor and analyse for the variety of compounds present and to control because they are emitted wherever fuel is handled or burned. Many of them are a key component in almost all atmospheric reactions. Gaseous hydrocarbons are, furthermore, toxic or causes of cancer, though not at the concentrations measured in city air. They can, by absorption or reaction, become fine particles capable of being inhaled.

According to OECD (Table 3.7) over the period 1970–75, total emissions of HC increased in Norway, Sweden and the United Kingdom and remained stable, or decreased slightly in Germany, the Netherlands, Canada, and the United States. The increases are explained by higher emissions from sources such as fuel handling and the use of solvents for which little or no control technology is or can be used. The motor vehicle with its emissions of unburnt HC constitutes an important source for this type of pollutant, particularly in areas with high traffic density, and certainly tighter control measures will be needed in the future.

In some parts of the world like Australia, Japan, south-eastern Canada and the western United States photochemical oxidants have become an important urban problem. The phenomenon, however, is also regional in scale, and observations in eastern North America and Europe indicate the need for control over wide areas in order to avoid the growth of photochemical smog. Photochemical oxidants are always formed in the atmosphere on sunny summer days, but may be formed in winter as well.

Networks of stations for monitoring ambient levels of NO, HC and oxidants are of recent origin and provide only limited information. Measurements made during the hot summer of 1976 showed that oxidant levels in major cities in most OECD countries were considerably above levels recommended for the protection of public health. The conviction is growing that nitrogen oxides and oxidants must be monitored and controlled more extensively.

CONCLUSIONS

Energy production and use and particularly combustion of fossil fuels is at the origin of most of the air pollution encountered within EEC member states. Therefore, policies destined to combat such pollution must consider what contribution to overall emissions is possibly coming from combustion processes, what types of fuel are used and where such use takes place. Some important questions need to be answered. What is the future pattern of energy production and use? Where is this energy to be used and within what equipment? Which pollutants are likely to be emitted and in what quantities? What impact will these emissions have once transformed into ground-level concentration?

Only a thorough analysis of the energy situation in terms of supply and demand can provide more or less precise answers to these questions on which policies can then be built against air pollution. Out of the many possible approaches one can choose those then likely to achieve, with minimum effort in terms of money invested and energy consumed, the desired air quality for specific pollutants at regional or overall scale.

In the light of forecast made on EEC energy use, and the ensuing air pollution, all efforts made to save energy will result in a net benefit to air quality, thus requiring less stringent control actions for the future. Despite such efforts emissions of pollutants like SO_2, particulates, NO_x, HC and CO_2 will increase in the medium term and more control will be needed in order to keep the air quality acceptable, particularly in large urban areas.

REFERENCES

[1] Brimblecombe, P. (1979). Von den spitzen Ecken der Atome, *Umwelt*, **1**.

[2] The World Energy Conference (1978). *World Energy Demand to 2020*, IPC Science and Technology Press, Guildford, UK.

[3] Thompson, T.J. (1971). IAEA-SM-146/4, Vienna.
[4] U.S. Department of the Interior (1972). *United States Energy—A Summary Review*, Washington DC.
[5] EEC Commission (1974). Preliminary report on the problems of pollution and nuisances originating from energy production, SEC(74)1150 final, Luxembourg, 3 April.
[6] EEC Commission (1975). Second report on the achievement of Community Energy policy objectives for 1985, COM(77)395 final, Brussels, 29 July.
[7] EEC Commission (1978). Energy objectives for 1990 and programmes of the Member States, COM(78)613 final, Brussels, 16 November.
[8] Environmental Resources Limited (1978). *Energy Strategies and the Environment 1976–2000*, study made for the EEC Commission, London.
[9] OECD (1978). *Clean Fuel Supply*.
[10] Ellis, R.J. (1977). An evaluation of economical consequences resulting from the application of directive proposal COM(75)681. The use of low sulphur fuel oils with the aim of decreasing sulphurous emissions, EUR-6011 EN. Study carried out for the EEC Commission.
[11] OECD (1979). *The State of the Environment* (restricted) April.
[12] ECE Geneve (1977). *Proceedings of the ECE Seminar on Fine Particles, Villach, Austria, 17–22 October*.
[13] OECD (1977). *Long Range Transport of Air Pollutants*.
[14] SRI International (1977). Study on socio-political impacts of CO_2 built-up in the atmosphere due to fossil fuel consumption. Prepared for the US Energy Research and Development Administration.

Pollutants and Their Ecotoxicological Significance
Edited by H. W. Nürnberg

CHAPTER 4

Physical Parameters of Pollutant Transport and Diffusion

P. E. RAVUSSIN

SUMMARY

The atmospheric turbulence is the main cause of dispersion of the pollutants, but the turbulence is one of the most difficult physicomathematical problems, which is not yet completely solved.

The problem of fluid movements can be approached in different ways. The Eulerian way considers the field of the speed vectors at fixed points of the space. This is described by the Navier–Stokes equation which is based on the d'Alembert (or Newton) principle. The Lagrangian approach follows a particle of fluid. It is based on the principle of least action. In addition, to close the system, we need two more equations: the equation for the conservation of the momentum and one of the thermodynamic equations, for instance the equation between density, pressure and temperature. In the atmosphere, the processes are mainly adiabatic.

In a laminar flow, the variables in the equations are 'good variables': at a given time they take only one value: but this is not so in turbulent flow which is the most usual case. We cannot use the above equations here. However, if the variation of the movement is stationary enough we can make the ergodic assumption and replace in the above equations the 'ordinary' variables by the 'statistical' variables and make a time average on it. This will lead to the tensor of Reynold which is one of the parameters of the turbulence.

PHYSICAL PARAMETERS OF TRANSPORT

The purpose of this chapter is not to describe all the mechanisms which occur in the atmosphere to transport and diffuse pollutants, but to explain the concept which leads to a general approach of the solutions on a local or regional point of view. These concepts can also be applied to water and ground.

Wind

The process of transport is driven by the vectorial field of the wind. A smoke which is released by a chimney at point A will be transported along a *stream line* of the vector field (which is in a steady state in the case of the Figure 4.1).

So one physical parameter of transport is indeed the *tri-dimensional field of tri-dimensional vector of wind speeds*. If the wind field is more or less homogeneous on a horizontal plane (on a flat country), this is generally not the case on a vertical profile. This simple scheme would be sufficient if the temperature of the atmosphere is constant and if the temperature of the smoke is the same, which is obviously not the case.

Temperature

Generally the smoke has a higher temperature at the output of the chimney. The plume will ascend through the combined effect of the initial speed and buoyancy. The pressure of the plume will rapidly reach its equilibrium point with atmosphere, but as it is warmer the smoke has a smaller density than the ambient air. So the plume will ascend through its Archimedes strength. It will cool mainly through conductive and turbulence processes until it reaches an equilibrium point.

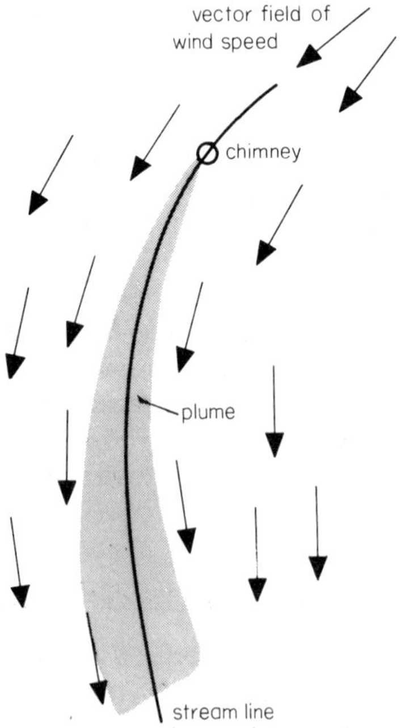

Fig. 4.1 Transport and diffusion of a plume of smoke

Because of the gravity field of the earth, the pressure of the atmosphere diminishes with the altitude.

$$\frac{\partial p}{\partial z} = \rho(p)g$$

where p = pressure ($\mathrm{N\,m^{-2}}$)
z = altitude (m)
ρ = specific mass ($\mathrm{kg\,m^{-3}}$) $\equiv$ ($\mathrm{N\,m^{-4}\,s^{-2}}$)
g = gravity acceleration ($\mathrm{m\,s^{-2}}$).

Neglecting the Coriolis strength of the earth's rotation, we can calculate at a thermodynamic equilibrium, the vertical profile.

$$\frac{\partial T}{\partial z} = -\frac{g}{C_p}$$

where T = temperature (K)
z = altitude (m)
g = gravity acceleration ($\mathrm{m\,s^{-2}}$)
C_p = specific heat at constant pressure ($\mathrm{m^2\,s^{-2}\,T^{-1}}$)

At the ground the normal temperature gradient $\partial T/\partial z$ is $\sim 1\,°\mathrm{C}\,100\,\mathrm{m}^{-1}$. But through the general circulation of the atmosphere, and also through the conductive heat exchange with the ground, the radiative absorption and emission processes, and the transformation of kinetic energy through the viscous processes, this gradient is reached very seldom. Sometimes it even occurs that an inversion layer of temperature exists. The plume can no longer reach its normal height and is blocked in the layer. It occurs if the difference of temperature between the plume and the atmosphere at the bottom of the inversion is smaller

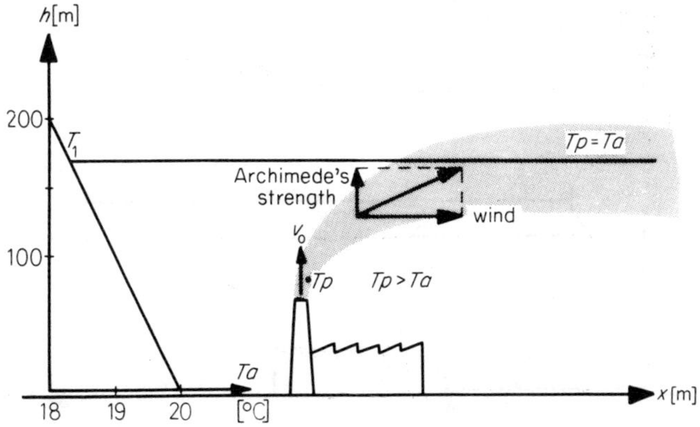

Fig. 4.2 Ascension of a plume of smoke

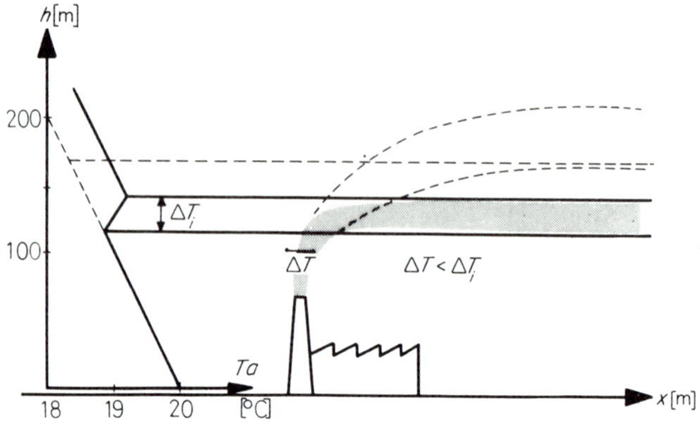

Fig. 4.3 Stagnation of a plume in an inversion layer

than the difference of temperature between the bottom and the top of the inversion.

In the other case the inversion is pierced and the plume reaches another equilibrium point. Sometimes the wind blows in the opposite direction over the inversion layer. There are many other thermal phenomena in the atmosphere. For instance it exists in winter as an islet of heat over a city.

Another important thermal phenomenon is the ascension of the vertical plumes of hot air in good weather conditions. So another important physical parameter of pollution transport is the *scalar field of temperature.*

In fact it is the difference of temperature between the cold polar air and the tropical hot air which leads to the formation of the big Atlantic depression. Because of the Coriolis acceleration, the depressions are turning counterclockwise in our hemisphere.

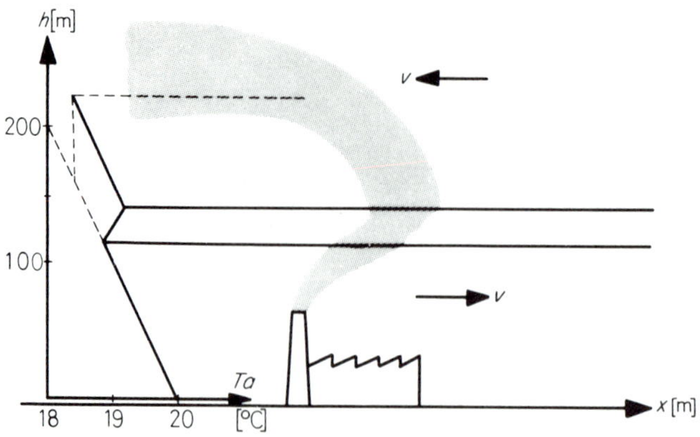

Fig. 4.4 The plume pierces the inversion layer

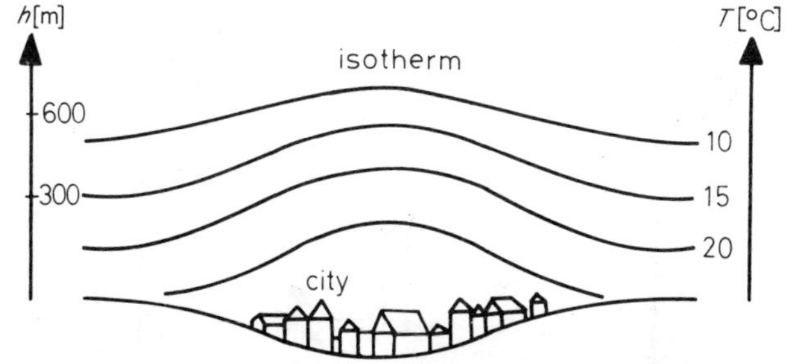

Fig. 4.5 Islet of heat

Pressure

The temperature field and the pressure field of the atmosphere can be expressed in terms of available potential energy (APE) of the atmosphere which is transformed to the kinetic energy of the atmosphere. This latter is in fact the wind and its turbulence.

So the third physical parameter for pollutant transport is the pressure field; which is the direct cause of the appearance of the wind. Unfortunately it is very difficult to measure, especially on the regional or local scales, because the horizontal pressure gradient or the deviation of the vertical gradient from the gradient of equilibrium is very small. Therefore we generally measure only the causes and the consequences of the pressure-gradient field which respectively are the temperature field and the wind field. To be complete we should also speak of the heating process of the ground or sea by the sun radiation and their cooling through the thermal radiation and of the waters latent heat exchange processes.

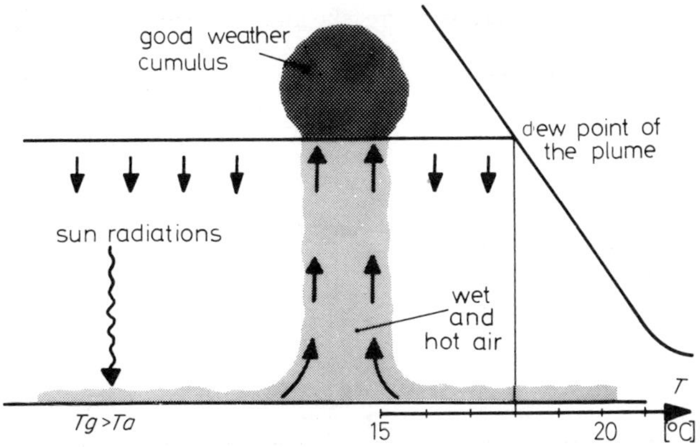

Fig. 4.6 Ascension of the hot-air plume

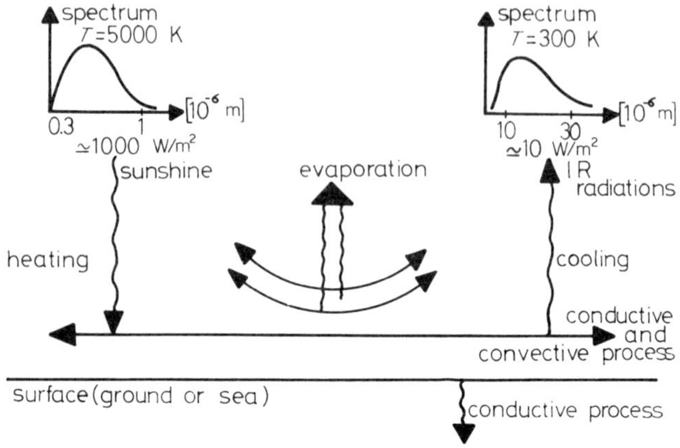

Fig. 4.7 Heat exchange

THE PHYSICAL PARAMETERS OF THE DIFFUSION

Until now, the parameters that we have seen are good, classical parameters. At one point and at a given time they take only one value, which can vary with the time and the location. But to explain the diffusion we need a statistical or *aleatory variable*; there are no more simple variables. At one point and at a given time, an aleatory variable represents an infinite number of values. We will see later what are the reasons to introduce this concept. We make here the distinction between the statistical variable and the aleatory variable, which corresponds to the general case.

Molecular Diffusion

We must consider two types of diffusion: the first one is *molecular diffusion*.

At normal pressure and temperature, the average speed of the molecules of the

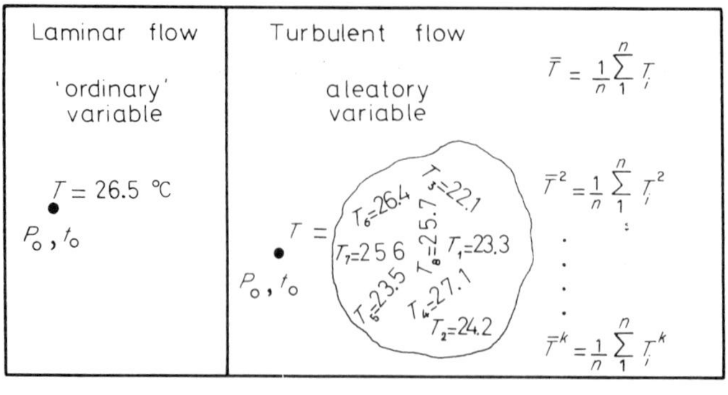

Fig. 4.8 Aleatory variable

air is about 480 m s^{-1}. If we consider a small group of marked molecules (for instance radioactive isotope of oxygen), we could imagine that they will spread out at this speed. This would be the case if the air is very diluted, and it could then be considered a *perfect fluid*. This could also be the case for another fluid: liquid helium. At very low temperatures (a few Kelvin degrees) one part of helium, for reasons explained by the quantum mechanic, becomes a perfect fluid. But in normal conditions, the numerical density of the molecule is very high and their size a few angström (1 Å $= 10^{-10}$ m), that they collide very often. The average free path between collisions is only $\sim 0.05\,\mu$ and this obviously considerably slows the molecular-diffusion process. Therefore, it is very seldom that molecular diffusion plays a role in the atmosphere, but it occurs occasionally. In practice the molecular diffusion speed in the atmosphere is of the order of magnitude of 0.1 mm s^{-1} and is proportional to the gradient of concentration.

For instance we have observed on the slope near Lake Léman (Lake Geneva), a very good phenomenon of molecular diffusion. In March, early in the morning, when the sun starts to heat the ground, and the lake remains cold, it appears as a very laminar light thermal wind. Also as the wine-growers burn the old wine-shoots, we can observe, (a) a turbulent cooling process, (b) a very smooth transport process with a molecular diffusion, and (c) a stagnation layer in the thermal inversion layer of the lake.

The conclusion is that situations can arise in the atmosphere where the pollutants are transported over long distances without appreciable dilution. Fortunately smoke is generally rapidly diluted through the turbulent-diffusion processes.

Let us now look at the concepts which lead to the parameters of turbulent diffusion. The movement of a laminar viscous fluid can be described by five equations. The first three are the vectorial Navier–Stokes equations.

Let us consider a small volume (control volume) of fluid, small enough in such a way that the general movement of the different parts of the volume is the same everywhere and big enough so that the fluctuation of the number of molecules in

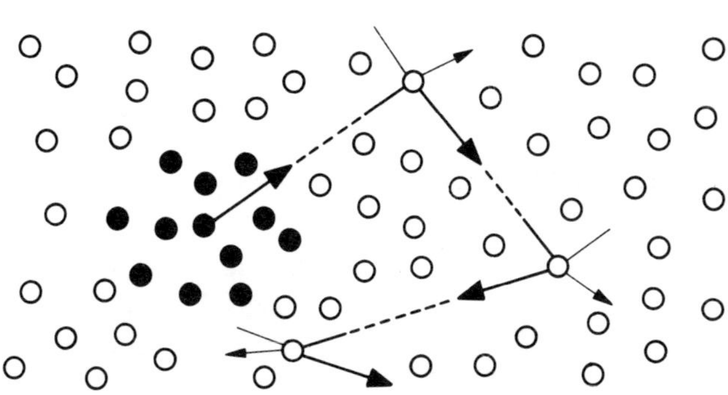

Fig. 4.9 Molecular diffusion

this volume is negligible. We have defined the *particle of fluid* which is connected with a *space scale* notion. For the material point of the mechanic, Newton's principle is expressed by

$$\vec{F} = M\vec{a}\,.$$

On a control volume, $\vec{F}$ is replaced by the difference of pressure between the two sides of the control volume, ΔV,

$$\frac{p_1 - p_2}{\Delta x} \equiv \overrightarrow{\text{grad}}\ p$$

and the mass, M, by the ratio between the mass of the molecules in the particle of fluid divided by the volume, ΔV

$$\frac{M}{\Delta V} = \rho$$

where ρ is the density of the fluid.

So our equation becomes

$$-\overrightarrow{\text{grad}}\ p = \rho\vec{a}\,.$$

It is the Eulerian equation of the perfect fluid. But we must also take into account all the collisions between the molecules, also called internal friction or viscosity. This leads to a diminished effective pressure in the equation.

As we will observe the phenomenon from a fixed point relative to the movement of the fluid, $\vec{a}$ is the total derivation of the speed, so the Navier–Stokes equation can be written

$$d_t\vec{v} = -\frac{1}{\rho}\ \overrightarrow{\text{grad}}\ (p - r_v).$$

The other equations are: an equation of continuity for the conservation of momentum

$$\frac{\partial \rho}{\partial t} + \text{div}\,(\rho\vec{v}\,) = 0$$

and, for instance, a thermodynamic relationship

$$\frac{\mathrm{d}\rho}{\mathrm{d}t} = 0 \qquad \text{adiabatic process}$$

or

$$\frac{\partial \rho}{\partial t} + \vec{v}\ \overrightarrow{\text{grad}}\ \rho = 0.$$

These equations describe the fluid at the Eulerian point of view: we consider the speed at fixed point of the space.

If we want to follow the fluid particle in its movements, we must apply the principle of least action

$$A = \int_{t_1}^{t_2} L(q_n, \dot{q}_n, t)\,\mathrm{d}t = \text{minimum}$$

where L is the Lagrangian function which expresses all the possible movements of the fluid. The movement of fluid is a straight line in the space of phases.

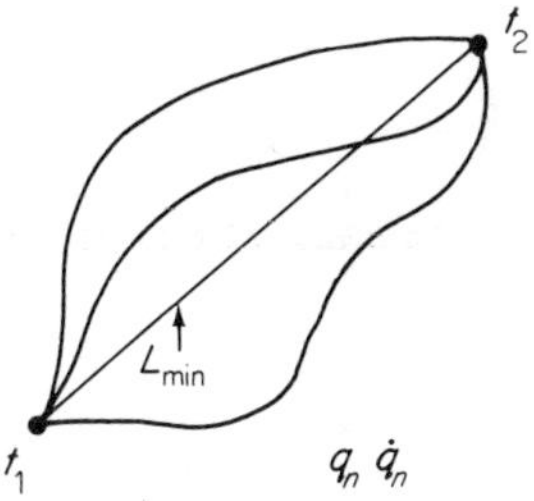

Fig. 4.10 Principle of least action

In a turbulent fluid the variables become aleatory functions. We cannot use these equations without further assumption. The new hypothesis is that you can use it *on average*. Of course we do not know the statistical averages of the aleatory variable, but we can use the ergodic theory which says that if the phenomenon is stationary, you can replace the *statistical average* by *time average*. But even with the stationary assumption, it appears in the equation's new terms like the Reynold's strength tensor

$$\rho\left(v_j \frac{\partial v_i}{\partial \lambda_j}\right)$$

and we need more equations to find the solution. The problem of the closure of the set of equations of turbulent fluid is *not yet completely solved.*

In the non-stationary case, which occurs most frequently in the atmosphere, we need more concepts. The one we are using is the concept of the aleatory mechanic.

1 The particle of fluid is replaced by the aleatory corpuscle which, as co-ordinate of position q and speed $\dot{q}$, has aleatory functions of time.

Thus the aleatory corpuscle $\mathscr{R}$ is an association of a field of probability density of appearance ρ and of an aleatory field of speed f.

$$\mathscr{R}(q, \dot{q}/t) = \rho(q/t) \otimes f(q, \dot{q}/t)$$

2 The principle of least action is replaced by a principle of least loosening of the field of probability density.

$$\rho = \int_{t_1}^{t_2} L(q, \dot{q}/t)\,\mathrm{d}t = \text{minimum}$$

Conclusion

The problem of turbulence is very complicated and not yet completely solved. However, it is possible to define two parameters of the diffusion. They are

1 for the turbulence of the wind

$$k_m = \frac{\bar{u}_1 \bar{u}_3}{\partial \bar{u}_1 / \partial x_3}$$

2 for the turbulence of the temperature

$$k_G = \frac{\overline{Tu_3}}{\partial T / \partial u_3}$$

These parameters are the mostly used for the problem of diffusion studies.

REFERENCES

1 Dedebant, G., and Wehrle, Ph. (1945). Mécanique aléatoire, *Portugaliae Physica.*

2 Dedebant, G. (1948). *Essai d'une Axiomatique Aléatoire*, Publicaciones de la Facultad de Ciencias fisicomatematicas, La Plata.

3 Dedebant, G. (1951). Sobre la prevision estadistica del tiempo, *Serie Meteorologica*, **7**, Buenos Aires.

4 Werhle, Ph. (1956). *L'Univers Aléatoire*, Editions du Griffon.

5 Baatard, F. (1963). *Structure Générale d'une Mécanique de la Diffusion*, Multi Office, Lausanne.

6 Favre, A.J. (1965). The equations of compressible turbulent gases, *Institut de Mécanique statistique de la turbulence*, Marseille.

7 Ravussin, P. (1971). *Mesure de l'Inconnexe d'une Fonction Aléatoire*, Multi Office, Lausanne.

8 Ravussin, P. (1973). *Théorie de la Prévisibilité*, publication EPFL No. 134, Lausanne.

Pollutants and Their Ecotoxicological Significance
Edited by H. W. Nürnberg

CHAPTER 5

Transfer of Industrial Pollutants to the Rural Environment: A European Problem

C. RONNEAU, J. L. NAVARRE AND P. PRIEST

SUMMARY

Heavy elements are emitted into the atmosphere together with sulphur dioxide which can be considered as a universal indicator of pollution events. These pollutants are transferred towards rural environments where they are eventually deposited to the ground by sedimentation and by the action of precipitations. In certain European regions the contamination rate due to airborne particles is much higher than the contamination rate due to the use of chemical fertilizers.

INTRODUCTION

This chapter intends to demonstrate the need for a thorough research on the contamination of the European agricultural soils by the mid- to long-distance transfer of various toxic elements from industrial and urban zones. The demonstration is based on two studies.

1 A regional study performed in Belgium in order to quantify the contamination of rural atmosphere by industrial or urban plumes and to quantify the contamination of agricultural soils by heavy elements.
2 The results of the regional study will be extrapolated to a European scale by using the data of the OECD programme *Long Range Transport of Air Pollutants* (LRTAP).[2]

THE TRANSFER OF AIR POLLUTANTS

Provided the observations are made in rural areas, it is possible to perceive sulphur dioxide (SO_2) immission episodes at distances up to a few hundred kilometres from industrial sources. These episodes are essentially intermittent, depending mainly on air-mass trajectories from the sources.

Because of its good detectability, its wide prevalence as an air pollutant and its very low natural level in air, SO_2 is used as an indicator of the occurrence of immission episodes in a typical rural area of Belgium. If, during an episode, air particles are sampled by filtration or by impaction, it becomes possible to determine the amount of toxic elements transferred together with SO_2 and, moreover, to estimate the proportions of a given element relative to sulphur. In order to simplify the discussion, we will choose one single element as an example; we will consider the case of arsenic.

In the course of 132 episodes observed in a Belgian rural area, the relative proportion in air of arsenic and sulphur have been found[1] to be of the order of

$$\frac{\mathrm{As}}{\mathrm{S}} = 0.21 \times 10^{-3} \tag{5.1}$$

Figure 5.1 shows the observed correlation: each point represents one single air pollution episode in the rural area. It can be seen that the concentrations of arsenic measured in our rural atmosphere are seldom a matter of concern from a toxicological point of view. But what about the fate of that metal and of the many other toxic elements emitted by our technology and by the combustion of fossil

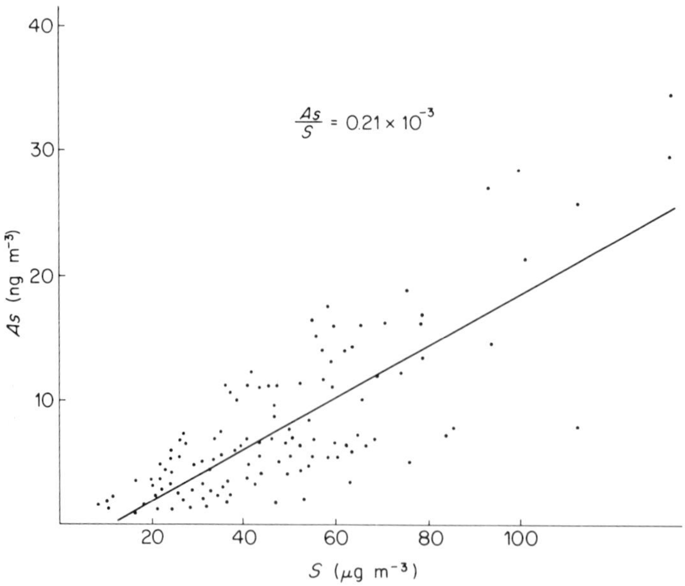

Fig. 5.1 Correlation between As and S in the rural atmosphere

fuels? It is estimated that Belgium emits 500×10^3 t of sulphur each year: if we assume that the proportions of arsenic and sulphur *in air* are an acceptable estimation of the proportions of arsenic and sulphur at the sources, it should mean that 105 t of arsenic are emitted each year by that country!

If this is really the case, concern should be oriented towards the agricultural soils which constitute the largest areas for disposal of these airborne pollutants.

THE DEPOSITION OF HEAVY ELEMENTS

It is estimated that about 200×10^3 t of sulphur are deposited each year on the Belgian ground. If we accept the hypothesis that arsenic is deposited at the same rate as sulphur, using the ratio given by (5.1) gives a figure of 42 t of arsenic being deposited each year on the Belgian ground.

We tried to verify this figure by a network of nineteen rain gauges installed in rural zones all over the country. Analysing rain samples collected by the network gives a figure of 45 t. This is of course a very lucky coincidence! Data relative to other elements differ more or less markedly from the theoretical forecast. However, except for sodium, the figures are of the same order of magnitude: Table 5.1 gives a comparison between the computed and the observed figures. The relative similarities between columns B and C allow us to speculate about the impact of heavy element deposition over a whole continent.

Table 5.1 Elemental deposition rates on Belgian soils

Element	A $\frac{\text{Element}}{\text{Sulphur}}$ observed ($\times 10^3$)	B Deposition rates calculated from column A $10^3\,t\,y^{-1}$	C Observed deposition rates $10^3\,t\,y^{-1}$
Na	4.5	0.90	30.0
Cr	0.16	0.032	0.066
Fe	12.0	2.5	18.0
Zn	9.5	1.9	11.4
As	0.21	0.042	0.045
Se	0.053	0.011	0.0064
Br	1.3	0.26	0.081
Cd	0.050	0.010	0.057
Sb	0.088	0.018	0.090
Pb	5.9	1.2	0.75

Column B calculated from sulphur deposition rates and from ratios observed in rural atmospheres (column A). Column C observed from our network of rain gauges.

EXTRAPOLATING TO EUROPE

Extrapolating results obtained in a small country to Europe is certainly erroneous; however, it permits us to emphasize the real concerns of environmentalists.

Regions in Europe are submitted to a sulphur deposition rate of more than $10\,g\,m^{-2}\,y^{-1}$, or $10 \times 10^3\,kg\,km^{-2}\,y^{-1}$.[2] Figure 5.2 gives an illustration of the deposition patterns.

Again, if we accept the relationships observed in Belgium, the figure should correspond to a deposition rate for arsenic of $2.1\,kg\,km^{-2}\,y^{-1}$. Such a figure is not explicit in itself, it should be compared to another source of contamination of the soils: the heavy elements present in chemical fertilizers which are a cause of great concern. The figure of $2\,kg\,km^{-2}\,yr^{-1}$ of arsenic is more than ten times higher than the contamination due to the use of chemical fertilizers in Belgium.

Fig. 5.2 Sulphur deposition pattern in Europe (LRTAP)

Table 5.2 Contamination of agricultural Belgian soils by heavy elements. Results are given in $kg\,km^{-2}\,y^{-1}$

Element	Fertilizers	Precipitations
Na	1030.0	1010.0
Sc	0.055	0.050
Cr	8.0	2.2
Fe	2540.0	590.0
Co	0.073	0.54
Zn	10.0	380.0
As	0.15	1.5
Se	0.096	0.18
Br	12.0	2.7
Cd	0.073	1.9
Sb	0.031	3.0
Cs	0.0050	0.080
Ba	21.0	11.0
La	1.4	0.47
Ce	0.90	0.50
Sm	0.27	0.050
Eu	0.034	0.084
Hg	0.027	0.054
Pb	0.30	25.0

Table 5.2 gives the results of a study, performed in Belgium, about the relative contamination of agricultural soils by fertilizers and by deposition of airborne particles.

It should be born in mind that regions in Europe are likely to be more exposed than Belgium where sulphur deposition rate is estimated to be of the order of $7\,g\,m^{-2}\,y^{-1}$, whereas rates of $10\,g\,m^{-2}\,y^{-1}$ have been calculated for areas of central Europe.

CONCLUSIONS

As far as toxic metals are concerned, deposition is likely to be more alarming than the use of chemical fertilizers: this justifies a legitimate anxiety about contaminants from the air! Furthermore, it is worthwhile noticing that, by an unfortunate geographical coincidence, the zones of highest sulphur deposition coincide with the zones of highest cereal production. In other words, the soils most intensively devoted to human food production are most intensively exposed to toxic metals contamination.

REFERENCES

[1] Navarre, J-L. (1979). *Contamination de l'environnement rural par les polluants atmosphériques, un exemple-type: Louvain-la-Neuve*. Doctoral thesis, Université de Louvain.

[2] OECD (1977). *Long Range Transport of Air Pollutants. Measurements and Findings*, OECD, Paris.

[3] Maas, G. (1979). *Contamination des sols agricoles belges par les éléments-traces des engrais et des précipitations atmosphériques*, Mémoire de fin d'études, Université de Louvain.

Pollutants and Their Ecotoxicological Significance
Edited by H. W. Nürnberg

CHAPTER 6

Plants as Bioindicators for Atmospheric Pollution

A. C. Posthumus

SUMMARY

The use of plants as indicators and/or accumulators for the monitoring of biological effects of air pollutants is the main topic of this chapter. Definitions of indicator and accumulator plants are given for a better understanding of biological effect monitoring. Both acute and chronic effects of air pollutants are described, and examples of indicator plants of differing degrees of sensitivity are mentioned. Great importance is attached to the standardization of plant material and growth conditions in biological effect monitoring. Examples and possible uses of plants as indicators and/or accumulators for monitoring the effects of air pollutants on a local, regional, national and even an international scale are explained. The problems caused by other harmful agents and by the combined effects of different air pollutants are discussed.

As an example of a national monitoring network for biological effects of air pollutants the nation-wide monitoring network with plants in the Netherlands is described. In this network different types of biological effect monitoring are used.

Suggestions are made for possible international programmes for monitoring networks for biological effects of air pollution. The expected difficulties and needs are indicated, and finally the ultimate goals and problems of monitoring biological effects of air-borne pollution are discussed.

INTRODUCTION

Air pollution in relation to vegetation may be characterized in different ways. First of all, the concentration levels of the pollutants may be measured and their possible effects may be estimated, if the dose-effect relationships in the prevailing conditions are known. But a second, more direct approach is the actual measurement of the effects of air pollutants on the plants in question.

For the first approach several physical and chemical methods have been developed and very sophisticated techniques are now available. But data on dose-effect relationships for many plant species and varieties with differing degrees of sensitivity are mostly lacking or are insufficient, and the influences of the external and internal environmental conditions are generally unknown. For this reason it is easier, in the present circumstances, to study the effects themselves on the naturally growing plants or cultivated crops, and it is even better to expose standardized indicator plants in well-known conditions.

Monitoring of effects of air pollutants on vegetation may be very important, for example to be aware of possible damage to natural or crop plants and also to raise the alarm for possible risks to men and animals from special air pollutants to which plants are more sensitive. In addition, continuous effect monitoring with indicator plants over not too large areas may help to study the presence and distribution of pollutants in time and place, to control air-pollution abatement measures, and to perform air-pollution trend research.

In this chapter examples of monitoring of air-pollution effects on plants will be discussed and it will be stressed that both levels and effects of air pollutants should be monitored simultaneously as they complement each other. Measurements of levels are rather accurate and objective and may be helpful in predicting and explaining the occurrence of effects, but they cannot replace measurements of effects, integrating the influence of concentrations of pollutants and the external and internal conditions of the plants.

As an example of a national programme, the Dutch national monitoring network for air-pollution effects on plants will be reviewed. Existing international co-operation will be mentioned, and the need for more international effect monitoring networks will be stressed.

MONITORING OF BIOLOGICAL EFFECTS OF AIR POLLUTANTS BY THE USE OF PLANTS AS INDICATORS AND/OR ACCUMULATORS

Because of the relatively high sensitivity of plants to several air pollutants and the sometimes rather specific symptoms accompanying the effects, some plant species may be used as indicators for the detection, recognition and monitoring of air-pollution effects. When the plants also accumulate the polluting compounds without changing their chemical nature by metabolism and the pollutants are easily analysed in plant samples, they may also be used as accumulators. And if the accumulation of air pollutants by plants is also considered as an effect of the pollution, plants are very suitable for detection, recognition and monitoring of air-pollution effects.

Several types of effects are known and these may be divided into acute effects of high concentrations during short time periods and chronic effects of low concentrations of air pollutants during long time periods. Acute effects are the clearly visible chlorosis and necrosis of leaf tissue, leaf, flower or fruit abscission, epinastic curvatures of leaf stems and others. Chronic effects may be shown in retardation or disturbance of normal growth and development (resulting in reduction of growth, yield or quality), but also slow discoloration (chlorosis) and even leaf-tip necrosis and total die-back of plant organs may be caused in the long run. In some cases the symptoms of acute and chronic effects may be rather specific for a special air pollutant or combination of pollutants.

Quite different plant species may be used as indicators and/or accumulators of air pollutants by showing special effects. For example species of lichens, mosses, ferns and higher, vascular plants (mostly phanerogames) have been used for this purpose. Biological effect monitoring may be performed by using the natural vegetation and crop plants present in the studied area, but differences in soil, water and other conditions may influence the effects and diminish the comparability of the results. For this reason it is better to use selected indicator and/or accumulator plants, cultivated in standardized conditions of soil and watering.

For this purpose mostly higher plants have been used till now, in addition to the use of transplanted lichens as used in the Ruhr area in the Federal Republic of Germany.[1] Several species and varieties of natural and cultivated plants, which are shown to be sensitive to one or more air pollutants, are in use in effect monitoring networks. Table 6.1 shows an example of a set of indicator plants of differing degrees of sensitivity, based on the experience of the national monitoring network for air pollution effects in the Netherlands.[2]

Indicator plants in standardized conditions in the open air may be used to detect, recognize and measure the acute effects of air pollutants and to study the presence, intensity and distribution of these effects in place and time. But chronic effects may also be determined by comparing growth, development, production and quality of indicator plants in pairs of small greenhouses, one of which is ventilated with charcoal-filtered air and one with unfiltered ambient air. Growth retardations and yield reductions have been found in the unfiltered, as compared to the charcoal-filtered, air. In this way possible crop losses in polluted areas may be indicated, although the effects on crop yield in practice are often quite different because of the different conditions. The use of open-top chambers instead of closed greenhouses, as practised in the United States of America,[3,4] is an improvement, but still different from the open-air situation. Plants grown in the open air are much more resistant to air pollution than plants cultivated in greenhouses. As a consequence, greenhouse crops are generally more vulnerable when air pollutants are also present in the greenhouses than crops in the open field.

Accumulator plants may also be cultivated in a standardized way at several locations to study the presence and distribution of air pollution. In particular the

Table 6.1 Review of indicator and accumulator plants, used for effect measurements of different air pollutants in the Netherlands

Plant species and variety	Air pollutant	Symptoms
Gladiolus (*Gladiolus gandavensis L.*) var. Snow Princess and Flowersong Tulip (*Tulipa gesneriana L.*) var. Blue Parrot and Preludium	Hydrogen fluoride (HF)	Leaf tip and marginal necrosis; F-concentrations in dry matter
Tobacco (*Nicotiana tabacum L.*) var. Bel W_3 Spinach (*Spinacia oleracea L.*) var. Subito and Dynamo	Ozone (O_3)	Leaf upper surface speckle necrosis
Small nettle (*Urtica urens L.*) Annual meadow grass (*Poa annua L.*)	Peroxyacetyl nitrate (PAN)	Leaf under surface band-forming necrosis
Lucerne (*Medicago sativa L.*) var. Du Puits Buckwheat (*Fagopyrum esculentum Mönch*)	Sulphur dioxide (SO_2)	Interveinal chlorosis; necrosis
Petunia (*Petunia nyctaginiflora Juss.*) var. White Joy	Ethene (C_2H_4)	Flower bud abortion; small flowers
Italian rye-grass (*Lolium multiflorum Lam.*) var. Optima	Fluoride and metal ions (F, Cd, Mn, Pb, Zn)	Ion concentrations in dry matter

non-convertible, persistent air pollutants such as fluoride and heavy metal ions may be measured relatively easily by physical/chemical analysis. This method is applied using standardized grass cultures in the Federal Republic of Germany,[5] and has already produced considerable results.[6] It has to be stressed, however, that the accumulation of the pollutants in the grass is strongly dependent on the growth of the plants and on the climatic conditions. So, in order to make comparisons between samples of different locations or periods it is necessary to know the internal and external conditions of the plants. Also other plant species may be used for this purpose, but with the same problems.

For a better understanding of the biological monitoring research with plants it could be useful to give the following definitions and descriptions.

Indicator plants are plants which may show clear symptoms, indicating the possible presence of some pollutant(s). These symptoms may be rather specific and lead to a qualitative determination of a pollutant, but mostly it is not a real identification and the presence of a specific pollutant has to be proven by other methods. The indicator plants have a detector function for recognition of the effects of the pollutants. But these effects may also be measured quantitatively to determine the effect intensities for monitoring (permanent measuring) of these effects. The permanent measurements of effect intensities may be used for surveys of air quality in relation to plants. The influences of the internal and external conditions of the plants are already integrated in the effects.

Accumulator plants are plants which may accumulate very easily specific air-polluting compounds. These compounds may be analysed in the plant material after some time by physical/chemical methods to recognize the pollutants and to obtain a quantitative measurement of the pollution burden (total amount of pollutants accumulated over some period). As the uptake of the pollutants by the plants (to be determined in washed samples of plant material) is looked at as an effect, this may give us a method for effect monitoring by accumulator plants. But the total pollution burden may also be monitored by measuring the pollutant concentrations in unwashed samples of plant material.

Sometimes the same plant species may act as both indicator and accumulator for a special pollutant, for example tulips and gladioli for fluoride. These species are also an example of plants reacting to the pollutant (HF) with both acute and chronic effects, dependent on the concentration. In the long run enough fluoride is accumulated to cause the same symptoms as in acute effects (leaf-tip and rim necrosis), and in addition bulb or tuber yield reduction may result.

Standardization of indicator and accumulator plants for air-borne pollution is a very important prerequisite, to be sure about the cause and intensity of the effects studied. As the effects of air pollutants on plants are, apart from the influence of the pollutant and the exposure time, dependent on plant species and variety, developmental and physiological stage of the plants and the physical environmental conditions, it is quite understandable that the selection of plant material and the growth conditions are very important. Selected uniform seeds or other plant material which is as genetically homogeneous as possible should be used (e.g. clones). Growth conditions ought to be optimal and the same at all the monitoring sites to be compared. This is only possible when totally artificial growth cabinets are used for the cultivation of the plants. Only differences in the quality of the air to be blown through the cabinets may influence the plants. This procedure has not been used in the practice of monitoring, because of its high costs. The use of plants grown in plain soil at different locations (with different soils) has been practised frequently, but is not to be encouraged. Standardization as far as possible of the soil and water conditions of the plants is advisable and possible at relatively low costs. Differences in climatic factors cannot be excluded in this way, but these parameters can be measured on the spot to be included in the comparison of effects.

Monitoring of biological effects of air pollutants by the use of plants as indicators and/or accumulators has been applied on the local, regional and national scale already. On a local scale, indicator and accumulator plants have been used to survey the possible unwanted effects of air pollution from a single source or a cluster of sources on special horticultural, agricultural or forestry crops, or on natural vegetation. The same indicators and accumulators may be used to sustain possible claims for compensation of economic losses. On a larger, regional or national scale the results of biological effect monitoring networks have been used to get some idea of the distribution of these effects in place and time. This may lead to conclusions about differences in air pollution effects burden between different regions of a country. In the long run it is also possible to get an idea about trends in incidences of effects, for example the occurrence of maximal values of ozone (O_3) effects in summer. On an international scale it would be possible to compare the air-pollution effects burden between adjacent regions of different countries and to trace back eventually some air pollutants to the source, if necessary also across the national frontiers. It could be a worthwhile source of information to have an extensive, biological-effects monitoring network, spread over all European countries to show the distribution of specific effects of air pollution on plants in place and time.

In every case the selection of plants and conditions is to be adapted to the specific aim of the study. When the pollutant to be studied is known, selection of specifically sensitive indicator plants is the first step for adequate results. In cases of unknown pollutants a series of plant species, sensitive to different compounds should be used simultaneously. For the comparison of the effects of air pollutants on plants at all locations of a European monitoring network, it is preferable to use identical indicator and accumulator plants in standard conditions. In all cases it is advisable to use more tolerant or insensitive varieties of the indicator-plant species next to the sensitive ones, to distinguish possible adverse effects of other biotic and a-biotic factors on the plants. These may be pathogens, like viruses, insect pests, frost, and possible nutritional deficiencies.

Another problem in biological effect monitoring with plants is the possible combination effects, when more than one pollutant is present (which is often the case). Two and more different polluting compounds may act on the plants additively, synergistically or antagonistically. Combined effects of mixtures of pollutants are hardly interpretable, when no information about the pollutants is available. The ideal indicator plant, for this reason, should be a plant sensitive to only one air pollutant. As this is impossible for all the different pollutants, it is worthwhile to look for indicator plants whose sensitivity is as specific as possible to one component of the total complex of air pollution.

EXAMPLE OF A NATIONAL MONITORING NETWORK FOR AIR-POLLUTION EFFECTS

Under the auspices of the Directorate-General for Environmental Hygiene of the Dutch Ministry of Public Health and Environmental Hygiene, from 1973 the Research Institute for Plant Protection at Wageningen has determined the effects of several air pollutants on indicator plants in one part of the country. By 1976 the biological-effects monitoring network was completed to forty experimental fields, regularly spread all over the country with a concentration in the area west of Rotterdam. This monitoring network is operated in close cooperation with the nation-wide automated monitoring network for air-pollution levels of the National Institute of Public Health at Bilthoven.

Biological-effect monitoring is performed during the vegetation period of the plants (from April until November) at thirty-seven stations of the national monitoring network for levels of air pollution and at three other locations. The methods and materials are the same at all forty experimental fields and these fields are visited weekly in the monitoring period, but not all on the same day. This should be taken into account when comparing the results of the concentration measurements and the effect measurements, resulting from one-week exposures.

The basic concept of effect measurement is the exposure of sensitive indicator and/or accumulator plants in a well standardized way. Use is made in the Netherlands of a special plant cultivation set, derived from a similar system developed and used in the Federal Republic of Germany.[7,8] Plants are grown in standard soil in plastic containers with ceramic filter candles in the soil, for automatic and regular watering. Three of these containers are put together in a larger container, serving as a water store. In these standardized cultivation sets all species of indicator and accumulator plants may be cultivated, by adapting the number of filter candles per container to the watering need of the plants.

Both acute and chronic effects of air pollutants are monitored. The acute effects are measured weekly on sensitive indicator plants in cultivation sets in the open air. Chronic effects are studied and measured by comparing the growth and production of sensitive plants after long-term exposures in cultivation sets in pairs of small greenhouses, one of which is ventilated with charcoal-filtered air and one with unfiltered air. Also standardized cultures of Italian rye-grass in the open air are used to accumulate several air-polluting compounds during 14-day periods for analyses of these compounds in the harvested plant material. Examples will be worked out from these three types of effect monitoring. A list of all indicator and accumulator plants used in our national monitoring network for air pollution is presented in Table 6.1.

Acute effects of O_3 on the sensitive tobacco variety, Bel W_3, are determined by estimating the percentage of leaf area damaged by the specific speckle necrosis for all leaves of four plants per location in the following classes: 0 per cent, 0–5 per cent, 5–10 per cent, 10–25 per cent, 25–50 per cent, 50–75 per cent and 75–100 per cent. From these ratings the ultimate mean value for all leaves is calculated for every location. Doing this for all weeks during the vegetation period, a distribution of the O_3 effect intensity in place and time is obtained. From this type of result it has appeared that the mean O_3 effect intensity is higher in the western half of the Netherlands than in the eastern half (see Figure 6.1), and that there are some special weeks every year with maximal O_3 effect intensity, correlated with a sunny type of weather.[9]

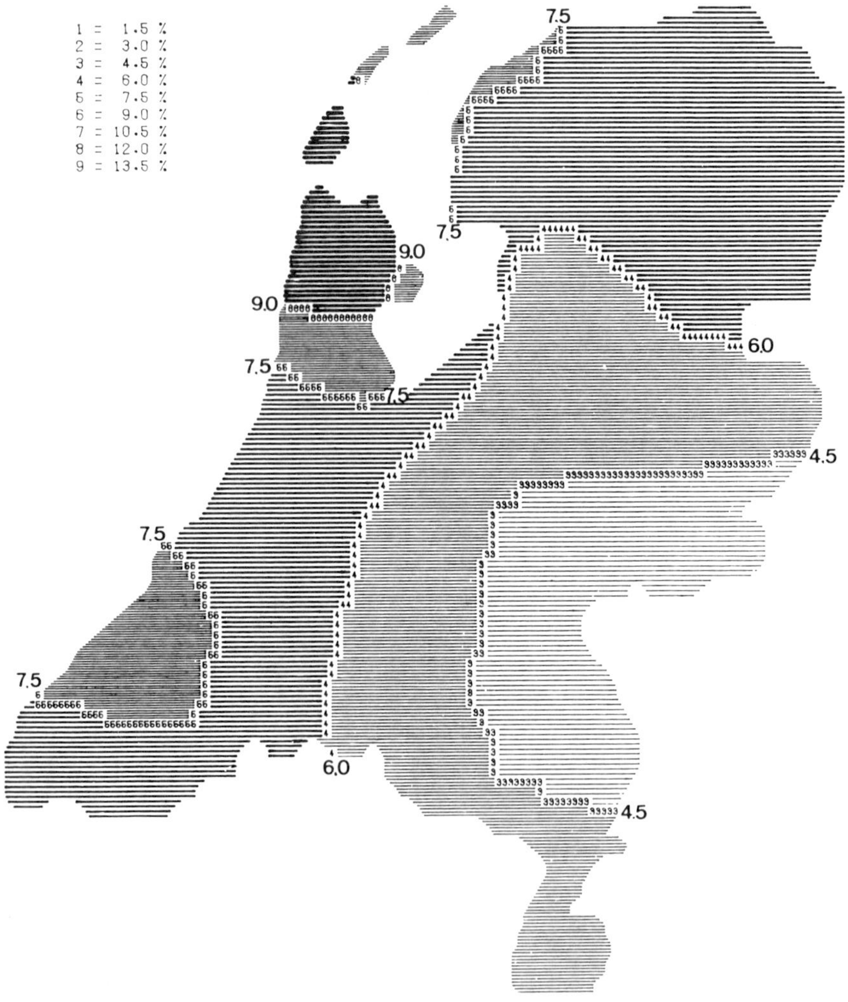

Fig. 6.1 Mean percentage of leaf area damaged by ozone on tobacco (Bel W_3) during the period from 6 June 1977 to 28 October 1977, in the Netherlands

Acute and also chronic effects of HF on sensitive monocotyledonous ornamental plants such as tulips and gladioli are studied. In the spring the tulip variety, Blue Parrot, and in the summer the gladiolus variety, Snow Princess, are cultivated in the cultivation sets outdoors. The leaf-tip and rim necrosis caused by air pollution containing fluoride is rather specific and may be assessed quantitatively by measuring the mean length of the necrotic tips, as a measure for the necrotic leaf area. The fluoride content of the 5.0 and 7.5 cm long leaf tips of tulips and gladioli appeared to be very well correlated with the leaf necrosis. From the fluoride effect monitoring network it was possible to show the distribution of fluoride pollution over the country, the south-western part being most polluted.

Chronic effects of air pollution as growth and yield reductions of horticultural crop plants have been studied in the industrial area west of Rotterdam. Pairs of small greenhouses, ventilated with filtered and unfiltered air, have been used to grow tomato, lettuce and other crops. Yield reductions of up to 20 per cent in the unfiltered air have been found. Differences in these reductions for separate locations indicated the differences in air-pollution effect burden. This type of result might be used for selection of areas well suited for horticulture, as far as the natural resource air is concerned.

Chemical/physical analysis of samples of standardized grass cultures for Pb, Cd, Zn and F have revealed possible differences in these air-polluting compounds in different regions of the Netherlands. This may be important for the productivity and/or quality of some crops, also when no acute effects are produced on the plants.

In the Dutch monitoring network for air pollution it should be possible to compare results of concentration measurements and effect measurements. However, up to now this has not been done for several reasons. Sulphur dioxide concentrations have been measured intensively but the concentrations have been rather low in the vegetation period, and practically no effects of SO_2 have been measured on the indicator plants. Clear-cut effects of fluoride have been found and measured quantitatively, but HF is not measured in the automated measuring network. For O_3 very good evidence for effects on indicator plants has been gained, but the O_3 concentration measurements in the network up to 1979 have been rather rare and unreliable. So, no general correlations between concentration and effect measurements have been established so far. This will surely be the aim of future research, but also other factors as for example climatic factors will have to be taken into account.

POSSIBLE INTERNATIONAL PROGRAMMES FOR MONITORING NETWORKS FOR AIR POLLUTION EFFECTS

Apart from the exchange of results of regional or national monitoring of air pollution effects, the establishment of international programmes for the monitoring of effects of air-borne pollution on vegetation should also be promoted. The need for this type of co-operation is quite clear, when trans-frontier air pollution accounts for a great deal of the total pollution burden.

Real international monitoring networks, set up and maintained under the auspices of some international authority or organization, would present great advantages for the participating countries. Good comparability of the results, by using the same measuring methods and standardized material, should be an important goal. For the purpose of comparison of the effects of air pollution on plants at all points of an international (European) air quality surveillance network, it is perhaps necessary to use identical indicator and accumulator plants at all sites. To this end standardized methods for cultivation, exposure, handling and observation of selected plants have to be elaborated. The personnel needed should also be educated to be able to fulfil this important task satisfactorily.

Differences in climate between countries participating in international programmes for monitoring the effects of air pollution on plants may be a problem for the selection of uniform collections of indicator and accumulator plants. But it should be borne in mind that the effects measured after some time of exposure of the plants are the integrated results of the influences of air pollutants and other (e.g. climatic) factors. This is the same in cases of damages in practice, so that the indicator plant effects may have a predictive value for the estimation of possible risks to the vegetation in general. These risks may be compared for the plants in different regions or countries, and threatened zones may be delimitated.

DISCUSSION AND CONCLUSIONS

The monitoring of levels and effects of air-borne pollution is a must if we are to control them. Without data on concentrations and effects we have no possibility of studying the occurrence and distribution of air pollutants and their effects in place and time, of setting standards for the protection of man and his environment, and of checking the results of abatement measures. We need an alarm system in order to prevent the occurrence of too high concentrations, resulting in hazardous effects on men, animals, plants and materials.

For these purposes both physical/chemical concentration measurements and biological effect measurements with plants are needed. The concentration levels measured by monitors will not be sufficient to predict all possible effects of air pollutants, and indicator plants will never be able to give information about the identity and concentration of polluting agents in the air. There is no question of replacing ambient air monitoring by physical/chemical methods with effect monitoring with plants. Both should be used jointly. Immission concentration measurements, effect intensity measurements on plants (including chemical/physical leaf analysis), and measurements of meteorological parameters may produce the total picture of the pollution situation.

Plants may be used as indicators and accumulators of air pollutants, both for recognition and monitoring purposes. An important problem remains and that is the representativity of the effects on the indicator/accumulator plants for vegetation as a whole of natural and crop plants. The comparability of effects on indicator/accumulator and other plants is not very well known, and a lot of information about dose–effect relationships for many plants in all possible conditions is still lacking. Also the occurrence of special effects of combinations of air pollutants is a great problem. Additive, synergistic, and antagonistic effects have to be taken into account as far as possible.

When plants are used for the monitoring of air-pollution effects, a high degree of standardization of the plant material and the physical conditions is a prerequisite. In international monitoring networks geographic and climatological differences should be taken into account. But it seems better to use uniform starting material and to accept an integrated effect of air pollutants and environmental conditions, than to use different and uncomparable plant material adapted to different local conditions. The risk for effects of air pollutants on vegetation is largely determined by environmental factors, so these should not be excluded in the effect monitoring research.

Although plants may have only an alarm function as to the potential risks of air pollution for other objects, the demonstration of the disastrous effects of this type of deterioration of the environment on vegetation may be important to stimulate possible prophylactic measures to prevent or diminish it.

REFERENCES

1. Schönbeck, H. *Staub-Reinhaltung der Luft*, **29** (1969) 14.
2. Posthumus, A.C. *Proceedings of the Kuopio Meeting on Plant Damages Caused by Air Pollution, Kuopio* (1976) 115.
3. Heagle, A.S., Body, D.E., and Heck, W.W. *J. of Environ. Quality*, **2** (1973) 365.
4. Mandl, R.H., Weinstein, L.H., McCune, D.C., and Keveny, M. *J. of Environ. Quality*, **2** (1973) 371.
5. Scholl, G. *Staub-Reinhaltung der Luft*, **34** (1974) 89.
6. Prinz, B., and Scholl, G. *Schriftenreihe der LIS*, **36** (1975) 62.
7. Scholl, G. *Zeitschr. für Pflanzenernährung und Bodenkunde*, **124** (1969) 126.
8. van Haut, H., Scholl, G., and van Haut, G. *Landwirtschaftl. Forschung*, **25** (1972) 42.
9. Floor, H., and Posthumus, A.C. *VDI-Berichte*, **270** (1977) 183.

Pollutants and Their Ecotoxicological Significance
Edited by H. W. Nürnberg

CHAPTER 7

*Effects of Air Pollution on Vegetation**

B. PRINZ AND C. J. BRANDT

SUMMARY

This paper outlines the impact of the principal atmospheric pollutants on vegetation. Along with sulphur dioxide, effects of hydrogen fluoride, nitrogen oxides, photochemical oxidants, hydrogen chloride and heavy metals are discussed. Where possible, threshold values are indicated for the protection of vegetation against effects of air pollution.

INTRODUCTION

Although scientific studies on the effects of air pollutants on vegetation have a long history, the general aspects of these studies have changed. At least in Germany, in the years around the Second World War extensive acute injury due to air pollution occurred quite often. At the present time, except for acute injury in the neighbourhood of specific sources, much more attention has to be paid to the chronic effects of long-term, low-level concentrations of air pollutants. These effects are not uncommonly caused by a combination of various interacting components.

These effects mainly impair the physiological strength of the plant without causing clearly visible symptoms. For example, photosynthesis may be decreased considerably or early senescense of leaves and conifer needles may occur. The aging of conifer needles can be so accelerated that needles older than two years are usually lacking on affected trees.

These effects are not only of economic importance. More relevant are the ecological consequences. Sensitive plant species may die out, the biomass *in toto*

* This chapter does not treat the recent significant damages reported on forests in Middle Europe. This subject which has become recently very significant has been treated by the authors in ref. 95.

is reduced with equal impairment of the function of vegetation as a sink for air pollution, and the resistance towards parasites and other stress factors is decreased.

When comparing the effects of air pollution on plants with the effects of air pollution on man the following distinctions can be made. Due to the very low degree of self regulation, plants in general react more sensitively than man. Clear symptoms of injury to plants may occur in a range of concentrations where it is difficult to derive health effects of any relevance. For this reason, the investigation of air pollution problems is very often induced by complaints of injury to vegetation, and epidemiological studies of possible harmful effects on man are carried out at a much later stage.

It is evident that experimental studies on plants are much easier to carry out than on man. On the other side, due to poor self regulation, reactions of plants to air pollutants are much more dependent on numerous endogenous and exogenous factors, such as temperature, humidity, soil condition etc. For this reason and because of the wide variety of plant species and their corresponding degrees of sensitivity and resistance, a large number of experimental variables have to be considered if dose–effect relations valid for each case are to be derived. This also means that protection of each individual element in the environment through setting of appropriate standards is principally impossible.

The pollutants most important for plants are sulphur dioxide, fluoride, chloride, nitrogen oxides, photochemical oxidants (including ozone), and some heavy metals, the latter equally important in connection with the food chain. The threshold for the toxic effects of these components on plants are, of course, a matter of on-going discussion. Within the VDI-Kommission, 'Reinhaltung der Luft', in Germany the following threshold values are presently being discussed (the range of values corresponds to different degrees of resistance):

Sulphur dioxide	annual mean concentration	50–120 $\mu g\ m^{-3}$
Fluoride	average over 24 h	2–4 $\mu g\ m^{-3}$
	average over 7 months (vegetation period)	0.3–1.4 $\mu g\ m^{-3}$
Chloride	annual average	60–150 $\mu g\ m^{-3}$
	average over 24 h	700–1800 $\mu g\ m^{-3}$
Nitrogen dioxide	average over 7 months (vegetation period)	250–500 $\mu g\ m^{-3}$
	average over 30 min	4000–9000 $\mu g\ m^{-3}$
Ozone	average over 4 h	50 $\mu g\ m^{-3}$
	average over 30 min	300–1000 $\mu g\ m^{-3}$

These figures represent a preliminary stage of discussion and may therefore still be corrected. In the following the effects of important phytotoxic pollutants are described in more detail.

SULPHUR DIOXIDE

Responses of vegetation to SO_2 have been by far more widely studied than the responses to other atmospheric pollutants and the symptomatology and dose–response relationships for SO_2 as a single pollutant are essentially complete. Reviews of the literature have appeared frequently in the past, for example Thomas,[1] Daines,[2] Heck and Brandt,[3] and they are recommended for further reference.

Natural background concentrations of SO_2 usually range between 0.28 and 2.8 $\mu g\ SO_2\ m^{-3}$ and whenever mean annual local concentrations of SO_2 exceed about 50–200 $\mu g\ SO_2\ m^{-3}$, extensive phytotoxic effects can appear in a number of species, although some are affected at lower concentrations.[4] Different species of plants vary widely in their sensitivity to SO_2 when exposed to this gas under conditions most favourable for its uptake by the plants. This fact must be taken into consideration when evaluating phytotoxic effects.

Symptomatology of SO_2 injury to vegetation varies considerably among plant groups. In dicotyledonous plants acute injury is expressed in the form of marginal or intercostal necrosis. Injured areas first appear dark green and soon become bleached to an ivory colour. The necrotic lesions are visible from both sides of the leaf. Chronic injury from the impact of SO_2 appears as general intercostal chlorosis. In monocotyledonous plants, acute injury appears as necrotic streaking, starting at the leaf tip and spreading toward the leaf base. Chronic injury may appear as a bleaching of the leaf tips. In conifers, injury initially appears as a banding around the needles and later the terminal portions turn reddish brown.

Epiphytes, such as lichens, are among the more sensitive organisms to SO_2. As early as 1866 Nylander[5] proposed his 'air pollution hypothesis' concerning the absence of lichens in the city parks of Paris. Sernander[6,7] described the central part of Stockholm as a 'lichen desert', completely devoid of lichens, surrounded by a zone in which some kinds of lichens could survive, although the thalli often had a different form and colour than healthy specimens. Field studies and fumigation experiments have shown that lichens show measurable characteristic responses to phytotoxic air pollutants including SO_2. This property may be used for the establishment of an effects monitoring system.[8–10]

Among the vascular plants, coniferous trees are the most sensitive to SO_2. In fumigation studies, Douglas fir (*Pseudotsuga menziesii*) seedlings showed necrosis after exposure to 812 $\mu g\ SO_2\ m^{-3}$ for 44 h and white pine (*Pinus strobus*) showed necrotic lesions after exposure to 280 $\mu g\ SO_2\ m^{-3}$ for only 8 h.[11] Knabe[12] found that in the Ruhr area, pine trees could not survive if the mean annual concentrations of SO_2 exceeded 196–224 $\mu g\ SO_2\ m^{-3}$ and that some necrosis of needles occurred at only 28 $\mu g\ SO_2\ m^{-3}$. Studies in Sweden[13] showed that mean annual SO_2 concentrations of 56 $\mu g\ SO_2\ m^{-3}$ led to a reduction in the general growth rate of forests.

Linzon[14] and Dreisinger[15] found that growth of white pine was reduced under average concentrations of 45 μg $SO_2\,m^{-3}$ during the growing season (April–October). Materna[16] found extensive chronic and some acute injury to spruce (*Picea abies*) stands after long-term (several years) exposure to average annual concentrations of 25–40 μg $SO_2\,m^{-3}$. Severe necrosis occurred under average concentrations of 50–70 μg $SO_2\,m^{-3}$ and peak concentrations of 320 μg $SO_2\,m^{-3}$ for 2.5 per cent of the monitoring time. Effects of other pollutants acting in combination with SO_2 cannot be excluded, however.

From 1959 to 1962 a lot of different plant species were exposed in open field at different distances from an iron ore foundry in Biersdorf/Westerwald[94] to receive quantitative relationships between SO_2 exposure and adverse effects. 60 μg $SO_2\,m^{-3}$ seemed to be a critical concentration for long lasting exposure (1 y and longer) where deciduous and coniferous trees showed clear indication of growth inhibition and other adverse effects. With 30 μg $SO_2\,m^{-3}$ the growth was even stimulated so that 50 μg $SO_2\,m^{-3}$ seemed to be a fairly safe level of SO_2 exposure where only slight effects on trees if at all would appear.

The field studies described here were carried out under considerably different pollutant conditions, characterized either by high peak concentrations and long pollutant-free times or by long-term action of low concentrations. Based on the average concentrations, however, the injury limits are in the same range. The extreme phytotoxicity of the pollutant type with high peak concentrations results from the progressive increase of acute injury with an increase in concentration, whereas injury from the pollutant type with long-term action of low concentrations rests on the cumulative effects of the pollutant and the absence of pollutant-free times allowing the plant to recover. This is especially true for long-term effects on fruit and forest cultures, as slight effects can add up over a period of years or even decades and cause a reduction of the useful value of the plant. Injury limits derived from fumigation experiments under conditions similar to natural conditions, tend to be slightly higher than those derived from field studies, usually because of the shorter exposure times in the fumigation studies. In extensive fumigation studies it could be shown that a 20 per cent reduction in yield occurred at 0.4 mg $SO_2\,m^{-3}$ with red clover and between a range of 1.0 and 1.5 mg $SO_2\,m^{-3}$ with radish, bean, cress and tobacco (see Figure 7.1a and b). All studies were carried out over an exposure period of two weeks. A summary of the effects of SO_2 on vegetation is presented in Table 7.1.

Environmental factors strongly affect the susceptibility of plants to SO_2. In general, conditions of high light intensity, moist soil, high relative humidity and moderate temperatures are conditions under which plants are most susceptible. This is due to the maximum aperture of stomata leading to optimum gas exchange. Oertli[88] found that conditions that cause moisture stress also induce resistance. Seasonal variability in sensitivity has been observed. The most sensitive period being spring or early summer, the least sensitive winter.[19,88] Berry (in Treshow[34]) demonstrated that nitrogen-deficient plants are much more

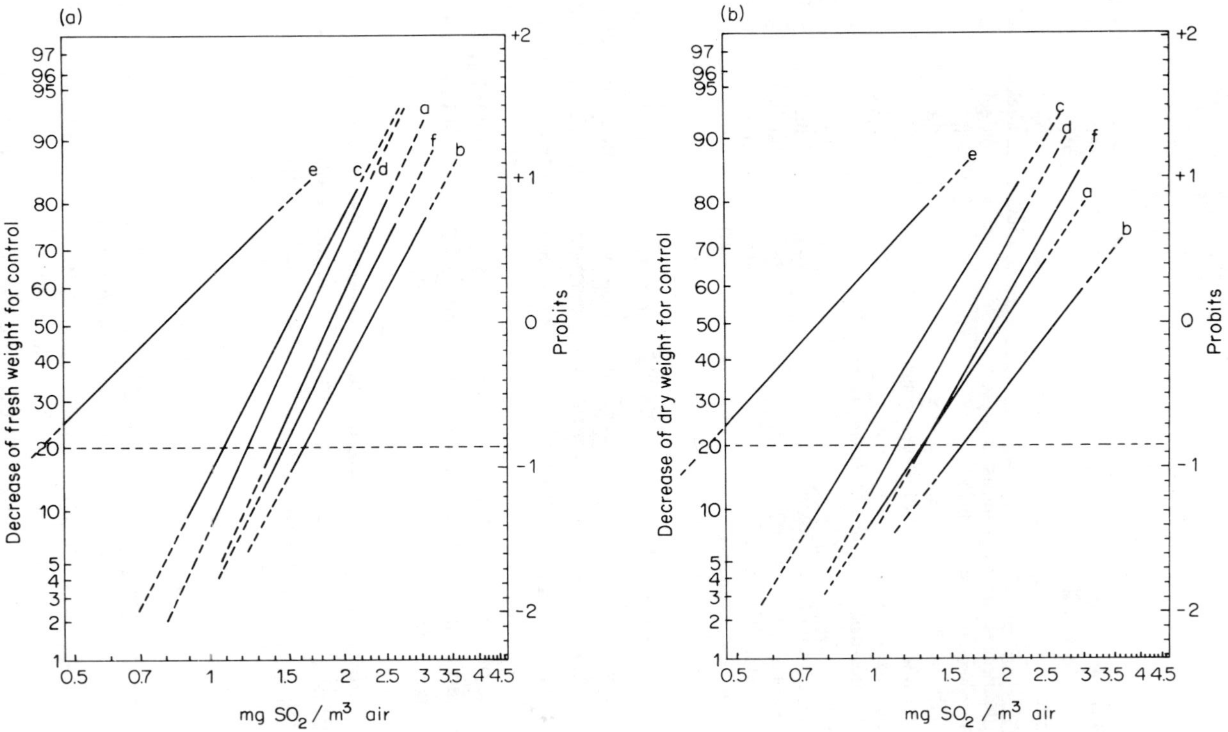

Fig. 7.1 Decrease in freshweight (a) and dry substance (b) of a: bean, b: radish-leaves, c: cress, d: radish-bulb, e: red clover, f: tobacco by SO_2

Table 7.1 Summary of effects of sulphur dioxide on vegetation

Effect	Concentration ($\mu g\, m^{-3}$)	Length of exposure	References
Weak necrosis on larch	840	8 h	19
Necrosis of needles of Douglas fir	812	44 h	11
Foliar injury to wheat, oats, spinach	560	150 h	20
Foliar injury to alfalfa, clover, vetch	420	150 h	20
20 per cent reduction in yield of red clover	400–1500	2 weeks	17, 18
Slight reduction in photosynthesis of alfalfa	364	6 weeks	21
Necrosis of needles of white pine	280	8 h	11
Death of the lichen *Hypogymnia physodes*	230	29 days	22
52 per cent reduction in growth of rye grass (*Lolium perenne*)	190	6 months	23
60 per cent necrosis of thallus of *Hypogymnia physodes*	80	68 days	22
Growth reduction of white pine	45	7 months	14
Growth reduction of oak, beech, Norway spruce, Scotch Pine, and Larch	60	One year and longer	94
Chlorosis of needles of spruce	25–40	several years	16

susceptible than normal plants. Pfeffer[90] found that plants growing on limestone and basalt were less severely affected than those growing on sandy soils.

Sulphur dioxide, as the most ubiquitous of air pollutants, is generally taken as the leading component in most pollutant types. That is, where other atmospheric pollutants, such as fluoride or photochemical oxidants, occur, SO_2 is almost always also present. Therefore, the impact on the vegetation of several pollutants in combination cannot be excluded. The present air-pollution situation is in fact characterized by persistent effects of low ambient concentrations and the simultaneous occurrence of several pollutants. Because of the infinite number of possible combinations of pollutants and concentrations, reliable dose–response relationships for the action of pollutants in combination are not available. Synergistic effects of SO_2 with ethylene and a number of phytotoxic metals have been documented by Krause (unpublished information).

HYDROGEN FLUORIDE

Among the fluoride-containing air pollutants, the gaseous, water-soluble compounds have the highest phytotoxicity. Aluminium, rare earth metals, steel, phosphorus, chemical and fertilizer factories, ceramic, brick and glass works are main sources of fluoride. Hydrogen fluoride is of the most practical importance. Hitchcock *et al.*[24] have shown that the phytotoxicity of other gaseous fluoride compounds, such as F_2, SiF_4 and H_2SiF_6, is comparable to that of HF.

Gaseous fluoride compounds are also indirectly responsible for air pollution

Table 7.2 Summary of effects of hydrogen fluoride on vegetation

Effect	Concentration ($\mu g\,m^{-3}$)	Length of exposure	References
Severe necrosis on gladiolus	0.08	5 weeks	49
Necrosis of needles of Ponderosa pine and leaves of gladiolus	0.4–0.7	2 weeks	38, 50
Severe tipburn on tulip	0.6	16 days	51
20 per cent reduction in photosynthesis of barley and alfalfa without visible injury	0.8	2 h	47
Inhibition in growth of orange trees	0.8–4.0	2 years	39
Reduction in growth of barley	3.3	12 days	41
Reduction in yield of alfalfa, orchard grass, lettuce	100	20 days	52

injury to animals,[25,26] that is after accumulation in fodder plants. For evaluation of the risks to plants and animals, two criteria should be considered: the level of gaseous fluoride compounds in the air, and fluoride accumulation in plants. The veterinary aspect is fairly straightforward.[27] For example, in order to avoid chronic fluorosis in milk cows, average fluoride content in fodder of 30–40 ppm fluoride in the dry matter should not be exceeded.[28–30]

Extensive studies on the relationship between fluoride accumulation in plants and atmospheric fluoride concentrations have been carried out and have been summarized by various authors, including Christiani and Gautier,[31] Hitchcock *et al.*,[32] Garber,[33] Treshow and Pack[34] and NAS.[30] Symptoms vary widely between plant groups. In broad-leaf dicotyledons, fluoride is translocated to the margins of the leaf where necrosis develops. A distinct reddish-brown band typically separates the necrotic tissue from healthy tissue. In monocotyledons, the tips of the leaves are the first areas to become necrotic with a red-brown band on the margin. In cereals, the necrotic tissue may appear almost white. Corn typically exhibits chlorotic flecking concentrated along the leaf margin and apex. In conifers, injury begins at the needle tip and spreads to the base. Initially chlorotic, the tissue turns reddish brown with age.

Natural background concentrations of fluoride in plant material generally range between 0.5 and 25.0 ppm fluoride ($\mu g\,F\,g^{-1}$ dry matter).[11,33] Accumulation in the range of 25–100 ppm fluoride is the critical range for sensitive plants.[30,35] Resistant species, however, may accumulate several hundred ppm before chlorosis or necrosis appear. Accumulation of fluoride by plants exposed to HF has been shown to be approximately a linear function of concentration and exposure time.[32,36,37]

In a field study near a fluoride source, gladiolus varieties and Ponderosa pine showed necrosis after exposure for 2 weeks on sites with average concentrations of 0.7–0.4 μg HF m^{-3}.[38] After continuous exposure to 0.8–4.0 μg HF m^{-3} for 2 years, orange trees showed an inhibition in stem and lateral growth, reduced leaf size with severe chlorosis, retardation in blossom formation, lower yields, and a reduction in external quality and vitamin C content of fruits.[39] Similar effects were found after exposure to 0.8 μg HF m^{-3} for 27 months.[40] In fumigation experiments with sensitive forest trees, such as spruce (*Picea abies*), white pine (*Pinus strobus*), Nordman's fir (*Abies nordmanniana*) and red beech (*Fagus silvatica*), such extensive leaf injury occurred after exposure to 1.3 μg HF m^{-3} for several days that long-term exposures would probably cause severe reduction in growth.[41] A significant reduction in growth was found in winter barley (*Hordeum vulgare*) after exposure to 3.3 μg HF m^{-3} for 12 days and on sensitive ornamentals such as tulips, daffodils, hyacinths, and crocus after exposure to average concentrations of 1.5–2.0 μg HF m^{-3}.

Since the fluoride ion is known to inhibit certain enzyme reactions, physiological injury without the necessary appearance of visible injury may be expected from the action of atmospheric fluorides. Studies on the effects of fluoride on gas exchange of plants have shown that, depending on the plant species, a given amount of fluoride may stimulate or reduce dark respiration.[42,37] McNulty and Newman[43] found an increase in respiration of the gladiolus variety Snow Princess after fumigation with HF. McNulty[44] also found that respiration of bean was increased by 25 per cent through action of sodium fluoride (NaF) in the nutrient solution.

On the other hand, McNulty and Newman[43] showed that respiration of *Prunus* spp. could be reduced by low concentrations of NaF. A reduction in respiration was also found in bean plants after fumigation with HF.[44,45]

Most studies on the effect of fluoride on CO_2 uptake by plants have been concerned with the effects of HF on the apparent photosynthesis. Thomas and Hendricks[21] found a correlation between the reduction in photosynthesis and the increase in leaf necrosis of gladiolus fumigated with HF. Short-term exposure, even to higher concentrations, caused only a temporary reduction in photosynthesis and the plants recovered after exposure.

In fumigation experiments with bean and tomato, Weinstein[46] found a reduction in the production of chlorophyll a and b, but no effect on photosynthesis. Bennett and Hill[47] exposed winter barley and alfalfa to six different air pollutants and found that exposure to 0.8 μg HF m^{-3} for 2 h caused a 20 per cent reduction in photosynthesis without visible injury. Based on the effects on photosynthesis, they were able to establish the following toxicity scale:

$$HF > O_3 > Cl_2 > SO_2 > NO_2 > NO$$

A review of the literature yielded the threshold values shown in Figure 7.2 for injury to various plant species from exposure to hydrogen fluoride.[48]

The influence of environmental variables on plant response to fluorides is difficult to establish. Greenhouse experiments have indicated that turgid cells are most susceptible to injury because of physiological activity, whereas field

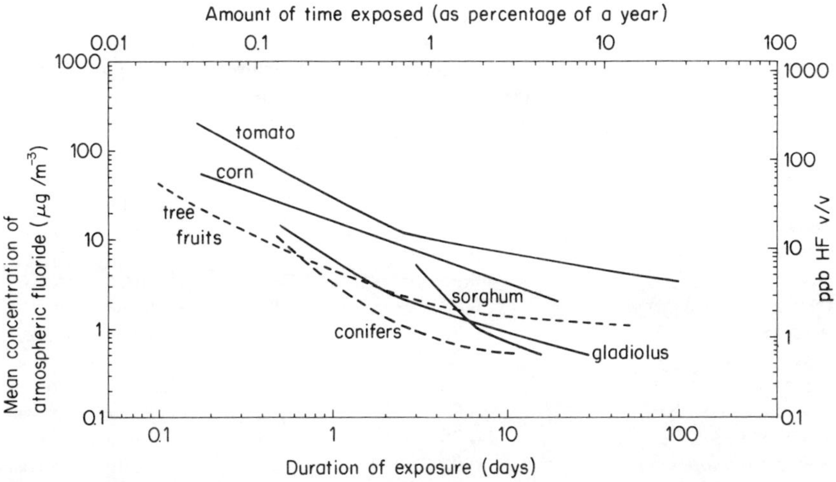

Fig. 7.2 Threshold concentrations for injury from atmospheric hydrogen fluoride, with reference to various plant species. *Source:* McCune (1969).

observations have shown that plants under moisture stress are frequently the most severely affected,[91,34] Daines *et al.*[92] found that plants with a water deficit were more resistant to fluoride than those receiving adequate water. They also found that plants fumigated in darkness show less injury than those fumigated in daylight. MacLean and Schneider[93] reported that gladiolus plants exposed to HF at a constant vapour-pressure deficit exhibited greater tip burn with increasing temperature. Foliar accumulation of fluoride decreased in gladiolus but increased in sunflower leaves. Suboptimal nutrient supplies of calcium, magnesium, potassium, phosphorus, and nitrogen have been associated with increased injury.[50,53–55] Accumulated fluoride in fodder plants is the causative agent in the debilitating illness of grazing animals, fluorosis. A limitation of the fluoride uptake rate in fodder plants for the protection of animals has been discussed.

Extensive investigation of the uptake of fluoride in standardized grass canopies[37,56,57] has led to the development of correlations between the uptake rate in grass and phytotoxic effects on other plants. Using this method of surveillance in the field, it could be shown that injury to very sensitive plants may occur when, after a fourteen-day exposure, fluoride accumulation in the standardized grass exceeds 30 ppm. Injury to sensitive and less sensitive plants may be expected when fluoride accumulation in grass exceeds 60 ppm and 80 ppm, respectively.

HYDROGEN CHLORIDE

Hydrogen chloride (HCl) and HCl aerosols are not considered to be of major importance except near isolated industrial processes such as chemical works and

Table 7.3 HCl concentrations that cause injury to vegetation

Plant species	Average concentration ($mg\,m^{-3}$ air)	Exposure time (h)	Type and extent of injury
Vitis vinifera	0.08	90	Slight necrosis Slight necrosis
Grape	0.28	114	Severe necrosis
Sorbus intermedia	0.08	90	Slight necrosis
Rowan tree	0.28	114	Medium necrosis
Aesculus hippocastanum	0.08	90	Slight necrosis
Horsechestnut	0.28	114	Medium necrosis
Populus sp.	0.08	90	Slight necrosis
Poplar sp.	0.28	114	Medium necrosis
Fagus silvatica	0.08	90	Very slight necrosis
European beech	0.28	114	Medium necrosis
Robinia pseudoacacia	0.08	90	Very slight necrosis
Black locust	0.28	114	Slight necrosis
Corylus avellana	0.12	140	Slight necrosis
Hazelnut	0.19	150	Medium necrosis
Betula verrucosa	0.11	229	Slight necrosis
Cut-leaved birch	0.18	228	Medium necrosis
Tilia platyphyllos	0.11	229	Slight necrosis
Basswood	0.18	228	Medium necrosis
Acer campestre	0.11	229	Very slight necrosis Slight chlorosis
Hedge maple	0.18	228	Slight necrosis
Gingko biloba	0.11	229	No injury
Ginko	0.18	228	Slight necrosis
Quercus pedunculata	0.11	229	No injury
English oak	0.18	228	Very slight necrosis
Pinus strobus	0.08	90	No injury
White pine	0.28	114	Slight necrosis

Source: Van Haut and Guderian.[59]

waste incineration. Observations of effects of HCl on vegetation in the field have been minimal because the concentrations necessary to produce injury rarely occur. Injury has been related to incineration of polyvinyl chlorides. Field injury occurred in vegetation within 1 km of the open burning source.[58]

Hydrogen chloride causes an acid-type necrotic lesion similar to the lesions caused by SO_2. Symptoms are generally lighter than necrosis caused by other pollutants, however. The reaction is thought to be a pH response in plant tissue. In the field, no chronic or growth effects have been reported for HCl.

In fumigation experiments, sensitive deciduous trees and shrubs showed significant foliar injury at an average concentration of only 0.08 mg $HCl\,m^{-3}$ after 90–230 h. This suggests that a reduction in growth and yield would occur after long-term exposure.[59] The extreme phytotoxicity of HCl is evident when it is considered that exposure was discontinuous for 6 h per day (Table 7.3).

NITROGEN OXIDES

Nitric oxide and nitrogen dioxide, the principal oxides of nitrogen, are primarily produced by high temperatures in combustion reactions, such as the burning of coal or in internal combustion engines. Nitrogen oxides are one of the important groups of chemicals found in photochemical smog. Nitric acid factories are secondary sources of nitrogen oxides.

Experimental results indicate that nitrogen dioxide (NO_2) is the most phytotoxic of the nitrogen oxides. Air-quality criteria and standards for the protection of vegetation from effects of NO_2, therefore, also guarantee protection from the other oxides of nitrogen.

Phytotoxic effects of nitrogen oxides have been summarized by van Haut and Stratmann[60] and by Taylor and MacLean.[61] Acute injury in dicotyledons is expressed as irregular necrotic lesions, that are intercostal and usually occur near the margins of the leaf. The necrotic tissue is white, tan, or brown, and it may closely resemble SO_2 injury. No visible chronic symptoms have yet been described for NO_2 in the field and no evidence is available that would indicate a chronic effect of NO_2 on plants.

Plants are uninjured by many mg $NO_2\,m^{-3}$ if exposures are for a few minutes. Continuous exposure to 0.095–0.19 mg $NO_2\,m^{-3}$ can result in physical injury, but this effect may be very difficult to detect unless careful comparisons are made, especially in the absence of appropriate epidemiological studies. In general, however, it can be said that phytotoxic nitrogen dioxide concentrations for long-term exposures lay in the range of 0.28–0.56 mg $NO_2\,m^{-3}$.

An important factor influencing the phytotoxicity of NO_2 is the action of NO_2 in combination with other pollutants, especially SO_2, as shown in recent experiments. Tingey *et al.*[62] reported that below 3.8 mg $NO_2\,m^{-3}$ or 1.3 mg $SO_2\,m^{-3}$ tobacco showed no injury when exposed to the pollutants alone, but plants exposed to mixtures of 0.09–0.47 mg $NO_2\,m^{-3}$ and 0.13–0.65 mg $SO_2\,m^{-3}$ were injured, indicating a synergistic effect. They observed that the levels of NO_2 and SO_2 that occur in urban areas may result in yield losses to plants grown under field conditions.

PHOTOCHEMICAL OXIDANTS

The primary phytotoxicants among the photochemical oxidants are ozone (O_3) and the peroxyacetylnitrates (PAN). These components have been studied extensively, especially on the west coast of the USA, where injury due to O_3 was first observed on vegetation in the early 1950s.[63,64] In Europe, the photochemical oxidants, or smog, are considered to be a fairly new pollution type.[65] Although

O_3 concentrations are increasing in Europe from year to year. Through production of HC and NO_x, contributing factors to O_3 production, from burning of fossil fuels, it is not expected that the European smog problem will reach the extreme conditions present on both coasts of the USA due largely to the prevailing climatic conditions. Although interest in the phytotoxic effects of oxidants is increasing in Europe, most of the information on dose–response relationships is taken from field observations and fumigation studies carried out in the USA. The presentation of the quantitative effects of oxidants has to be limited to laboratory and controlled field exposures because, under ambient conditions, the effects of these compounds cannot be differentiated.

Research has shown that additional phytotoxicants may be present in the photochemical complex.[66–68] Combined effects of the toxicants discussed and other atmospheric pollutants may also produce injury on sensitive plant species.[62,68,70–72] On the basis of available information, ozone is the most important phytotoxicant of the photochemical complex.

Although ozone is an important constituent of the upper atmosphere, Berry[73] has found that the background concentrations in the lower portion of the atmosphere usually range between about 20 and 40 $\mu g\ O_3\ m^{-3}$. In metropolitan areas the concentration of O_3 ranges to above 2000 $\mu g\ O_3\ m^{-3}$, which is well above the phytotoxic level.

Ozone-induced symptoms in plants have been summarized by Treshow[74] and Hill *et al.*[75] In most species, O_3 characteristically causes a necrotic stipple or flecking of tissue on the upper surface of the leaf adjacent to small veins. Internally, the palisade tissue is the most sensitive to O_3, and these cells become non-functional and collapse before the lower portion of the mesophyll becomes injured. The affected areas initially appear bleached from the upper surface. After several days, the colour may change to tan or a parchment colour. A darker colour will appear if new pigments are produced. At higher concentrations or long-term exposures, O_3 injury will extend to other leaf tissues, and consequently the necrotic areas will expand and coalesce, making injury visible from both surfaces. Newly matured leaves are usually the most susceptible to O_3. In conifers, the stipple initially appears at the tips of new needles, but at the base of old needles. The most sensitive species are injuries by O_3 fumigations between 100 and 200 $\mu g\ O_3\ m^{-3}$ for 2–4 h, while resistant species tolerate concentrations as high as 2000 $\mu g\ O_3\ m^{-3}$. Not only does susceptibility vary between species[76] but also among varieties of a species.[77,78]

The acute effects of O_3 have received sufficient study to permit the construction of preliminary profiles to relate time, concentration and response for certain plant species. An adequate graphical expression has been developed for pinto bean and Bel W_3 tobacco exposed to O_3.[79] Based on this work, a number of plants were exposed to a series of times and concentrations and a model was derived as an empirical relationship between concentration, time and response.[80]

An extensive review of the literature was made for critical information on dose

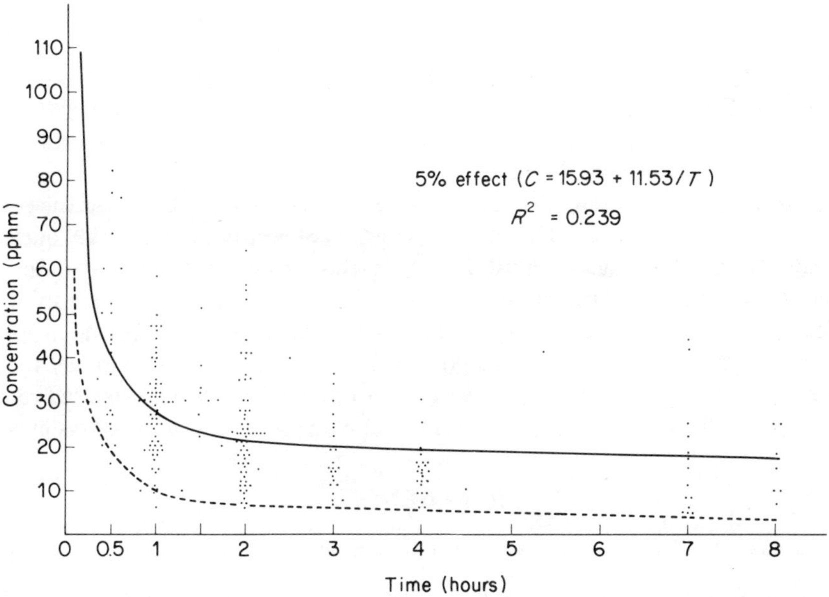

Fig. 7.3 Ozone concentration time scatter diagram for a 5 per cent acute plant response from over 100 species and varieties tested using data from the reference list. The data were used to calculate an average 5 per cent plant response curve using the equation $C = a + b/T$ and plotting C v. T. (*Source:* Linzon *et al.*, 1975)

and response.[81] Curves of concentration response and time response were developed for over 100 species and varieties obtained from the literature review. From these curves, the scatter diagram for a 5 per cent response was drawn (Figure 7.3).

HEAVY METALS

Classically, heavy-metal pollution has been associated with ore-smelting operations and mining. However, heavy-metal pollution also results from many other sources, such as electrical power generation, additives in gasoline, fertilizers and various fungicides and insecticides.[82]

Heavy-metal pollution can affect the vegetation directly through uptake by the leaves and shoots, as well as indirectly through deposition and accumulation in the soil and uptake by the roots. Heavy-metal pollutants that occur in easily soluble form, such as sulphate or chloride, can be taken up quickly through leaves and shoots. Near heavy-metal sources, however, the pollutants usually occur in a less soluble form, such as oxides or in metallic form. These pollutants may still enter the plant to a certain extent through the leaves[83] but the major portion is taken up through the roots after accumulation and break-down into more soluble form in the soil.

Heavy metals taken up from the soil accumulate in the plant organs and influence growth, yield and quality of the plant. Chlorotic symptoms induced by heavy metals frequently resemble nutrient deficiency symptoms of essential elements. The specific mechanism of injury is not known for most heavy metals, but competition with various ions, especially iron[84] and phosphorus and magnesium[85,86] have been proposed.

Through their accumulation in plants, heavy metals can also enter the food chain of animals and man. Therefore, evaluation of limiting values or air-quality standards for toxic metals must also take this aspect into consideration. A specific example is the recent case in Westfalia (Germany) where thallium was released from iron oxide used in the production of cement.[87] It is evident that further studies on the effects of air pollutants on vegetation can no longer focus on effects on single plant species or varieties but will have to be extended to an analysis of effects on entire plant communities, food chains and ecosystems.

REFERENCES

1 Thomas, M.D. (1961). Effects of air pollution on plants, in, *Air Pollution*, WHO, Geneva.

2 Daines, R.H. (1968). Sulfur dioxide and plant response, *J. Occup. Med.*, **10**, 516–524.

3 Heck, W.W., and Brandt, C.S. (1977). Impact of air pollutants on vegetation: Crops, forest, native, in, *Air Pollution* (Ed. A.C. Stern), 3rd edn, Academic Press, New York.

4 Nash, T.H. (1973). The effects of air pollution on other plants, particularly vascular plants, in, *Air Pollution and Lichens* (Eds B.W. Ferry, M.S. Baddeley, and D.L. Hawksworth), The Athlone Press, University of London.

5 Nylander, W. (1866). Les lichens du Jardin du Luxembourg, *Bull. Soc. Bot. Fr.*, **13**, 364–372.

6 Sernander, R. (1912). Studier öfvar lafvarnes biologi. I. Nitrofila lafvar, *Svensk bot. Tidskr.*, **6**, 803–883.

7 Sernander, R. (1926). *Stockholm Natur*, Alingvist and Wiksells, Uppsala.

8 Schönbeck, H., and van Haut, H. (1971). Exposure of lichens for the recognition and evaluation of air pollution. *International Symposium on Identification and Measurement of Environmental Pollution, Ottawa.*

9 Prinz, B., and Scholl, G. (1975). Erhebungen über die Aufnahme und Wirkung gas- und partikelförmiger Luftverunreinigungen im Rahmen eines Wirkungskatasters. (Mitteilung der Ergebnisse für die Erhebungsjahre 1972–1974). *Schriftenr. Landesanst. Immissionssch. d. Landes NW*, **26**, 25–59.

10 Prinz, B., and Scholl, G. (1978). Erhebungen über die Aufnahme und Wirkung gas- und partikelförmiger Luftverunreinigungen im Rahmen eines Wirkungskatasters (II. Mitteilung; Erhebungszeitraum: Januar 1975 bis Dezember 1976). *Schriftenr. Landesanst. Immissionssch. d. Landes NW*, **46**, 25–77.

11 Treshow, M. (1970). *Environment and plant response*, McGraw-Hill, New York.

12 Knabe, W. (1971). Air quality criteria and their importance for forests, *Mitt. Forstl. Bundesversuchsanst., Wien*, **92**, 129–150.

13 Swedish Royal Ministry of Agriculture (1971). Air pollution across national boundaries. The impact on the environment of sulfur in air and precipitation, Sweden's case study for the UN conference on the human environment. *Swed. Prop. Comm. for UN Conf. on Human Environ.*, Stockholm.

[14] Linzon, S.N. (1971). Economic effects of sulfur dioxide on forest growth, *JAPCA*, **21**, 81–86.
[15] Dreisinger, B.R. (1970). Monitoring atmospheric sulfur and correlating its effects on crops and forests in the Sudbury area. *Impact of Air Pollution on Vegetation Conference, Toronto*.
[16] Materna, J. (1972). Einfluß niedriger SO_2-Konzentrationen auf die Fichte, *Mitt. Forstl. Bundesversuchsanst., Wien*, 219–232.
[17] van Haut, H., Prinz, B., and Höckel, F.E. (1979). Ermittlung der relativen Phytotoxizität von Luftverunreinigungen im LIS-Kurzzeittest—verschiedene organische Komponenten und Ammoniak—*Schriftenr. Landesanst. Immissionssch. d. Landes NW*, **49**, 29–65.
[18] van Haut, H., and Prinz, B. (1979). Beurteilung der relativen Pflanzenschädlichkeit organischer Luftverunreinigungen im LIS-Kurzzeittest, *Staub-Reinh. Luft*, **39**, 408–414.
[19] Katz, M., and McCallum, A.W. (1952). The effect of sulfur dioxide on conifers, in, *Air Pollution*, vol. I, pp. 401–443, Academic Press, New York.
[20] Zahn, R. (1961). Wirkungen von Schwefeldioxid auf die Vegetation. Ergebnisse aus Begasungsversuchen, *Staub*, **21**, 56–60.
[21] Thomas, M.D., and Hendricks, R.H. (1956). Effects of air pollutants on plants, in, *Air Pollution Handbook*, McGraw-Hill, New York.
[22] Schönbeck, H. (1968). Einfluß von Luftverunreinigungen (SO_2) auf transplantierte Flechten, *Naturwiss.*, **55**, 451–452.
[23] Bell, J.N.B., and Clough, W.S. (1973). Depression of yield in ryegrass exposed to sulfur dioxide, *Nature*, **241**, 47–49.
[24] Hitchcock, A.E., Zimmermann, P.W., and Coe, R.R. (1963). The effects of fluorides on milo maize, *Contr. Boyce Thompson Inst.*, **22**, 175–206.
[25] Shupe, J.L., Miner, M.L., Harris, L.E., and Greenwood, D.A. (1962). Relative effects of feeding hay atmospherically contaminated by fluoride residue, normal hay plus calcium fluoride, and normal hay plus sodium fluoride to dairy heifers, *Am. J. Vet. Res.*, **23**, 781.
[26] Oelschläger, W. (1971). Problematik der Immissionsmessungen von Fluor hinsichtlich der Wirkungen auf Pflanze und Tier, *Staub Reinh. Luft*, **31**, 457–459.
[27] Brandt, C. S. (1971). Ambient air quality criteria for hydrofluorine and fluorides, *VDI-Berichte*, **164**, 23–27.
[28] Shupe, J.L. (1969). *Fluorosis in livestock*, Air Quality, Monographs, pp. 69–74, American Petroleum Institute, New York.
[29] Suttie, J.W. (1969). Air quality standards for the protection of farm animals from fluorides, *JAPCA*, **19**, 239–242.
[30] NAS (1971). *Fluorides, Biologic Effects of Atmospheric Pollutants*, National Academy of Sciences, Washington, D.C.
[31] Cristiani, H., and Gautier, R. (1925). Emanations fluorées des usines: étude experimentale de l'action du fluor sur les vegetaux, *Ann. Hygiene Publ.*, **3**, 49–64.
[32] Hitchcock, A.E., Zimmermann, P.W., and Coe, R.R. (1962). Results of ten year's work (1951–1960) on the effect of fluorides on gladiolus, *Contr. Boyce Thompson Inst.*, **21**, 303–344.
[33] Garber, K. (1966). Die Beeinflussung der Pflanzenwelt durch fluorhaltige Immissionen, *Angew. Bot.*, **40**, 12–21.
[34] Treshow, M., and Pack, M.R. (1970). Fluoride, in, *Recognition of Air Pollution Injury to Vegetation: A Pictorial Atlas* (Eds J.S. Jacobson and A.C. Hill), APCA, Pittsburgh.
[35] Bolay, A., and Bovay, E. (1965). Observations sur la sensibilité aux gaz fluorés de quelques espèces végétales du Valais, *Phytopath. Z.*, **53**, 289–298.

[36] McCune, D.C., and Hitchcock, A.E. (1970). Fluoride in forage: Factors determining its accumulation from the atmosphere and concentration in the plant. *Second International Clean Air Congress, Washington, D.C.*
[37] Brandt, C.J. (1980). Untersuchungen über Wirkungen von Fluorwasserstoff auf Lolium multiflorum und andere Nutzpflanzen, PhD Dissertation, University of Bonn.
[38] Adams, D.F., Shaw, C.G., Gnagy, R.M., Koppe, R.K., Mayhaw, D.J., and Yerkers, W.D. (1956). Relationship of atmospheric fluoride levels and injury indexes on gladiolus and ponderosa pine, *J. Agric. Food Chem.*, **4**, 64–66.
[39] Brewer, R. R., Sutherland, F.H., Guillemet, F.B., and Creveling, R.K. (1960). Some effects of hydrogen fluoride gas on bearing navel orange trees. *Proc. Am. Soc. Hort. Sci.*, **76**, 206–214.
[40] Thomas, M.D., and Alther, E.W. (1966). The effects of fluoride on plants, in, *Handbuch der experimentellen Pharmakologie*, vol. XX/1, pp. 231–366, Springer Verlag, Berlin.
[41] Guderian, R., van Haut, H., and Stratmann, H. (1969). Experimentelle Untersuchungen über pflanzenschädigende Fluorwasserstoff-Konzentrationen, *Forsch. Ber. des Landes NW*, 2017.
[42] Yu, M-H., and Miller, G.W. (1967). Effect of fluoride on the respiration of leaves from higher plants, *Plant and Cell Physiol.*, **8**, 483–493.
[43] McNulty, I.B., and Newman, D.W. (1957). Effects of atmospheric fluoride on respiration rate of bush bean and gladiolus leaves, *Plant Physiol.*, **32**, 121–124.
[44] McNulty, I.B. (1959). The influence of fluoride on leaf respiration. Proceedings of the Ninth International Botany Congress, vol. 2, p. 245.
[45] Applegate, H.G., and Adams, D.F. (1960). Effect of atmospheric fluoride on respiration of bush beans, *Bot. Gazette*, **121**, 223–227.
[46] Weinstein, L.H. (1961). Effects of atmospheric fluoride on metabolic constituents of tomato and bean leaves, *Contrib. Boyce Thompson Inst.*, **21**, 215–231.
[47] Bennett, J.H., and Hill, A.C. (1973). Inhibition of apparent photosynthesis by air pollutants. *J. Environ. Qual.*, **2**, 526–530.
[48] McCune, D.C. (1969b). *On the Establishment of Air Quality Criteria with Reference to the Effects of Atmospheric Fluorine on Vegetation*, Air Quality Monographs, 69-3, American Petroleum Institute, New York.
[49] Compton, O.C., and Remmert, C.F. (1960). Effect of air-borne fluorine on injury and fluorine content of gladiolus leaves, *Proc. Am. Soc. Hort. Sci.*, **75**, 663–675.
[50] Adams, D.F., and Sulzbach, C.W. (1961). Nitrogen deficiency and fluoride susceptibility of bean seedlings, *Science*, **133**, 1425–1426.
[51] Spierings, F.H., and Wolting, H.G. (1971). Der Einfluß sehr niedriger HF-Konzentrationen auf die Länge der Blattspitzen-Schädigung und den Zwiebelertrag bei der Tulpenvarietät 'Paris', *VDI-Ber.*, **164**, 19–21.
[52] Benedict, H.M., Ross, J.M., and Wade, R.W. (1964). The disposition of atmospheric fluorides by vegetation, *Intl. J. Air Water Pollut.*, **8**, 279–289.
[53] MacLean, D.C., Roark, O.F., Folkerts, G., and Schneider, R.E. (1969). Influence of mineral nutrition on the sensitivity of tomato plants to hydrogen fluoride, *Environ. Sci. Technol.*, **3**, 1201–1204.
[54] McCune, D.C., Hitchcock, A.E., and Weinstein, L.H. (1966). Effect of mineral nutrition on the growth and sensitivity of gladiolus to HF, *Contr. Boyce Thompson Inst.*, **23**, 295–299.
[55] Pack, M.R. (1966). Response of tomato fruiting to HF as influenced by calcium nutrition, *JAPCA*, **16**, 541–544.
[56] Scholl, G. (1971). Ein biologisches Verfahren zur Bestimmung der Herkunft und Verbreitung von Fluorverbindungen in der Luft, *Landw. Forsch.*, **26**, 29–35.

[57] Scholl, G. (1976). Vorschläge für die Begrenzung der Aufnahmerate von Fluorid in standardisierter Graskultur zum Schutz von Pflanzen und Weidetieren, *Schriftenr. d. LIS*, **37**, 129–132.
[58] Lacasse, N.L. (1968). Open burning and our forest; a new threat, *Forest Notes*, Summer, 23–25.
[59] van Haut, H., and Guderian, R. (personal communication).
[60] van Haut, H., and Stratmann, H. (1967). Experimentelle Untersuchungen über die Wirkung von Stickstoffdioxid auf Pflanzen. *Schriftenr. LIS*, **7**, 20–70.
[61] Taylor, O.C., and MacLean, D.C. (1970). Nitrogen oxides and the peroxyacyl nitrates, in, *Recognition of Air Pollution Injury to Vegetation: A Pictorial Atlas* (Eds J.S. Jacobson and A.C. Hill), APCA, Pittsburgh.
[62] Tingey, D.T., Reinert, R.A., Dunning, J.A., and Heck, W.W. (1971). Vegetation injury from the interaction of NO_2 and SO_2, *Phytopath.* **61**, 1506–1511.
[63] Middleton, J.T., Kendrick, J.B., and Schwalm, H.W. (1950). Injury to herbaceous plants by smog or air pollution, *Pl. Dis. Reptr.*, **34**, 245–252.
[64] Middleton, J.T., Darley, E.F., and Brewer, R.R. (1958). Damage to vegetation from polluted atmospheres, *JAPCA*, **8**, 9–15.
[65] Knabe, W., Brandt, C.S., van Haut, H., and Brandt, C.J. (1973). Nachweis photochemischer Luftverunreinigungen durch biologische Indikatoren in der Bundesrepublik Deutschland. *Proceedings of the Third International Clean Air Congress, Düsseldorf.*
[66] Haagen-Smit, A.J. (1952). Chemistry and physiology of Los Angeles smog, *Indus. Chem.*, **44**, 1342–1346.
[67] Heck, W.W. (1966). The use of plants as indicators of air pollution, *Intl. J. Air Water Pollut.*, **10**, 99.
[68] Hindawi, I.J., Dunning, J.A., and Brandt, C.S. (1965). Morphological and microscopical changes in tobacco, bean, and petunia leaves exposed to irradiated auto exhaust. *Phytopath*, **55**, 27–30.
[69] Taylor, O.C. (1968). Effects of oxidant air pollutants, *J. Occup. Med.*, **10**, 53–60.
[70] Applegate, H.G., and Durrant, L.C. (1969). Synergistic action of ozone–sulfur dioxide on peanuts, *Environ. Sci. Technol.*, **3**, 759–760.
[71] Menser, H.A., and Heggestad, H.E. (1966). Ozone and sulfur dioxide synergism: Injury to tobacco plants, *Science*, **153**, 3734, 424–425.
[72] Dochinger, L.S. (1968). The impact of air pollution on eastern white pine: The chlorotic dwarf disease, *JAPCA*, **18**, 814–816.
[73] Berry, C.R. (1964). Differences in concentrations of surface oxidant between valley and mountaintop conditions in the southern Appalachians, *JAPCA*, **14**, 238–239.
[74] Treshow, M. (1970). Ozone damage to plants. *Environ. Pollut.*, **1**, 155–161.
[75] Hill, A.C., and Bennett, J.J. (1970). Inhibition of apparent photosynthesis by nitrogen oxides, *Atmos. Environ.*, **4**, 341–348.
[76] Davis, D.D., and Wood, F.A. (1972). The relative susceptibility of 18 coniferous species to ozone, *Phytopath*, **62**, 14–19.
[77] Brennan, E., Leone, I.A., and Daines, R.H. (1964). The importance of variety in ozone plant damage, *Pl. Dis. Reptr.*, **48**, 923–924.
[78] Feder, W.A., and Campbell, F.J. (1968). Influence of low levels of ozone on flowering of carnations, *Phytopath.*, **58**, 1038–1039.
[79] Heck, W.W., Dunning, J.A., and Hindawi, I.J. (1966). Ozone: Nonlinear relation of dose and injury in plants, *Science*, **151,** 577–578.
[80] Heck, W.W., and Tingey, D.T. (1970). Ozone, time-concentration model to predict acute foliar injury, *Proceedings of the Second Clean Air Congress*, Washington, D.C.

81 Linzon, S.N., Heck, W.W., and MacDowall, F.D.H. (1975). Effects of photochemical oxidants on vegetation, in, *Photochemical Air Pollution: Formation, Transport, Effects*. Nature Research Council, Canada.

82 Vandergrift, A.E., Shannon, L.J., Salkee, E.E., Gorman, P.G., and Park, W.R. (1971). Particulate air pollution in the United States, *JAPCA*, **21**, 321–328.

83 Krause, G.H.M. (1974). Zur Aufnahme von Zink und Cadmium durch oberirdische Pflanzenorgane. PhD Dissertation, University of Bonn.

84 LEH, H.O. (1969). *Handbuch der Pflanzenkrankheiten*, vol. 1, 7th edn, Verlag Paul Parey, Berlin and Hamburg.

85 Suchodoller, A., and Wanner, H. (1969). Die Absorption von Blei und Mangan durch Wurzeln von Hordeum vulgare, *Z. Pflzphysiol.*, **61**, 122–128.

86 Suchodoller, A., and Wanner, H. (1970/71). Die Absorption von Blei und Kupfer durch Wurzeln von Hordeum vulgare, *Ber. Schweiz. Bot. Ges.*, **80**, 334–340.

87 Prinz, B., Krause, G. H. M., and Stratmann, H. (1979). Thalliumschäden in der Umgebung der Dyckerhoff Zementwerke AG in Lengerich, Westfalen, *Staub-Reinh. Luft*, **39**, 457–462.

88 Oertli, J.J. (1959). Effects of salinity on susceptibility of sunflower plants to smog. *Soil Sci.*, **87**, 249–251.

89 Leone, I.A., Brennan, E., and Daines, R.H. (1965). Factors influencing SO_2 phytotoxicity in New Jersey. *Pl. Dis. Reptr.*, **49**, 911–915.

90 Pfeffer, A. (1965). *The Effect of Air Polluted with Sulfur Dioxide on the Countryside*, Liblice House of Scientific Workers, CSSR.

91 Zimmermann, P.W., and Hitchcock, A.E. (1956). Susceptibility of plants to hydrofluoric acid and sulfur dioxide gases, *Contrib. Boyce Thompson Inst.*, **18**, 263–279.

92 Daines, R.H., Leone, I., and Brennan, E. (1952). The effect of fluorine on plants as determined by soil nutrition and fumigation studies. Proceedings of the U.S. Technological Conference on Air Pollution, McGraw-Hill, New York.

93 MacLean, D.C., and Schneider, R.E. (1970). Fluoride phytotoxicity: Its alteration by temperature. *Proceedings of the Second Clean Air Congress*, Washington, D.C.

94 Guderian, R., and Stratmann, H. (1968). Freilandversuche zur Ermittlung von Schwefeldioxidwirkungen auf die Vegetation, III. Teil. Forsch. Ber. des Landes NW, 1920.

95 VDI-Kommission Reinhaltung der Luft (1983). Säurehaltige Niederschläge—Entstehung und Wirkungen auf terrestrische Ökosysteme. Düsseldorf.

Pollutants and Their Ecotoxicological Significance
Edited by H. W. Nürnberg

CHAPTER 8

Modelling the Formation of Secondary Pollutants in the Atmosphere Over North-West Europe

R. G. DERWENT

SUMMARY

The phenomenon of photochemical air pollution has recently been observed over large areas of north-west Europe. Modelling the formation and long-distance transport of photochemically generated secondary pollutants is a complex problem involving knowledge of precursor emissions, atmospheric chemistry and meteorology. At present the modelling problem can only be tackled using simple numerical techniques which are based on a detailed knowledge of photochemical air-pollution formation. One such approach is described in this paper which aims to illustrate some of the basic aspects of secondary pollutant formation, transport and destruction in Europe.

INTRODUCTION

Most people are aware of the local effects of atmospheric pollutants, such as smoke and grime, damage to health and vegetation and reduction in visibility. By the application of the Clean Air Acts and other legislation, the disastrous fogs of London and other urban areas are almost a thing of the past and industrial pollution is steadily being reduced. Although much still remains to be done in the field of primary pollution, attention is now also being directed at the more distant effects of some atmospheric pollutants. Of particular interest are those new classes of pollutants, not directly emitted into the atmosphere, but formed there by atmospheric chemical reactions, which may turn out to have unexpected effects in parts of the environment which are unaffected by their precursors.

In north-west Europe much attention has been directed to studies of those atmospheric conversion processes, driven by sunlight, which give rise to photochemical air pollution. Systematic measurements carried out in the Netherlands since 1968[1] and in the UK since 1971,[2] have shown that hourly mean ozone (O_3) concentrations may reach values approaching 0.25 ppm (250 ppb). Measurements in France, Germany[3] and Scandinavia[4] from 1974 onwards have shown hourly mean values of between 0.1 and 0.2 ppm (100 and 200 ppb). In addition to the formation of O_3, there is also evidence for the formation of other secondary air pollutants such as peroxyacetylnitrate[5] and visibility-reducing aerosol species.[6] Table 8.1 contains a summary of the main photochemical air pollutants and their environmental targets.

Table 8.1 The photochemically-generated secondary pollutants routinely monitored in the UK and their environment targets

Secondary pollutant	Environmental target
Ozone	Human health
	Plant life
	ornamental plants
	trees
	vegetables
	other cash crops
	Materials
	rubber and painted surfaces
Peroxyacetylnitrate	Human health
	Plant life
Nitric and sulphuric acids	Visibility reduction
ammonium nitrate and sulphate	Freshwater ecosystems
	Human health

The time scales for the formation and decay of the individual photochemical air pollutants are the subject of this review. Observations show dramatically that the time scales are long enough to allow significant long-range transport inside north-west Europe. Transport over distances approaching 1000 km has been illustrated with visibility-reducing aerosols[6] and ozone.[7]

Photochemical air pollutants arise by reactions involving hydrocarbons and oxides of nitrogen driven by sunlight.[8] The atmospheric degradation of hydrocarbons by hydroxyl (OH) radicals plays a central role in the oxidation of nitric oxide (NO) to nitrogen dioxide (NO_2) and hence in O_3[9] and other secondary pollutant formation. The secondary pollutants may have both chemical (photolysis and thermal decomposition) and physical sinks (dry deposition and aerosol scavenging). Their overall behaviour in the atmosphere, leading to characteristic diurnal variations, represents a subtle interplay between photochemical production and physical and/or chemical destruction.[10] Representation of this behaviour is a complex modelling problem requiring sophisticated techniques.

In the present review the competition between photochemical production and destruction processes is represented by the simple two-step consecutive reaction scheme:

$$\text{Hydrocarbon} \xrightarrow{k_1} \text{secondary pollutant} \xrightarrow{k_2} \text{sink}$$

Here k_1 and k_2 represent first-order rate coefficients, whose reciprocals, t_1 and t_2, represent the time scales for hydrocarbon degradation (and hence secondary pollutant formation) and the time scale for the subsequent loss process for the secondary pollutant respectively.

The purpose of the present review is to define the time scales t_1 and t_2 and to examine photochemical air pollution in north-west Europe using the simple two-step model.

PRODUCTION AND DESTRUCTION OF THE SECONDARY POLLUTANTS

Ozone

There is only one important source of O_3 in polluted atmospheres,[8] viz. the photolysis of NO_2 in reaction which is usually followed by the rapid reaction (8.2).

$$NO_2 + h\nu = NO + O \tag{8.1}$$

$$O + O_2 + M = O_3 + M \tag{8.2}$$

These reactions when coupled with the rapid oxidation of NO by O_3 in (8.3), serve to maintain an overall balance between oxidizing and reducing agents in the atmosphere.

$$NO + O_3 = NO_2 + O_2 \tag{8.3}$$

The production of peroxy radicals (RO_2) by the degradation of hydrocarbons through oxidation of NO to NO_2, lead to net O_3 formation, viz.

$$RO_2 + NO = RO + NO_2 \tag{8.4}$$

$$NO_2 + h\nu = NO + O \tag{8.1}$$

$$O + O_2 + M = O_3 + M \tag{8.2}$$

net result $$RO_2 + O_2 + h\nu = RO + O_3 \qquad (8.4) + (8.1) + (8.2)$$

The overall reaction which occurs as a rapid set of consecutive reactions leads to net O_3 formation, leaving the oxides of nitrogen unchanged.

Peroxy radicals are formed by the attack of OH radicals on hydrocarbons of all types and by photolysis or OH attack on the intermediate aldehydes and ketones produced in their degradation. Demerjian *et al.*[11] have demonstrated how peroxy radicals (RO_2) are formed from methane, *n*-butane, ethylene and propylene. They also characterized the fate of the alkoxy (RO) radicals formed in (8.4) as elimination of a hydroperoxyl (HO_2) radical with the formation of an aldehyde in a reaction such as (8.5).

$$CH_3O + O_2 = HO_2 + HCHO \tag{8.5}$$

The structure of the hydrocarbon dictates how many peroxy radicals can be formed during its degradation, and hence its stoichiometry, i.e. O_3 molecules produced per molecule of hydrocarbon oxidized. In fact, most hydrocarbons produce between 2 and 4 O_3 molecules per hydrocarbon oxidized under typical United Kingdom conditions.[12]

The potential for O_3 production is fairly evenly distributed over the whole spectrum of hydrocarbons. However, this is not to say that all hydrocarbons contribute evenly to O_3 production. Two further criteria are more significant than the stoichiometry factor. These criteria involve the reactivity of the hydrocarbon and that nature of the emission sources themselves. The latter is the simplest to deal with and covers the variability of the emission rates of the individual hydrocarbons from source to source. Hence the amount available of each hydrocarbon varies from one source area to another.

The criteria of reactivity is important because two different hydrocarbons with similar O_3 production potentials and emission rates would contribute differently to O_3 production depending on the rates at which they are attacked by OH radicals. The rate coefficients for OH attack on most hydrocarbons of atmospheric interest are now available[13] and reactivities have been quantitatively assessed.

In the United Kingdom, from a consideration of the reactivity, stoichiometry and emission factors, it appears that most of the O_3 production involves ethylene, propylene and *o*-xylene. In the framework of the simple two-step model O_3 production can be represented by $k_1 = 0.1\,h^{-1}$ (on the basis of $k_{OH+C_2H_4} = 2.2 \times 10^{-12} \exp(385/T)$ and $[OH] = 3 \times 10^6$ molecule cm^{-3} and an α (O_3 produced per hydrocarbon oxidized) of about 4. Having qualitatively identified the time scale for O_3 production, attention is turned to its destruction.

The sink processes for O_3 in the lower atmosphere involve both chemical and physical destruction. In urban areas, local sources of NO may scavenge O_3 and act as an efficient local O_3 sink, particularly during night-time. However, during daylight since the NO_2 produced is readily photolysed back to O_3 and NO, this sink is only temporary. In rural areas, the NO concentrations are typically much lower than the O_3 concentrations so that chemical reactions with NO is unimportant during both daytime and night-time. Other chemical sinks involve reactions with olefinic hydrocarbons or hydroperoxyl radicals. These are of some significance in urban and background atmospheres, respectively, but generally do not compete effectively with the main physical sink, viz. dry deposition.

Several workers[14,15] have shown that land surfaces absorb O_3 from the atmosphere. Ozone shows a marked diurnal variation even in rural atmospheres with concentrations much lower at night-time than during the daytime. Balloon observations[16] and numerical simulations[10] have shown that depletion occurs below the nocturnal inversion. The morning increase of concentration is due to downward mixing of air from aloft following the disappearance of the inversion.

For the simple consecutive reaction model, the decay time constant for O_3 was estimated on the basis of dry deposition alone. A value of t_2 of $0.02\ h^{-1}$ was taken for daytime on the basis of the observed v_g of $0.58\ cm\ s^{-1}$ and a mixing height of 1 km for O_3.

Peroxyacetylnitrate

Early chemical studies of photochemical air pollution revealed the presence of a previously unknown compound, peroxyacetylnitrate (PAN).[8] Subsequent studies[17] have confirmed that this substance is formed by the reaction of peroxyacetyl radicals with NO_2 in reaction (8.6).

$$CH_3CO(O_2) + NO_2 = PAN \qquad (8.6)$$

Peroxyacetyl radicals are formed in the atmosphere by OH attack on acetaldehyde (8.7 + 8.8), and by photolysis of molecules containing the CH_3CO group, (8.8 + 8.9), such as biacetyl ($CH_3COCOCH_3$):

$$OH + CH_3CHO = H_2O + CH_3CO \qquad (8.7)$$

$$CH_3CO + O_2 = CH_3CO(O_2) \qquad (8.8)$$

$$CH_3COCOCH_3 = CH_3CO + CH_3CO \qquad (8.9)$$

The ability of hydrocarbons to produce PAN is, therefore, strictly limited to those with the correct molecular structure which leads either to acetaldehyde or acetyl formation. Whereas the ability to produce some O_3 was shared by most hydrocarbons this is not true of PAN. The stoichiometry factor, α, viz. PAN molecules produced per hydrocarbon molecule oxidized, may be in fact very small for some hydrocarbons.

In the United Kingdom, from a consideration of the stoichiometry, reactivity and emission factors, it appears that most of the PAN production involves the degradation of *o*-xylene and propylene.[12] For the simple two-step model values of k_1 of $0.3\ h^{-1}$ (on the basis of $k_{OH+C_3H_6} = 4.1 \times 10^{-12} \exp(545/T)$ and $[OH] = 3 \times 10^{-6}$ molecule cm^{-3}) and α of 0.6 were taken.

The destruction processes for PAN in the lower atmosphere, as with O_3, involve both chemical and physical destruction processes. PAN is readily decomposed by surfaces and a deposition velocity of about one-third of the value for O_3 has been reported.[18] However, laboratory investigations[19] have shown that PAN is thermally unstable near room temperature with a lifetime of about 40 min for decomposition to NO_2 and peroxyacetyl, viz.

$$PAN \rightarrow NO_2 + CH_3CO(O_2) \qquad (8.10)$$

Table 8.2 Typical lifetimes of PAN in rural and urban atmospheres during daytime and night-time due to thermal decomposition and deposition, hours

	Daytime (25 °C)	Night-time (10 °C)
Urban	5 (thermal decomposition)	10 (thermal decomposition)
Rural	5 (thermal decomposition)	10 (deposition)

Notes: $v_g = 0.25\,cm\,s^{-1}$.[18]
$[O_3] = 160$ ppb.[2]

In the atmosphere, the overall lifetime of PAN depends on whether the peroxyacetyl radicals are scavenged by NO or NO_2. Whereas the former would lead to PAN decomposition the latter would not. Table 8.2 shows the typical lifetime calculated for PAN in various atmospheric regimes during daytime and night-time.

Visibility-reducing Aerosol Species

The formation of visibility-reducing aerosol species is often readily apparent during photochemical air-pollution episodes. These aerosol species consist of nitric and sulphuric acids and their ammonium salts.[20] The free acids are the main precursors which are produced by OH attack on nitrogen and sulphur dioxides.

The mechanisms of gaseous nitric acid formation is relatively clear cut via reaction (8.11),

$$OH + NO_2 = HNO_3 \tag{8.11}$$

The nitric acid must then be scavenged by the atmospheric aerosol by processes that are not yet quantitatively understood. Nevertheless, the processes appear quite efficient and the time scale, t_1, for nitrate formation in the simple two-step model was estimated as 8.3 h on the basis of $K_{OH+NO_2} = 1.1 \times 10^{-11}\,cm^{-3}$ molecule$^{-1}\,s^{-1}$.

The mechanism of sulphate formation is not so well understood. Many photochemical conversion routes have been proposed but only the $OH + SO_2$ route can account for a significant fraction of the conversion of SO_2 to sulphate during sunlit daytime conditions. The attack of OH on SO_2 leads to the formation of sulphur-oxide radical species. The actual mechanism of sulphuric acid production has not yet been elucidated.[21] However, the time scale for daytime sulphuric acid production can be estimated on the basis of $k_{OH+SO_2} = 1.1 \times 10^{-12}\,cm^{-3}$ molecule$^{-1}\,s^{-1}$ [21] as 84 h. Atmospheric oxidation during daytime competes readily for SO_2 against dry deposition which has a time scale of about 35 h.[22] The overall behaviour of sulphate aerosol formation in the simple two-step model is therefore represented by a time scale, t_1, based on both deposition and photochemistry but a stoichiometry factor, α, reflecting the competition between them. Accordingly k_1 is taken to be $0.04\,h^{-1}$, and α, 0.3. This behaviour provides a contrast to that for nitrate formation where OH oxidation is the dominant fate for NO_2 since dry deposition is much less efficient than for SO_2.

Deposition processes provide the important removal processes for aerosol particles. Dry deposition of sub-micron particles is inefficient and a lifetime of 300 h is estimated on the basis of the observed particle size distribution.[22] In contrast to dry deposition, wet deposition is highly efficient for nitrate and sulphate aerosols. This is an episodic process and makes little contribution to the overall behaviour during photochemical episodes. The dry deposition rate is, therefore, employed in the simple two-step model shown below.

APPLICATION OF THE SIMPLE TWO-STEP REACTION SCHEME

The Model Equations

The time dependence of the hydrocarbon precursor HC is given by (8.12)

$$\frac{d}{dt}[HC] = -k_1[HC] \qquad (8.12)$$

where $k_1 = k_{OL+HC}[OH]$, and [OH] is the OH concentration. Equation (8.12) may be integrated to give (8.13),

$$[HC] = [HC]_{t=0}\exp(-k_1 t) \qquad (8.13)$$

The time dependence of the secondary pollutant, (SP), is given by (8.14)

$$\frac{d}{dt}[SP] = \alpha k_1[HC] - k_2[SP] \qquad (8.14)$$

which may be integrated to give (8.15)

$$[SP] = \frac{\alpha k_1}{k_1 - k_2}[HC]_{t=0}[\exp(-k_2 t) - \exp(-k_1 t)] \qquad (8.15)$$

The integrated equation (8.15) shows that the secondary pollutant reaches a maximum, SP_{max}, at a time, t_{max}, given by (8.16) and (8.17)

$$[SP]_{max} = \alpha[HC]_{t=0}\exp(-k_2 t_{max}) \qquad (8.16)$$

$$t_{max} = \frac{1}{k_1 - k_2}\ln\left(\frac{k_1}{k_2}\right) \qquad (8.17)$$

Photochemical Episodes

Under normal, cloudy conditions for most hydrocarbons and secondary pollutants k_1 and k_2 are fairly comparable. Hydrocarbon degradation produces

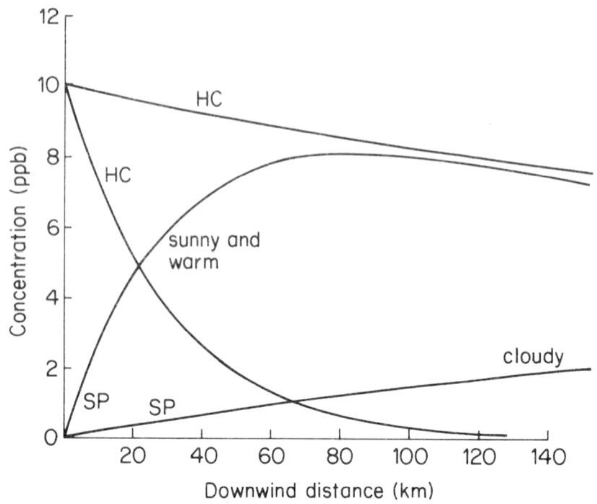

Fig. 8.1 The effect of [OH] on secondary pollutant formation. Sunny and warm, high [OH]; cloudy, low [OH]

the secondary pollutants over a long time scale giving them maximum exposure to their respective sink processes. The resulting secondary pollutant concentrations are accordingly quite trivial.

Under photochemical episodic conditions the higher photoloysis rates stimulate OH radical production and increase proportionately the k_1 values leaving the k_2 values unaffected. The secondary pollutants now build up very much quicker than they can decay away and they are left in the atmosphere close to the source region of their precursors. Figure 8.1 illustrates diagrammatically the effect of k_1 on secondary pollutant formation assuming the same hydrocarbon, HC, and secondary pollutant but changing the OH concentration by an order of magnitude.

Effect of Hydrocarbon Reactivity

The equations (8.15), (8.16) and (8.17) also show the importance of the reactivity of the hydrocarbon on secondary pollutant formation. The more reactive hydrocarbons liberate their secondary pollutants closer to their point of emission, minimizing their exposure to their destruction processes. Figure 8.2 illustrates the effect of changing hydrocarbon reactivity from highly reactive HC_1 (with k_1) to low reactive HC_2 (low k_1) on the formation of the same secondary pollutant, SP, with a constant α value, and [OH].

Effect of Lifetime of Secondary Pollutant

The equations also show the effect of the lifetime of the secondary pollutant. If the secondary pollutant has a short lifetime (k_2 large), then the maximum in its

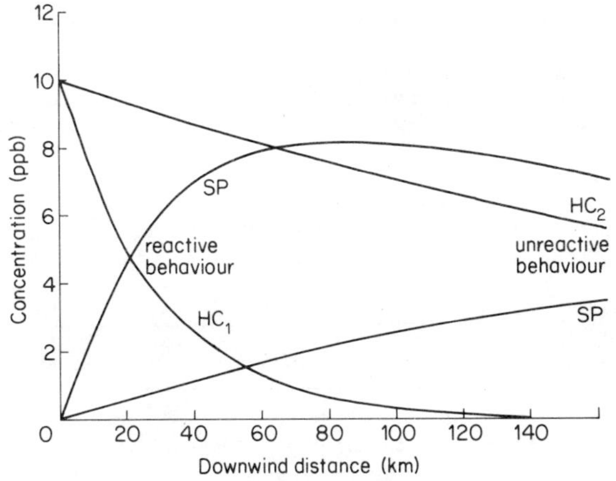

Fig. 8.2 The effect of k_{OH+HC} on secondary pollutant formation. HC_1 is highly reactive and HC_2 is an order of magnitude less reactive

concentration appears closer to the source of the hydrocarbons. Hence the distribution of the shorter-lived secondary pollutants gives a clearer indication of the precursor source areas. Ozone PAN and sulphate aerosol have markedly different lifetimes with PAN the most reactive. The observed distribution of PAN is much more uneven than for O_3 for example, with urban concentrations much higher than rural values.[23] With O_3 there is little difference.[2] Sulphate

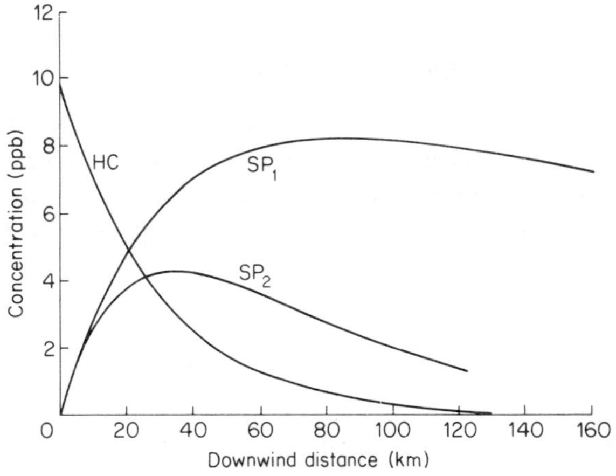

Fig. 8.3 The effect of secondary pollutant lifetime on the location and magnitude of its peak downwind concentration. SP_1 has a long lifetime and SP_2 an order of magnitude shorter lifetime

Table 8.3 Typical behaviour of the secondary pollutants in south-east England

Secondary pollutant	Precursor emission rate (molecule s^{-1})	α	Initial precursor concentration (ppb)	k_1 h^{-1}	k_2 h^{-1}	Time of maximum (h)	Downwind distance of maximum (km)	Maximum secondary pollutant concentration (ppb)
Ozone	4.6×10^{25}	4	24	0.1	0.02	20	172	64
PAN	1.4×10^{25}	0.6	7	0.3	0.2	4	35	1.8
Nitrate aerosol	3.3×10^{25}	1	17	0.1	0.003	36	310	15
Sulphate aerosol	3.4×10^{25}	0.3	18	0.04	0.003	70	605	4

aerosol concentrations shown much less variation than O_3 over the southern British Isles.

Figure 8.3 illustrates the effect of changing the secondary pollutant lifetime, k_2, by over an order of magnitude keeping k_1, OH and α constant. The movement of the maximum concentration downwind and the increase in the magnitude of the maximum are clearly apparent.

Typical United Kingdom Conditions

In Table 8.3 the simple two-step model is used to calculate the behaviour of O_3, PAN, nitrate and sulphate aerosols in south-east England. A mean windspeed of 2.4 ms^{-1} is assumed[10] and the effective length of the source region is taken as 50 km.

ACKNOWLEDGEMENTS

This work was performed as part of an air-pollution research programme sponsored by the United Kingdom Department of the Environment. The author wishes to express his thanks to Oystein Hov, Institute of Geophysics, University of Oslo for his assistance with the calculations and stimulating discussions.

REFERENCES

1 Guicherit, R. (1975). Photochemical smog formation in the Netherlands, *TNO Report*, Delft, the Netherlands.
2 Derwent, R.G., McInnes, G., Stewart, H.N.M., and Williams, M.I. (1976). Warren Spring Laboratory Report LR 227 (AP), Stevenage, Herts, UK.
3 Guicherit, R., and Van Dop, H. (1977). *Atmos. Environ.*, **11**, 145–155.
4 Schjoldager, J., Sivertsen, B., and Hanssen, J.F. (1978). *Atmos. Environ.*, **12**, 2461–2468.
5 Sandalls, F.J., Penkett, S.A., and Jones, B.M.R. (1974). AERE Harwell Report—R7807, HMSO, London, UK.

[6] Lovelock, J.E. (1972). *Atmos. Environ.*, **6**, 917–926.
[7] Cox, R.A., Eggleton, A.E.J., Derwent, R.G., Lovelock, J.E., and Pack, D.H. (1975). *Nature*, **255**, 118–121.
[8] Leighton, P.A. (1961). *Photochemistry of Air Pollution*, Academic Press, New York, USA.
[9] Calvert, J.G. (1976). *Environ. Sci. Technol.*, **10**, 256–262.
[10] Garland, J.A., and Derwent, R.G. (1979). *Quart. J. Roy. Met. Soc.*, **105**, 169–183.
[11] Demerjian, K.L., Kerr, J.A., and Calvert, J.G. (1974). *Advances in Environmental Sci. and Technol.*, **4**, 1–262.
[12] Derwent, R.G., and Hov, O. (1979). AERE Harwell Report—R9434, HMSO, London, UK.
[13] Atkinson, R., Darnall, K.R., Lloyd, D.C., Winer, A.M., and Pitts, J.N. (1979). *Advances in Photochemistry*, **11**, 375–488.
[14] Galbally, I.E. (1971). *Quart. J. Roy. Met. Soc.*, **97**, 18–29.
[15] Regener, V.H. (1957). *J. Geophys. Res.*, **62**, 221–228.
[16] Galbally, I.E. (1968). *Nature*, **218**, 456–457.
[17] Louw, R., Van Ham, J., and Nierboer, H. (1973). *J. Air Poll. Control Ass.*, **23**, 716.
[18] Garland, J.A., and Penkett, S.A. (1977). *Atmos. Environ.*, **10**, 1127–1131.
[19] Cox, R.A., and Roffey, M.J. (1977). *Environ. Sci. and Technol.*, **11**, 900–906.
[20] Atkins, D.H.F., Cox, R.A., and Eggleton, A.E.J. (1972). *Nature*, **235**, 372–376.
[21] Calvert, J.G., Su, F., Bottenheim, J.W., and Strausz, O.P. (1978). *Atmos. Environ.*, **12**, 197–226.
[22] Garland, J.A. (1978). *Atmos. Environ.*, **12**, 349–362.
[23] Penkett, S.A. (1978). *Environ. Poll. Management*, Jan./Feb., 8–12.

Pollutants and Their Ecotoxicological Significance
Edited by H. W. Nürnberg

CHAPTER 9

The Effect of Man-made Halocarbons on Stratospheric Ozone

D. H. EHHALT AND U. SCHMIDT

ABSTRACT

The chemistry of stratospheric chlorine and the impact of man-made halocarbons on stratospheric ozone are discussed. If the present release of chlorofluoromethanes is continued, one-dimensional models predict stratospheric ozone to decrease by about 10 per cent. The new steady-state concentration is reached with a relaxation time of about 60 years. The relative error in this prediction is rather large, larger than 40 per cent.*)

INTRODUCTION

For about 10 years it has been recognized that man influences, not only his immediate atmospheric environment, but also regions as remote as the stratosphere. In the meantime four pathways have been identified by which man can perturb chemistry and climate of the stratosphere. Most of the impact on the stratosphere centres around the fact that in every case the concentration of ozone (O_3) is changed. Listed in order in which they were published, the human activities altering stratospheric ozone are:

1 The emission of nitrogen oxides (NO_x) into the stratosphere in the exhaust of high-flying (supersonic) aircraft (1970).
2 The emission of halocarbons at the earth's surface (1974).
3 The enhanced release of nitrous oxide (N_2O) from soil after application of fertilizer (1976).

* Since this paper was submitted, a number of reaction rates have been remeasured and changed. This changes also some conclusions. In particular the steady state ozone depletion based on CFM emission has been reduced to 5%. For details see ref. 29.

4 The emission of carbon dioxide (CO_2) from fossil-fuel burning. The latter causes primarily a decrease in stratospheric temperature which in turn causes a slight increase in stratospheric O_3 (1977).

Presently, the halocarbons are thought to have the strongest impact on stratospheric O_3.

The man-made halocarbons of atmospheric importance are mainly the chlorofluoromethanes (CF_2Cl_2, $CFCl_3$ and CCl_4) to a less degree also methyl chloroform (CH_3CCl_3). They all have important commercial uses, as refrigerants, aerosol propellants, solvents, or foam-blowing agents. In all their uses these gases are eventually released to the atmosphere. Since the chlorofluoromethanes (CFM) are chemically relatively inert, they are virtually not destroyed in the troposphere. They accumulate and are eventually mixed by vertical transport into the stratosphere, i.e. the atmosphere above about 10 km altitude.

In the stratosphere the ultraviolet component of the solar radiation becomes increasingly stronger with altitude and is shifted to shorter wavelengths. Above about 25 km solar photons are encountered of sufficient energy to photolyse the CFM, injecting Cl atoms into the stratosphere. In a catalytic chain reaction these Cl atoms destroy O_3, causing a perturbation of the natural O_3-balance. To appreciate the impact of this perturbation we have to understand first the natural O_3-cycle—at least in its salient features.

THE NATURAL OZONE BALANCE

The presence of O_3 in the stratosphere is due to the photodissociation of molecular oxygen (O_2) and the subsequent combination of the oxygen atoms with O_2

$$O_2 + h\nu \rightarrow O + O \qquad \lambda \leq 240\,\text{nm} \tag{9.1}$$

$$O + O_2 + M \rightarrow O_3 + M \tag{9.2}$$

Since ultraviolet radiation with wavelengths below 240 nm cannot penetrate into the lower atmosphere, O_3 production by reactions (9.1) and (9.2) is limited to the stratosphere. Moreover, since the reactions involve only radiation and molecular oxygen, which is the second most abundant molecule in the atmosphere, stratospheric O_3-formation is little affected by man. Integrated over the globe reactions (9.1) and (9.2) produce 5×10^{31} O_3-molecules sec^{-1} or 1.3×10^{11} t of O_3 per year. This is an enormous amount and dwarfs any chemical produced by man including CO_2 emitted from fossil-fuel burning.

The O_3-production is balanced by a number of destruction mechanisms. Most of the O_3-destruction, incidentally, takes place within the stratosphere, since less than 1 per cent of O_3-produced in the stratosphere reaches the troposphere and is destroyed there. As is well known, O_3 is rapidly photodissociated

$$O_3 + h\nu \rightarrow O + O_2 \tag{9.3}$$

However reaction (9.3) does not provide a net sink or loss of O_3, since the O atom thus formed reacts back to O_3 through reaction (9.2). Reactions (9.2) and (9.3) establish a rapid photochemical equilibrium between O and O_3. Because of this, O_3 and O are often lumped together as odd oxygen. The first loss reaction of odd oxygen was identified by Chapman[5]

$$O_3 + O \rightarrow O_2 + O_2 \tag{9.4}$$

Reaction (9.4) has a relatively high activation energy (4.4 kcal) and therefore has a low rate constant at stratospheric temperatures (200 K). Thus, despite high concentrations of O_3 and O, reaction (9.4) proceeds at a low rate in the stratosphere. As a consequence a number of chain reactions which catalyse reaction (9.4) become quite important in the destruction of O_3. The chains are carried by several free radical families. The first such radical family introduced into stratospheric chemistry was $HO_x[OH, HO_2]$ (Bates and Nicolet,[4] Nicolet[19]). They react via

$$OH + O_3 \rightarrow HO_2 + O_2 \tag{9.5}$$

$$HO_2 + O \rightarrow OH + O_2 \tag{9.6}$$

$$\text{net } O_3 + O \rightarrow O_2 + O_2 \tag{9.4}$$

to destroy O_3. The OH-radicals required for reaction (9.5) are produced by following reactions

$$O_3 + h\nu \rightarrow O(^1D) + O_2 \qquad \lambda \leq 310\,\text{nm} \tag{9.3a}$$

$$O(^1D) + H_2O \rightarrow OH + OH \tag{9.7}$$

$$O(^1D) + CH_4 \rightarrow OH + CH_3 \tag{9.8}$$

$$O(^1D) + H_2 \rightarrow OH + H \tag{9.9}$$

A similar catalytic chain is provided by the nitrogen oxides, $NO_x[NO, NO_2]$. Most of the NO_x in the stratosphere is introduced by the reaction

$$N_2O + O(^1D) \rightarrow NO + NO \tag{9.10}$$

The catalytic O_3 destruction by the NO_x takes the form

$$NO + O_3 \rightarrow NO_2 + O_2 \tag{9.11}$$

$$NO_2 + O \rightarrow NO + O_2 \tag{9.12}$$

$$\text{net } O + O_3 \rightarrow O_2 + O_2 \tag{9.4}$$

We note that the production of excited O atoms, $O(^1D)$, from the photolysis of O_3 is rather important for the production of HO_x and NO_x radicals in the stratosphere (reactions (9.7), (9.8), (9.9), (9.10)).

Finally there is the ClO_x-destruction cycle of O_3. Cl atoms are also naturally present in the stratosphere. They are transported upward from the troposphere in the form of methylchloride (CH_3Cl) which reacts also with OH and eventually releases a Cl atom. The Cl atom reacts with O_3 in a catalytic cycle (Molina and Rowland[15])

$$Cl + O_3 \rightarrow ClO + O_2 \tag{9.13}$$

$$ClO + O \rightarrow Cl + O_2 \tag{9.14}$$

$$O + O_3 \rightarrow O_2 + O_2 \tag{9.4}$$

in complete analogy to the NO_x and HO_x cycles.

Of course there are many more reactions involving HO_x, NO_x, ClO_x. These radicals react also among themselves, in ways which reduce the destruction of O_3. Their overall effect on O_3 is summarized in Figure 9.1 which shows the O_3-destruction rates of the various catalytic cycles at different altitudes. They were calculated from a one-dimensional, diurnally averaged model. It is clearly seen that the direct reaction (9.4) of O with O_3, characterized by the dash-dotted curve, provides only a minor fraction of the O_3 destruction at all altitudes. In fact in 1979, the date for which these calculations were made, the catalytic destruction by ClO_x ((9.13) and (9.14)), given by the dashed curve, is dominant at most altitudes. Only at high altitudes it is exceeded by the destruction through the

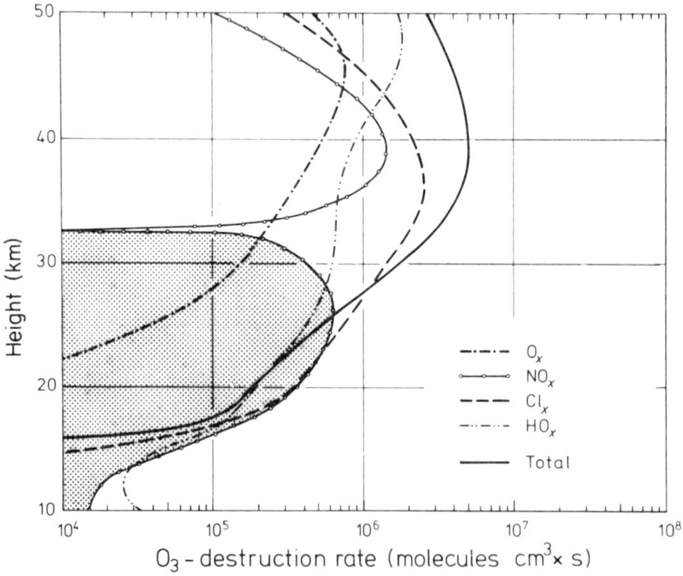

Fig. 9.1 Vertical distribution of the ozone destruction due to the different catalytic chain reactions. Note that NO_x-chemistry leads to a production of ozone at altitudes below 33 km, as indicated by the stippling. The calculations were made by a 1-D, diurnally averaged model with rate constants from NASA[18] and total chlorine of 2.5 ppb. (*Source:* Röth *et al.*[22])

HO_x-cycle. Due to various coupling reactions between the radical families there is also some O_3 produced. For reasons which will become clear below, this production is assigned to the NO_x cycle. Thus in the stratosphere below 33 km altitude the presence of NO_x causes a production of O_3 as indicated by stippling in Figure 9.1. Above that altitude it contributes to the destruction in an important way.

Figure 9.1 shows that most of the stratospheric O_3 is destroyed by the catalytic radical cycles. Because of the catalytic nature of the cycles the radicals themselves are not destroyed—only recycled. Thus one radical can destroy many thousands of O_3 molecules. Therefore, radicals present with mixing ratios of 1 ppb or less can significantly alter the amount of O_3 which is present with mixing ratios of a few ppm in the stratosphere. (The mixing ratio is the number of molecules of a specific gas to the total number of molecules in dry air: ppm = part per million, indicates a mixing ratio of 10^{-6}; ppb = part per billion, indicates a mixing ratio of 10^{-9}; ppt = part per trillion, indicates a mixing ratio of 10^{-12}.) Since man's activities can increase the stratospheric loading of ClO_x or NO_x on the order of a few ppb, there is definitely a possibility of an anthropogenic influence on the stratospheric O_3 layer. Moreover it is noted that the radicals are derived from gases like H_2O, CH_4, N_2O, CH_3Cl or the CFM all of which are emitted at the earth's surface and transported by atmospheric motions into the stratosphere. Thus no direct emissions into the stratosphere are required. Emission at the

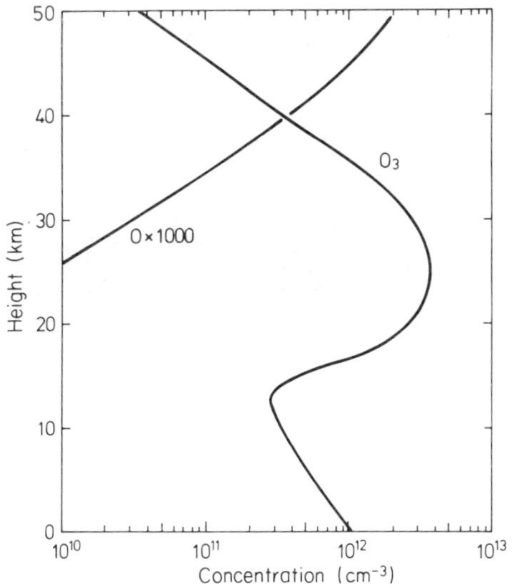

Fig. 9.2 Average vertical profile of O_3- and O-concentrations in the stratosphere. The profiles were calculated using a 1-D, diurnally averaged model. (*Source:* Röth *et al.*[22])

earth's surface of these or other source gases carrying the elements, H, Cl and reactive N is sufficient.

The balance between the O_3 destruction rates shown in Figure 9.1 and the O_3-production rates ((9.1) and (9.2)) yields a stratospheric O_3-profile which is shown in Figure 9.2. It was also calculated by a one-dimensional diurnally averaged model and describes the average O_3 distribution in the stratosphere. The O_3 concentration exhibits a broad maximum in the stratosphere around 25 km altitude, the so-called O_3-layer. Also shown is the average concentration of O atoms. It is much lower than the O_3 concentration, by a factor of 1000 at 40 km altitude, but increases rapidly with altitude. This is due to the fact that the photolysis of O_2 (reaction (9.1)) increases with altitude and that reaction (9.2), the three body recombination of O-atoms with O_2 molecules to O_3, decreases with the second power of the air density. At 60 km altitude the O-atom concentration reaches the same value as the O_3 concentration.

If we divide the O_3 concentration (Figure 9.2) by the total destruction rate of O_3 we obtain the chemical lifetime of O_3 at any given altitude. At 30 km altitude

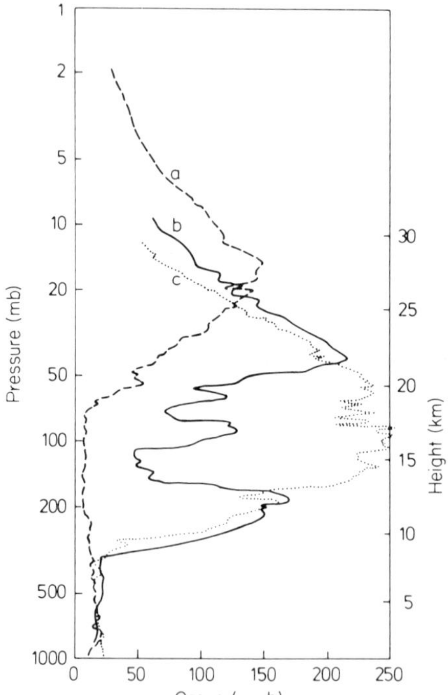

Fig. 9.3 Measured O_3-profile in the stratosphere. Curve (a) represents a tropical profile, curve (b) a profile at mid-latitudes, curve (c) an arctic profile. (*Source:* Dütsch.[6])

this lifetime is about 2 weeks and decreases rapidly with higher latitudes. Therefore the distribution of O_3 is little influenced by horizontal air motions. Below, however, the chemical lifetime of O_3 increases to a few months. In this region the photochemical equilibrium of O_3 is reached so slowly that a redistribution of O_3 by horizontal air motions cannot be compensated. The effect of transport is clearly visible in measured O_3 profiles (Figure 9.3). Because in the stratosphere poleward air motions are always associated with subsidence the arctic O_3 maximum is displaced to much lower altitudes than the tropical one. The profile at mid-latitudes exhibits a typical, layered structure caused by the interweaving of tropical and polar air masses. Only the superposition of many measured profiles averages to a smooth curve like the one predicted by the one-dimensional models. With the sinking motion the O_3 concentration maximum also intensifies. This leads to a latitudinal dependence of the O_3 overburden.

The overburden or O_3 column is the amount of O_3 above a unit area and is

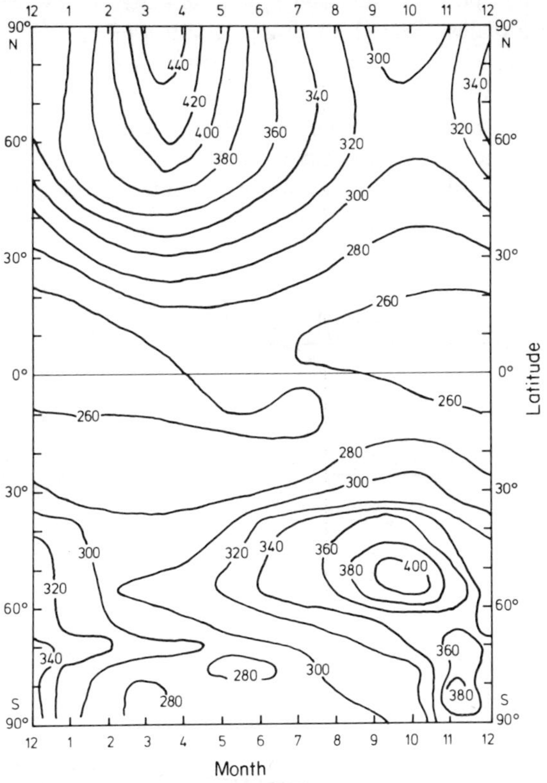

Fig. 9.4 Total ozone overburden as function of latitude and season. The isopleths are labelled in Dobson units (10^{-3} cm atm ozone). (*Source:* Düsch.[6])

obtained by integration over the vertical concentration profile. The overburden is shown as a function of latitude and season in Figure 9.4. A vertical straight line through the isopleths yields the latitudinal variations at a given month. The latitudinal variation exhibits the largest gradients during spring. Similarly the crossing points of a horizontal line with the isopleths allows to read the seasonal variation of O_3 at a given latitude. It is largest at high latitudes with a total variation of about 30 per cent. This seasonal variation is due to the seasonal variation of the horizontal transport by the general circulation of the stratosphere.

THE STRATOSPHERE CHLORINE CHEMISTRY

After this rough outline of the major features of the chemistry and distribution of stratospheric O_3, we can begin to look more closely at the impact on stratospheric O_3 by man-made perturbation like that of the CFM.

The main Cl compounds and their reactions are summarized in Figure 9.5. This flow diagram describes the average conditions at 30 km altitude. The symbols in the boxes give the major Cl compounds; the numbers give their concentration in molecules cm^{-3}. The symbols in the arrows give the reactants by which the Cl-compounds are interconverted. The respective numbers give the reaction rates or fluxes of Cl in molecules $cm^{-3}\ s^{-1}$. The numbers are rounded to one digit for simpler drawing. They were calculated using a one-dimensional,

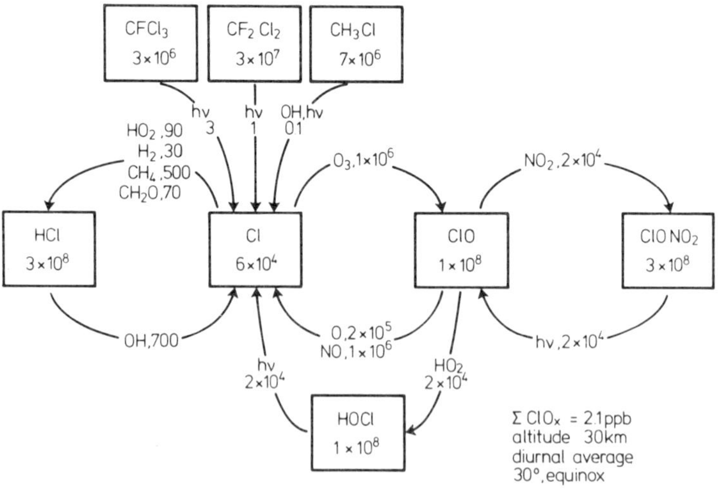

Fig. 9.5 Flow diagram of Cl-chemistry. The numbers in the boxes indicate the concentrations (molecules cm^{-3}), the numbers in the arrows the Cl-conversion rates (Cl-molecules $cm^{-3}\ sec^{-1}$). Concentrations and reaction rates were calculated for 30 km altitude using a 1-D diurnally averaged model.

diurnally averaged, steady state model. Figure 9.5 allows us to trace out the major pathways of Cl reactions.

We notice that the man-made source gases of Cl, $CFCl_3$, CF_2Cl_2, have already decreased substantially in concentration at 30 km altitude. Therefore the injection rates of Cl-atoms from their photolysis is rather small. The primary step of photolysis is the dissociation of one Cl atom as in

$$CFCl_3 + h\nu \rightarrow CFCl_2 + Cl$$

The details of the further reaction of $CFCl_2$ have not been resolved in full detail. But is is likely that they proceed fast and that all three Cl atoms are injected as Cl or ClO close to the altitude of primary photodissociation.

By far the fastest reaction of Cl atoms is that with O_3 which starts the catalytic cycle formed by reactions (9.13) and (9.14). However, reaction (9.14) is not the most likely ClO reaction to occur. The fastest reaction of ClO is that with NO. This has an interesting consequence since reaction (9.15) produces NO_2

$$ClO + NO \rightarrow Cl + NO_2 \qquad (9.15)$$

$$NO_2 + h\nu \rightarrow NO + O \qquad (9.16)$$

$$O + O_2 + M \rightarrow O_3 + M \qquad (9.2)$$

$$Cl + O_3 \rightarrow ClO + O_2 \qquad (9.13)$$

$$\text{net} \qquad \emptyset \rightarrow \emptyset$$

The most likely reaction of NO_2 is photolysis (reaction (9.16)), which yields an O atom, which in turn recombines to O_3 (reaction (9.2)). Up to this step there is a production of O_3. Including reaction (9.13) completes a full Cl-reaction cycle, and destroys the O_3 generated from NO_2. Consequently the net result of the cycle consisting of reactions (9.15), (9.16), (9.12) and (9.13) is zero. It is a so-called 'zero' or 'do nothing cycle'. Nevertheless it is quite important, because five out of six Cl atoms completing the inner cycle between Cl and ClO do not destroy O_3. That is, through this coupling between the NO_x and ClO_x families the potential O_3 destruction by Cl is reduced by a factor of 5.

This cycle also explains why the presence of NO_x can produce O_3 and does so in the lower stratosphere. In Figure 9.1 the NO_x cycle was credited with the O_3 production from reactions (9.15), (9.16) and (9.12), the equivalent O_3 loss through reaction (9.13) was assigned to the Cl_x-cycle. Both of course cancel.

Figure 9.5 shows another way in which ClO_x and NO_x interact. It is through the formation of chlorine nitrate, $ClONO_2$.

$$ClO + NO_2 + M \rightarrow ClONO_2 \qquad (9.17)$$

Although $ClONO_2$ is rapidly photolysed,

$$ClONO_2 + h\nu \begin{cases} \nearrow Cl + NO_3 \\ \searrow ClONO + O \end{cases} \qquad (9.18)$$

its formation (9.17) is sufficiently fast to allow the build up of substantial concentrations, $3 \times 10^8\,cm^{-3}$ of $ClONO_2$. Since $ClONO_2$ does not react with O_3, substantial amounts of ClO_x and at the same time of NO_x are inactivated and the potential of O_3 destruction is reduced further. It should be noted, however, that the products of $ClONO_2$ photolysis are still uncertain. If the upper channel of reaction (9.18) dominates, a minor destruction cycle of O_3 through $ClONO_2$ can be constructed.

Another and even more important inactive reservoir of chlorine is the hydrochloric acid molecule (HCl). It is formed by a number of hydrogen abstraction reactions of the Cl atom. The most important is the reaction with methane, CH_4.

$$Cl + CH_4 \rightarrow HCl + CH_3 \qquad (9.19)$$

As indicated by the small fluxes the forward reactions are not so fast. However, the back reaction of HCl with OH is also a slow process

$$HCl + OH \rightarrow Cl + H_2O \qquad (9.20)$$

Thus large concentrations of HCl are built up. In fact HCl is the most abundant product of Cl chemistry at all stratospheric altitudes. This inactive reservoir further reduces the impact of the Cl perturbation on O_3.

A particularly interesting compound is hypochloric acid (HOCl). It is formed via

$$ClO + HO_2 \rightarrow HOCl + O_2 \qquad (9.21)$$

with a fairly strong rate of $2 \times 10^4\,cm^{-3}\,s^{-1}$.

Since HOCl does not react with O_3 and is present in substantial amounts it also forms a reservoir of inactivated chlorine. But this is not its major effect. The major effect is given by another reaction cycle. Photolysis of HOCl forms Cl and OH (9.22).

Since Cl reacts further by reaction (9.13), and OH by reaction (9.5), reactions (9.21), (9.22), (9.13) and (9.5) form a catalytic cycle which destroys O_3.

$$HO_2 + ClO \rightarrow HOCl + O_2 \qquad (9.21)$$

$$HOCl + h\nu \rightarrow OH + Cl \qquad (9.22)$$

$$Cl + O_3 \rightarrow ClO + O_2 \qquad (9.13)$$

$$OH + O_3 \rightarrow HO_2 + O_2 \qquad (9.5)$$

$$\text{net} \quad O_3 + O_3 \rightarrow O_2 + O_2 + O_2$$

This process turns out to be quite important, because it circumvents the necessity of the pressure of O-atoms, which is required in all other catalytic cycles. Thus it becomes quite important in the lower stratosphere where, as we have seen, the concentration of O-atoms is low.

We note that the reactions leading to HOCl and back from HCl couple the concentration of the ClO_x radicals to that of the HO_x radicals. As with NO_x this coupling is quite strong. As a consequence any changes in the reactions controlling the concentration of OH will not only change OH, but via reaction (9.20) also the concentration of ClO_x, and therefore the impact of the CFM on stratospheric O_3. This fact has caused some of the difficulties in the prediction of the future stratospheric O_3 depletion.

The discussion just presented provides as much insight as one can hope to obtain from a semiquantitative system analysis of the stratospheric Cl chemistry. To go beyond that, model calculations for the full stratospheric chemistry are required. Such models can then also be used to estimate the future decrease of stratospheric O_3 from anthropogenic perturbations. The discussion of the models is well beyond the scope of this paper. We will simply present their results. In doing so we use the predictions of a one-dimensional diurnally averaged model. Such a model treats the atmosphere as horizontally uniform and considers only vertical variations and transport, moreover it averages over the diurnal variations of the solar radiation as well as the concentrations of radicals and other short-lived trace species.

COMPARISON OF CALCULATED AND MEASURED TRACE-GAS PROFILES

Before we present the model prediction of a future O_3 depletion, we would like to test how well the model describes today's trace-gas concentrations in the stratosphere by comparing calculated vertical trace-gas profiles with the measured ones. To remain brief, we concentrate on the chlorine species. Figure 9.6 summarizes the measured vertical profiles of $CFCl_3$, one of the important man-made Cl source gases in the stratosphere. The measured $CFCl_3$ mixing ratios (indicated by the symbols) show a steep decrease with altitude above the tropopause (10–15 km altitude). Measurements have been made by various investigators and at different latitudes (see review by Ehhalt[8]). Like O_3 the measured vertical CFCl profiles vary with latitude, with a steeper gradient at higher latitudes. The measurements are compared to a theoretical profile, which represents the global mean vertical profile and corresponds closest to 30° N latitude. The agreement between theoretical profile and measured ones is quite reasonable, especially when we consider that the tropopause altitude decreases with increasing latitude. Using the actual tropopause altitude of 10 km in the calculation of the theoretical profile at 40–55° N instead of 15 km, would shift the theoretical profile down by 5 km and bring it into quite good overlap with the experimental data.

A similar picture is obtained for CF_2Cl_2 (Figure 9.7). A slower decrease with altitude is shown with CF_2Cl_2, because its photolysis rate is slower than that of $CFCl_3$. This is well exhibited by both the measured and the theoretical profiles.

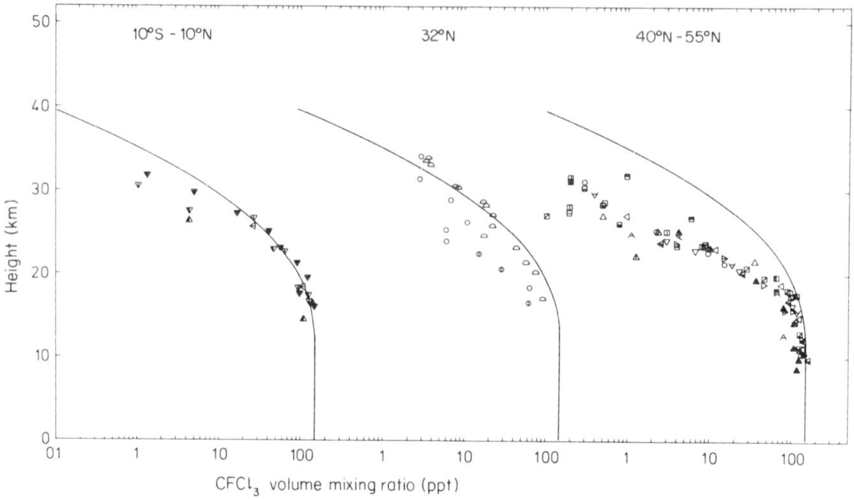

Fig. 9.6 Measured vertical profiles of the $CFCl_3$-mixing ratio in the stratosphere. The symbols represent data by various investigators. The full curves represent a profile calculated for 30° N by a 1-D, diurnally averaged model. (*Source:* Ehhalt.[8])

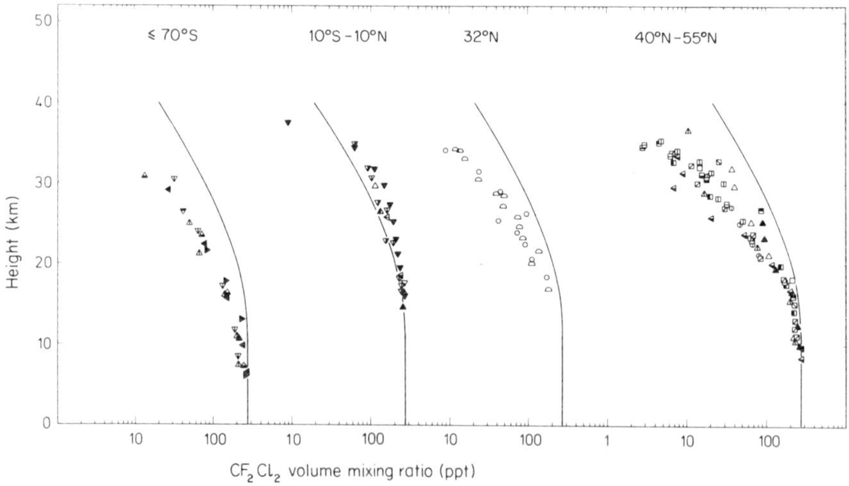

Fig. 9.7 Measured vertical profiles of the CF_2Cl_2-mixing ratio in the stratosphere. The symbols are the same as in Fig. 9.6. The full curve represents a profile calculated for 30° N by a 1-D, diurnally averaged model. (*Source:* Ehhalt.[8])

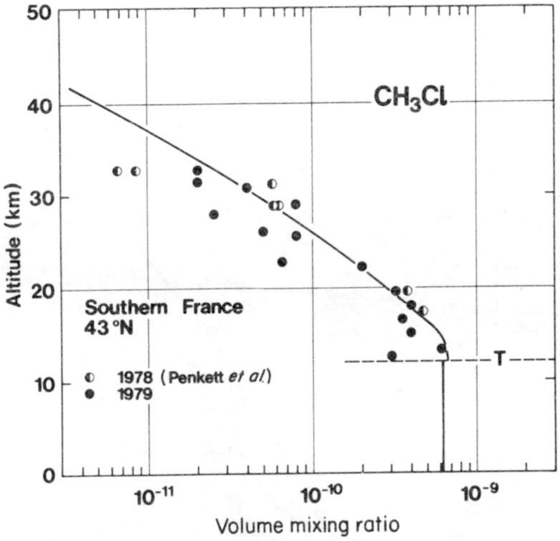

Fig. 9.8 Measured vertical profiles of the CH_3Cl-mixing ratio in the stratosphere. The full curve represents a profile calculated for 30° N by a 1-D, diurnally averaged model. (*Source:* Schmidt *et al.*[24])

A third source gas, CH_3Cl, is presented in Figure 9.8. Natural production of CH_3Cl is responsible for the natural background of Cl in the stratosphere. It is, however, more difficult to measure than $CFCl_3$ and CF_2Cl_2 and only few measurements exist (Penkett *et al.*;[20] Schmidt *et al.*[24]). As expected for a source gas which is destroyed in the stratosphere, its mixing ratio also decreases rapidly with altitude above the tropopause. Within the scatter of the experimental data the agreement between theoretical mean curve and the data is good.

This comparison between measurements and calculations for the source gases tests two aspects of the model. It shows that the model describes the mean vertical transport in the stratosphere and the destruction rates (mainly by photolysis) reasonably well.

To test the details of the Cl chemistry, we have to look at the products of this chemistry: HOCl, HCl and $ClONO_2$. No measurements are yet available for HOCl. The measurements for HCl and $ClONO_2$ are shown in Figure 9.9 and 9.10. The measurements are more difficult than those of the CFM, so there are fewer results with higher experimental errors. The HCl measurements agree reasonably well with the calculations. The few $ClONO_2$ measurements that exist lie at the upper range of the model predictions which is indicated by the two full lines in Figure 9.10.

This agreement can be viewed as quite a success for the modelling, because both HCl and $ClONO_2$ were predicted to be present before they were measured.

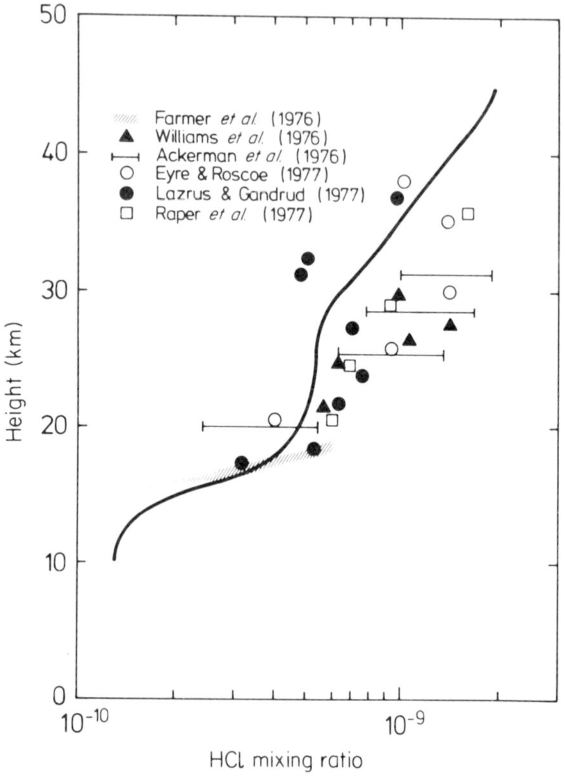

Fig. 9.9 Measured vertical profiles of the HCl-mixing ratio in the stratosphere. The full curve represents a profile calculated for 30° N by a 1-D, diurnally averaged model. (*Source:* NASA.[18])

Moreover, the shape of the profiles as well as the absolute values of the mixing ratio were predicted in reasonable agreement with the data.

Of course the crucial test for the model, as far as its capability to predict the O_3 depletion is concerned, are its correct predictions of radical concentrations for Cl and ClO. Both have been measured in the stratosphere mainly by Anderson *et al.*[2,3] The comparison for Cl atoms is shown in Figure 9.11. The direct measurements of Cl atoms between 40 and 35 km altitude agree well with the model calculations. However an upper limit for Cl concentration derived from ethane measurements (Rudolph *et al.*[23]) shows values up to a factor of 5 lower than the calculated Cl concentrations below 30 km altitude. A similar trend is observed for ClO (Figure 9.12). At upper altitudes the agreement between the data points and the range of theoretical predictions indicated by the two full curves is reasonable, especially when one considers the large scatter in the

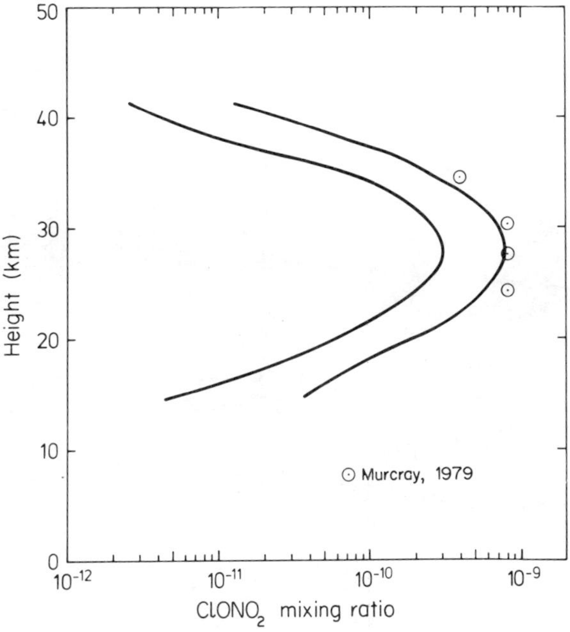

Fig. 9.10 Measurements of the $ClONO_2$-mixing ratio in the stratosphere. The full curves indicate the range predicted by 1-D, diurnally averaged models. (*Source:* NASA.[18])

experimental data. Below 30 km the majority of the measurements fall below the calculated ClO concentrations—again up to a factor of 5. It is noted that the very high ClO values measured on 14 July, 1977, are somewhat doubtful because they exceed the accepted value for the total amount of all chlorine compounds in the stratosphere at least at the upper altitudes. Again the fact that the presence of Cl and ClO was correctly predicted is a great success of the models, even more so as the prediction above 30 km agrees well with the measurements. However, this comparison also indicates some difficulties in the prediction of Cl and ClO in the lower stratosphere.

This is in a way to be expected. At higher altitudes, above the major part of the O_3 layer, the ultra violet of the solar radiation is so intense that only the simpler Cl molecules, like Cl, ClO and HCl, escape rapid photolysis. The more complicated Cl-containing molecules, like HOCl, $ClONO_2$, or possibly others, are rapidly photolysed and hardly present. Below the peak of the O_3 layer, where solar ultra violet is attenuated, these molecules live longer and are able to participate in the Cl chemistry. Thus the Cl chemistry in the lower stratosphere is more complex, maybe even more so that hitherto thought.

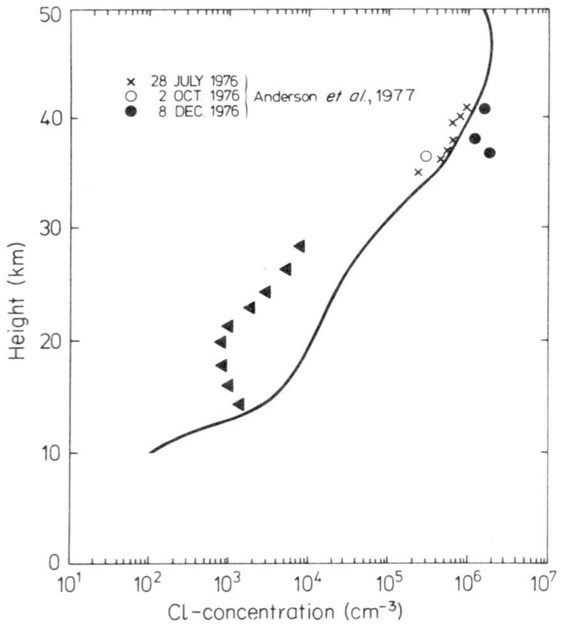

Fig. 9.11 Measurement of the Cl-atom concentration in the stratosphere. The triangles represent an upper limit derived from ethane measurements. The full curve represents a profile calculated for 30 °N by a 1-D, diurnally averaged model. (*Source:* Rudolph *et al.*[23])

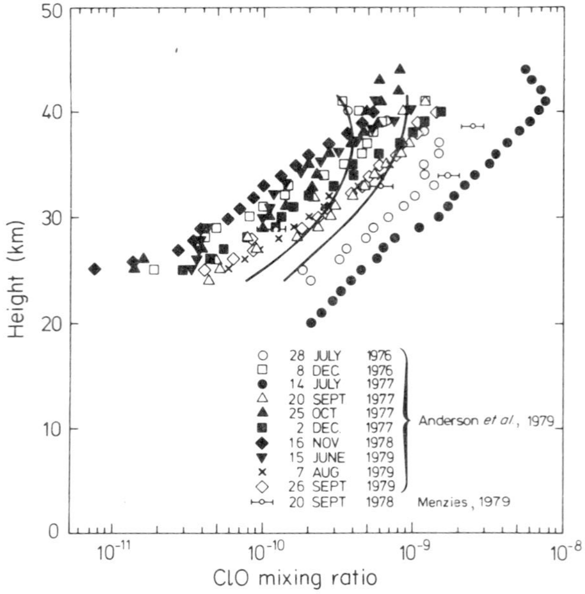

Fig. 9.12 Measured vertical profiles of the ClO-mixing ratio in the stratosphere at 32 °N latitude. The full curves indicate the range predicted by 1-D, diurnally averaged models (from NASA.[18])

On the whole it can be stated that our understanding of the stratospheric Cl chemistry has come a long way and seems satisfactory in the upper and middle stratosphere. There are still some uncertainties in the lower stratosphere, which may have implications for the predicted O_3 reduction.

PREDICTED OZONE REDUCTION

The same models were also used to predict the future O_3 depletion in the stratosphere due to the continued release of man-made CFM. To do that one has to define a release scenario. For the following predictions we assume a constant future release of the CFM at the 1977 rates. Release rates and resulting perturbations are shown in Figure 9.13.

Up to 1978 figures on the $CFCl_3$ and CF_2Cl_2 release rates have been published by the industry. They are accurate to ± 5 per cent. Until 1974 the release rates increased more or less exponentially. The decrease since 1974 reflects the economic recession, the attempts of government regulation of production and release and possibly also a negative attitude among the consumers. Taking note of this—at least temporary—decrease, the future release rates were assumed to remain constant at the 1977 values (NAS[17]). It is of course quite possible that future production and release of CFM will accelerate again.

In any case continued release leads to further accumulation of the CFM in the atmosphere. The increase in the average tropospheric CFM mixing ratio corresponding to the chosen release scenario is also shown in Figure 9.13. The concentrations increase is controlled by the constant release rate and the increasing stratospheric destruction. The destruction rates are essentially determined by the time it takes to transport the CFM to the altitudes where they are photolysed. The resulting atmospheric lifetimes are 50 years for $CFCl_3$ and 80 years for CF_2Cl_2. The latter is somewhat larger because CF_2Cl_2 absorbs at shorter wavelengths and has, therefore, to be transported to higher altitudes for efficient removal. As a consequence the steady-state concentration of CF_2Cl_2 is enhanced and approached more slowly than that of $CFCl_3$. The steady-state mixing ratios are 0.7 ppb for $CFCl_3$ and 1.5 ppb for CF_2Cl_2. The present mixing ratios are 0.14 and 0.26 ppb respectively. Consequently we expect a future increase in CFM concentration by about a factor of 5. This increase is of course uncertain—not only because of the projected release rates, but also because of the errors in the estimated lifetimes which alone amount to about 20 per cent.

Finally, the lowest panel shows the resulting mean stratospheric O_3 depletion as a percentage of the present column density as predicted by a one-dimensional chemical model of stratospheric chemistry. The perturbed steady-state O_3 column density is about 10 per cent lower than the present one which averages about 0.3 cm STP O_3. The new state is approached with a characteristic time which is essentially a weighted mean of the $CFCl_3$ and CF_2Cl_2 lifetimes, i.e. about 65 years. Due to uncertainties in a number of model parameters, mostly in

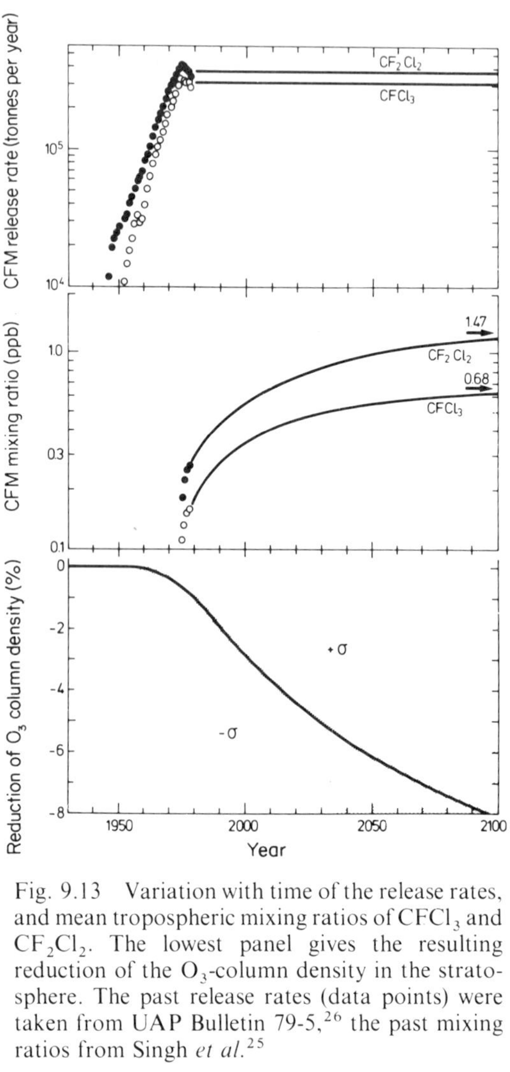

Fig. 9.13 Variation with time of the release rates, and mean tropospheric mixing ratios of $CFCl_3$ and CF_2Cl_2. The lowest panel gives the resulting reduction of the O_3-column density in the stratosphere. The past release rates (data points) were taken from UAP Bulletin 79-5,[26] the past mixing ratios from Singh *et al.*[25]

chemical kinetics, the predicted O_3 depletion is uncertain with a relative error, $\sigma/\Delta O_3 \sim 0.4$, as indicated by the hatched area (cf. NAS 1979,[17] Ehhalt *et al.*[7]). The presently predicted O_3 reduction (10 per cent) is nearly a factor of 2 lower than that of last year. The O_3 depletion is not uniform with altitude. Figure 9.14 shows that the major impact is at 35 km altitude (Wine *et al.*[28]). Thus not only the total column density of O_3 changes but also the vertical structure of the O_3 profile. It should be noted that much of the depletion at lower altitudes is due to the inclusion of the catalytic destruction cycle involving HOCl.

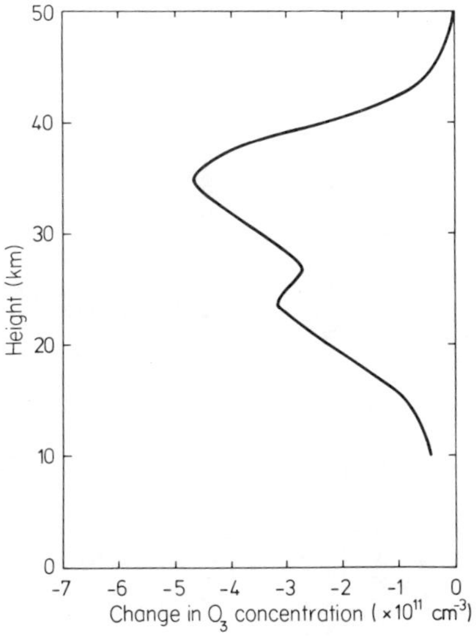

Fig. 9.14 Vertical profile of the reduction in O_3-concentration at steady state from continued release of CF_2Cl_2 and $CFCl_3$ at 1977 rates. *Source:* Wine *et al.*[28])

Finally a few words need to be said about the accuracy of the prediction. In Figure 9.13 a 1-σ error was shown. This error, however, includes only the estimated errors of the input parameters of the model. In addition there could be unknown systematic errors, for example due to an incomplete representation of the stratospheric chemistry in the model. It is obviously impossible to estimate such an error. But we can look at the past history of O_3 predictions and check how often corrections had to be made to accommodate new insights in stratospheric chemistry. The major events in this record are shown in Figure 9.15.

The first predictions in 1975 scattered around an O_3 reduction of 15 per cent. This value changed by about a factor of 2, to 7 per cent, when in 1976 $ClONO_2$ was introduced into the model chemistry. Incidentally the connecting lines are somewhat misleading, as the changes in the prediction did not vary gradually but rather in form of sudden jumps. Interestingly enough, $ClONO_2$ had been considered before 1976, but the laboratory data then available suggested a photolysis rate much too fast, which rendered the previously calculated $ClONO_2$-concentrations insignificant. In 1976 it was found that its absorption spectrum was lower especially at the higher wavelengths which increased its calculated photolytic lifetime and made it important in the stratosphere.

Also most of the other changes were introduced by improved laboratory data

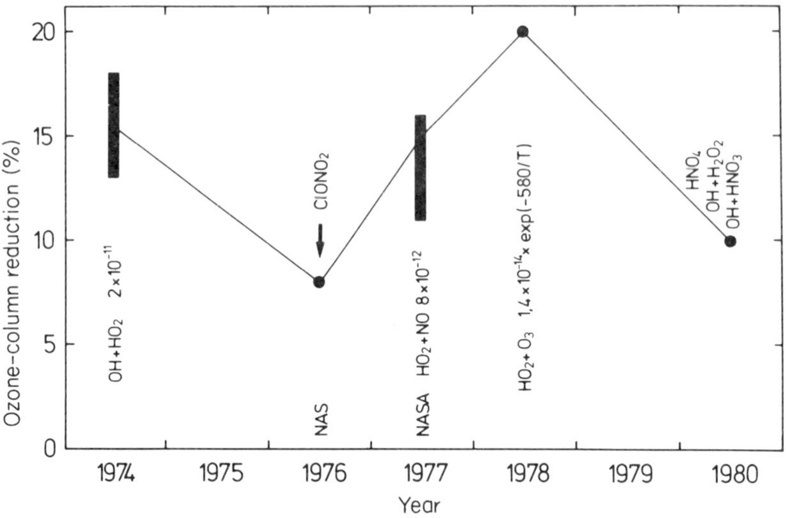

Fig. 9.15 Historical record of the prediction of O_3-column reduction from continued release of CFM. Only the major changes are shown

on reaction rate constants or absorption spectra. Thus when Howard and Evenson[11] remeasured the rate of

$$HO_2 + NO \rightarrow OH + NO_2 \tag{9.23}$$

and found a 40 times higher-rate constant than previously measured. This faster-rate constant increased the calculated OH-concentrations in the stratosphere by a factor of about 2. Since the OH concentration regulates the Cl concentration mainly because of its reaction with HCl (9.20) the calculated Cl atom concentration was also raised and the impact of the CFM on O_3 became larger.

The same happened for the reaction

$$HO_2 + O_3 \rightarrow OH + 2O_2 \tag{9.24}$$

Its new rate constant became a factor of 6 faster at stratospheric temperature than before, with similar effects as for reaction (9.23). During 1979 HOCl was introduced in the stratospheric chemistry. This caused less of a change in the column density, but a large effect in the vertical structure of the O_3 reduction.

The reactions indicated in 1980 all have the inverse effect, they destroy OH radicals

$$OH + H_2O_2 \rightarrow H_2O + HO_2 \tag{9.25}$$

$$OH + NHO_3 \rightarrow H_2O + NO_3 \tag{9.26}$$

Thus when the reaction rate constants were found to be larger than previously measured (Keyser;[12] Wine *et al.*[28]), they tended to lower the calculated OH concentration with a concurrent decrease in Cl and the predicted O_3 depletion. As a result the present prediction of the O_3 depletion caused by CFM is lowered to 10 per cent, as indicated in Figure 9.15.

Figure 9.15 indicates that there have been significant additions to the list of substances and reactions important to the stratospheric O_3 depletion at regular intervals over the past 10 years. Extrapolating from Figure 9.15 we suspect that our understanding of stratospheric chemistry is still not complete and we must expect further systematic changes. If we take the measured low Cl and ClO data in the lower stratosphere (Figures 9.11 and 9.12) as guide lines we would expect that the predicted O_3 depletion is more likely to decrease further than to increase in the near future.

Another conclusion implicit in the record of Figure 9.15 is that in some cases the errors of the laboratory data had been greatly underestimated. It is quite possible that some of the errors in the set of rate constants presently used in modelling are still underestimated. Consequently the relative error of 0.4 estimated for the O_3 depletion in Figure 9.13 rather presents a lower limit.

In summary: if the present CFM-release is continued, one-dimensional models predict stratospheric O_3 to decrease by about 10 per cent. This new steady state is reached with a relaxation time of about 60 years. The relative error in this prediction is rather large, larger than 40 per cent.

REFERENCES

1. Ackerman, M., Frimont, D., Girard, A., Gottignier, M., and Muller, C. (1976). Stratospheric HCl from infrared spectra, *Geophys. Res. Lett.*, **3**, 81–83.
2. Anderson, J.G., Margitan, J.J., and Stedman, D.H. (1977). Atomic chlorine and the chlorine monoxide in the stratosphere: Three *in situ* observations, *Science*, **198**, 501–503.
3. Anderson, J.G., Grassl, H.J., Shetter, R.E., and Margitan, J.J. (1980). NO, HO_2, ClO, and $O(^3P)$ in the earth's stratosphere: Do the rates of odd oxygen production and destruction balance, *J. Atmos. Science*.
4. Bates, D.R., and Nicolet, M. (1950). The photochemistry of atmospheric water vapor, *J. Geophys. Res.*, **55**, 301–327.
5. Chapman, S. (1930). On ozone and atomic oxygen in the upper atmosphere, *Phil. Mag.*, **10**, 369–383.
6. Dütsch, H.U. (1974). The ozone distribution in the atmosphere, *Can. J. Chem.*, **52**, 1491–1504.
7. Ehhalt, D.H., Chang, J.S., and Butler, D.M. (1979). The probability distribution of the predicted CFM-induced ozone depletion, *J. Geophys. Res.*, **84**, 7889–7894.
8. Ehhalt, D.H. (1980). The observation of long lived trace gases in the stratosphere, *Proceedings of the International Ozone Symposium, Boulder, Colorado, USA, 4–9 August*. Vol. II, 728–737.
9. Eyre, J.R., and Roscoe, H.K. (1977). Radiometric measurement of stratospheric HCl, *Nature*, **266**, 243–244.
10. Farmer, C.B., Raper, O.F., and Norton, A.H. (1976). Spectroscopic detection and vertical distribution of HCl in the troposphere and stratosphere, *Geophys. Res. Lett.*, **3**, 13–16.

[11] Howard, C.J., and Evanson, K.M. (1977). Kinetics of the reaction of HO_2 with NO, *Geophys. Res. Lett.*, **4**, 437–440.
[12] Keyser, L.F. (1980). Absolute rate constant of the reaction $OH + H_2O_2 \rightarrow HO_2 + H_2O$ from 245 to 243 K, *J. Phys. Chem.*
[13] Lazrus, A.L., Gandrud, B.W., Greenberg, J., Bonelli, J., Mroz, E., and Sedlacek, W.A. (1977). Midlatitude seasonal measurements of stratospheric acidic chlorine vapor, *Geophys. Res. Lett.*, **4**, 587–589.
[14] Menzies, R.T. (1979). Remote measurement of ClO in the stratosphere, *Geophys. Res. Lett.*, **6**, 151–154.
[15] Molina, M.J., and Rowland, F.S. (1974). Stratospheric sink for chlorofluoromethanes: Chlorine atom catalyzed destruction of ozone, *Nature*, **249**, 810–812.
[16] Murcray, D.G., Goldman, A., Murcray, F.H., Murcray, F. J., and Williams, W.J. (1979). Stratospheric distribution of $ClONO_2$, *Geophys. Res. Lett.*, **6**, 857–859.
[17] NAS (1979). *'Stratospheric Ozone Depletion by Halocarbons': Chemistry and Transport*, National Academy of Sciences, Washington, D.C.
[18] NASA (1979). *The Stratosphere: Present and Future* (Eds. R.D. Hudson and E.I. Reed), Publication 1049, NASA, Washington, D.C., December.
[19] Nicolet, M. (1970). Ozone and hydrogen reactions, *Ann. Geophys.*, **26**, 531–546.
[20] Penkett, S.A., Derwent, R.G., Fabian, P., Borchers, R., and Schmidt, U. (1980). Methyl chloride in the stratosphere, *Nature*, **283**, 58–60.
[21] Raper, O.F., Farmer, C.B., Toth, R.A., and Robbins, B.D. (1977). The vertical distribution of HCl in the stratosphere, *Geophys. Res. Lett.*, **4**, 531–534.
[22] Röth, E.P., Tönnißen, A., and Janssen-Schmidt, Th. (1980). *Report from Study Group III (Modelling) on the Ozone Depletion Problem* (Ed. G. Brasseur), Commission of the European Communities.
[23] Rudolph, J., Ehhalt, D.H., and Tönnißen, A. (1980). Vertical profiles of ethane and propane in the stratosphere, *Proceedings of the International Ozone Symposium, Boulder, Colorado, USA, 4–9 August*. Vol. II, 809–815.
[24] Schmidt, U., Rudolph, J., Johnen, F.J., Ehhalt, D.H., Volz, A., Röth, E.P., Borchers, R., and Fabian, P. (1980). The vertical distribution of CH_3Cl, $CFCl_3$, and CF_2Cl_2 in the midlatitude stratosphere, *Proceedings of the International Ozone Symposium, Boulder, Colorado, USA, 4–9 August*. Vol. II, 816–823.
[25] Singh, H.B., Salas, L.J., Shigeishi, H., Smith, A.J., Scribner, E., and Cavanagh, L.A. (1979). Atmospheric distributions, sources and sinks of selected halocarbons, hydrocarbons, SF_6, and N_2O, *Stanford Res. Inst. Report Proj.* 4487.
[26] UAP (1979). Upper Atmospheric Programs Bulletin, UAP 79-5, Federal Aviation Agency, Washington, D.C.
[27] Williams, W.J., Kostus, J.J., Goldman, A., and Murcray, D.G. (1976). Measurements of the stratospheric mixing ratio of the using an infrared absorption technique, *Geophys. Res. Lett.*, **3**, 383–385.
[28] Wine, P.H., Ravishankara, A.R., Kreutter, N.M., Shah, R.C., Nicovich, J.M., Thompson, R.L., and Wuebbles, D.J. (1980). Rate of reaction of OH with HNO_3, preprint.
[29] WMO (1981) 'The Stratosphere 1981, Theory and Measurements' WMO Global Ozon Research and Monitoring Project, Report No. 11, World Meteorological Organization, Geneva, Switzerland, May.

Pollutants and Their Ecotoxicological Significance
Edited by H. W. Nürnberg

CHAPTER 10

The Carcinogenic Components of Automobile Exhaust

G. GRIMMER, A. GLASER, A. HILDEBRANDT AND H. BÖHNKE

SUMMARY

The carcinogenic effect of automobile exhaust condensate and fractions thereof by dropping onto mouse skin is described. The result of fractionating and discriminating these fractions by the biological assay shows that

1 Almost the total carcinogenity of exhaust condensate results from polycyclic aromatic hydrocarbons (= PAH) with 4–7 rings. This fraction is about 4 % by weight of the total condensate.
2 The amount of benzo(a)pyrene present in the condensate accounts for only 10 % of its carcinogenicity.
3 A mixture of 15 PAH occurring in the exhaust condensate, explain about 50 % of the carcinogenic activity of the exhaust if applied epicutaneously onto the mouse skin.

The analytical investigation of the conditions which control the emission of PAH shows that

1 A total amount of 2 t benzo(*a*)pyrene annually emitted in Germany can be estimated from the average fuel consumption of 100 passenger cars (410 g per Europa test) and the benzo(*a*)pyrene amount emitted under these conditions (41 μg BaP per Europa test).
2 An excess of fuel in the fuel–air ratio beyond the stoichiometric ratio results in a considerable increase of the PAH emitted (10- to 40-fold more than under normal conditions).
3 A 'cold' cycle of the 12-federal-7-mode test emits about 1500 μg BaP per gallon, whereas a 'warm' cycle produces only about 50 μg BaP per gallon.

4 The variation of the content of aromates in the fuel (0 per cent, 26 per cent, 32 per cent, 42 per cent, 48 per cent) significantly influences the PAH amount emitted during the Europa-test, e.g. in case of BaP from 4.7 up to 28.2 μg BaP per test.
5 Europa-test using PAH-containing or analogous PAH-free fuel did not result in any differences of the PAH profile emitted. Thus, obviously all PAH present in the fuel are completely combusted, or retained in the motor oil, respectively.
6 Mileage and resulting PAH accumulation in the lubricating oil do not significantly influence PAH emission. Dropping of a solution of diesel exhaust condensate onto the skin of mice provokes similar carcinoma incidences as it has been demonstrated with gasoline engine exhaust.

INTRODUCTION

A direct inhalation of automobile exhaust gas by experimental animals would be the best simulation model to test its carcinogenic burden in man. Since a carbon monoxide-content of 2–5 per cent in the exhaust of spark-ignited combustion engines, however, is too toxic, animals can only survive if the exhaust is considerably diluted. This, on the other hand, results in exhaust concentrations which do not provoke carcinogenic effects in the respirative tract of animals. Thus, only the particle phase of exhaust can be tested for carcinogenic activity. The favoured animal experiment to test carcinogenic effects is the epicutaneous dropping of substances onto the mouse skin. The carcinogenic effect of automobile exhaust condensates in this model has been described repeatedly,[1–4] and the assay therefore seems to be fairly convenient. In total, 53 mg condensate administered onto the mouse skin for a year provoke carcinomas in 1 per cent of the animals, 158 mg provoke carcinomas in 30 per cent and 478 mg in 72 per cent of the animals tested. An average of about 0.5 g exhaust condensate results from 1.3 l fuel under city driving condition. Generally, this condensate is produced in a so called 'Europa-test' (following the EEC rule 15) which simulates city driving on a chassis dynamometer. It allows the production of larger amounts of condensate representing city driving, and which is demanded for the animal test.

How can we isolate carcinogenic compounds from this complex mixture? The concept for this is as follows: The condensate is separated chemically by liquid–liquid distribution or chromatography techniques into a carcinogenic and non-carcinogenic part. The efficiency can be provoked by the animal test, in this case by epicutaneous application of both fractions onto the mouse skin in several dosages, respectively. For this we should have known which carcinogens occur in the condensate or which class of compounds provokes the biological effect. Since, however, we do not know this in advance, we suppose a certain class to be active, and then have to prove whether our hypothesis is correct or not by bioassay.

FRACTIONATION AND CARCINOGENICITY OF FRACTIONS

We suppose the polycyclic aromatic hydrocarbons (PAH) to be the predominant or exclusive carcinogens of the automobile exhaust condensate. Thus, the exhaust has been separated into PAH-containing and PAH-free fractions. If the PAH-free fraction would show biological activity also, we have to conclude that other classes of compounds contribute to the total carcinogenic effect of automobile exhaust, e.g. nitrogen-containing compounds (carbazoles or acridines).

The condensate has been separated in two steps into PAH-containing and PAH-free fractions and a third step separates the PAH according to their molecular size. The fraction containing PAH with four to seven rings was then separated by means of chromatography. The PAH of this fraction suspected of being carcinogenic were combined again by mixing fifteen pure PAH according to their percentages in which they originally occurred in the condensate, and this mixture was tested in the mouse-skin assay. The biological effect was compared with that resulting from the original condensate. The condensate was separated according to the following criteria: (a) hydrophobic and hydrophilic substances; (b) aromatic and aliphatic compounds; (c) PAH with two to three rings and PAH with four to seven rings. Finally the biological effect of the automobile exhaust condensate was compared with that of, (a) pure benzo(*a*)pyrene and (b) of the mixture of 15 pure PAH.

Hydrophobes and Hydrophiles

The first step is a partition between methanol:water:cyclohexane (9:1:10) in which the hydrophilic phase is extracted repeatedly with cyclohexane. It results in an almost complete removal of the PAH from the methanol water phase (only about 1–2 per cent of the PAH remain in this phase). The biological effect of these fractions in the epicutaneous test (with 80 CFLP mice per dosage group) is given in the Table 10.1. The table shows that the condensate and hydrophobic

Table 10.1 Carcinogenic effect of automobile exhaust condensate (epicutaneous dropping test). Separation into hydrophobic and hydrophilic fractions

Weight (%)	Fraction	PAH-content	Tumour incidence (%)	
			Dosage I	Dosage II
100	Automobile exhaust	+	15	60
67	Methanol–water phase	–	1	3
33	Cyclohexane phase	+	14	61

Dosage I represents 1.05 mg condensate twice weekly, dosage II 3.15 mg. The tumour incidence of the condensate results from a probit transformation.

Table 10.2 Carcinogenic effect of automobile condensate. Separation of the hydrophobic fraction into aromatic and aliphatic compounds

Weight (%)	Fraction	PAH-content	Tumour incidence (%)	
			Dosage I	Dosage II
100	Exhaust condensate	+	15	60
15.3	Cyclohexane phase	–	1	0
17.7	Nitromethane phase	+	15	65

Dosage I (II) = 1.05 mg (3.15 mg) of the condensate, twice weekly.

parts thereof result in comparable tumour incidences in the animals. Only a small part of the carcinogenic effect of the condensate results from non-PAH.[4–6]

Aromatic–aliphatic Compounds

In the next step the residue of the hydrophobic phase was redissolved in cyclohexane and extracted repeatedly with nitromethane (five times) in order to remove the PAH from the cyclohexane phase. The result is given in Table 10.2. The dosages of both fractions correspond to the concentration in the condensate. The Table 10.2 shows that the original condensate and the aromatic fraction thereof provoke comparable tumour incidences, whereas the aliphatic parts do not have any effect.[4–6]

PAH with Two and Three-rings—PAH with Four to Seven Rings

A third step separates the aromatic compounds according to their molecular weight using Sephadex LH 20 chromatography with *i*-propanol as eluant. It results in a fraction with aromates possessing two or three rings and a fraction with four to seven ring aromates. The Table 10.3 shows the results of the animal

Table 10.3 Carcinogenic effect of automobile exhaust condensate. Separation into PAH with different ring numbers (molecular size)

Weight (%)	Fraction	Tumour incidence	
		Dosage I	Dosage II
100	Automobile exhaust condensate	9	61
13	PAH with 2–3 rings	0	0
4	PAH with 4–7 rings	6	61

Dosage I (II) = 1.05 mg (3.15 mg) condensate, twice weekly.

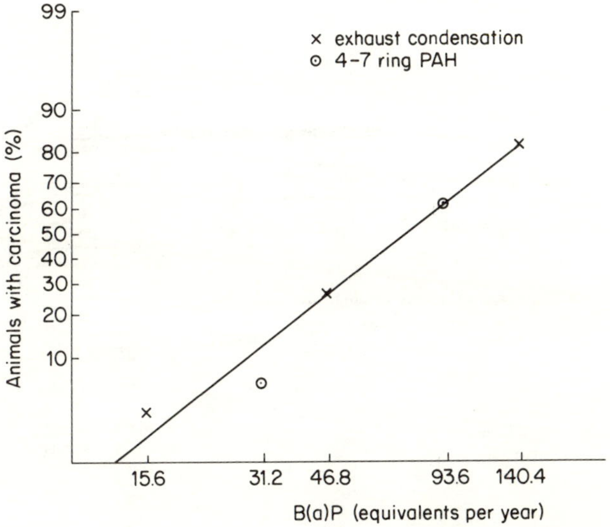

Fig. 10.1 Dose-response relation of automobile exhaust and PAH-fraction thereof

experiment after 83 weeks treatment. Dosages are proportional to the condensate.

Table 10.3 demonstrates that the carcinogenic effect of the total condensate results from PAH with four to seven rings. Figure 10.1 shows this once again graphically. The original S-formed dosage-response curve is probit transformed into a straight line. The dosage on the abscissa (given in $\mu g\, y^{-1}$) is standardized to the benzo(*a*)pyrene (= B(*a*)P) amount actually present in the condensate or in the fraction, not to the condensate itself or the weight of the fraction. Thus, the four to seven rings containing PAH are equivalently dosed; this means 100 mg exhaust condensate are compared with 4 mg of the PAH fraction which both contains about 50 μg B(*a*)P, respectively. The ordinate shows the tumour incidence. The cross-marked points are B(*a*)P amounts of the condensate, the open circles those of the PAH fraction. Since this fraction counts for only 4 per cent of the weight of the condensate, one-twenty-fifth of the amount of the condensate is necessary to get the same tumour incidence. The five points are approximately on a straight line; this means the carcinogenic activity is focused in this small fraction. On the other hand, the ten-fold amount of B(*a*)P must be applied to provoke the same tumour incidence using pure B(*a*)P. We, therefore, have to assume that 90 per cent of the carcinogenic activity of the condensate results from other PAH, since only 10 per cent of the effect can be correlated with B(*a*)P content.

How does this fraction look? Figure 10.2 shows a section of the gas chromatogram of this fraction for which a 20 m glass column with OV 101 and about 50 000 HETP was used. Some of these signals have been identified by

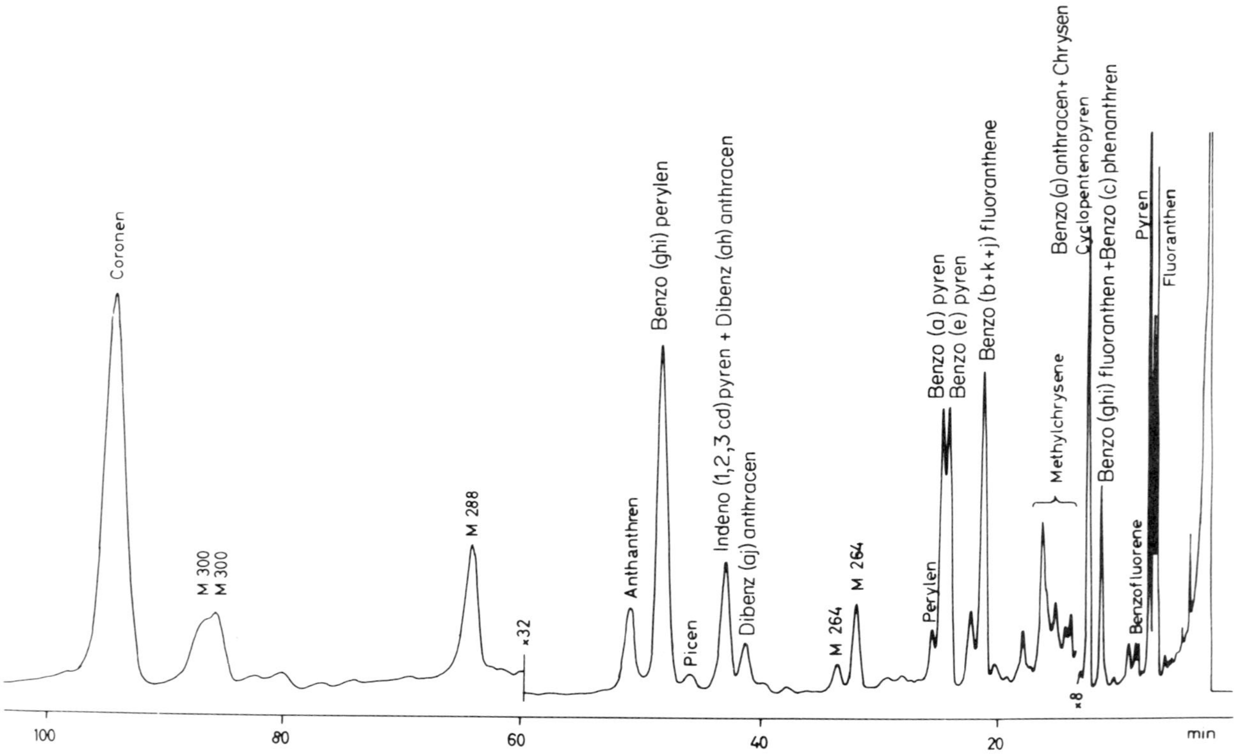

Fig. 10.2 Gas-chromatographic separation of the PAH-fraction of automobile exhaust

comparison with the retention times and mass spectra of those from authentic substances, eleven of which are well-known carcinogens: benzo(*c*)phenanthrene, benz(*a*)anthracene, cyclopenta(*cd*)pyrene, chrysene, benzo(*b*)fluoranthene, benzo(*j*)fluoranthene, benzo(*a*)pyrene, dibenz(*a*, *j*)anthracene, ideno(1,2,3-*cd*)pyrene, dibenz(*a*, *h*)anthracene and anthanthrene. Four unknown compounds occur in considerable concentrations also, namely: methylene-benzo(*e*)pyrene, methylene-benzo(*a*)pyrene, and two PAH with a molecular weight of 300 (PAH-M 300 A and B).[7]

We have simulated the carcinogenic activity of exhaust condensate by a mixture of these eleven PAH plus the four non-identified PAH. If these substances are mixed in the ratio in which they occur in the exhaust, about 50 per cent of the effect produced by the original condensate results. Though not fully satisfactory, it is more than the 10 per cent resulting from benzo(*a*)pyrene. To summarize

1 The total carcinogenic effect of exhaust condensate results from the four to seven ring PAH. This fraction is only about 4 per cent of the weight of the total condensate.
2 The amount of benzo(*a*)pyrene in the condensate may explain only 10 per cent of its carcinogenicity.
3 A mixture of 15 PAH occurring in the exhaust condensate, explain about 50 per cent of the carcinogenic activity of the exhaust, if administered epicutaneously onto the mouse skin. It is supposed that there are additional carcinogenic PAH in the exhaust condensate besides the fifteen individuals tested.

Other Animal Test Systems as Detectors

Fractions are investigated in other animal test systems according to their biological activity, e.g. by subcutaneous injection into mice.[8] In this model, however, the carcinogenic activity of exhaust condensate from gasoline engines is very small. Only after removal of the hydrophilic constituents from the condensate does the remaining fraction show a strong carcinogenic effect. Thus PAH must be considered to be the predominant carcinogens of automobile exhaust.

WHAT AMOUNT OF PAH IS EMITTED BY VEHICLES?

Vehicles with Spark-ignited Combustion Engines

The annual fuel consumption of the FRG is about 20 million t (normal + super).

Kind and amount of PAH emitted

About 150 different PAH have been detected in the emission from two commercially available vehicles during the Europa-test (simulating city

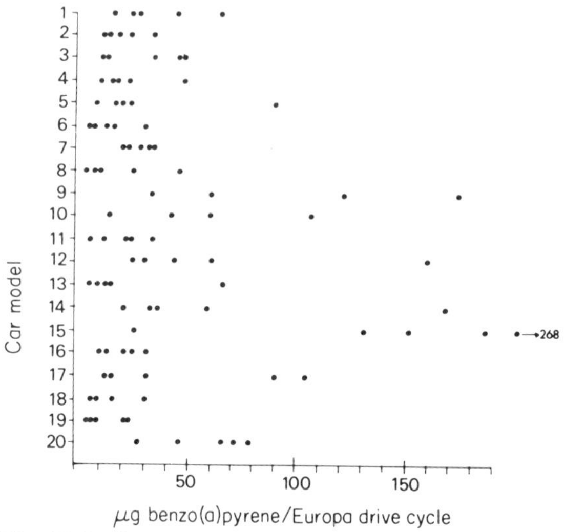

Fig. 10.3 Benzo(*a*)pyrene emitted during the Europa Drive Cycle; comparison of twenty different vehicle models, five cars per model

driving), 73 of which could be identified by comparison with authentic samples (GC retention time and mass spectra). The profile of the four and more rings containing PAH is shown in Figure 10.2.[7]

Range of emission

Five of each of the twenty most abundant vehicles have been investigated in cooperation with the TÜV in Essen[9] and the profiles of fourteen selected PAH were determined. Figure 10.3 demonstrates the range of benzo(*a*)pyrene emitted during an Europa test.

For this study the following major components were selected: phenanthrene, anthracene, 3-methylphenanthrene, 2-methylphenanthrene, fluoranthene, pyrene, chrysene + benz(*a*)anthracene + M266, benzofluoranthene, benzo(*e*)-pyrene, benzo(*a*)pyrene, indenol(1,2,3-*cd*)pyrene, benzo(*ghi*)perylene anthanthrene, and coronene.

The reproducibility of technical condition and the homogeneity of the driving cycles for a single car were determined by several repeated Europa tests. For this purpose one car was driven for five Europa tests by the same driver. The margin of error for any single PAH was 6.3–10.9 per cent (variation coefficient).

Benzo(a)pyrene amounts annually emitted by vehicle traffic

From the average fuel consumption of 100 passenger cars (410 g per Europa test) and the benzo(*a*)pyrene amount emitted under these conditions (41 μg benzo(*a*)pyrene per Europa test), a total amount of 2 t benzo(*a*)pyrene emitted per year can be estimated for the FRG. This corresponds fairly well with the 10 t annually emitted by vehicles in the USA.[10]

Influences on the PAH amount of exhaust by motor condition, fuel and lubricating oil

To summarize, it can be stated that only the fuel–air ratio and the temperature of the engine influence strongly the amount of PAH emitted. There are contradictory results as far as the influence of various factors on the amount of benzo(*a*)pyrene and other PAH are concerned. It can be seen from a survey of publications until 1971 (Particulate Polycyclic Organic Matter, 1972) and another one dealing with investigations until 1974[9] that sporadic experiments can not contribute to the answer of this question, and that only serial experiments result in relevant information. According to our experiences a minimum of five Europa tests, under identical conditions, is necessary to answer specific questions.

Fuel–Air Ratio

An excess of fuel in the fuel–air ratio beyond the stoichiometric ratio ($\lambda = 1.00$) yields a considerable increase of PAH emitted. A mixture rich in fuel can not be avoided when the cold engine is started and yields a large PAH emission; e.g. a badly adjusted commercial vehicle (8 vol. per cent CO during idling) emits up to 0.4 mg benzo(*a*)pyrene during the first Europa cycle (in 195 s) and consumes about 170 g of fuel with a cold motor. This is about forty-fold more than during a normal Europa test with warm engine.[5]

Under these conditions the concentration of all PAH increases also, so that the profile remains constant. This has been found with American vehicles also, where normal emissions of 15 μg benzo(*a*)pyrene per gallon (0.9–1.4 vol per cent CO) increases up to 160 μg benzo(*a*)pyrene per gallon (2.85 vol per cent CO).[11]

The authors reported a significant influence of the temperature of the engine. For instance, a 'cold' cycle of the 12-federal-7-mode test emits about 1.500 μg benzo(*a*)pyrene per gallon, whereas a 'warm' cycle produces only about 50 μg benzo(*a*)pyrene per gallon. From analytical difficulties other PAH were not included in this investigation. It is, however, likely that the results are comparable with those on German vehicles.

Fuel, content of lower aromates

The variation of the content of aromates in the fuel (0 per cent, 26 per cent, 32 per cent, 42 per cent, 48 per cent) significantly influences the PAH amount emitted (positive regression coefficient). For this a commercial vehicle (1.6 l, driven for approximately 15 800 km) has been driven four times in the Europa test with each of the fuels and the averages of the emissions regarding the eighteen most abundant PAH (fluoranthene-coronene) were compared. Increasing content of lower aromates in the fuel correspond to linear increase of the PAH emission; in case of benzo(*a*)pyrene 4.7 up to 28.2 μg benzo(*a*)pyrene per Europa test (average consumption about 420 g).[42]

Fuel, PAH-content

Several times repeated tests using PAH-containing and analogous PAH-free fuel did not result in any differences of the PAH profile emitted. Thus, obviously all PAH present in the fuel are completely burned, or retained in the motor oil, respectively. All PAH in the exhaust have been formed during the combustion.[13] Meyer and Grimmer[13] did not find any difference in the amounts of ten of the most abundant PAH before and after the removal of PAH with boiling points higher than 300 °C by distillation.

Mileage of the engine

In a huge investigation in which 240 Europa tests have been evaluated (entitled, Einflüße der Betriebsart und Betriebszeit des Motors sowie des Motorschmieröles auf die Emission von PAH aus Kraftfahrzeugen mit OTTO-motor. BMI-DGMK-Project 110), two brand-new, commercially available vehicles were driven 20 000 km in a simulated city driving programme and the exhausts analysed. The emissions of both cars during the Europa test (six tests beginning with cold engine, subsequently twelve tests with warm engine) were measured at the beginning, after 5000 km, 10 000 km, and 20 000 km, respectively. After the aforementioned mileages the eighteen single tests were performed with fresh and used lubricating oil and the PAH emission analysed.

Mileage, deposit formation

Only a small increase of the PAH emission from a brand-new engine was detected after 10 000 km of city driving (e.g. about 10 per cent). After a subsequent 1000 km of high-speed driving the PAH emission decreased to about the same rate.

PAH Concentration in Lubricating Oil

There occurs a considerable mileage-dependent accumulation of PAH in the lubricating oil. Under the aforementioned conditions a PAH-profile develops which is very similar to that of the fuel (amounts of about 20–200 mg B(*a*)P kg^{-1} oil).

PAH-emission with fresh or used lubricating oil

Mileage and resulting PAH accumulation in the lubricating oil do not significantly influence PAH emission; details of this investigation are given in the Scientific Project 110 of the DGMK.

Vehicles with a Diesel-engine

The annual consumption of diesel fuel by vehicles presently is about 8 million t.

Kind and amount of PAH emitted

No systematic investigation on the diesel-exhaust condensate has been published so far; according to our own investigations (Grimmer, unpublished information), the PAH profiles do not differ significantly from those of spark-ignited combustion engines. The concentration of PAH emitted during an Europa test, however, is about 50 per cent if compared with spark-ignited combustion engines of the same efficiency. This investigation was performed with diesel-engine vehicles; larger engines have not yet been tested. Thus, it is not possible to estimate the benzo(*a*)pyrene amount resulting annually from truck traffic. Investigations on American diesel-driven trucks showed similar PAH emissions as those of comparable gasoline engines (particulate Polycyclic Organic Matter, 1972). From this, the benzo(*a*)pyrene amount annually produced by diesel vehicles, can be expected to be significantly lower than that produced by OTTO-engines, according to the minor fuel consumption.

Carcinogenic effect of diesel-exhaust condensate

Dropping of a solution of diesel-exhaust condensate onto the mouse skin provokes carcinoma incidences, similar to that demonstrated by gasoline-engine exhaust condensate. If related to the benzo(*a*)pyrene amount, the tumour incidences of both condensates are equally high.

HOW LARGE IS THE CONTRIBUTION OF VEHICLE TRAFFIC TO THE ATMOSPHERIC POLLUTION WITH CARCINOGENIC PAH?

Table 10.4 shows the various sources of benzo(*a*)pyrene emission or the annually produced benzo(*a*)pyrene amount by various sources in USA, respectively. According to these figures the predominant sources are coal heating, open fires

Table 10.4 Estimated benzo(*a*)pyrene emission in the USA

	$t\,y^{-1}$	
Automotive exhaust		
Gasoline powered		
Automobiles	10	
Trucks	12	
Diesel-fuel powered		
Trucks and buses	0.4	
		22 = 1.7 %
Heating		
Coal		
Hand-stoked residential furnaces	420	
Intermediate units (chain-grate and spreader-stokers)	10	
Coal-fired steam power plants	1	
Oil		
Low-pressure air atomizer and others	2	
Gas	2	
Wood	40	
		475 = 38 %
Refuse burning		
Commercial, industrial, institutional and apartment refuse burning	33	
Open burning		
Forest and agricultural	140	
Vehicle disposal	50	
Coal refuse fires	340	
		563 = 45 %
Industrial plants		
Petroleum cracking	6	
Asphalt air blowing	< 1	
Coke production	192	
		200 = 16 %
		1260 $t\,y^{-1}$

and coke plants. The contribution to the pollution by vehicular traffic, however, remains comparatively low with about 1.7 per cent.

REFERENCES

[1] Kotin, P., Falk, H.L., and Thomas, M. (1954). *A.M.A. Arch. Indust. Hyg.*, **9**, 164.
[2] Wynder, E.L., and Hoffmann, D. (1962). *Cancer*, **15**, 103.
[3] Brune, H. (1977). *IARC Sci. Publ.*, **16** (Lyon), 41.
[4] Brune, H., Habs, M., and Schmähl, D. (1978). *J. Environ. Path. Toxicol.*, **1**, 737.
[5] Grimmer, G., and Böhnke, H. (1978). *J. Environ. Path. Toxicol.*, **1**, 661.
[6] Misfeld, J., and Timm, J. (1978). *J. Environ. Path. Toxicol.*, **1**, 747.
[7] Grimmer, G., Böhnke, H., and Glaser, A. (1977). *Zbl. Bakt. Hyg. I. Abt. Orig. B*, **164**, 218.
[8] Pott, F., Tomingas, R., and Misfeld, J. (1977). *IARC Sci. Publ.*, **16** (Lyon), 79.
[9] Grimmer, G., and Hildebrandt, A. (1975). *Zbl. Bakt. Hyg., I. Abt. Orig. B*, **161**, 104.
[10] Hangebrauck, R.P., von Lehmden, D.J., and Meeker, J.E. (1967). Public Health Service Publication 999-AP-33. Cincinnati: US Dep. Health Education and Welfare, p. 48.
[11] Jacobs, E.S., Brandt, P.J., Hofman, C.S., jr., Patterson, G.H., and Willis, R.L. (1971). Presentation to the Am. Chem. Soc., 31 March 1971, Los Angeles. Printed from Du Pont Petroleum Chemicals, E. I. Du Pont de Nemours.
[12] Grimmer, G., and Voigtsberger, P. (1980). *Erdöl und Kohle—Erdgas-Petrochem.*, **33**, 226.
[13] Meyer, J.P., and Grimmer, G. (1974). DGMK Forschungsbericht 4547 II (Hamburg 1974).
[14] Grimmer, G., Brune, H., Deutsch-Wenzel, R., Naujack, K.-W., Misfeld, J., and Timm, J. (1983). *Cancer Letters*, **21**, 105.
[15] Grimmer, G., Brune, H., Deutsch-Wenzel, R., Dettbarn, G., and Misfeld, J. (1984). *J. Nat. Canc. Inst.*, **72**, 733.

Note Added in Proof

More recent investigations[14,15] have confirmed these results showing that

(1) PAH containing more than three rings account for 84–91 % (epicutaneous dropping test/mouse skin) or 81 % (implantation into the rat lung) of the total carcinogenic effect of vehicular exhaust condensate;

(2) The benzo(*a*)pyrene content accounts for 6–7.6 per cent (epicutaneous dropping test/mouse skin) or 2.4 per cent (implantation into the rat lung) of the total effect;

(3) A mixture of 15 pure PAH may explain only 41 per cent of the total carcinogenicity of vehicular exhaust condensate, if this mixture is dosed in concentrations as occurring in the exhaust.

Pollutants and Their Ecotoxicological Significance
Edited by H. W. Nürnberg

CHAPTER 11

Influences of Human Activities on a Regional Climate

ANDRÉ JUNOD AND PIERRE LEHMANN

SUMMARY

Human activities interfere with natural energy fluxes. On the one hand, man-made energy transformations end up generally as heat dissipates in the environment. Large heat sources such as towns and power stations can influence the regional climate when many of them are concentrated in the same region. Modifications of the local/regional air circulations as well as triggering of convective phenomena like storms are some effects of concern. On the other hand, man's activities and settlements modify the ground surface and produce emissions of gases and particulates in the atmosphere. This in turn alters both the natural radiative budget in the lower atmospheric layers and the water budget through changes in runoff and evaporation conditions.

Project CLIMOD is an attempt to study man's impact on the regional climate in a section of the Rhine Valley, east of Basle. While man's activities in this region have already reached a high level, plans exist to increase anthropogenic energy releases still further through the operation of three nuclear power plants with cooling towers along the Rhine. A detailed description of the present and past climate and special field measurements during meteorological episodes likely to be perturbed by major energy releases provide the data base which is needed. Besides physical and mathematical modelling techniques are used to describe and predict quantitatively the influences of the activities under consideration on the sensitive meteorological processes. Finally, it will be necessary to interpret such effects in terms of climatic variables.

INTRODUCTION

Human activities interfere with natural energy fluxes and thus have the potential to modify what is generally termed 'climate'. This term could be appropriately

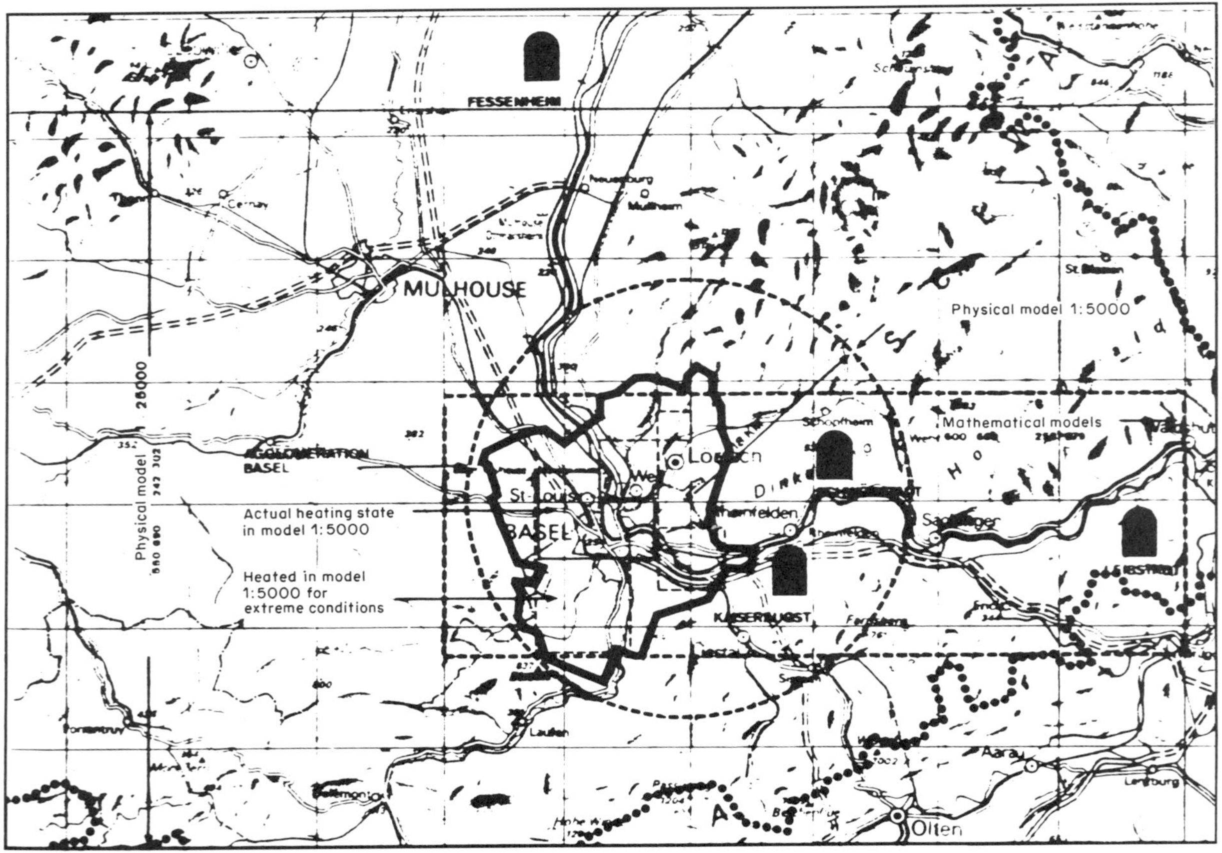

Fig. 11.1 Study region of the CLIMOD project

defined for our purpose as representing the statistical distributions in space and time of those parameters which describe the state of the atmosphere, ocean and land masses. Such parameters are, for instance, temperature, moisture, aerosol concentration, rainfall, etc. Any activity that causes significant changes in the natural energy fluxes is bound to modify the distributions of these climatic parameters.

As we are concerned with distributions, particularly spatial distributions, we immediately realize that we will be in a different situation if we consider the problem on a global scale for which averages for instance can be defined in a meaningful way, than if we look at it on a regional scale. Here the averaging process is not straightforward as the regions outside also have an influence on the phenomena inside the region of interest. Furthermore proper boundaries and boundary conditions must be defined.

With these difficulties in mind, project CLIMOD (for *CLI*mate *MOD*ification) was launched in 1975. The study region is the section of Rhine Valley that runs east–west between Waldshut (Aar–Rhine junction) and Basle about 50 km downstream. It is limited in the north by the Black Forest mountain range and in the South by the Jura (Figure 11.1). The major impetus for the study came from the realization that the three planned nuclear power stations in this section of the Rhine Valley (Leibstadt, Schwörstadt, Kaiseraugst) would increase by almost an order of magnitude, the already not negligible anthropogenic heat and moisture releases which, in turn, could possibly lead to unwanted modifications of local meterological processes.

In what follows we shall first discuss in a general way the influence of man on natural-energy fluxes in the atmosphere and at the earth's surface. Then we will consider the possible effects the human influences can have on regional meteorological processes. Next we shall describe the meteorological approach followed in the CLIMOD project to assess and predict such influences.

THE NATURAL ENERGY FLUXES

The motor of climate is sunshine. The sun's energy arrives at the upper level of the atmosphere in the form of radiation, the spectrum of which is closely approximated by the emission of a black body at 6000 K. The power received on a surface perpendicular to the sun's rays is about 1.35 kW m^2.

As the radiation goes through the atmosphere it will be partially absorbed and partially reflected back to space. These interactions are not identical for all wave lengths so that the solar spectrum reaching the earth's surface is not identical with the one at the top of the atmosphere. In particular a good deal of ultraviolet radiation is held up in the stratosphere causing it to be warmed up.

The radiation reaching the ground will be partially reflected and the rest is absorbed by the earth's surface. The earth and the atmosphere re-radiate the energy they receive from the sun at wave lengths which depend on their

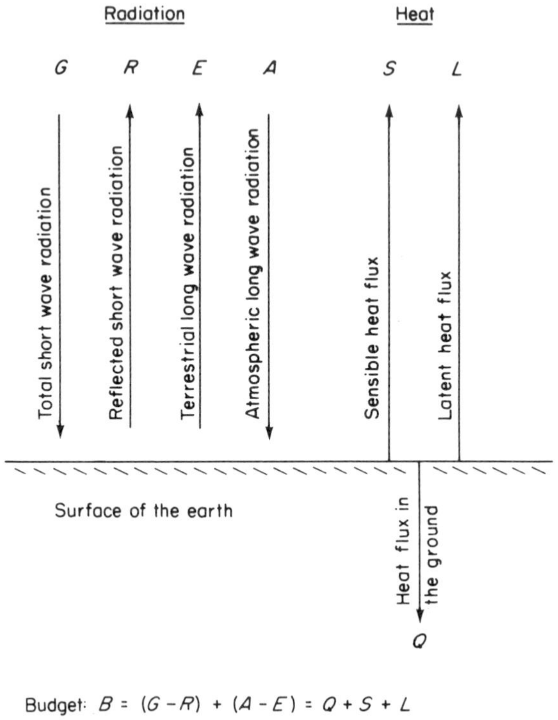

Fig. 11.2 Radiation and heat flux budget at earth surface. *Note:* The overall average of Q is zero since the temperature of the ground averaged over 1 year remains essentially constant

temperature. Clearly the earth as a whole must re-radiate back to space all the energy it receives as otherwise its mean temperature would change in time.

The radiation budget at the earth's surface is shown in Figure 11.2. This budget is composed of four terms. Two terms concern the short-wave (solar) radiation and the other two refer to the infra-red radiation of the earth and the atmosphere. Human interferences can affect all four components of the budget as will be discussed in the next section.

If the radiation budget at the earth's surface is positive the ground will warm up (daytime). If the budget is negative it will cool (night). The temperature of the earth's surface in turn controls the fluxes of heat to the atmosphere and into the ground. The heat flux to the atmosphere has two components, sensible and latent. Human activities can influence the heat fluxes either directly by injection of heat in the atmosphere or in the ground or water bodies, or indirectly by modifying the earth's surface.

INTERFERENCES WITH THE RADIATION BUDGET

The global short-wave radiation can be influenced by those human activities which modify the composition and the aerosol load of the atmosphere. The burning of fuels of all kinds produces gases and particulates. One well-advertised problem is the increase of the carbon dioxide (CO_2) content of the atmosphere as a consequence of the burning of fossil fuels. It is worthwhile to note that the burning of renewable fuels such as wood or biogas does not alter the CO_2 content of the atmosphere, as long as the equilibrium of biomass is maintained, and thus does not contribute to the increased greenhouse effect that CO_2 is causing.

The aerosol load is of course influenced by a variety of sources and human inputs are not necessarily the most important on a global scale. Regionally, however, the aerosols of human origin can be quite important. They change the turbidity of the atmosphere mainly because they help increasing the haze. This can result in a decrease at global radiation. The aerosols can also serve as condensation nuclei thus increasing cloud cover and possibly precipitation.

It is important to realize that the gases and aerosols that man introduces in the atmosphere not only affect the radiative budget at the surface but also change the thermal structure of the atmosphere because they modify the pattern of radiative absorption and reflection.

The amount of radiation that is reflected at the earth's surface depends on the nature of this surface and also on solar elevation. The reflection also varies somewhat with wave length. The main point for our purpose is that any modification man causes to the surface will change the amount and composition of the backscattered radiation. Clearly the effect is not easily predictable in detail and it will in general be necessary to measure the albedo (ratio of reflected to incident radiation) of the various types of surface (forest, grass, towns, etc.). The highest albedo values are of course those of ice and snow for which it can reach 0.8 or even more (for the solar radiation as a whole; in the infra-red region snow has a low albedo). A smooth sea has a very low albedo <0.1 and other surfaces range between these extremes. Towns have typically a smoothed value of about 0.15.

The direct human influence on the terrestrial long-wave radiation is probably not very large. Since the surface radiates more or less like a black body and the emissivities of most substances are in the range 0.9 to 0.99 (a notable exception is quartz with an emissivity of 0.71) the infra-red radiation from the surface will primarily depend on its temperature. Man's influence will, in general, be towards an increase of this component of the radiative budget to increased surface temperatures, particularly in built-up areas.

The atmospheric infra-red radiation is depending on the composition of the atmosphere and its temperature. Here human influences are more of an indirect nature via CO_1, H_2O and an impurity content.

To sum up, the human interference with the radiation budget is a complicated process since all the components of the budget can be affected so that there can be compensations: for instance a warm-up due to increased atmospheric infra-red radiation may be compensated by a cooling due to increased albedo.

INTERFERENCES WITH THE HEAT FLUXES

Human activities result in energy transformations which practically all end up as heat-dissipates in the environment. These heat fluxes add directly to the natural ones, sensible and latent. For instance the operation of a cooling tower causes both sensible and latent heat (in a wet cooling tower the latent part is by far the largest: about 70 per cent) to be introduced in the atmosphere. Burning of fuel will cause sensible heat to be released but it will also result in CO_2 and H_2O the latter being in the form of vapour and thus corresponding to a latent-heat flux. Heating of water bodies will also increase the latent-heat flux.

Heat releases of all kinds will be transferred to the atmosphere by turbulent mixing. Atmospheric turbulence is a complicated process which is influenced, among other things, by the heat fluxes themselves.

When the ground is warm and heats the atmosphere from underneath the turbulence will be high in the lower layers and the turbulent transfer is large. On the contrary, when the ground cools off, the temperature structure of the lower layers becomes stable and turbulence is reduced.

These effects are important mainly in the planetary boundary layer and more so in the lower part of it, the so-called surface layer. But clearly any heat transferred to the atmosphere must go through these layers and consequently the heat fluxes will be largely controlled by the turbulence. Turbulence also depends on the surface conditions, in particular its roughness. So land utilization is also a factor influencing the heat fluxes.

The latent-heat flux depends on the presence of moisture and on the surface available to exchange the water vapour with the atmosphere. Land use will influence both terms. A concrete or asphalt surface increase runoff and quickly dries out while a forest canopy will retain precipitation water thus making it available for evaporation later on. Any human interference with the hydrological cycle will modify the latent-heat flux.

So it is seen that human activities can influence the heat fluxes both directly by releases to the atmosphere and indirectly by changing the factors that affect the natural heat fluxes. As with the radiative budget, the net result of the human interference is not easy to quantify.

INTERACTIONS AND FEEDBACKS

The separation of energy exchanges in radiative and heat components should not hide the fact that these components interact. Clearly the radiation from the sun is the primary input. But as heat fluxes are generated they modify temperatures which in turn change the infra-red radiative fluxes. On the other hand, evaporation results in clouds which in turn reflect short-wave radiation to space while preventing the terrestrial radiation to escape. Pollutant releases resulting from fuel burning will cause smog and haze which in turn reduces the incident solar radiation. Horizontal temperature gradients produced by differential heating between slopes and flat land or between surfaces with different heat capacities and albedos will cause air motions which in turn may be responsible for changes in the radiation field. Human interferences with one particular flux can thus have consequences on all the others and the net result may be difficult to assess. As a result major modelling effort has been undertaken worldwide. To quote a publication of the WMO (GARP 16): 'There seems to be no clear alternatives to the modelling approach for understanding climate sensitivity. This is a venture in which the chances of success are still uncertain and perhaps modest.' This statement which more specifically refers to the global climate is certainly also valid for modelling climate on a regional scale.

THE REGIONAL PROBLEM

The weather over a region is quite obviously the result of the worldwide meteorological situation. Conversely any region contributes its share of inputs to the weather evolution worldwide. When we consider the human effects on a regional climate we are of course not concerned with these global aspects but rather with the perturbations that the human activities may cause to the climate of the considered region or its immediate surroundings. For example, we may ask if the release of heat and pollution over a town may cause clouds to precipitate downwind of the town but we shall consider the advection of clouds to be independent of regional influences.

Disregarding this particular case of induced precipitation for the time being, one can say that the regional influence on the regional climate will obviously be most important when the direct large-scale influences are reduced, i.e. when advection is small. These situations may be subdivided in three main classes: (a) high-pressure situations with subsidence and inversions; (b) situations with weak horizontal pressure gradients and weak winds; (c) situations with weak winds and overcast skies.

A common feature of these three situations is the absence of strong synoptic winds, i.e. winds due to the large-scale pressure distribution. This does not mean, however, that there can be no strong wind on a smaller scale. This brings us right in the centre of the difficult problem of choosing the scales of the phenomena we want to deal with and the relation of this choice with the geographical extent of the region to be considered. Obviously for phenomena which have scales larger than the region of interest there will be a problem of boundary conditions at the limits of the region. On the other hand phenomena with scales smaller than the grid to be used for computation or, in other words, smaller than the smallest variation we want to consider in a deterministic way, will have to be 'parameterized'. This means that their effect is simply expressed by a parameter entering the equations describing the phenomena at larger scale. Such small-scale parameterized phenomena are described as turbulence.

So we see that what we must attempt to do is to look at reality through a kind of binocular which selects a band in the spectrum of scales which influences the phenomena we want to analyse. But we know that the scales we have cut out may have an important influence on what we are looking at. We thus take them into account by parametrization on the small-scale end of the spectrum and by boundary conditions at the large-scale end. It is important to realize that this procedure cannot be fully justified *a priori* since it makes the implicit assumption that there is no interaction between scales.

The characteristic time scales of organized atmospheric phenomena vary between about 1 month and a few minutes. These time scales are related to the period of earth rotation, the Brunt–Vaisala period, and characteristic depth or length of the concerned air masses. The larger time scales depend on latitude. Clearly the horizontal scale will vary in accordance with the time scale. If we assume that a region has linear dimensions of, say, 100 km, and if we consider a grid spacing in the order of, say, 1 km, than the typical meteorological phenomena which qualify as regional, include deep convection, urban heat islands, thunderstorms, low-level jets, mountain and lake disturbances, gravity waves. Things such as thermals and small vortices, plumes, etc. will have to be included in the turbulence and parameterized, while fronts, cyclones and anticyclones and their displacements in the general circulation will influence boundary conditions but cannot be treated explicitly.

In cases when clouds are advected over a region by the general atmospheric circulation, precipitation releases can possibly be helped by human releases of aerosols and by lifting of air due to an urban heat island or other artificial heat inputs. This is more a problem of cloud and precipitation physics of a general nature, not necessarily associated with any particular region.

PERTURBATIONS OF REGIONAL METEOROLOGICAL PROCESSES

With the assumption that there is little or no driving due to the large-scale weather evolution, the regional air motions in the planetary boundary layer will depend on the energy inputs the air receives in the region itself. A typical case is the valley wind which is due to the heating or cooling of the air on mountain slopes. Such winds will develop in particular with situations of type (a) and (b) when daytime warming and night-time cooling are effective and there is no strong advection to overwhelm the regional winds. If the air masses are close to unstable, as might be the case in situations of type (b), regional heating of the air by proximity to the ground can trigger deep convection. With situations of type (c), where the natural energy input is strongly reduced due to overcast skies, the local circulations will in general be weak, of small horizontal extension and uncorrelated.

What kind of influences can human activities have on such processes? As we have seen, the primary effect of man is with the radiative and heat fluxes. Since local winds depend on differential heating of air masses any human influence that modifies the spatial distribution of the energy budget will have consequences for these winds. A well-known case is the urban heat island. The heating of the air provided by the town will cause air to rise over it. As a consequence, air is drawn from the town surroundings and a circulation sets in. If the town is in a valley, the heat island may be superimposed on the valley wind and a more complicated circulation pattern will result. As a limit case the town may block the valley wind or cut it in separate cells. The effects we have here attributed to a town may of course be caused by other human disturbances of the natural environment. Power parks and power stations with cooling towers may have similar effects. Replacement of one type of vegetation by another, the ploughing of fields and all activities which change the radiative and heat-flux characteristics of a large enough surface can have an effect on the regional circulation pattern.

When air masses are conditionally unstable, that is if the vertical temperature gradient is in between the dry and the wet adiabatic lapse rates, deep convection may be triggered if air warmed up by the surface reaches the condensation level. Such a triggering may be helped by large human heat and moisture releases like those produced by cooling towers (note that the heat release alone is sufficient; dry cooling towers are probably as effective for triggering as wet ones). Deep convection does not necessarily lead to thunderstorms but of course the probability for having a thunderstorm over a region will increase if convection occurrences increase.

In situations of type (c) with calm air and overcast skies, the human heat releases may become energetically comparable to the natural radiation and heat fluxes. They may thus become the main driving force for local circulations. It should be emphasized, however, that such circulations cannot be of very large extent nor very strong. The reason is that, considered on a regional scale, the man-produced heat fluxes are small compared to the natural radiation fluxes that cause regional circulations.

REGIONAL METEOROLOGICAL PROCESSES AND REGIONAL CLIMATE

So far we have considered in general terms how man's activities may interfere with regional meteorological processes. In the next sections we will discuss the ways and means to assess, in a quantitative way, the importance of these perturbations. Assuming that this assessment is possible, the problem remains to interpret such results in a climatological sense. This means that one has to appreciate the modifications that the human influences will cause to the statistical distributions describing the climate from the climatologist's point of view. For instance if it can be shown that precipitations downwind of a town increase as a consequence of its heat and pollution releases, one will have to assess not only at what frequency such additional precipitations occur but also how the pattern of rainfall in the region is modified and how the statistical distribution of rainfall at various places in the region is affected; and here it will not be sufficient to estimate changes in mean values. From a climatological point of view the likelihood of extreme cases is an important consideration. It is not irrelevant that a given amount of precipitation falls gently over 1 month or suddenly in a few minutes. The average may be the same but the effects certainly not. This climatological analysis is not an easy task and, so far, has not been completed in our CLIMOD project.

PRESENT SITUATION

In the preceding sections we have spoken of meteorological processes and climate that may be modified by man's activities. In order to be able to describe a modification one has first to describe in a suitable way the thing that may be modified. This requires a methodology because meteorological events and climate cannot be described in a mechanistic way like, say, the revolution of planets around a star.

Climate will have to be described by looking at long records of climatological parameters such as temperature, sunshine duration, precipitation, humidity, wind, visibility, etc. But what is a long record? In general good and comparable records exist for a few tens of years, in some cases for a few hundred years (in Basle temperature records date back to 1755). One will have to decide, somewhat arbitrarily, that the climate is defined over a number of years for which a good enough data base is available. The 30 year period 1931–1960 is often used as reference even though it is realized that it was by no means typical for say the last 300 years and may not be adequate for our climate today. We are confronted here with a problem that is somewhat similar to the scale problem discussed under the regional problem: time scales larger than a few tens of years cannot be included in our statistics and will be taken care of, if at all, by changes in 'boundary' conditions.

The situation is somewhat easier, at least in principle, when considering a particular meteorological process like a valley wind for instance. Here one will have to set up a measuring network that will enable a physical description of the process with an accuracy deemed sufficient for perturbations to be unequivocally identified. But the statistical and scale problem is not by-passed for all that. Turbulence is present in all measurements and since measurements are limited in time and in resolution the whole spectrum cannot be taken into account. This has the important consequence that meaningful measurements leading to meaningful averages and other statistical moments can only be made if the situation is stationary or close to it.

Human inputs are already present today. Consequently whatever meteorological process we consider may already be influenced to some extent. A proper description of the present situation must, therefore, include an estimate of man-made heat, moisture and pollutant releases as well as an adequate description of the present state of land use.

MODELLING

Since we want to know what effects additional human inputs will have on the climate before these effects are discernable, there is no choice but to deduce them from physical reasoning which, as already stated above, implies modelling. There are two main modelling alternatives, mathematical and physical.

1 Mathematical modelling assumes that the phenomena can be correctly deduced from basic differential equations expressing the conservation of mass, momentum and thermodynamic energy. As is well known these equations are basically non-linear and do not have easy analytical solutions except in very simplified cases. As a consequence it is necessary to solve these equations by finite difference methods. The continuum of reality is replaced by a three-dimensional set of discrete grid points at which the values of the variables describing the state of the atmosphere are calculated. The grid extends over the volume of interest. Boundary conditions must be imposed at the outward-lying points of the volume. If time is an open variable initial conditions must be prescribed at all grid points.
2 Physical modelling assumes that it is possible to correctly reproduce the phenomena on a scaled-down relief of the region. This requires that similarity conditions are satisfied. There are quite a few similarity requirements. They are expressed by the conservation of dimensionless numbers (Reynolds, Froude, etc.). Unfortunately the conservation of some of these numbers conflicts with the conservation of others and it is not possible to have a perfect similarity. The choice of the compromise is predicated on the phenomena one is most interested to simulate properly.

Models require inputs and these have to be established as much as possible on the basis of measurements actually made *in situ*. To test a model it will be necessary to compare its outputs to experimental results. This interplay between models and observations is a very important step if one is to have confidence in modelling results describing the effect of perturbations that cannot be tested against reality.

THE CLIMOD PROJECT

Figure 11.1 shows the region concerned by the CLIMOD project. The major human impact today is the town of Basle. At Leibstadt a nuclear power plant is now being constructed. A project exists to build another one at Kaiseraugst, but public opposition has so far prevented its construction. On the German side a project exists at Schwörstadt. Leibstadt and Kaiseraugst are in the 1000 MWe range, Schwörstadt could be twice as large, perhaps even more. All three nuclear power stations will be equipped with wet, natural draft, cooling towers.

Meteorological data collected for the project can be subdivided in three main classes.

1 Data collected on a routine basis by existing stations in the region. These consist mainly of hourly values of temperature, humidity and wind. Wind is usually measured at 10 m above ground. A notable exception is the meteorological towers at the sites of the planned power stations. Leibstadt has a tower 60 m high equipped with sensors at 10 and 60 m. Kaiseraugst: 110 m high with levels 10, 60, 110. Schwörstadt: 150 m high, levels at 10, 40, 100 and 150. Some stations give additional information such as hourly values of sunshine and precipitation. This data material is used for both climatological studies and to document particular meteorological episodes. Data were collected from about thirty stations, most of which are located in the Swiss part of the region. But the relevant German (in particular Feldberg which is very important, being the highest measuring point available) and French measurements were also taken into account.
2 Continuous fixed-point measurements made specially for CLIMOD at the Observatory of Basle-Binningen. These include the direct determination of the radiative budget and of the latent heat flux (lysimeter).
3 Measurements made specially for CLIMOD on particular days when the weather situation was thought to be of a kind where human influences could be important (see above). Such measurements include microaerological soundings with a tethered balloon system, giving vertical profiles of temperature, humidity and wind, horizontal flights at different altitudes with a motorized sailplane to supply horizontal profiles of temperature, humidity and, to some extent, turbulence. In addition remote sensing of ground temperature was carried out from an aircraft.

In addition to these measurements direct observations of the weather evolution were made with 8-mm cameras located at suitable observation points.

Parallel to meteorological data gathering and observations a comprehensive emission inventory was prepared and conceivable scenarios of future urban development and power generation developed.

A modelling effort was undertaken. Mathematical modelling has so far concentrated on two approaches.

1 A comprehensive three-dimensional approach aimed at a correct simulation of the phenomena in the planetary boundary layer and, particularly, at the ground/air interface. The mathematical model was developed originally by the

Center for Environment and Man (CEM) as an air/sea interaction model and adapted later to ground/air. It contains in particular very elaborated calculations of the radiative fluxes in the short and long wave domain.

2 An elaborate wet-plume calculation program named FOG. This program is able to reproduce the behaviour of plumes from cooling towers taking into account the actual profiles of temperature and wind. However it cannot be used to assess the influence of the cooling-tower plumes on the meteorological processes themselves. This program was used in particular to estimate the increase in humidity and, occasionally, of cloudiness when plumes from several cooling towers are superposed in the same stream.

A major physical modelling effort has been started in the EPFL. A first problem was to choose between a liquid and a gas for the simulation fluid. In spite of anticipated difficulties, air was finally preferred to water because it was thought to be more favourable for proper simulation of turbulent effects.

Two relief models of the region were constructed and set in a large wind tunnel. One model is a reduction 1:25 000 which covers the whole study region including the highest ridges of the Black Forest to the north and of the Jura to the south. The other model 1:5000 scale is circular with an equivalent radius of 18 km and centred close to Basle. It covers only one part of the study region. Its main purpose was to investigate the heat island effect of Basle and its interaction with the cooling tower plumes from Kaiseraugst and Schwörstadt. A second 1:5000 model may be constructed at a later stage to cover the part of the region extending between Schwörstadt and Leibstadt.

What have been results so far? To tell it straight, the problem of man's influence on the regional climate has by no means been solved in a general way and more questions remain as were answered.

Difficulties have been encountered in all phases of the project. For one thing it is not easy, even with a fairly large number of measuring stations, soundings, flights etc. to construct an accurate description of the air flow over the topography. Of course the flow pattern is understood in a general way. But this may not be sufficient for detailed comparisons with computer outputs.

The modelling difficulties are great. The three-dimensional mathematical modelling effort has not, so far, succeeded to give clear answers. The physical modelling has shown interesting phenomena, like the blocking of the valley wind by the heat island of Basle. Its results are, however, difficult to quantify because there are important factors which cannot be taken into account. For instance, adiabatic warming or cooling of air is of necessity absent in such a model. Furthermore the interaction of the radiation field with the field of motion is not reproducible.

From results obtained so far one may conclude that the town of Basle is an important factor for the local wind systems and that cooling towers at Kaiseraugst and Schwörstadt have the potential to generate cloud cover over

Basle if there is a strong inversion to block the ascent of their plumes. Such situations do occur and one was documented in December 1977. The effect of man's activities on regional precipitation should not be large but this is to be considered as a tentative answer.

The CLIMOD effort will continue until 1981. Aside from the improvement of the techniques used so far to investigate the CLIMOD problems, attention will also be paid to more direct approaches to atmospheric energetics.

REFERENCE

[1] Eidgenössische Kommission Meteorologie des schweizerischen Gebietes Hochrhein/Oberrhein (1978). *Möglichkeiten regionaler Klimaveränderungen durch menschliche Einwirkungen, 2, Zwischenbericht über das Projekt CLIMOD*, September.

SECTION II:
AQUATIC ENVIRONMENT

Pollutants and Their Ecotoxicological Significance
Edited by H. W. Nürnberg

CHAPTER 12

Basic Ecological Parameters, Monitoring and Biological Monitors in the Aquatic Environment

RAMÒN MARGALEF

SUMMARY

Any research project must be an attempt to provide a sensible answer to a relevant question, and the information it provides must be organized using a scientific framework. The variables selected for monitoring should be operationally defined, suitable for automatic measurement, and should have a certain built-in conservatism that allows comparisons to be extended over time and space. Organisms are very sensitive indicators and are convenient for putting together historical sequences of events, but it is inadvisable to rely on them very much because of frequent mistaken identification, incomplete lists and inadequate knowledge of their requirements. The variables used for monitoring rarely coincide with those preferred by ecologists, which are based on more elaborate levels of conceptualization. It seems more constructive to compare networks of interrelated variables at different levels than to look for one-to-one correspondences. Precise definition of the values of variables related to public or legal action (tolerated maxima, thresholds) is approached with reluctance on the part of ecologists since single variables are never a fair expression of the state of an ecosystem. Practical problems involve situations that usually fall into one of two classes: (a) the flow and eventual accumulation of specific pollutants (radioactive isotopes, heavy metals, organic substances) along the normal pathways of an ecosystem that has not changed very much, and (b) generalized phenomena, such as eutrophication, in which an ecosystem is force-fed, a state accompanied by an acceleration in its internal cycles and the development of external loops that remove, at least temporarily, some of the materials.

INTRODUCTION

The aim of this introduction is twofold. First, to comment in very broad terms on some general principles of ecology that might be relevant in connection with monitoring and, secondly, to identify some of the areas where contacts and exchanges between ecologists and those in authority (or with powers of decision in environmental matters) usually generate misunderstanding and conflict. The particular issues discussed are selected according to my own experience, which is not very extensive. Nevertheless, I have been particularly impressed by a frequent reluctance on the part of ecologists to expose the degree of ignorance that exists generally about ecology and their reluctance, too, to produce precise ecological statements—for instance about allowable limits for the values of critical variables. This position leads to behaviour that appears uncooperative and, as a result, ecologists have less influence on decisions and not more.

Quantitative data are valuable only in a well-defined and appropriate setting. Bare numbers mean different things to different people. The gathering of information whose only purpose is an enlargement of the store of collected data is perhaps worse than the exercise of science for science's sake. This last phrase has pejorative overtones only when used by environmental managers and not, of course, by ecologists, who see no harm in science for science's sake. (This is not intended to taunt or provoke environmental managers, although I would welcome any reaction that might be turned to constructive ends.)

From the point of view of ecology, an inflow of information can be confusing if the intellectual foundation lacks structure: that is, if there is no scientific basis or frame of reference into which the incoming data are assimilated and integrated so that many of them become, one hopes, redundant as well as being illustrations of some general principle. If a large organization is involved in collecting and processing the information, bureaucratic inertia can develop and can lead to graveyards of numbers into which it is difficult to breathe any life. The existence of a frame of reference that is soundly based on general scientific principles facilitates understanding between people dealing with similar problems. This is desirable at any level, but at the lowest of them we must be prepared to concern ourselves with the comparability and compatibility of data obtained by different people and different apparatus at different places and different times. All these preliminary questions, and others, pose themselves at the beginning of any survey or monitoring procedure.

For the record, I summarize here the substance of my contribution on general ecology to a past summer school in Venice.[1] Everything appears to consist of matter and energy, together with organization or information. Information or form is the way matter and energy are structured. This is the result of a historical process in which usable energy is exchanged and degraded as material things change or as information accumulates. Ecology deals with natural systems and focuses on a level at which the elementary units of action are whole organisms, plants, animals and bacteria, interacting inside a physical matrix. Measurements involving matter and energy in the ecosystem pose no problems over and above technical ones in the field of physics. However, any attempt to measure organization or structure raises special questions for two reasons: (a) the discontinuous nature of the individuals in the sense that ecosystems are formed by quanta with a wide range of sizes and (b) the fact that organization is always a function of size, and that it has to be expressed in terms of a whole range of reference spaces. Consider, as an example, diversity, which expressed how the complete set comprising the whole ecosystem is distributed among different subsets, each equivalent to one of the various unispecific populations. Diversity needs a spectral expression: in other words, any quantitative value of it will refer to a definite volume, and an appropriate expression for it must offer a series of values giving average diversities for different ranges of some chosen parameter.

It is a truism to say that an ecosystem is a *system*, but such a statement must be kept in mind when collecting data and when modelling. The law of diminishing returns must be expected to operate with the costs of any study that is carried out. As the study progresses, the amount of new knowledge per unit of exploratory effort decreases. The knowledge of the entire universe provided by the study of a part of it is out of all proportion to the relative volume of the studied part compared with the total volume of the universe. Such convergent behaviour or closure is a property of any system. In a closed system, the only sort that can be properly mapped on to a model, prediction would be possible. The imperfect models used in practice, however, have to be tailored to the nature, the properties and the degree of closure of the respective systems. Since ecosystems are size-limited and open, it is easier to translate this to the level of models by conceiving of them as open and allowing for stochastic inputs. We call these inputs stochastic because they are not determined from the interior of the system under scrutiny, although they may be strictly determined as part of the interactions between *different* systems or subsystems.

I am not overimpressed by the predictive capacity of ecologists: like students of evolution, they are never short of *post facto* explanations. The problem is not merely an academic one, since it is very much present in the practical application of ecology and in justifications of the way ecology is founded. Practical realism would, in my opinion, often demand that a flat 'no' be given to a number of projects and works that would introduce changes in the environment when obvious adverse effects are to be expected. The same realism would look with rather benign indifference upon many other projects without the need to base such a decision or attitude on verbose and often scientifically ridiculous assessments of ecological impact. The only benefit of these assessments is in delaying action and allowing time for second thoughts. Cost–benefit analysis always cheats in relation to social costs. It would seem to me more sensible to require that every important project having an impact on the environment should have a small percentage of its costs set aside for research into the environmental consequences (similar to the way in which important public works in many countries must include in their estimate of costs a small percentage for artistic or especially creative features). In other words, what I would ask for is that any proposed predictions of environmental responses to the impact of new developments should be followed up and used as a control. If considered acceptable, this would seem to be the only way to bring the whole subject back into the field of standard science.

Such an approach would require a more sensible and less bureaucratic method of dealing with the environment. I feel particularly unhappy at the way things usually go and much of the blame falls on ecologists. What happens briefly is that politicians and administrators are naturally driven to dividing any environmental problems into different parts or aspects, each allocated to a different group of specialists for study, reporting and an eventual proposal of solutions. All the reports from these groups agree in strongly advising a continuation of the research for at least a further year (ecologists, like everybody else, try to keep their jobs for a while). Usually, however, any synthesis eludes the individual workers, and it eludes the decision-makers as well, who have to go ahead and come to a decision at a particular moment, unfortunately without having benefited from complete and exhaustive reports from the specialists. I am sure that it would be better if there were some person having wide experience and some naturalistic flair who could sort out the flow of numerical data or else test, with sound scientific methods, the reliability of the outcomes predicted by some 'ecological study'.

What is probably the most important decision has to be taken at the beginning, and that consists of asking relevant questions: What is our problem? What do we want to know? Defining the question conditions the terms of the answer, and requires the setting of limits in space and time and the selection of a number of variables to be measured. The selection must be made with reference to a realistic model of the ecosystem and not by simply duplicating the forms used in some previous work. The model must be open to allow for stochastic inputs, and the resultant uncertainty will be transmitted through to the answers or predictions.

A number of variables are selected and these are periodically measured. Their behaviour is interpolated and extrapolated statistically, or else a model is adopted that represents some hypothesis concerning the functional relationships between the variables. The model should be able to predict how the set of variables will change if some particular input from the outside acts on a definite region of space and time, directly influencing a subset of the variables. One difficulty is that the model is usually tailored *ad hoc* to fit the particular problem, and the hypotheses it embodies have not been adequately validated in anticipation. In other words, success in prediction, usually poor, is quite often the only way to check the model, and in many predictions involving matters of practical consequence any eventual follow-up studies may be too short. This over-hasty jump from an untested hypothesis to a practical goal not only gives ecology a bad name but is socially unacceptable. In these pages we shall be concerned only with very general, uncomplicated models, not expecting too much from them. In addition, discussion will be restricted to problems connected with surveying and monitoring by the selection and measurement of variables.

The approach will be different according to the sort of ecosystem to be studied. In coastal areas, which are subjected to sizeable movements of water and to continuous shifts of hydrographic structures and of populations, it is more realistic and appropriate to think in terms of processes that generate patterns, since function and structure are clearly interdependent. Planktonic ecosystems can be compared to clouds. Other systems, like forests and some benthic systems, have a more persistent or rigid organization which is a manifestation of past history, and here the initial conditions in our study are much more significant.

BUILDING A MONITORING SYSTEM CAPABLE OF EVOLUTION

Some variables, like temperature, have a precise meaning and are included in a programme as a matter of course. The same is true of other variables such as the concentrations of chemical elements: although methods for their analysis undergo changes and improvements, they can be described at any time and each variable is operationally defined. There exist other groups of variables, however, for which the methods by which they are understood and measured have a larger range of uncertainty and a correspondingly greater potential for evolving. This is because they are expressions of ecological concepts. These concepts (primary production is an example) require rather complicated descriptions and great detail, which are difficult to encapsulate in an operational definition that is simple enough to allow nearly perfect reproducibility. Such concepts are usually approached through parallel and converging measurements, and are defined in rather abstract terms based on theoretical representations. They are continually being reviewed and criticized, and people not in the field find themselves uneasy when working with them. In addition to primary production, most expressions for populations, whether in numbers of cells, volume, or quantity of photosynthetic pigments, belong to this same group. It is difficult, for instance, to provide simple and sharp criteria for distinguishing between active and non-active chlorophyll, or for moving confidently around the various ways of expressing numbers and biomasses.

Such variables as those just mentioned are not to be recommended for monitoring purposes. They are unattractive because they change and evolve rapidly, and are often subject to error when used both by ecologists and technicians. For monitoring, it is better to select variables that can be measured in an unequivocal way and, if possible, automatically. It is acceptable to use a selected measurement as a good approximation to an ecological concept (such as relying on fluorescence as a fair indication of the amount of chlorophyll), and then to look for an approximate conversion factor if it is deemed necessary to express the results in the more commonly used ecological unit.

This is the standard procedure, and I have developed rather strong views about it which I would like to put forward in some detail. It is my opinion that a good approach would be to measure not only two parallel variables, but two sets of them, at least over a certain time. On the basis of theoretical or idealized concepts at the ecological level, we could adopt as variables the concentration of chlorophyll, the population density of each species and the total added density, primary production, and so on. Suppose that it is these three variables that have been selected with reference to phytoplankton. It is not only the sets of values of the three variables themselves that are significant in understanding the properties of the system under scrutiny: the ratios between each pair (chlorophyll per cell, carbon fixed per unit time per amount of chlorophyll or per cell) are equally significant. They are important in themselves and no mere 'conversion factors'. The measurement of plant pigments was introduced into planktology to save time by replacing cell counts and other estimates of biomass. This proved unacceptable since the quantities of plant pigment per cell varied widely according to the taxonomic composition and physiological condition of the phytoplankton. Later, it was assumed that plant pigments might be better related to primary production (notice how this goes with the shift of interest from biomass to production in ecology), but the wide range of assimilation numbers proved that this was not so either. It has now been recognized that quantities of pigment cannot provide good estimates either of biomass or primary production. It is obvious that a ring or network of ecological concepts, and the corresponding variables, undergo shifts in their definitions and in the interpolation of any possible relationships among them. Such relationships are causal, and the chains of dependence linking them find an appropriate expression more in regression than in simple correlation. This means that most relationships have to be considered asymmetrical, or at least that an asymmetrical logic has to be included in the analysis of the model, introducing selection and irreversibility.

No serious objections can be raised against devising comparable networks of interconnected variables at another level, variables that are defined operationally without much theoretical bias. A conservative attitude may be advisable here in order to facilitate comparisons between different places and different operators, and over significant periods of time. Examples of this kind of variable could be any measurement of fluorescence with a flow fluorimeter, the counting of particles in a Coulter counter, or the measurement of concentration of a selected nutrient (phosphate, for example) using a standardized procedure and maybe an automatic analyser. A set of variables like that can form an ensemble of interrelated measurements which also allows meaningful ratios to be used (such as that of the intensity of fluorescence under standard conditions to the density of particles with a fixed size range). It seems (although it may only appear to be so) that it is more difficult to make such relative measures rational and acceptable

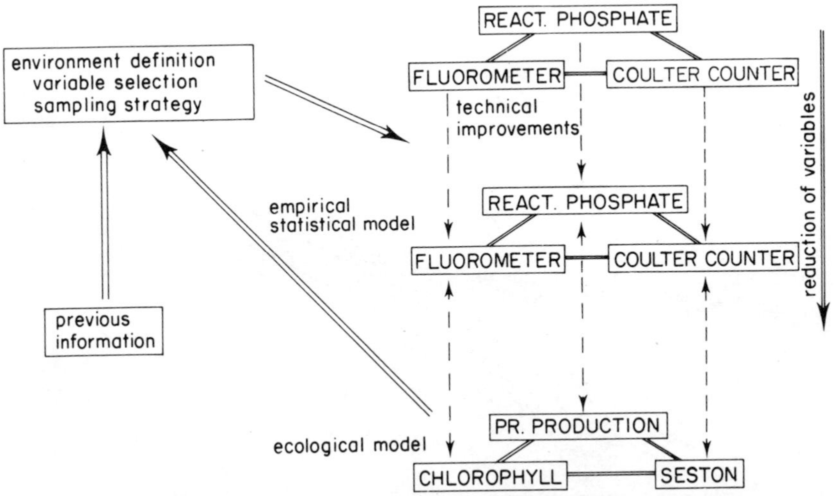

Fig. 12.1 Development of a monitoring system using a network of interrelated variables. The usual variables in the study of phytoplankton would be (1) plant pigments, (2) cell counts or particulate materials (seston), and (3) primary production, as sketched in the lower right-hand corner. Comparison among these variables gives a fair idea of the physiological status and production capacity of the plankton. A project for continuous monitoring of a body of water can adopt three variables that, in principle, are more or less parallel to the previous ones. They also recommend themselves because their measurement can be automated. It is useless to claim a one-to-one correspondence between the two sets of variables, but a simultaneous comparison of the ensembles of both sets will be meaningful. Two feedback loops for improvement can be considered, one at the technical level related to changes in technology (although for large projects a certain conservatism is to be recommended to make comparisons easier). Another feedback loop runs through the two levels and may help to discard variables that do not add much information, and this can also be achieved through the study of the covariance among the variables of any set

than it is in the case of ratios between the 'true' ecological variables mentioned before. These empirical or technical variables are defined operationally; the study of correlations between them and, if highly correlated, the attempt to reduce the number of variables, are useful statistical approaches to their analysis.

It might be appropriate to establish the relationship between both levels as a whole: that is, between the level of empirical and operational measures, and the level of more sophisticated variables based on ecological theory. In other words, not to establish a one-to-one relationship between pairs of variables (fluorescence to chlorophyll in Figure 12.1) but rather to project the whole set of relationships at one level on to the other. For example, it is possible to find and express a satisfactory relationship between quantity of plant pigment and fluorescence, but a closer look reveals that the ratio varies according to the time of day or night (Figures 12.2 and 12.3), since fluorescence is higher if the incoming photons are

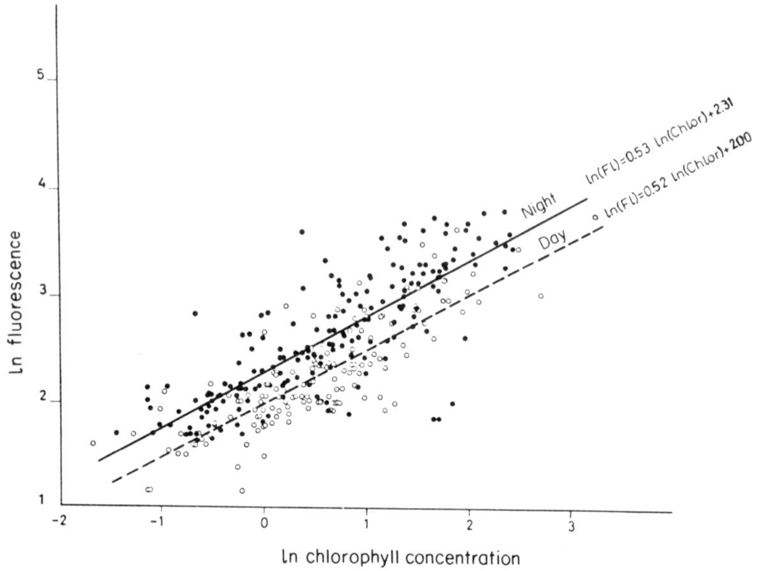

Fig. 12.2 Relationship between concentration of chlorophyll a (mg m^{-3}), and fluorescence measured with a Turner 111 fluorometer. Both variables have been transformed and are expressed as natural logarithms of the original values. The graph is based on many measurements made in western Atlantic, in and around the upwelling area of north-west Africa and the data have been distributed in two groups, according to time, if day (open dots) or night (full or black dots). Regression has been computed separately, and assuming a logarithmic relation between the two variables. The result can be accepted as suggesting that the information provided by a fluorometer gives a fair estimate of the amount of chlorophyll, and of phytoplankton, acceptable for the purpose of controlling eutrophication. Surely it is wiser to think that the 'conversion factors' are not always the same, and in this case are related to the physiological condition of phytoplankton in function of diel rhythmicity. (*Source:* Estrada, 1974)

not used in the acquisition of reducing power by the cell. In fact, dichlorodiphenyl-dimethylurea (DCMU, a herbicide and inhibitor of photosynthesis) apparently enhances fluorescence, thus providing the basis of a new method being developed for the estimate of primary production. Following similar lines of thought, a useful correspondence can be found between the number of particles in certain size ranges counted with an electronic counter and the density of the populations of definite groups of organisms (Figure 12.4). It should be noted that better correlations can be found, at least for certain sets of data, if different channels are combined, providing, so to speak, a better image of the whole ecosystem (even though it may be one that is difficult to define). As I have already said, it is my view that the most useful correspondences to work with will be found in comparing whole sets or nets of relationships.

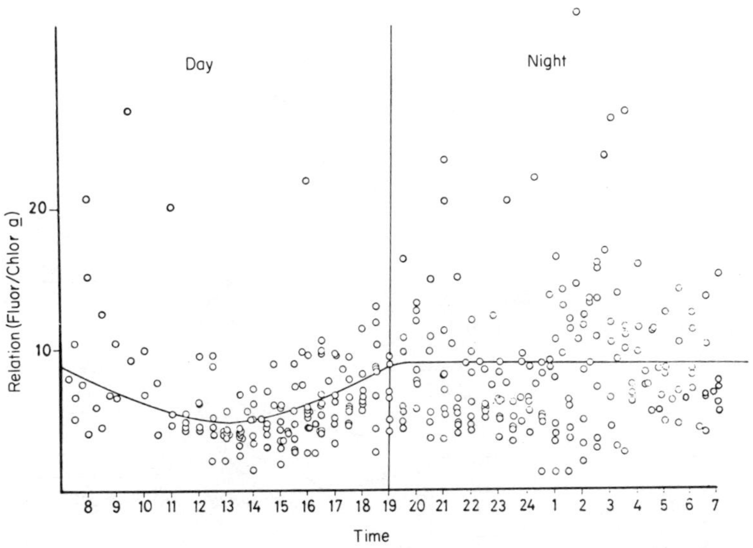

Fig. 12.3 Simple ratio between fluorescence as measured with a Turner fluorimeter, model 111, and chlorophyll a in mg m^{-3} for the same set of data as in Figure 12.2, and distributed according to the time of sampling. In daytime, the fluorescence/chlorophyll ratio is depressed

Some theory is needed, of course, in relating all the variables, in helping with their selection, in guiding their evolution and consolidation, in choosing the most significant variables appropriate to automation for further use, and in dropping redundant ones. This process should result in a steady feedback so that the whole system of monitoring is improved. To bring that about, it is probably useful, and even necessary, to have a backing of sound ecological concepts. This does not mean that all the variables mentioned have to be measured for ever and ever, but only that a necessary background must be provided to get the most from the appropriate variables chosen for monitoring.

Automation leads to the ready accumulation of large amounts of information, but this is an advantage and none of the information should be discarded. Continuous analysis along transects is carried out on the way. Of course, a more even distribution of the sampled points could result in a better understanding of ecosystems, but a receptively open mind can obtain valuable hints from the analysis of transects. Three-dimensional distributions have to be inferred from the sequences provided by transects, or else the transects have to be used to validate or falsify hypotheses about distributions in three dimensions. The pattern of the community on a small scale is in itself an expression of relevant ecological processes. As suggested at the beginning, plankton populations are almost independent of past situations, and the current pattern is a reflection of processes that are actively in operation. Small structures, or the spectra of the variance of populations in relation to space, are useful indicators of the

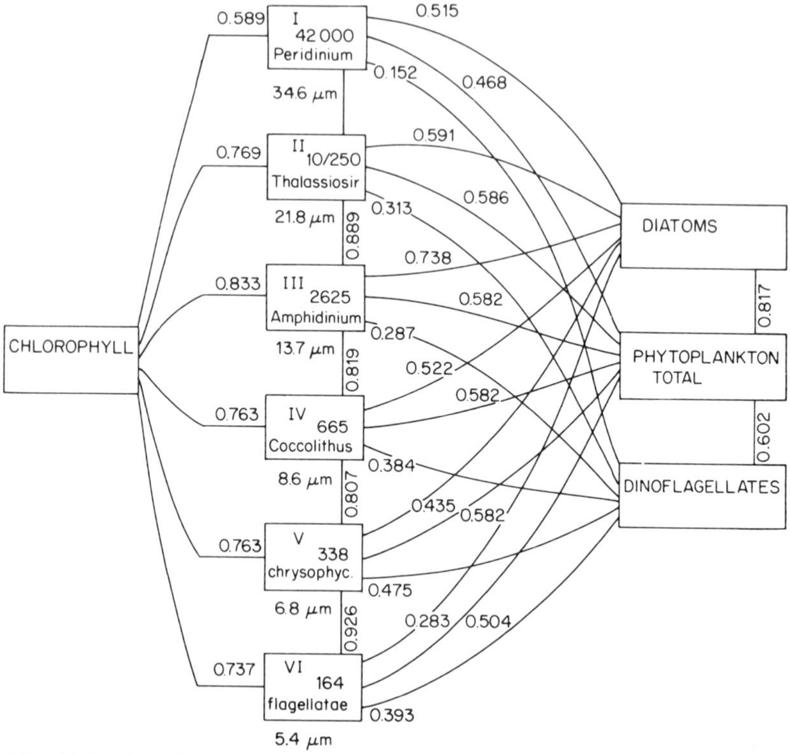

Fig. 12.4 Correlations among a number of variables referring to more than 200 samples of Atlantic plankton (area of north-west Africa). Correlations have been computed on the logarithms of the original data. *Chlorophyll* was measured spectrophotometrically on plankton extracts; *fitoplancton total* refers to the total number of cells; *diatomeas* and *dinoflageladas* refer to these groups, as comprised in the total plankton. The six rectangles in the central column refer to six channels of a Coulter Counter to which an aliquot of the same sample was fed. Inside each rectangle is the name of a typical organism counted in such a channel, as an example to provide orientation to the reader, and the average volume in μm^3 of the organisms counted in the channel; between adjacent rectangles is the average diameter of the particles lying in the boundary. On the basis of the set of correlations among the different variables it is feasible to develop expressions for the computation of some of them from the values of the others, that can be useful for interpolation in monitoring water bodies in which an automatic counter is routinely used

properties of ecosystems and of the underlying processes, and are important from the point of view of practical problems such as the diffusion of pollutants.

The usual approach to the study of transects computes power spectra and cross correlations between different variables, with or without lags. A serious inconvenience is that the detail of the analysis along transects is not normally matched by equivalent information about evolution with time. The required mathematics needs to be more appropriate and more complex than the pedestrian approach which uses and compares means and variances. I have expressed the view in other papers that, in the study of patterns of distribution, one senses that biological amplification of stochastic distributions seems to be very important and that some appreciation of the topology of the pattern is perhaps more significant than the assignation of some size to the dubious structures. The most common pattern is one of disconnected patches of high density with a predominance of organisms of rapid turnover, dispersed over another phase shaped more like a honeycomb or a net in which active secondary producers are prevalent.

I should like to mention that automatic analysis of pattern, useful in establishing relationships between pollution and the response of communities, can also be applied to the study of benthos on hard or rocky bottoms. They are highly characteristic, can be reproduced in colour, and the pictures can be scanned and reduced to some type of abridged quantitative representation. Much interesting work is anticipated in this field of study.

INDICATOR SPECIES

The appreciation of communities as indicators of physico-chemical or synoptic properties of the environment is coming full circle. In the 'good old days' of ecology, communities were always considered to be the best expression of the basic properties of the environment. A well-known example of a practical application was the 'Saprobiensystem' introduced in Germany at the beginning of this century[2] in connection with freshwater pollution. It claimed to provide basic guidelines for the assessment of the quality of water according to the presence or absence of certain indicator species. Various *faux pas* and mistakes occurred along the way: pollution was assumed to be a simple environmental factor and treated accordingly, unnecessary and impossible divisions and subdivisions were proposed, and the whole affair was suspect in that it appeared to be a ruse to provide jobs for biologists. Surely, smelling the water or carrying out some uncomplicated analysis would provide better and cheaper information than the critical appraisal of long lists of organisms. Taxonomy is a subject always requiring much care, taxonomists are increasingly scarce, and it is not uncommon to find an amazingly high proportion of mistaken identifications in lists of 'indicators'. At any rate, the method never fared very well far from its breeding ground in Central Europe. The sophistication and ease of physico-chemical analysers has rightly displaced the use of biological indicators for many purposes.

However, a careful consideration of the significance of the composition of communities emerging from modern ecology gives renewed emphasis to the role of organisms as very sensitive indicators of properties of an environment. Organisms are very sensitive to inputs, each species representing a particular filter in the partition between information and noise, and being capable of detection as the best available equipment. There is no longer any doubt that the presence of particular species is the best detector and indicator of the general properties of each volume of the environment. So we have come full circle in the appreciation of biological analysis and on the way we have perhaps discarded a few prejudices. In connection with this new appreciation of the use of indicator organisms in general, it has to be said that communities are not an expression of instantaneous conditions, but that they integrate environmental properties over a certain length of time. As a result, the distribution of organisms, including possibly those appropriate for use as indicators, reflects past events as well as current conditions. For this reason, the distribution of organisms is the best detector of the effect of industrial effluents that are discharged in an irregular way or at weekends when controllers are not working.

The position can be seen very clearly if we prepare a comparative and critical list of properties of an environment, some of which may have a special interest for us, and pause for a moment to reflect on the way such qualities are seen at different levels. As personal observers, we tend to discriminate and isolate factors according to our senses: we measure light, temperature, salinity, and so on. Sophisticated apparatus has amplified our senses and has removed the definition and measured values of traditional variables far from the biological field while retaining their physical significance. Think only of the concentrations of substances given to the nearest nanomol or picomol, or of temperatures given to hundredths of a degree, and so on. Such precision might be significant and useful (and the proof of this lies in 'artificial' pollutants), but it can be described as unbiological in the sense that it is difficult to extract any significance from eventual correspondence found between the distribution and activities of organisms and the variables measured at the highest precision. Ecology usually operates with coarser scales. At this stage, too, organisms are filters of the properties of the environment. Some inputs are made irrelevant, are screened out and cannot be a source of information. A homoiotherm is indifferent to changes of a small fraction of a degree in the temperature of the environment and can usually accommodate variations in the chemical composition of water. Other species may be sensitive to those variables but indifferent to stimuli that strongly affect a bird or mammal. Each species uses a different method for sorting out relevant from irrelevant information, the latter also being classifiable as noise in a specific sense. As an illustrative analogy, note that we do the same at a personal level: what is information for one person is noise for another, and vice versa—in the end everybody is content, just as in nature a number of species manage to live together without excessive strife.

The distribution of species and the way organisms react to environments make it likely that our sorting out of different environmental factors is probably sensible, since it is based on our senses (which are biological). However, perhaps this is not the way in which the environment is used as a reference in the process of the organization of a community. Perhaps the organisms practise some sort of equivalent of our principal component analysis, in the sense that survival is allowed or responses triggered by recurrent and definite combinations of factors. This would provide the theoretical background for another way of continuing the unending and confusing discussions about niche space, segregation and coexistence of species. Take bird migration, an example as good as any other, in which the animals simultaneously or alternately make use of several mechanisms of detection and guidance involving different 'senses', with the result that the only recognizable basic mechanism is a trend towards the recovery of a harmonic system of inputs, characteristic of the species: in other words, towards pattern matching.

These thoughts are mentioned here not only because of their general interest but also because it is common nowadays, when data are abundant, to proceed to a canonical analysis of the information, and this often produces the desirable result of combining the first set of variables into new ones that are more convenient or reduced in number. The ecological space defined by the possible combinations of all the species in our records is large and confusing, and it depends also on the dimension of the samples studied. Many combinations are never found or are extremely improbable, but a reduction in the number of variables that suffice for a fair description is not easy: sometimes because of computational problems, too, as happens with the influence of double absences on correlations. Judging by my own experience in the field of phytoplankton, the analysis of communities along such lines (that is, with the purpose of recognizing core communities of fairly precise composition that could reflect precise combinations of environmental parameters) is an almost hopeless task, and the existence of groups of species that have among them a high propensity to occur together is hard to prove. Nevertheless, the principal factors or components that emerge from such attempts do quite often make sense, as do components obtained in the study of physico-chemical variables or other variables, such as particle counts at various size ranges. This can be seen in Figures 12.5 to 12.7, which show some distributions in surface waters along the Spanish Mediterranean coast near Barcelona in a more or less eutrophicated area. Some components that emerge from the multivariate analysis, both of particle distributions and of complete phytoplankton counts species by species, agree with distributions of simple environmental factors, such as temperature and salinity, and with the distribution of selected species.

Nevertheless, an exercise of that sort is less objective than it seems. Voluntary selection, in anticipation of routine work, of variables that are expected to be more or less correlated among themselves, marks out the analysis and automatically fixes the first principal component. For instance, in the study of fresh water a large number of inorganic ions are routinely analysed: the first component reflects mineralization. It is clear that, in a situation where we imagine that it is easy to measure the concentrations of all types of organic matter, the

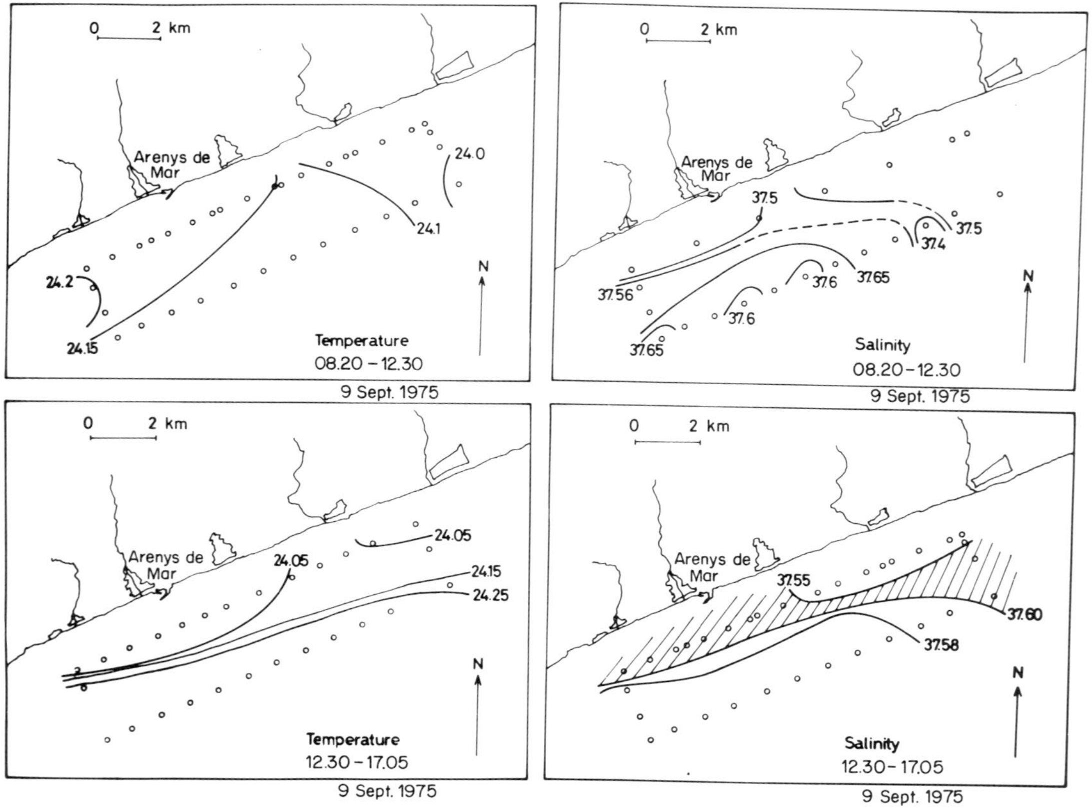

Fig. 12.5 Temperature and salinity at two different periods of a day (9 Sept. 1975) in the surface water of a coastal area close to Barcelona, north-west Mediterranean. (*Source:* Estrada, 1974)

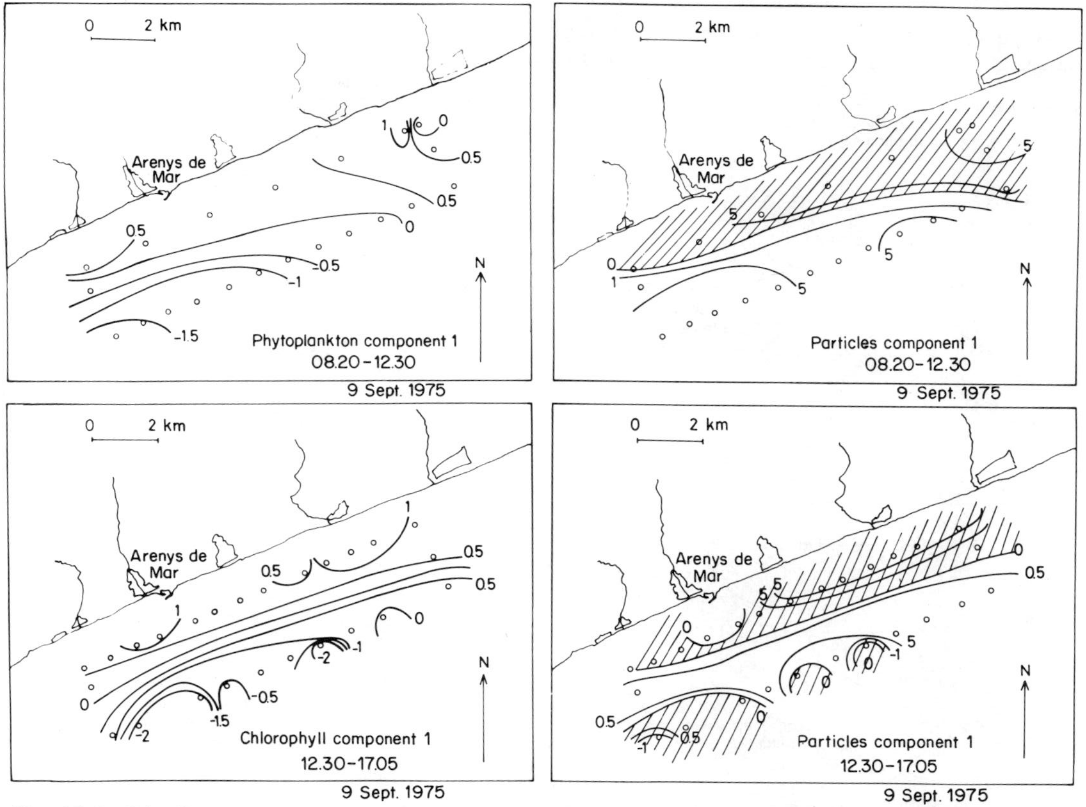

Fig. 12.6 Distribution of the weight of principal component 1, in two analyses referring respectively to phytoplankton composition, and to particle counting (with Coulter Counter) and in the same situations depicted in Figure 12.5. (*Source:* Estrada and Flos.[8])

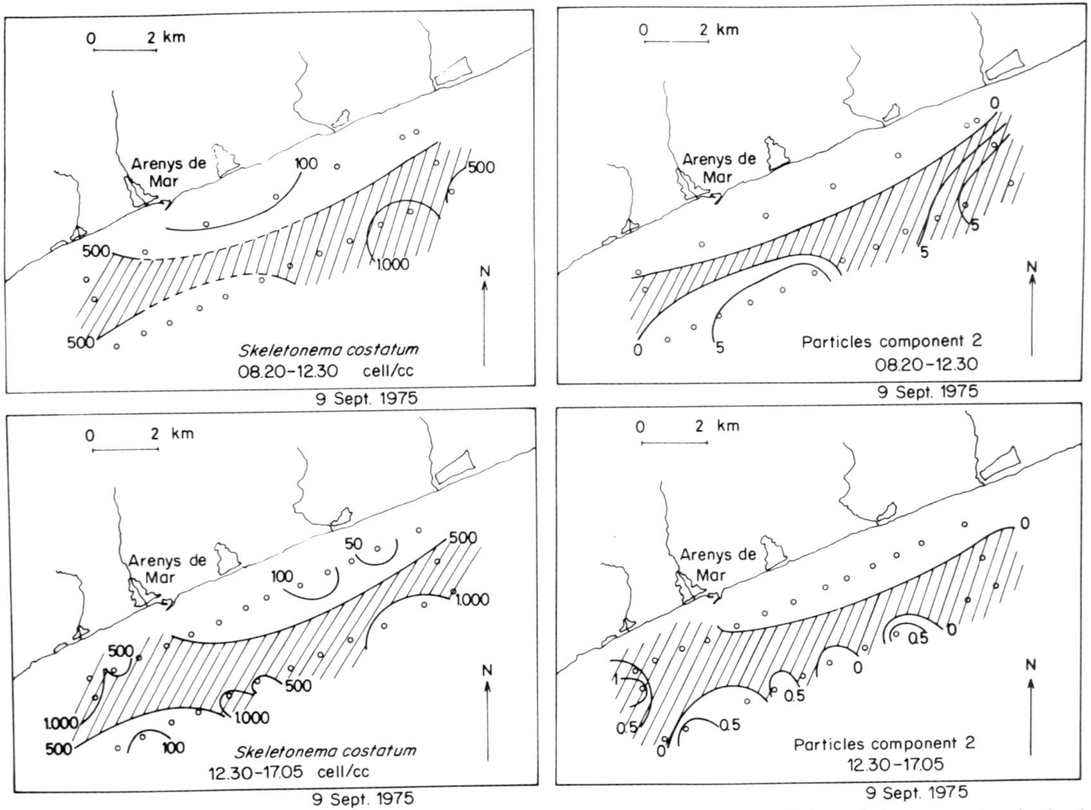

Fig. 12.7. Distribution of the diatom *Skeletonema costatum* and of the weight of the second principal component in the analysis of particulated material (with Coulter Counter), in the situation depicted in Figure 12.5. (*Source:* Estrada and Flos.[8])

most important component would then be related instead to total organic matter content. It is the evolution and adapatation of species in sets of environments exhibiting a necessary physico-chemical continuity that give significance to the different procedures which map or project together the distributions of species and combinations of selected factors. Such procedures represent a modernization of the old idea of indicator species without adding anything substantially new, unless it be the glamour of the computer.

The use of indicator organisms also includes that form of biological analysis in which known strains of the organisms are used as reagents, by examining how they grow in the medium provided by the suspect water. It is better to use mixed populations,[4] in which the way competition goes provides a much finer estimate of environmental properties than unispecific cultures, where the growth and survival of one species is the only available criterion.

It is once again necessary to stress that correct identification is essential in biological analysis. Taxonomic work is difficult and time consuming, and sloppy taxonomic work is a waste of time. It is perhaps a good general policy not to rely heavily on the identification of organisms for monitoring purposes. A special case occurs in programmes like the mussel watch, where individuals of a certain species are selected and analysed as concentrators of definite elements or other pollutants. The need for correct identification must then be extended, if appropriate, to the level of taxonomic units below that of species. This is because much genetic variation within the species is often highly relevant from the standpoints of ecology and indicator value, for example in cases where differences exist in resistance to pesticides or in response to temperature variations.

HOW TO SELECT LIMITS OR REFERENCE VALUES IN ENVIRONMENTAL VARIABLES

Many environmental problems might look as simple as involving merely the quotation of the limiting value of some variable in the environment or ecosystem. This limit is then adopted in law: the quantity of phosphate in water, or the amount of a certain chemical substance in food or drugs, must not exceed a precise value. If a substance has deleterious effects on health, the safest decision is to ban it altogether, or else set the tolerated limits at levels corresponding to the ever-decreasing limits of detection. In the real world, this is almost impossible and some compromise has to be reached. However, the sensitivity of the methods used and the time and space covered by monitoring is naturally very much related to legal questions, and there is a feedback between legislation (or its feasibility) and progress in the study of the environment. Any realistic approach must go through an appropriate and complicated representation of the ecosystems involved. This is very clear, for instance, in the distinction between pollutants that flow and those that accumulate at certain levels through the workings of the ecosystem.

In general, the law claims to make a classification of ecosystems: they fall to one side or the other of a fixed limit and, according to which side they fall, a different action, or no action, is taken. Professionals, such as ecologists, are then asked to propose a limit in relation to which these decisions are to be made. Understandably enough, ecologists are reluctant to recommend that some limit be not exceeded with reference to a single variable (for example, in the allowable input of phosphorus for a lake, or the allowable concentration of a heavy metal in a discharge). It is often assumed that a particular input is safe if it does not exceed a certain value, or that it is possible to define a boundary between an area of stability (whatever that means) and the breakdown of the system. The simplest answer is that it is impossible to recognize any discontinuity, and the best that can be done is to find a function that relates inputs and outputs. Moreover, the structure of this function will depend on other variables that are not referred to in the law—and perhaps cannot be referred to, either because of ignorance or because it does not make sense to turn a book of law into a treatise on ecology. The following example is as good as any other: the concentration of phosphorus in the water flowing out of a reservoir is a monotonic function of its concentration in the inflowing water, but the precise form of the function relating the two depends on the fact that the sediment accumulates phosphorus from the water and gives out phosphorus to it. Furthermore, this property of the sediment is a function of the concentration of phosphorus in the incoming water, of the properties of the sediment itself and of the degree and type of mixing and stratification (that is, whether it uses natural energy (wind) or artificial energy (aeration of the deeper layers of the lake)).

Once the ecological background has been clarified, ecologists can and must understand that practicalities require some value to be stated in legislation, and they may be cajoled into providing a figure. Of course, they must play safe and here there occurs another point of deep disagreement with the lawmaker. The ecologist can cast him or herself into a role similar to that of a bridge designer and use a high safety factor. In that case, nobody would object to the suggestion of a number several times larger or smaller than any reasonable estimate according to the demands of the situation: there is then a fair assurance that, even if other parameters change, it will provide a safe reference point. However, with the size of safety factors commonly used in the design of bridges and buildings, almost no environmental modification would be allowed. It must be clearly understood from the outset that the limits to be set in environmental problems can never have a large safety factor, but are rather more like the closer limits adopted in aircraft design or the construction of nuclear plant. Because we are working close to the limit in these cases, we need a superb technology. However, although small safety factors are accepted in ecology as a matter of course, the usual science and technology employed in the work is less than superb: no wonder there are

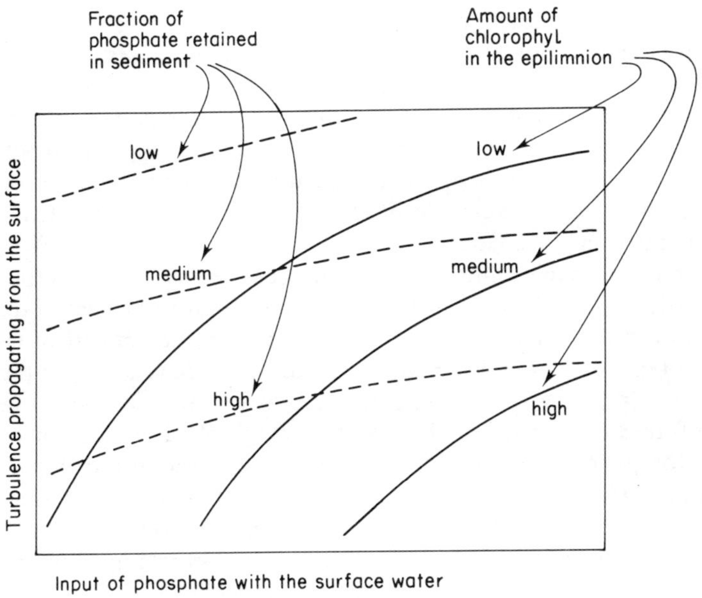

Fig. 12.8 An ideal example, not drawn to scale, of the kind of relationships between functionally related variables in a lake. The amount of chlorophyll in the surface water, an expression of the degree of eutrophication, is proportional to the amount of phosphate entering the lake, and to the concentration of phosphate in the water of the lake. In a strongly stratified lake, the concentration of chlorophyll close to the surface is much higher, and in such a situation much O and N is driven off to the atmosphere while reduced carbon compounds plus phosphorus and other elements are deposited in the sediments. Turbulence propagating from the surface mixes the water, creating different conditions: the total amount of chlorophyll may be lower and its concentration in the surface is certainly much lower, as are the losses from the system across the boundaries to the atmosphere and the sediment. Thus, chlorophyll in the surface is not related in a straightforward way to the input of phosphate. The percentage of the phosphate input that finds a way to the sediment, and is temporarily stored in it, is often a monotonic function of the concentration of phosphate in the input water, but under conditions of very high production, a larger fraction of the phosphate stored in the sediment may leak back into the water in a non-linear way

ceaseless arguments about the various aspects of policy-making in connection with the environment. Once the problem has been clarified and the responsibilities allocated, it should be easier to elicit a response from ecologists. Scientists should not be reluctant to advise about limits: if they don't then somebody else will probably do so much closer to the danger area. It is obvious that more responsibility and more care are necessary in the application of recommended precautions. It is also obvious that any system of survey or monitoring must always keep a close relationship to the practical problem and bear in mind the criteria for decision that are already incorporated or are being incorporated into the law.

Within reasonable limits, the ratio between the increases in output and input can provide a useful criterion. The same could apply to such general concepts as the ratio of the 'quality of life' to the increase in the amount of energy used. Incidentally, such an index might be expected to define a limit to the growth of cities, but it does not do so because other factors come into play that belong to other frames of reference, in this case the fact that the accumulation of people is a basis for political power. The criterion is, however, perfectly valid in other situations: for example, with the increase of chlorophyll in the surface water of a lake versus the increase in phosphorus input into the same lake, where the relationship is a function of the degree of vertical mixing in the water (Figure 12.8).

As technology improves, it becomes possible to detect pollutants in minimal concentrations, and a transition is made from the undetectable to the detectable. The core of the problem here is usually transferred to the possibility of the accumulation of the pollutant in a part of the system: in other words, the outcome is related more to the internal workings of the ecosystem than to the input. Good predictions are then very much dependent on our understanding of the processes going on in the ecosystem, a dynamic study that usually lies outside the possibilities of simple monitoring.

POLLUTANTS AND POLLUTION IN TRANSPORT SYSTEMS

Pollution is not an ecological factor and the word has a sense that is more legal than scientific. Pollution is what makes some resource inappropriate for a certain purpose. It changes according to knowledge, needs and fashions. In the old days, most bottled mineral water had labels boasting of its radioactive properties. Today, even anti-nuclear fanatics have forgotten about that. Natural radioactivity of the building materials in some areas is relatively high, but nobody seems to care. If aesthetic aspects are disregarded, the whole spectrum of undesirable changes divides into two large categories of ecological problem: (1) the presence and accumulation of elements or specific substances (such as heavy metals, radioactive isotopes, or stable organic compounds) which can be recognized as such and which can accordingly be monitored in the different compartments of an ecosystem, and (2) a certain malfunctioning of ecosystems, usually in the form of an irregularity or 'sickness' in the system of transport, or even a change in the whole structure of an ecosystem. Neither sort of change is ever quite independent of the other: a specific pollutant can greatly modify the primary production of an ecosystem; and a plume of sewage discharged at depth can introduce pollutants, increase the consumption of oxygen in deep water, and significantly modify the pattern of transportation.

The first group of problems includes the contamination of trophic paths with alien substances that eventually reach lethal concentrations. In ecology, this occurrence is not altogether negative. A prepared diet can provide all the necessary materials to raise animals of a given species quite successfully, but animals raised in nature have access to other substances that make them more tasty—'biological' chickens are more contaminated than farm chicken. A similar process has played an important role in evolution. The pollution of trophic chains amounts to the development of a defence mechanism. Insects get toxic substances from plants, molluscs get them from planktonic dinoflagellates, and they all then become toxic to their respective predators. The next step in evolution is to make the process more regular, ensuring a minimum concentration of the toxic substance and thus gaining the bonus of a reliable defence which often carries with it the coevolution of advertising signals. The process has been nicely documented by the study of the relationship between aposematic insects (such as the monarch butterfly) and plants. The evolutionary interpretation becomes more complicated because some insects synthesize protective substances identical to the products of metabolism of plants they receive as food, thus raising the final concentration of the materials on which they depend for defence. It is not farfetched to suggest that the mechanism of the accumulation of mercury along marine trophic chains which end with tuna and other large fish might provide a certain defence for such animals against their major predator, man, who recognizes the danger of eating large amounts of tuna because of the high concentration of mercury. It may be significant for the survival of tuna populations that man is discouraged from eating such polluted food. The concentration of mercury in tuna and hake is very much related to their age and to the latitudes in which they occur, always subject to the condition that concentrations of the pollutant increase more readily whenever and wherever turnover is slower. This also applies to the fractionation of the heavier isotopes of biogenetic elements. The distribution of fatty acids and their relative concentrations in different trophic levels show regularities that are based on different motilities and differences in metabolism. It is not irrelevant to mention them here since unwelcome fat in the composition of food can be considered as a pollutant.

With many pollutants, migration and eventual concentration is less dependent on biological mechanisms. The concentration of heavy metals in natural waters reflects the operation of complex equilibria. If the composition of seawater is considered as an equilibrium state under the conditions that reign on the earth today, it is clear that the addition of an extra quality of an element at some point must activate a chain of reactions, with the result that the same amount must be segregated somewhere else. This means that such pollutants can only be dangerous over a rather restricted range of time and space, and the perturbation will eventually be expected to iron itself out. The problem is to define the time necessary for a given decrease in the excess local concentration. Any mechanism leading to a new equilibrium has to be activated by a certain excess or 'head', so that perhaps we do not need to be unduly sensitive to some local increase in concentration. In fact, one important purpose of monitoring ecosystems is to detect the evolution of the field of perturbation created around a point source input.

The partition of CO_2 between air and water, and the consequences of an increased input of this compound, is another good example along the same lines. A column of sea water contains about one hundred times more inorganic carbon than an atmospheric column of the same horizontal cross-section. A small change in the depth at which foraminifera shells are dissolved could absorb a large amount of CO_2 and return the composition of the atmosphere to a former value. This process, however, requires time and is greatly influenced by other processes going on in the terrestrial ecosystem as well as being highly dependent on the conditions of mixing in the surface layers of the seas and on the activities of organisms. This is once again a good example of the fact that simple concentrations are not very revealing. The important things are the processes and the ratios between concentrations coming from such processes. It is important to know the time necessary for a perturbation to be reduced to a fraction (usually a half or 5 %) of its initial value, and this must be considered a major purpose in any monitoring project. An exponential decrease is expected: at the heart of highly controversial approaches to the notion of stability is the common belief that any perturbation will activate mechanisms which tend to nullify it, thus slowing down the speed of recovery as time goes on.

Pollutants do not necessarily act as nutrients, toxic agents or by any other biochemical path. They can also act as chelators, surfactants, precipitators of some elements, scavengers of other materials in the precipitates, or they can change the pH. In all these ways they have tremendous importance in the biology of populations, and their effects may be particularly apparent at interfaces such as at the surface of the water or the sediment. There is also synergism between different pollutants, or between pollutants and the intensity of other ecological factors. For instance, high temperature accelerates biological processes and can change the significance of certain elements or molecules as pollutants. In water birds, mortality is greatly increased by contamination with hydrocarbons if at the same time the salt gland is under stress because only saline water is available.

EUTROPHICATION AND OTHER FORMS OF GENERALIZED STRESS

Specific pollutants, especially if they are substances alien to the evolution of organisms, may inhibit the development of particular species or of larger sections of the trophic web. Often, however, this is not detrimental to the whole ecosystem, either because other species are available, or because the energy flow is bypassed through a detritic path, or simply because the trouble is passed on to another ecosystem where dilution finally takes care of the problem. Other changes are more generalized and are seen as a malfunction of the transport system. They are undesirable, but do not represent pollution attributable to a single substance that can be analysed and followed by simple monitoring. Eutrophication of natural waters is a good example. It results from overfeeding a lake with an excess input of nutrients and also with organic matter which brings in chemical energy but uses up oxygen. A discussion of the boundary between pollution and eutrophication, assuming for example that the introduction of organic matter into the deep layers of a lake should be considered as pollution, is immaterial. It might also be unsound to make a distinction between natural processes and eutrophication due to man. To place eutrophication and pollution in the framework of a general theory does not exonerate man. I do not go along with such an academic line: pollution is a real problem that has to be solved no matter at which door responsibility is laid. Nevertheless, a better perspective is achieved if relations are seen in an ecological context.

Urban areas import materials such as food and other organic substances, use them, process them, and then dump the residues in restricted spaces nearby. The transport system is active in one direction only since nobody is willing to pay for the return journey. The accumulation of organic residues activates other ecological mechanisms; the former communities are replaced by others with a higher proportion of bacteria and other decomposers. They use materials inefficiently in the sense that the biomass is kept small in relation to the energy flow. This might be seen as a positive outcome but many side-effects are undesirable and it would seem as if returning the used materials to their place of origin would cause most of the negative effects to disappear. If this were done, we should have added to the original ecosystem a loop of horizontal transportation: an extra loop, but a closed one.

Pollution and eutrophication are the result of an open loop. A moment's reflection reveals that most horizontal and some vertical cycles in nature are more or less open as well. Water and nutrients flow from land at a higher level to lower lying country and to the sea, and the process is kept alive for an indefinite time as long as orogenesis and a water cycle exist. In aquatic ecosystems, phytoplankton have a finite probability of sinking out of the photic zone and the effect of this on the transport of material is enhanced by animals. Zooplankton, for sure, recycle a small fraction of the nutrients on the photic zone, but the net result of their activity is an acceleration of the process of transporting organic material to deep water. Most components of zooplankton migrate daily up and down, and they excrete at a level deeper than the average one at which they feed. Vertical migration acts as a conveyor belt that diminishes reducing power. Copepods comprise more than half the biomass of zooplankton and have the peculiarity of compressing digested material into faecal pellets that sink at a relatively high speed and have time to travel downwards for several hundred metres before they burst and disperse their contents. As a result of several parallel or convergent mechanisms, a large proportion of the material becomes involved in a decelerated cycle, and the return to the photic zone is dependent on the operation of local or seasonal mechanisms: upwelling in the sea, for example, or overturning in a lake. All the phenomena mentioned above, including the behaviour of modern man, obviously belongs to the same class: in relation to cycles within ecosystems, man does not behave in a way that is different from that of copepods. When ecologists choose to become moralistic, they cannot recommend a more 'natural' behaviour, but can only explain that in a pelagic animal it would be more 'sensible' to feed down and excrete up. However, such ideal behaviour poses problems from the point of view of the energy budget, unless the animal in question could tap a source of energy that was external or not endometabolic. Man could do that by using external energy to accelerate returns and counteract imbalance in the flows.

Problems of generalized eutrophication and pollution, including aspects of unbalanced transportation, can be treated within the framework of a theory of interactions between ecosystems. Exchange across selected boundaries in the biosphere can be described using different terms. If one ecosystem receives from another more energy (chemical energy) than it gives, it can be said to exploit the other system. Exploitation implies some control on the part of the exploiter. However, the flow of material is one thing and control of the flow is quite another. If a lake receives organic matter and is turned into a cesspool, it would seem to be contrary to current usage to say that the lake exploits the source of organic material: it is preferable to say that the system is force-fed or stressed from outside. Similar apparent complications are found in other areas of ecology. Predators feeding on small prey, and small parasites living on the surface of large animals as well, get a supply of matter and energy from other organisms and can be considered as exploiters. Nevertheless, the respective systems are controlled more by the predator in the first case and more by the host in the second. In this special sense a system that is controlled can be said to be under stress.

Stress can develop in succession or in cascade. A fertilized agricultural area may be a source of stress for rivers and lakes (eutrophication). Freshwater systems are very much controlled by terrestrial ecosystems and the biology of running water is characterized more by the inputs of organic matter formed in terrestrial ecosystems than by autochtonous primary production. Only very

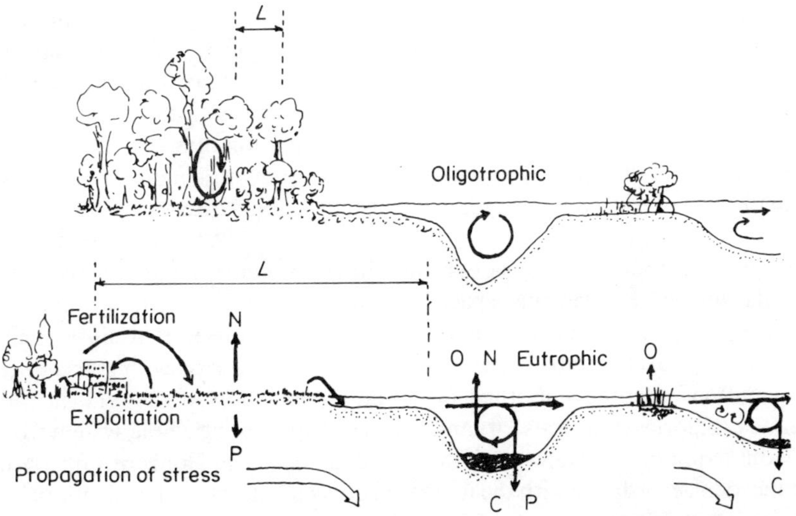

Fig. 12.9 A comparison between natural, local, cycles (upper sketch) and intensified horizontal transportation as a result of human activity (lower), leading to manifestations of exploitation and stress. L is the average horizontal dimension over which the biogeochemical cycles extend

mature terrestrial ecosystems such as the Amazonian region excrete almost distilled water. Rivers are the kidneys of the land and the best indicators of malfunction. Rivers, in their turn, act as sources of stress upon estuaries and coastal marine waters. It is important to realize that monitoring of the condition of fresh water provides a source of information about the state of conservation of terrestrial ecosystems.

If the concept seems trivial and matter-of-fact, a more detailed consideration of the processes involved may prove helpful when we need to select appropriate measures to quantify eutrophication or any other equivalent phenomena. In a force-fed system, the speed of cycles is usually increased, raising rates of turnover and production/biomass quotients at different levels. A proportion of the elements and compounds cycling within the system, or added to it, is removed. In a eutrophic lake,[5] oxygen and nitrogen go to the atmosphere, and phosphorus and reduced carbon compounds to sediments. This is absolutely general and is the best characterization to be made of the eutrophic lake. The same processes are to be observed in oceanic upwelling zones of high production and even on land. The response of cropland to fertilization is of course the anticipated one of an accelerated cycle, but there is also the denitrification of some of the nitrogen compounds and the immobilization of some of the added phosphate because it becomes insoluble. In fact, the ratio between the 'highly insoluble' and 'normal' fractions of the phosphate in a soil is an index of the intensity of human occupation and influence.[6]

I have dealt with these matters at some length because they are essential to an understanding of how difficult it could be to design appropriate combinations of variables for the detection and monitoring of such situations of generalized stress. The accelerations lead to characteristic changes in the structure of the community. One consequence might be the elimination of many species and the simplification of the ecosystem, even if primary production increases, as is usual. This effect may be important in relating the richness and diversity of species to the state of the system.

A general approach considers the different components of the system (populations of the various species, concentrations of nutrients and other dissolved materials) to be moving towards a steady state and, because of effects inherently connected with the accumulation of biomass or with storage, small perturbations become smoothed out. The introduction of material in certain definite categories (nutrients, organic matter) displaces the whole system from its previous equilibrium state, and the speed of reaction along certain paths is enhanced in accordance with the increased concentrations of the reactants. Some head is necessary to generate such activation and to cause other mechanisms involving accelerations and decelerations to function. One result to be expected is a decrease of biotic diversity, which needs to be characterized under many conditions and studied spectrally in relation to the reference space. Turbulent diffusion brings potential reactants closer together and thus accelerates the various processes. Keeping things apart makes their interaction impossible. The coincidence of light, nutrients and algae produce a bloom which lasts until the nutrients disappear into the darker layers and leave the photic zone once again very poor in them. Such simple principles are very helpful.

The condition of an ecosystem that is force-fed or stressed from outside can be recognized and quantified by applying such simple principles to a number of variables. Characteristically, it is unsafe to use single variables and the appropriate measure has the form of a ratio or quotient (Table 12.1). For practical purposes, what has been said in a previous section is applicable: that is, one might prefer ratios between variables that can be measured automatically or devise new ratios between variables defined operationally which fulfil the said condition.

Some of the indicator ratios in Table 12.1 are significant in any estimate of the degree of stratification as against turbulence (dinoflagellates/diatoms, etc.). Any aquatic environment that is very well stratified develops a vertical gradient that is reflected in several properties: perhaps the most significant are oxygen concentration and redox potential. The more pronounced the vertical gradients in an ecosystem, the higher the proportion of material driven out of its cycles, other things being equal. For this reason the accumulation of reduced carbon in sediments is not strictly proportional to primary production alone, but also to the vertical gradient (or redox potential and other variables). A stagnant environment runs more slowly but eliminates a larger fraction of the charge. Such

Table 12.1 Some composite variables suitable for quantifying the degree of eutrophication and other forms of stress in aquatic ecosystems[7]

Total biomass/primary production
Total biomass/biomass of bacteria
Zooplankton biomass/phytoplankton biomass
Biomass of carnivores/biomass of herbivores
Phytoplankton biomass/chlorophyll
Oligotraphent bacteria/bacteria that require a high concentration of easily degradable material
Average size of individual organisms = biomass/numbers (not always)
Reciprocal of the nutrient concentration at which speed of assimilation is maximum
Diversity (not always) and steepness of the diversity spectra
Degree of stratification and organization of the vertical migration and vertical transport
Dinoflagellates/diatoms (an indicator of turbulence)
Rounded dinoflagellates/flattened dinoflagellates (in stratified water)

principles serve as a background against which one can view the practice of injecting oxygen into the hypolimnion of lakes as a stopgap measure to avoid the consequences of eutrophication. The quality of the water may improve and reactions are accelerated, but more material is kept in the cycle and more trouble can be expected in the future.

Turbulence is very important as a mechanism of transport that operates mainly in the vertical dimension as far as the functioning of ecosystems is concerned. However, horizontal transport in humanized systems is closely comparable. Two alternative possibilities are: (1) accumulating used materials in a local source of pollution or (2) paying for the return of the materials to their place of origin, which means accelerating the cycle as well as getting rid of the pollution. There are options possible between these two, and they belong to a set of possible solutions realized in various natural ecosystems.

I should like to insist once more that none of these processes can be monitored using single variables, and that the selection of the group of variables to be measured needs careful consideration, both of the phenomena involved and of what we are concerned about.

REFERENCES

1. Margalef, R. (1977). Troisième Session de E4, Univ. de Venise, **1**, 71.
2. Kolkwitz, R., and Marsson, M. (1909). *Int. Rev. ges. Hydrob.*, **2**, 126.
3. Estrada, M. (1974). *Tethys*, **6**, 247.
4. Mosser, J.L., Fisher, N.S., and Wurster, C.F. (1972). *Science*, **176**, 533.
5. Margalef, R. (1975). *Schweiz. Z. Hydrol.*, **37**, 102.
6. Eidt, R.C. (1977). *Science*, **197**, 1327.
7. Margalef, R. (1975). In *Marine Pollution and Marine Waste Disposal* (Eds Pearson and Frangipane), Pergamon Press, Oxford and New York, p. 301.
8. Estrada, M. (1979). *Inv. Pesq.*, Barcelona, **43**, 637.

Pollutants and Their Ecotoxicological Significance
Edited by H. W. Nürnberg

CHAPTER 13

Some Selected Examples of Eutrophicated European Lakes

O. RAVERA

SUMMARY

Eutrophication means the enrichment of a water body with substances (nutrients) which increase the growth rate of plant populations, i.e. macrophytes and phytoplankton. The most various effects of eutrophication are the following: (a) degradation of water quality, with consequent higher cost of the supply; (b) modification of the community structure; (c) negative influence on the fishing quality; (d) reduction of tourist and sporting activities; (e) increasing frequency of pathogenic microorganisms, and in some cases, of toxic algae; (f) oxygen depletion in deep layers.

Eutrophication is the natural ageing of the lakes, but the influence of man may greatly accelerate this process giving rise to so-called 'cultural eutrophication'. Sewage, agriculture, industries, soil leaching and erosion and dry and wet fall-out are the most important sources of the nutrients which enrich lakes. Because the most common limiting factor is phosphorus, to stop trophic development, reduce its rate and/or to decrease the trophic degree of a lake, the phosphorus discharged into the lake must be reduced. Phosphorus removal from sewage or the diversion of waste waters outside the watershed are the most efficient methods of recovering eutrophicated lakes. In some small lakes the siphoning of the hypolimnetic waters has produced appreciable results. In the areas with high population density and considerable rural and industrial activities, such as the countries of Western Europe, the nutrient enrichment of the lakes is more and more great. The efforts devoted to recovering inland waters are, however, not proportional to their degradation. Consequently, the trophic evolution of these water bodies is accelerated. As an example, some European lakes with different characteristics have been chosen to illustrate the effects obtained from different intervention work. Future topics of the research on lake eutrophication are discussed.

CAUSES OF EUTROPHICATION

As time passes a lake is progressively enriched with mineral salts and organic matter. In most lakes, phosphorus is generally the most common limiting factor of plant growth, nitrogen may be limiting in some cases and very occasionally some other elements (e.g. carbon, silicium) are involved. One of the first symptoms of such enrichment is the increase of primary production, i.e. the production of living material by macrophytes and algae. An increase in primary production leads to an increase in bacterial activity. Beyond a certain level, which varies from one water body to another, the breakdown process cannot balance the primary production. As a consequence, there is a continuous accumulation of organic material in the water and above all in the sediment. These processes cause a series of modifications of the lake, which passes through successive stages called oligotrophic, mesotrophic and eutrophic. This evolution is common to all lakes and is called eutrophication. Eutrophication is the effect of the natural ageing of the lake and the time required to attain this stage depends on several factors (e.g. morphometric characteristics of the lake, hydrological retention time, size, morphology and lithology of its catchment area, vegetation cover of the catchment area and climatic characteristics of the zone). Man and his activities may greatly accelerate the eutrophication rate producing so-called 'cultural eutrophication'. If the time necessary for eutrophicating a medium-sized lake is of the order of some thousands of years, the interventions of man in the catchment area may reduce this time to a few years. Sewage, agricultural practices, soil leaching and erosion, some industrial activities (e.g. breweries, food products) and dry and wet fall-out are the main sources of nutrient enrichment of the lake. The importance of each of these sources obviously varies with the region; for example, one lake may be eutrophicated mainly by untreated sewage and another by agricultural activities and cattle.

In the report by Vollenweider[1] for the OECD (Organization for Economic Cooperation and Development), the source of nutrients and the evaluation of their loads in the lake were discussed. The author attributes to phosphorus and nitrogen compounds a fundamental role in eutrophication and justifies it with a mass of data. On this basis, in 1971, the Water Management Sector Group of the OECD initiated a cooperative study on the relation between nutrient loads in lakes and their trophic evolution. This study has been divided into four projects: (a) the Alpine Project (Switzerland, Germany, France, Austria and Italy; (b) the Nordic Project (Scandinavia); (c) the Shallow Lakes and Impoundments Project (Germany, Netherlands, Belgium, United Kingdom, Spain, Japan and Australia); and (d) the North American Project (USA and Canada). This research programme concerns about 200 lakes and reservoirs. In 1980 all the results of this study will be published. The report on U.S. lakes has been published by the USA Environmental Protection Agency.[2,3]

EFFECTS OF EUTROPHICATION

Several detrimental effects due to eutrophication have been observed; the most obvious are the following:

1 *Degradation of the water quality* and consequent increase of the cost of the supply, if the water is used for drinking or for some industries (e.g. electronics). This degradation is due to the high content of organic material dissolved and particulated. These organic substances, sometimes toxic, are the cause of the bad smell and taste of the water. During chlorination of this water (one of the most common and inexpensive treatments of drinking water) toxic halogenic products (chloroform and bromoform) are synthesized. If ammonia and amino compounds are present, chlorination produces chloramines which have a low toxicity for the microorganisms. The high concentration of nitrites, originally present in the water of the eutrophic basins, or produced by oxidation of ammonia during the chlorination, may produce diseases. In addition the high concentration of particles (e.g. zooplankton, algae, bacteria, detritus) may quickly clog the filters.

2 *Variation of the community structure* by elimination of several species and the increase in population density of others more able to survive in the modified environment. The most common effect of this selection is the occurrence of 'algal blooms' and particularly of blue-green algae.

3 *Decrease in fishing quality.* This is one of the consequences mentioned in 2. In eutrophic lakes the total yield of fish generally increases, but because the populations of the Salmonids decrease and those of Cyprinids increase, the commercial value of the fishing is reduced. Oxygen depletion in deep waters is one of the most important causes of the decrease or elimination of Salmonids.

4 *Reduction of tourist and sporting activities.* This is due principally to the variations of some organolectic characteristics of the water; smell, taste, colour and transparency. In addition, algal blooms and the massive growth of macrophytes have a negative influence on the recreational activities.

5 *Increase of the frequency of pathogenic microorganisms* and in some cases blooms of more or less toxic algae occur in eutropic lakes. An important contribution to eutrophication is given in most cases by domestic effluents which are generally rich in pathogenic microorganisms (e.g. *Salmonella*, *Leptospira*, *Escherichia*, *Pasteurella*, *Vibrio*).

6 *In deep water the oxygen concentration is very low or equal to zero.* This situation may be limited to a well-defined period (e.g. during summer stagnation, in winter if the lake is completely covered by ice) or may be permanent. The absence of oxygen in the water in contact with the sediment

decreases the mineralization rate of the organic substances, reduces nitrates to nitrite and ammonia and sulphates to hydrogen sulphide and allows the transfer of phosphorus, iron and manganese from the sediments to the water.

METHODS FOR REDUCING THE EUTROPHICATION OF LAKES

At present, the effort devoted to recovering the eutrophicated lakes is small when compared to the increase in the number of eutrophic lakes and in the level of trophy in lakes which are becoming eutrophic. The methods most commonly used are the following:

1 Diversion and canalization of the effluents. Satisfactory results have been obtained by the canalization of the effluents, which, after treatment, are discharged in the lake outlet.
2 Reduction of the amount of nutrients in effluents. The removal of phosphorus from the effluents by chemical precipitation is largely performed in modern treatment plants. Good results may be obtained if the size and the efficiency of treatment is proportional to the phosphorus load.
3 Modification of agricultural methods. Some modifications of the quantity, quality and distribution period of chemical fertilizers may reduce the amount of nutrients (phosphorus, nitrogen and potassium) washed down to lakes. In addition, a more rational use of detergents and the decrease in the concentration of the phosphates in them could result in a great reduction in the phosphorus concentration in municipal and domestic waste waters.
4 Removal of nutrients from the lake water and/or sediment and dilution of nutrients. The removal of superficial sediments decreases the amount of nutrients in the lakes. This method may only be applied with great care in small lakes. By uprooting and removing macrophytes, great accumulators of phosphates and nitrates, some temporary benefit has been obtained. A trophy of some small and shallow lakes has been decreased by siphoning the hypolimnetic waters, which are very rich in nutrients and poor in oxygen during the summer stagnation. In small eutrophicated lakes the input of large volumes of water with a low nutrient content has produced satisfactory results.
5 Reduction of the eutrophicating effect of the nutrients. In a few shallow lakes, the eutrophication action of the nutrients has been reduced by flocculating them by adding chemical substances to the lake waters. Satisfactory results have been obtained in some environments where the total water volume has been made to circulate during the summer stratification. This prevents the formation of the thermocline.

EXAMPLES

The following lakes have been chosen to illustrate the effects obtained by some methods applied to their recovery (Table 13.1). The information on Lake

Table 13.1 Morphometric characteristics of the lakes considered

Lake	1 Surface (km^2)	2 Mean depth (m)	3 Max. depth (m)	4 Drainage basin (km^2)	4 ÷ 1
Mauensee	0.51	3.9	6.8	4.3	8.4
Lugano	48.9	130	322	615	12.6
Annecy	27.0	41.5	64.7	163	6.0
Greifensee	8.49	17.7	32.6	160	18.8

Lugano is the result of our research,[4] that on Lake Annecy is from the 'I.N.R.A., Station d'hydrobiologie, Lucustre', Thonon-les-Bains,[5] that on Greifensee is by Ambühl and Thomas[6-8] and that on Mauensee is by Gächter.[9]

Lake Lugano

In the past this lake was probably oligotrophic, but after the Second World War a progressive increase of its trophy was evident.[10-13] Our studies carried out from 1972 up to now demonstrate that the eutrophication of this lake is so advanced that Lake Lugano may become included among the most eutrophic lakes of the alpine region.

The most important source of phosphorus is the domestic and municipal waste waters produced by the great size of the population living in the lake watershed. Indeed, in recent decades the population has increased very rapidly, particularly in the area near the lake border. The most obvious effects (direct and/or indirect) of the nutrient enrichment are: (a) a progressive decrease of the oxygen concentration in deep waters and an increase in the epilimnion (oversaturation); (b) several algal blooms, including blue-green algae (e.g. *Oscillatoria rubescens*); (c) mass mortality of fish; (d) decrease of the water transparency; (e) quantitative and qualitative modifications of the planktonic and benthonic community structure; and (f) obvious alteration of the organoleptic properties of the lake water.

Until 1976 no serious work was done to recover Lake Lugano and waste-water treatment of the effluents began only in July of that year. A plant was established near Lugano (Bioggio), the town which discharges the greatest part of the sewage emitted into the lake (Figure 13.1). This plant can treat a volume of waste waters of 120 000 equivalent inhabitants.

The phases of treatment are as follows: (a) filtration; (b) primary sedimentation; (c) biological mineralization; (d) and final sedimentation. The waste waters treated are discharged into the Agno basin through the Vedeggio river. In 1977 the third stage was added, i.e. phosphates are precipitated by aluminium salts introduced during the first phase. At the same time, a second treatment plant has been established near Riva San Vitale on the south-east

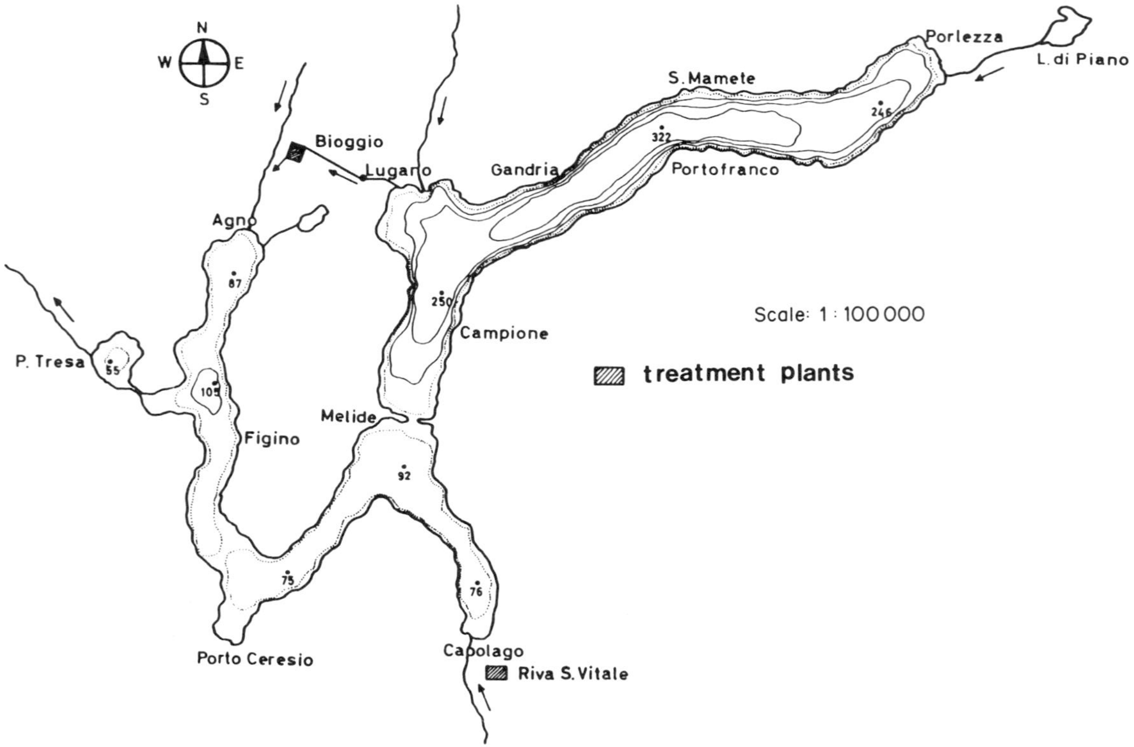

Fig. 13.1 Bathymetric map of Lake Lugano and sewage treatment plants. (*Source:* Ravera, 1973–78)

border of the lake. In this plant the removal of phosphates is obtained by precipitation with iron salts. So far the effects of these treatments are not yet evident.

Lake Annecy

Since 1970, because of the increasing trophy of this lake, principally due to human activity, sewage has been diverted by 'ring' canalization. Lake Annecy is divided into two basins: Grand Lac and Petit Lac. The effluents of the first are collected by a 'ring' and discharged out of the catchment area. Those of the latter are discharged untreated into the lake. As a consequence, the effects of the decrease of the nutrient load are more evident in Grand Lac than in Petit Lac. The studies carried out from 1971 up to now are very interesting and useful because they identify some causes of the delay in the lake recovery.

From 1971 up to now the effects of sewage diversion were evident in the epilimnion. For example, the transparency increased, chlorophyll concentration and primary production decreased, Diatoms populations increased, phosphorus concentration decreased and oxygen oversaturation disappeared. In the hypolimnion the improvement was less evident; the concentration of the oxygen was very low and the concentrations of phosphorus and other nutrients were high. The high concentration of phosphorus in deep water was principally due to the release of this element from the sediment when the oxygen concentration was too low, i.e. during the summer stagnation. In spite of the stratification period which lasted for several months, a certain exchange of nutrients between deep waters and epilimnion occurred and, as a consequence, the improvement of the surface water was delayed. Today this lake is judged mesotrophic.

Greifensee

Before 1919 Greifensee was oligotrophic, but from 1917 its trophic level was continuously increased by domestic waste waters and the phosphorus load from rural areas. At present there are eleven sewage treatment plants and the phosphorus removal from the effluents is obtained by chemical precipitation. An important improvement of the water quality has been obtained; for example, the phosphate concentration in lake water ranged from 380 to 590 $\mu g\,l^{-1}$ in 1973, but was about 260 $\mu g\,l^{-1}$ in 1977. In the near future (in addition to the phosphate precipitation), flocculation and filtration techniques will be added to the most important plants. By this method phosphorus in the treated effluent may be reduced to 0.1 $mg\,l^{-1}$.

Mauensee

Mauensee is a shallow and eutrophicated Swiss lake. To restore this lake, the hypolimnetic waters, which were the most rich in nutrient during the stratification period, have been siphoned according to the Olszewski's method.[14] In a short time satisfying results were obtained. The oxygen concentration decreased in the euphotic zone and increased in hypolimnetic water, transparency increased and the phosphorus and nitrogen concentrations decreased. At the same time the population density of *Oscillatoria rubescens* decreased and that of *Ceratium hirundinella* increased. An important increase of the thickness of the epilimnetic layer has been recorded. Another important result of the siphoning was the reduction of the mean residence time of the hypolimnetic waters.

During the warm season the amount of phosphorus exported through the outlet was greater than that immitted from external sources. This was due to the release of phosphorus from the sediments during this season, when the oxygen concentration in the hypolimnion was lower than 1 $mg\,l^{-1}$. During the rest of the year the oxygen concentration was higher than 3 $mg\,l^{-1}$ and, consequently, an important amount of phosphorus was accumulated in the sediments. The increase of phosphorus taken out by the siphoning reduced the accumulation of this element at the sediment surface. As a result, a decrease in the flux of phosphorus from the sediment during the summer stagnation has been noted. These modifications resulted in a decrease of the primary production and then of the trophic level.

DISCUSSION AND CONCLUSIONS

From the comparison of the methods adopted to recover the four lakes, considered in relation with the results obtained, the following points may be made.

Several characteristics of Lake Lugano combine to delay its recovery: this great lake is made up of different basins, thermal stratification occurs for several months and in the deep basin (Gandria-Lugano) full circulation occurs only in very cold winters and the mean hydrological renewal time is very long (12.5 years). In addition, the sewage treatment started when the eutrophication of Lake Lugano was very advanced. The first effect of the sewage treatment should be observed in the basins of Lugano, Agno and Capolago. Indeed, the effluents immitted in Lugano basin, after 1976, have been diverted to the treatment plant, treated and discharged, through the Vedeggio river, into the Agno basin. After 1976, the Capolago basin has received only treated waste waters. In any way, it is evident that the time needed to recover this lake will be long and will be influenced by the meteorological characteristics of a number of years.

A noticeable improvement in Lake Annecy has been obtained by ring-canalization, but to obtain a satisfying recovery of this lake, the canalization must include also the effluents of Petit Lac. In addition, the oxygenation of the hypolimnion could both increase the mineralization rate of the organic material in deep waters and reduce the release of phosphorus from the sediments to the water.

In spite of the expensive and sophisticated treatment planned for Greifensee, it is not probable that the water quality of this lake will be acceptable in the near future. This prediction is based on the high content of nutrients accumulated in the sediments for several years and released during the period of oxygen depletion in the hypolimnion. In addition, the phosphorus charge from diffused sources (e.g. agriculture) is important and will be reduced with difficulty.

The satisfying results obtained in Mauensee by siphoning the hypolimnetic waters show that this method may be applied with success in the shallow and small lakes. Indeed, appreciable results have been obtained using this inexpensive method by Olszewski,[15,16] Eschmann[17] and Pechlaner.[18]

In conclusion, each lake may be considered as a particular case in relation to its natural characteristics and the level of trophy reached. As a result, the recovery of a lake, or at least an improvement in its water quality, must be based on the knowledge of the lake metabolism in relation to the nutrient charge from external and internal sources, i.e. the nutrient flux from the sediment. The importance of the internal enrichment varies with the lake and season.

For most eutrophicated lakes, phosphorus is the most important cause of plant growth. As a result the reduction of the load of phosphorus is generally considered the most efficient way of stopping the trophic evolution, reducing its rate and/or decreasing the trophic degree of a lake.

There are different ways of reducing the phosphorus load, for example, the removal of phosphates from the sewage by chemical precipitation, the diversion of effluents by canalization, the siphoning of hypolimnetic waters and the reduction of the phosphorus load from diffused sources (e.g. fertilizers) by an efficient environmental policy. The input of phosphorus from the sediment may also be reduced by siphoning the hypolimnetic waters as well as oxygenating the anoxic layer. The latter method must be combined with the reduction of the external phosphorus load. If the effluents are the main source of phosphorus, the chemical precipitation of this element reduces the phosphorus load into the lake. On the other hand, this method gives few results if the phosphorus load is principally due to the diffused sources (e.g. fertilizers, soil leaching).

FUTURE TRENDS OF STUDY

Several researchers have carried out studies on the diversity and stability of eutrophicated ecosystems. Most of these studies are based on the concept that a decrease of the diversity reduces the stability of a community. Odum's hypothesis[19] on the relationship between energy, diversity and stability is undoubtedly of interest. According to this author, the stability of an ecosystem does not depend on its diversity level, but on the balance between the quantity and the quality of the energy flowing into the system and its diversity.

Consequently, systems are unstable either because they receive a low energy flow and have a low diversity, or because they are maintained by a rich flow of energy (or of nutrient substances) and have a high diversity. Important results from the research on diversity and stability in ecosystems with different level of trophy and receiving different charges of toxic pollutants should be obtained. Indeed, the combined effects of the eutrophication and toxic substances is not well known. From the little information available it seems that the effects produced by the same amount and type of toxic pollutants are minimized with the increasing trophy of the ecosystem. This is evident for heavy metals which are chelated and adsorbed in great quantity by the organic material which is very abundant in eutrophic lakes.

The use of more or less partially treated sewage in agricultural and aquicultural practices is an undoubtedly interesting research subject. It is obvious that rational proposals in this field must be based on the knowledge of the energy flow and the material cycle in the environment considered. In addition, the risk that pathogenic microorganisms may be transferred from the sewage to the yield obtained (e.g. agricultural products, fish) must be accurately evaluated.

Another important problem arises from the effects produced by treated effluents in the lake receiving them. Because the removal of nutrients from the waste waters is never 100 per cent, a certain aliquot of nutrients and chemicals used to precipitate phosphorus is concentrated in the discharge point. If there are also toxic substances in the waste waters (e.g. heavy metals), noxious effects may be occurred. In Lake Geneva deleterious effects on the benthonic fauna living in the area near Lausanne, influenced by the discharge of treated waste waters, were evident.[20,21] In fact, in the sediments of these areas, high concentration of nutrients, organic substances, iron, mercury and other heavy metals have been measured.

REFERENCES

1 Vollenweider, R.A. (1971). OECD, Paris.

2 United States Environmental Protection Agency (1977). Ecological Research Series, EPA-600/3-77-086.

3 United States Environmental Protection Agency (1978). Ecological Research Series, EPA-600/3-78-008.

4 Ravera, O. (1973, 1974, 1975, 1976, 1977, 1978). EURATOM Reports.

5 Institute Nationale de la Recherche Agronomique (1971, 1972, 1973, 1974, 1975, 1976, 1977, 1978). Station d'Hydrofiologie Lacustre, Thonon-les-Bains.

6 Ambühl, H. (1975). *Schweiz. Zeit. Hydrol.*, **37**, 1.

7 Ambühl, H. (1977). *Wasser, Abwasser, Abfälle*, **78**, 218.

8 Thomas, E.A. (1976). *Vierteljahresschrift*, **121**.

9 Gächter, R. (1976). *Schweiz. Zeit. Hydrol.*, **38**, 1.

10 Baldi, E., Pirocchi, L., and Tonolli, V. (1949). Ispettorato Federale Svissero per la Pesca, Berna.

[11] Gigax, F., and Nydegger, P. (1964). Tipografia Grafica Bellinzona.
[12] Vollenweider, R.A. (1965). *Mem. Ist. Ital. Idrol.*, **19**, 213.
[13] Jaag, O., and Marki, E. (1970). Eutrophication in large lakes and impoundments, Uppsala Symposium, OECD, Paris.
[14] Olszenstri, P. (1967). *Blatt*, **14**, 87.
[15] Olszewski, P. (1961). *Verh. Internat. Verein. Limnol.*, **14**, 855.
[16] Olszewski, P. (1971). *Verh. Internat. Verein. Limnol.*, **18**, 1792.
[17] Eschmann, K.H. (1969). *Gesundheitstechnik, Zürich*, **3**, 125.
[18] Pechlaner, R. (1971). *Corinthia II*, **31**, 97.
[19] Odum, E.P., van Dobber, W.H., and Lowe, R.H. (1975). McConnell Eds. W. Junk Publ., The Hague, 11.
[20] Lang, C. (1975). *Verh. Internat. Verein. Limnol.*, **19**, 1182.
[21] Lang, C. (1978). *Arch. Hydrobiol.*, **83**, 158.

Pollutants and Their Ecotoxicological Significance
Edited by H. W. Nürnberg

CHAPTER 14

A Regional Case Study of the Pollution of Natural Waters, Soils and Plants by Lead, Cadmium and Zinc

J. FAUCHERRE, A. M. PINART, J. PINART AND A. DUTOT

SUMMARY

It has been demonstrated:

1 That the pollution of the river run-off from mine workings has two sources: the open-cast mine during dry weather and the slag-heap during wet periods. These two sources cause pollution by different processes:

(a) Pb, Zn and Cd pollution from surface mining: this is mainly due to oxidation of pyrites and the resultant acidification (pH = 2) of the waters. This decreases regularly each year. The neutralization of waters at a junction with a non-polluted river is manifested by the precipitation of $Fe(OH)_3$, a process which takes out Zn, Pb and Cd by adsorption or coprecipitation and thus acts as an important purifying mechanism. The $Fe(OH)_3$ suspension, rich in polluting elements, is largely decanted along irrigation canals so that the soil of meadows and vegetable plots is only slightly contaminated.

(b) Pb, Zn and Cd pollution from the slag-heap: sands from this source, rich in pyrites, are physically transported by the river during wet periods and are deposited on the soil of meadows and vegetable plots with thicknesses of up to 20 cm. Oxidation of FeS_2 then takes place and causes an *in situ* acidification of infiltrating water with the activation of Zn^{2+}, Pb^{2+} and Cd^{2+}. As a result, the alkaline reserve of the soil is neutralized over a thickness of about 5 cm for a sand deposit of 20 cm depth. The survival of plants with shallow roots (meadow grasses) is severely jeopardized. Plants with deep roots, on the other hand, do survive although they are highly polluted.

In the case of vegetable plots, the sandy deposits are smaller but are dispersed in the soil by digging or ploughing. The acidification of the soil is therefore not sufficient to neutralize the alkaline reserve, the pH remains about 6.5 and the alteration of the pyrites produces only thiosulphates which form soluble complexes with Pb, Zn and Cd. In this case, the plants are highly polluted but their survival is not endangered.

2 The extent of the contamination of commonly consumed plants by Pb, Zn and Cd:

The first mechanism (pollution from surface mining) leads to only a small amount of Pb pollution in plants (2 to 3 times the permissible limits set by the World Health Organization), whereas pollution by the second mechanism (physical pollution from the slag-heap) is much more significant (2 to 16 times the WHO limits).

These standards are exceeded much more frequently for Pb than for Cd. Those parts of the plants above ground are affected to a much greater extent than the non-chlorophyllous parts (tubers, roots, fruits).

The maximum pollution occurs during the summer, when evapotranspiration is greatest.

3 That the zinc and cadmium concentrations in the annual growth rings of conifers are good chronological tracers of pollution in water courses. In this way, we can roughly reconstruct the local microclimate, the changes in mining activity and the background value of Zn and Cd which indicates anthropogenic pollution. On the other hand, Pb cannot be used as a tracer: it can be used as such to detect atmospheric pollution close to main roads, but only in a very limited way.

4 The effects of the pollution of commonly consumed plants on birth rate.

From 1926 to 1968, the number of children per family, about 2 on the national scale, is 2.1 for a non-polluted community. It falls to 1.1 for a community irrigated by a river polluted with Pb and Cd.

From 1886 to 1906, before the appearance of this pollution, the birth rate for both communities was normal (2.3 and 2.4). It therefore appears that a correlation exists between the extent of Pb contamination in the river and the birth rate.

To the east, south and west of the granitic Massif Central, there is a sedimentary plateau (Les Causses) containing more than a score of galena and blende deposits which have been, or are currently being, exploited either in drift or open-cast mines. The subject of this study is one of these mine workings located at Carnoules near Saint-Sébastien d'Aigrefeuille (Gard, France). In particular, we describe the geochemical mechanism involved in the weathering of the open-cast mine and of the slag-heap associated with it, and the effect of that weathering on the pollution both of the river waters and of the soils and plants they irrigate. This work is the subject of two research contracts, one with the Commission of the

European Economic Community[1] and the other with the Ministère de l'Environnement,[2] to both of which we wish to express our gratitude.

GEOGRAPHICAL AND GEOLOGICAL SETTING

The run-off from the open-cast mine at Carnoules (Figure 14.1) is collected by three channels R_1, R_2 and R_3 which flow into the River Reigous, itself a tributary of the Amous. The Reigous also receives the run-off from the slag-heap

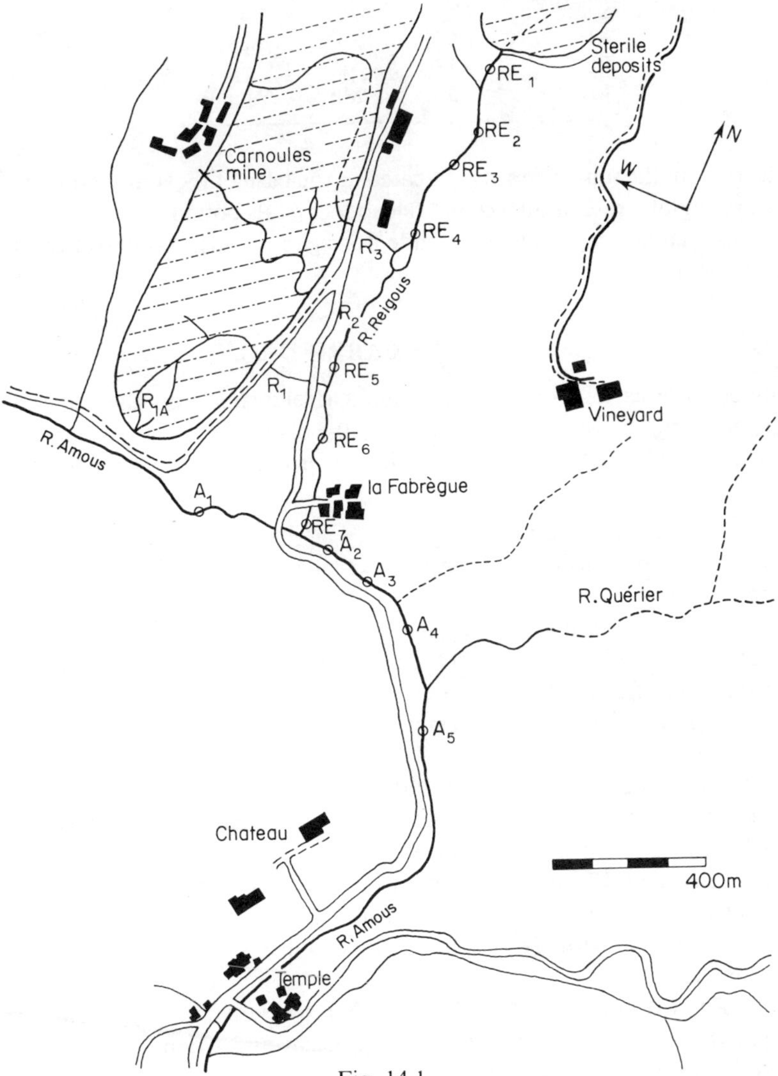

Fig. 14.1

associated with the mine workings ('sterile deposits' in the figure). Some 7 kilometres to the south, the Amous flows into the Gardon, a tributary of the Rhône. Downstream from its confluence with the Reigous, the Amous is polluted by a copious reddish precipitate; the aquatic fauna have completely disappeared.

Table 14.1 Mean chemical composition of the rock in the deposits

SiO_2	70%	Ca	2.0%
S	7%	Pb	3.5%
Al	1.5%	Zn	0.25 ppm
Fe	5.7%	As	500 ppm
Mg	2.5%	Mn	29 ppm

The rock in the mine workings, located at the base of the Trias, consists of an alluvial conglomerate of quartz and feldspar crystals bound by a greyish cement of pyrites, marcasite, galena and blende. Its mean chemical composition is given in Table 14.1.

ANALYSES CARRIED OUT

Analyses were carried out on the filtered river waters, on the red sediment transported by the river, on the sands of the slag-heap, on the soils and their interstitial water, and on the vegetation. Determinations were made of

1 pH using a glass electrode.
2 Redox potential using a platinum electrode.
3 Resistance.
4 Dissolved oxygen and temperature.
5 Total reducing ions Fe^{2+}, AsO_2^-, SO_3^{2-}, $S_2O_3^{2-}$ by permanganate titration.
6 Total iron after reduction of Fe^{3+} by ascorbic acid.
7 Total reducing anions by permanganate titration after passing through a cation exchange resin.
8 SO_4^{2-} by a potentiometric method (titration with $Pb(ClO_4)_2$ using a lead-selective electrode in 80 per cent ethanol) or by nephelometry of $BaSO_4$ at a $pH < 2$ and spectrophotometry at 750 nm.
9 S^{2-}, Sn^{2-}, SO_3^{2-} and $S_2O_3^{2-}$ by a potentiometric method: S^{2-} and Sn^{2-} titrated with Hg^{2+} at $pH = 13$ using a Hg^{2+}-selective electrode; then $SO_3^{2-} + S_2O_3^{2-}$ with Hg^{2+} at $pH = 7$ to 8; and $S_2O_3^{2-}$ alone after addition of formaldehyde.
10 Pb, Zn, Cd, Fe, Mn, Ca, Mg, K, Na by flameless atomic absorption spectrometry or, in some cases, for Pb, Zn and Cd by anodic stripping voltammetry.

DEGREE OF POLLUTION OF THE WATER, SOIL AND VEGETATION

Filtered River Water

Table 14.2 gives values for the pH and redox potential of the waters of the Reigous, the Amous and the lateral channels. It can be seen that:

–the lateral channels R_1, R_2 and R_3, which collect the run-off from the open-cast mine, are very acidic;

–the pond R_1A has the lowest pH of all those measured, no doubt because the stagnation of its waters gives time for redox equilibria to become established;

–the high redox potential at the level of the Reigous and of the lateral channels decreases sharply at their confluence with the Amous, a decrease accompanied by the precipitation of the red sediment ($Fe(OH)_3$).

Table 14.2

Location of sample	pH	Redox potential (V)	Location of sample	pH	Redox potential (V)
RE_1	3.0	0.56	R_1A	1.6	0.72
RE_2	3.4	0.57	R_2	2.0	0.64
RE_3	3.4	0.57	R_3	3.1	0.74
RE_4	3.5	0.57	A_1	7.9	0.38
RE_5	2.1	0.59	A_2	3.5	0.51
RE_6	2.0	0.66	A_3	5.5	0.30
RE_7	2.1	0.66	A_4	6.7	0.30
R_1	1.9	0.75	A_5	7.0	0.37

Table 14.3

	Fe (%)	Zn (ppm)	Cd (ppm)	Pb (ppm)
Non-polluted soil	2.25	268	1.5	148
Soil polluted by irrigation	2.95	780	2.9	782
Soil polluted by the deposit of sand from slag-heap	2.86	2201	8.7	5802
Slag-heap sand (90–95% SiO_2)	2 to 3	1000 to 5000	15 to 25	4000 to 10000

Soil Irrigated by the Amous

Table 14.3 gives some values of the concentrations of Fe, Zn, Cd and Pb in the soil of the vegetable plots irrigated by the Amous and located downstream from its confluence with the Quérier. It can be seen that:

–soil irrigated by water from the Amous is richer in Zn, Cd and Pb than non-polluted soil;

–soil reached by the water when the river is in flood is even richer in the polluting elements (part of the material from the slag-heap is carried away at that time and deposited on the soil).

Plants Irrigated by the Amous

Enrichment of the soil in Zn, Cd and Pb means that vegetables planted in it are similarly enriched, but it is noticeable that the effect is much more pronounced in the chlorophyllous parts of the plants (Table 14.4). The storage organs of these vegetables thus have little tendency to retain toxic metals, and the same phenomenon can be observed in potatoes and turnips. As a result, the consumption of lettuce or beet is more dangerous than that of potatoes.

It is noticeable, moreover, that the accumulation in the plants is greater in summer than in autumn (Table 14.5), probably because of evapotranspiration. The permissible weekly doses fixed by the World Health Organization (WHO) are 3 mg for Pb and from 0.4 to 0.5 mg for Cd. We have compared these norms

Table 14.4 Accumulation of heavy metals in the chlorophyllous parts of vegetables (in mg/kg fresh matter)

	Fe	Zn	Cd	Pb
Beet: green part	64	37	0.42	24
Beet: white part	11	10	0.14	8
Carrots: leaves	86	13	0.1	14
Carrots: roots	2.8	3.4	4×10^{-3}	2.1

Table 14.5 (Concentrations in mg/kg of fresh matter)

	Fe	Zn	Cd	Pb
Lettuce: July 1976	302	28	1.4	4.8
Lettuce: Nov 1976	18	19	0.12	1.6

Table 14.6 (Concentrations in mg/kg of fresh matter)

	Pb	Cd
Lettuce: non-polluted soil	0.56	0.05
Lettuce: polluted soil	4.6	0.9
Leeks: polluted soil	11.5	0.22
Beet: polluted soil	24	0.42

with the concentrations found in various vegetables planted in soil which has received deposits of sand from the slag-heap during periods of high water in the river. Several significant examples are to be found in Table 14.6. As can be seen, the WHO norms are clearly and systematically exceeded for lead, and in one case for cadmium. In the case of vegetable plots not reached by flood waters but irrigated by water from the Amous containing the $Fe(OH)_3$ sediment in greater or lesser amounts, the concentrations of Pb and Cd are not so high but do nevertheless exceed the permissible lead levels by a factor of two or three.

Effect of the Pollution of Commonly Consumed Vegetables on Birth Rates in the Local Population

We have carried out an investigation into the birth rate in the population by comparing the ratio (number of children/number of families) for the commune of Générargues, irrigated by the Amous, with that of the neighbouring commune of Mialet, supplied by a non-polluted stream. The standard of living of these two communes is similar and they are both predominantly Protestant. Figures taken from national censuses of the population (archives départmentales du Gard, Nîmes) are given in Table 14.7.

From 1926 to 1968, the ratio of the number of children to the number of families, around 2 on a national scale, was 2.1 in the non-polluted commune (Mialet), while it fell to 1.1 in the commune of Générargues which is irrigated by the Amous polluted with Pb and Cd. From 1886 to 1906, before the appearance of any pollution (the mine workings were established in 1922), the birth rate in both communes was normal. It would be possible to conclude from this that a correlation exists between the degree of contamination of the river by Pb and Cd

Table 14.7 Birth rates in the local population
From 1886 to 1906

	Number of families	Number of children	Number of children / Number of families
Mialet	24	56	2.3
Générargues	24	58	2.4

From 1926 to 1968

	No. families	No. children	No. children / No. families	No. families with ≤ 3 children	No. families with ≥ 3 children
Mialet	28	59	2.1	3	10
Générargues	27	29	1.1	9	2

and the birth rate, but this result cannot, in our view, be more than an assumption since the size of the sample is statistically insufficient: formal proof of a link between the concentrations of Pb and Cd and birth rates has not yet been rigorously established.

PHYSICO-CHEMICAL MECHANISM OF THE POLLUTION OF THE RIVER WATER AND THE SOIL

The pollution of the river by the run-off from the mine workings originates from two sources: the open-cast mine during dry weather and the slag-heap during wet periods. These sources cause pollution by two different mechanisms which we now study in turn.

Pollution by the Open-Cast Mine

The acidification of the water in the channels and rivers (Table 14.2) is due to the oxidation of S_2^{2-} ions in pyrites and marcasite by oxygen dissolved in the water:

$$S_2^{2-} + 8H_2O \rightleftharpoons 2SO_4^{2-} + 16H^+ + 14e^-$$

$$(\tfrac{1}{2}O_2 + 2H^+ + 2e^- \rightleftharpoons H_2O) \times 7$$

$$S_2^{2-} + (7/2)O_2 + H_2O \rightleftharpoons 2SO_4^{2-} + 16H^+$$

This oxidation is slow because of the irreversible nature of the two redox pairs, but it is greatly speeded up under natural conditions by thio-oxidant bacteria. Note particularly that monomeric sulphide ions S^{-2} do not create any H^+ ions through oxidation.

The overall equation representing the oxidation of FeS_2 depends on the pH.[3] For successively decreasing pH we have:

$$\underset{\downarrow}{\underline{2FeS_2}} + (15/2)O_2 + 7H_2O \rightarrow \underset{\downarrow}{\underline{2Fe(OH)_3}} + 4SO_4^{2-} + 8H^+ \qquad (1)$$

$$\underset{\downarrow}{\underline{2FeS_2}} + (15/2)O_2 + 3H_2O \rightarrow 2FeOH^{2+} + 4SO_4^{2-} + 4H^+ \qquad (2)$$

$$\underset{\downarrow}{\underline{2FeS_2}} + (15/2)O_2 + 1H_2O \rightarrow 2Fe^{3+} + 4SO_4^{2-} + 2H^+ \qquad (3)$$

$$\underset{\downarrow}{\underline{2FeS_2}} + (15/2)O_2 + 3H_2O \rightarrow 2Fe^{3+} + 4HSO_4^- + 2OH^- \qquad (4)$$

Evidence in support of the model above

We have applied three experimental tests which provide evidence in support of this model.

First test

In an environment of very low pH, it can be anticipated from Equation (4) that the phenomenon of acidification will be limited through this very reaction because of the pKa_2 value for H_2SO_4 (2.0). Moreover, at the end of the whole process, reactions (3) and (4), one acidifying and the other producing alkalinity, will lead to a limiting value for the pH. This limiting value, depending on the reaction coefficients, is seen to come down to a determination of the pH of a solution of ferric iron at a concentration C and of SO_4^{2-} ions at a concentration $2C$. Thus, the following system of equations has to be solved:

$$\begin{cases} [SO_4^{2-}] + [HSO_4^-] = 2C \\ [Fe^{3+}] + [FeOH^{2+}] = C \\ \dfrac{[SO_4^{2-}][H^+]}{[HSO_4^-]} = 10^{-2} \\ \dfrac{[Fe(OH)^{2+}][H^+]}{[Fe^{3+}]} = 10^{-2.2} \\ 2[SO_4^{2-}] + [HSO_4^-] = 3[Fe^{3+}] + 2[FeOH^{2+}] + [H^+] \end{cases}$$

The solution of this system of five equations with five unknowns leads by successive approximation to a limiting pH of 1.6, making corrections to ionic activities with the help of the approximate empirical coefficients of Davies.[4] This value is in agreement with the lowest pH that we have found experimentally (the pond R_1A, p. 193).

Second test

It can be seen from Equations (1) to (4) that throughout the process there is a relation between the pH and the quantity of SO_4^{2-} going into solution. This relation can be calculated, and is illuminated in Figure 14.2, where it is clear that the curve does tend towards a limiting pH of 1.6 and that the experimental points are distributed along the theoretical curve in a satisfactory way.

Third test: observed and calculated redox potentials

Three redox pairs are involved in the course of the whole transformation process: Fe^{3+}/Fe^{2+}, O_2/H_2O and As^{V}/As^{III}. Using Nernst's equation and the normal potentials of these three pairs, we have calculated the theoretical potential for different ox/red ratios. In the case of the irreversible pair O_2/H_2O, we have carried out the calculation using the 'reversible' potential $E = 1.22–0.06\,pH$ and the 'irreversible' potential defined empirically:[5] $E = 0.85–0.06\,pH$.

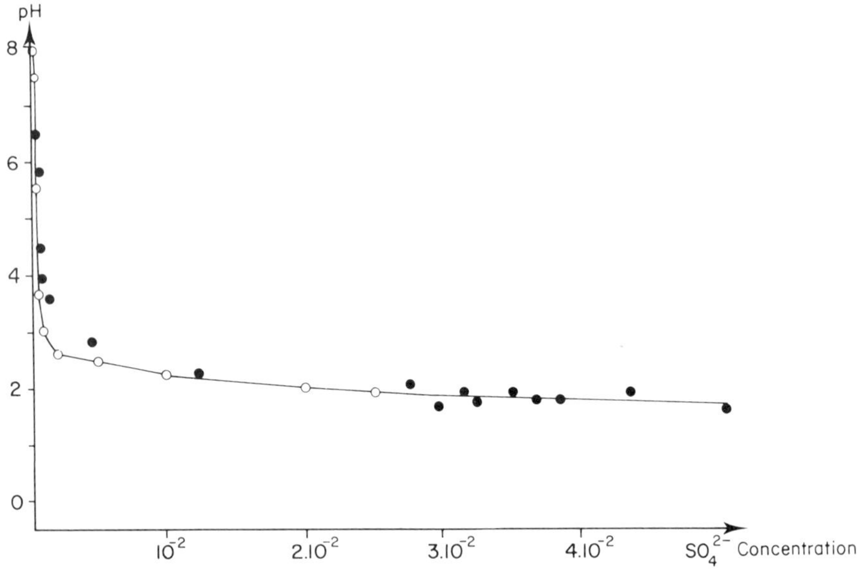

Fig. 14.2 Relation between pH and quantitites of sulphate carried into solution through oxidation of pyrites

Table 14.8 shows the various values thus calculated from these normal potentials for each of the redox pairs as well as the values actually measured using a platinum electrode at different points in the Reigous and the lateral channels.

It can be seen that:

–the calculated and experimental values are in complete disagreement for the As^{V}/As^{III} pair. The measured potential therefore allows no prediction of the state of oxidation of As in solution;

–the values are also in disagreement in the case of the 'reversible' pair O_2/H_2O but are in rather more satisfactory agreement for the 'irreversible' pair. Nevertheless, even in this latter case, the differences between measured and calculated potentials still vary quite randomly from -0.11 V to $+0.03$ V;

–the agreement in the case of the Fe^{3+}/Fe^{2+} pair is satisfactory, the differences varying from 0 to 0.04 V.

We are thus verifying the rule according to which it is always the fast redox system (Fe^{3+}/Fe^{2+}) that imposes its potential in a mixture of slow and fast systems.

Conclusion

The three experimental tests described above are in good agreement with our theoretical model (Equations 1 to 4) and thus provide excellent confirmation that

Table 14.8 Comparison of calculated and measured redox potentials for different redox pairs

Location of sample	*E* measured	*E* calculated			
		E_{Fe}	$E_{O_{2irrev}}$	$E_{O_{2rev}}$	$E_{As_{max}}$
RE_1	0.56	0.56	0.67	1.04	0.36
RE_2	0.57	0.61	0.66	1.03	0.34
RE_3	0.57	0.61	0.66	1.03	0.33
RE_4	0.57	0.61	0.66	1.03	0.33
RE_5	0.59	0.61	0.67	1.04	0.35
RE_6	0.66	0.67	0.67	1.04	0.35
RE_7	0.66	0.68	0.73	1.10	–
R_1	0.75	0.75	0.74	1.11	–
R_2	0.64	0.64	0.75	1.12	–
R_3	0.74	0.76	0.71	1.08	–

the acidification of the waters during the weathering of the open-cast mine is indeed linked to the passing of pyrites into solution through oxidation.

Transport of ions in solution and their trapping

We have seen (p. 193) that from the point of confluence with the Amous, the pH of the water rapidly increases, the redox potential falls and there is a precipitation of $Fe(OH)_3$ as a red sediment. The precipitation of $Fe(OH)_3$ is accompanied by a progressive removal of Pb, Cd, Zn, As and Mn ions. We have therefore attempted to make a quantitative evaluation of this phenomenon.

In order to eliminate the effects of dilution due to the addition of water from the Amous, the Quérier, etc., we have normalized the concentrations of the elements in relation to that of sulphates, whose absolute quantity remains constant after the last of the lateral channels. The values plotted in Figure 14.3 are thus those of the quantity $\bar{X}$, where

$$\bar{X} = 100x[\text{ion X}]_{\text{sample}}/[SO_4^{2-}]_{\text{sample}}$$

Figure 14.3 shows that the order in which the elements are precipitated or trapped is: As, Fe, Zn, Mn:

–arsenic is precipitated abruptly before the confluence with the Amous and hence at pH ≤ 2. We have shown[3] that this phenomenon is due to the fact that $FeAsO_4$ is less soluble than $Fe(OH)_3$ at pH < 2.5, whereas the reverse is true at pH > 2.5;

–iron is precipitated abruptly in steps at each confluence because the redox pair Fe^{3+}/Fe^{2+} is a fast one. The precipitation is caused by the increase in pH at each of the confluences and by the increased supply of dissolved oxygen;

–zinc seems to follow the precipitation of iron. However, since all compounds

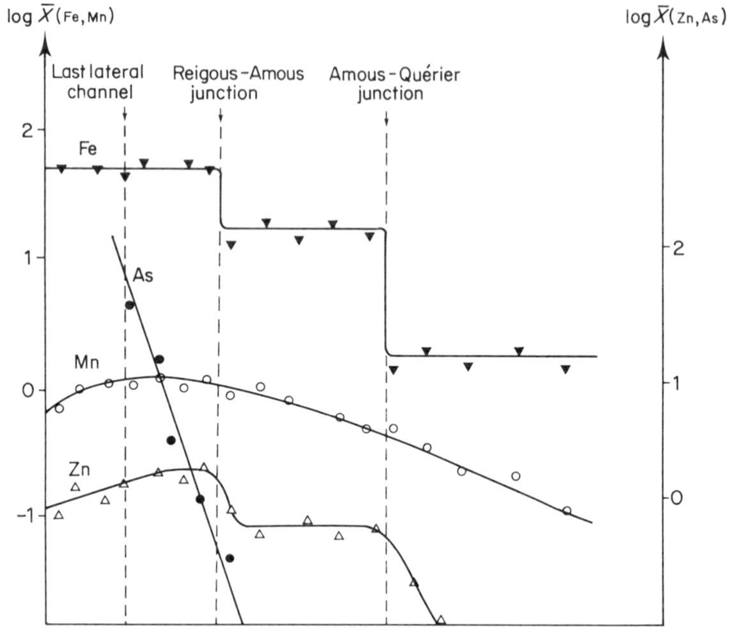

Fig. 14.3 Successive precipitation of metallic elements

of zinc are completely soluble at the pH values found in the river, this can only be caused by the Zn^{2+} ions being carried away through adsorption on $Fe(OH)_3$;

–manganese is also removed from solution but continuously and without any apparent connection with the precipitation of iron. The redox pair Mn^{2+}/Mn^{4+} is less rapid than Fe^{2+}/Fe^{3+}, so that manganese is precipitated more slowly than iron in the form of MnO_2.

We thus maintain that the neutralization of the waters of the Reigous by those of the Amous is manifested by the precipitation of $FeAsO_4$, MnO_2 and, above all, of $Fe(OH)_3$, which, by adsorption or coprecipitation, carries away not only Zn but Cd and Pb as well. The $Fe(OH)_3$ thus plays the role of a powerful purifying agent, so that it is not the river which is toxic but the suspension of $Fe(OH)_3$ that it carries along. This suspension, rich in polluting elements, is deposited on the soil of the meadows and vegetable plots irrigated by the river. There it is slowly remobilized by combination with oxalic, citric or malic acid liberated by plant litter in the soil.

Pollution by Deposition of Sand from the Slag-heap

During wet periods, some of the sand from the slag-heap is carried away by the river and deposited on the surface of flooded land. The pollution processes which then occur are different for undisturbed soil (in meadows) and for soil that is cultivated by digging or ploughing (in vegetable plots).

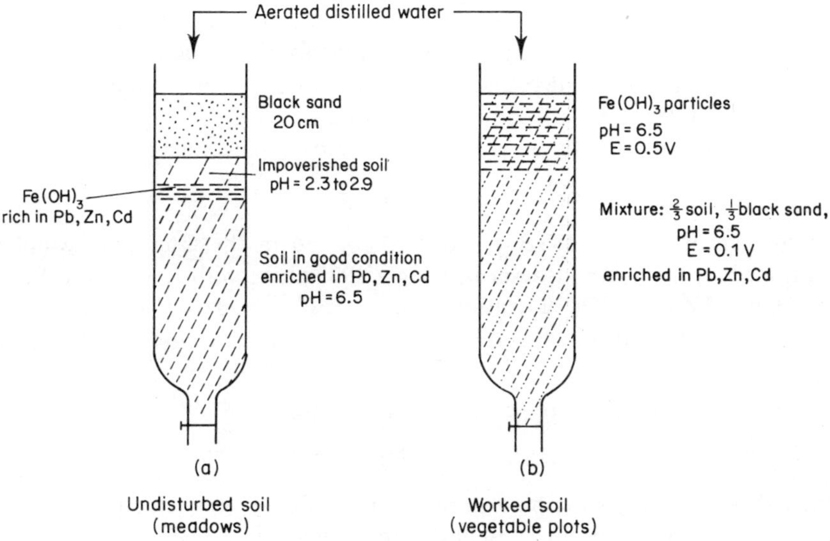

Fig. 14.4.

Meadows

Sands rich in pyrites are deposited here to a thickness of between 5 and 20 cm and the process of oxidation of the pyrites with liberation of H^+ ions then begins (see p. 196). This acidification activates Zn, Cd, Pb and neutralizes the alkaline reserve in the underlying soil. The laboratory experiment shown in Figure 14.4(a), carried out in a 10 cm diameter glass column, is designed to simulate this effect by varying the respective quantities of soil in good condition and black sand from the slag-heap. In this way, it is observed that, for a 20 cm layer of black sand, the alkaline reserve of the soil is neutralized over about 5 cm (pH = 2.3 to 2.9) and that this acidified portion is separated from the soil in good condition by a layer of $Fe(OH)_3$ enriched in Zn, Pb and Cd. This phenomenon must be similar to processes that take place in the formation of what soil scientists call the B horizon of podzols. When the whole of the sandy black layer has become whitish and the migration towards the bottom of the $Fe(OH)_3$ layer has come to an end, the transformation is complete.

It can be seen that, under these conditions, the survival of plants with shallow roots (meadow grass) is severely jeopardized; plants with deeper roots, on the other hand, will survive but will be enriched in Cd, Zn and Pb.

Vegetable plots

Here, the less plentiful sandy deposits, unlike those occurring in the meadows, are incorporated in the soil when it is dug or ploughed over. At any location, the proportion of soil is then much greater than that of the dispersed pyrites, so that the acidification of the soil that takes place is not enough to neutralize its alkaline reserve: the pH of the interstitial water remains fixed at around 6.5.

The redox potential decreases gradually after several centimetres of infiltration. Under those conditions, the oxidation of FeS_2 cannot reach the SO_4^{2-} stage and it stops at the intermediate thiosulphate stage:

$$FeS_2 + (3/2)O_2 \rightarrow S_2O_3^{2-} + Fe^{2+}$$

The concentration of the $S_2O_3^{2-}$ ions so formed in the interstitial water is determined by the method described on p. 192, along with that of the Fe^{2+}, Pb^{2+}, Cd and Zn ions. The activation of the toxic Pb^{2+} and Cd^{2+} ions is brought about by the formation of stable soluble compounds whose constants are known:[6,7]

$$Pb(S_2O_3)_3^{3-}: \log \beta = 6.34$$
$$Cd(S_2O_3)_3^{-}: \log \beta = 6.05$$

Under these conditions the plants are heavily polluted but their survival is not jeopardized because the environment is not sufficiently acidic.

The phenomenon could thus be summarized as follows:

–in a highly oxidizing environment, the acidification produced kills the plant;
–in a less oxidizing environment, the formation of $S_2O_3^{2-}$ ions activates toxic ions by complexation and the latter are then concentrated in the plant.

RECONSTRUCTION OF THE HISTORY OF POLLUTION IN THE REGION USING THE CONCENTRATIONS OF ZINC AND CADMIUM IN THE ANNUAL GROWTH RINGS OF CONIFERS[8]

We chose 6 pines (*Pinus sylvestris*) located between 5 and 25 m from the banks of the Reigous:

–P_1, P_2, P_3, situated downstream from the junctions with the lateral channels R_1, R_2, R_3 from the open-cast mine;
–P_4, P_5, P_6, upstream from these junctions;
–P_0, sufficiently far (300 m) from the Reigous to be considered as unpolluted.

From each of these 7 pines we took core samples using a core drill and picked out the growth rings perpendicular to the axis of the core by X-ray radiography. Each ring was then divided up for the determination of Pb, Zn and Cd ions by atomic absorption spectrometry. Results obtained for zinc are shown in Figure 14.5.

The peaks that can be observed in 1958 and 1963 for P_1, P_2 and P_3 coincide with two great floods that are well known in the climatic history of the region. In contrast, the pines P_4, P_5 and P_6 located upstream from the junctions with the channels from the open-cast mine show no comparable peaks.

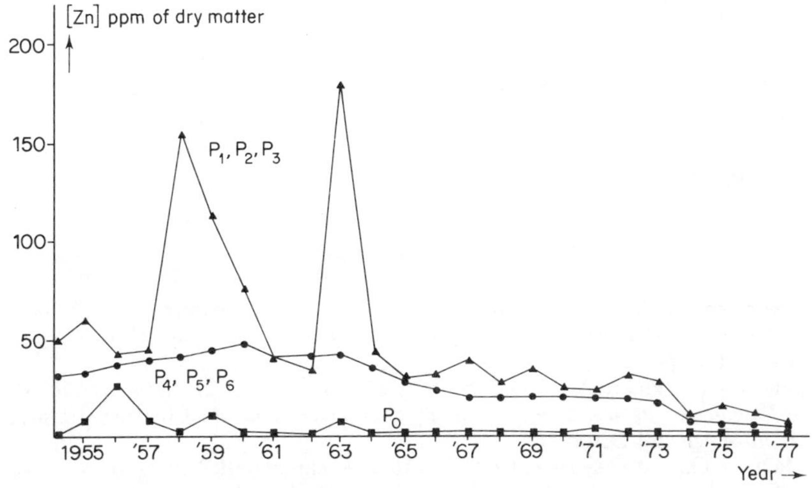

Fig. 14.5

The concentration of Cd shows similar variations. Lead, on the other hand, with a less soluble sulphate, shows no significant variations over that period.

In conclusion, it can be seen that since 1965, when mining activity ceased, the concentration of zinc has slowly decreased. An interpretation of these phenomena has been given in a recent publication.[9] It should be remembered that zinc and cadmium can be excellent indicators recording the history of local pollution and of fluctuations in regional water flow due to climatic variations (the dates of great floods and mine closures).

CONCLUSION

We have explained what physico-chemical processes are involved in the passage of Pb^{2+}, Zn^{2+} and Cd^{2+} ions into solution through the weathering of a mine working, by what means they are carried into the waterways, the method by which they are captured in the soils, and the degree of pollution in plants growing in these soils, as well as the effects of this pollution on birth rates in the local population. We have shown, finally, that the concentration of zinc in the annual growth rings of conifers irrigated by the polluted rivers forms an excellent chronological indicator of local pollution and microclimate. Two effects, however, would merit a deeper study: first, the nature of the phenomenon responsible for the carrying away of Pb^{2+}, Zn^{2+} and Cd^{2+} ions by iron hydroxides. Is it a precipitation, an adsorption, or a coprecipitation of basic salts or of mixed hydroxides? Such a study would probably make it possible to control the parameters responsible for decontamination, whether spontaneous or caused by natural waters. Secondly, the relation between the decrease in the birth rate of the population and pollution by Pb and Cd, since our investigation, as we have seen, forms no more than a strong suggestion and is certainly not a formal proof that such a relation exists. In particular, it would be desirable to carry out experiments similar to those of Howard *et al.*[10] on the litters of rats and mice whose food contained concentrations of Pb or Cd similar to those we have encountered in the present study, and above all to examine the long-term effects of both Pb and Cd together.

REFERENCES

1. Faucherre, J., Pinart, A.M., and Dutot, A. (1979). *Mécanisme biogéochimique de contamination des végétaux par le plomb, le cadmium et le zinc.* C. R. Contrat C. Comm. Europe, January.
2. Faucherre, J., Pinart, A.M., and Dutot, A. (1978). *Contamination des eaux, des sols et des végétaux par le plomb, le cadmium et le zinc.* C. R. Contrat Ministère de l'Environnement, October.
3. Michard, G., and Faucherre, J. (1970). Etude géochimique de l'altération des minerais sulfurés, *Chem. Geol.*, **6**, 63.
4. Davies, C.W. (1962). *Ion Association*, Butterworth, London.
5. Baas Becking, L.G.M., Kaplan, I.R., and Moore, D. (1960). Limits of the natural environment in terms of pH and oxidation–reduction potentials, *J. Geol.*, **63**, 243.
6. Desideri, P.G., and Penani, F. (1959). *Gazzetta*, **89**, 1349.
7. Pryszczewska, M. (1959). *Roczniki chem.*, **33**, 755.
8. Faucherre, J., and Dutot, A. (1976). Le zinc et le cadmium des anneaux de croissance des végétaux en tant que traceurs chronologiques de pollutions d'un cours d'eau, *C.R. Acad. Sci.*, **282**, 187.
9. Faucherre, J., and Dutot, A. (1979). Le plomb et le zinc des anneaux de croissance des conifères en tant que traceurs chronologiques de la pollution atmosphérique, *C. R. Acad. Sci.*, **288**, 1283.
10. Howard, D. *et al.* (1971). Reproductive ability and progeny of F1 lead-toxic rats, *Fertility and Sterility*, **27**, 785.

Pollutants and Their Ecotoxicological Significance
Edited by H. W. Nürnberg

CHAPTER 15

Toxic Metal Levels in the River Rhine

R. BREDER, H. W. NÜRNBERG, J. GOLIMOWSKI AND
M. STOEPPLER

SUMMARY

The levels of the heavy metals, cadmium (Cd), lead (Pb) and mercury (Hg), were determined in water and sediments of the River Rhine and the main tributaries from Lake Constance to the Dutch/German border within rather a dense sequence of sampling stations. The samples were taken under particular precautions to avoid accuracy deficiencies of the data due to contamination. Depending on the element, its concentration and matrix differential pulse anodic stripping voltammetry (DPASV) and flame or flameless atomic absorption spectrometry (AAS) were applied. The accuracy of the results was confirmed by using reference materials for comparison measurements and by the application of DPASV and AAS for the same sample. Both methods yielded within a satisfactory range coinciding data.

The anthropogenic heavy metal pollution increased downstream from Lake Constance to the Dutch/German border with some superimposed peaks. At least in one of the two years, 1977 or 1978, the highest Cd, Pb and Hg concentrations in River Rhine sediments occurred at Emmerich. The metal enrichment factors in the sediment from Emmerich in 1978 in comparison with the geochemical background levels were: Cd, 48.8; Hg, 22.4; and Pb, 17.6. The sediment data revealed a downstream shift of the maxima from 1977 to 1978. It could not be ascertained that the levels decreased in general in 1978. With regard to the tributaries the Lippe, the Wupper and the Main showed a high Hg pollution of the sediments. Also in the water of the Main exceptional high total Hg concentrations up to 401 $ng\,l^{-1}$ were observed. The background level for total Hg was about 5 $ng\,l^{-1}$. For Cd in sediments the Ruhr tributary showed a maximum in 1977 (30 $mg\,kg^{-1}$, DW), for Pb also Lahn and Moselle (600–750 $mg\,kg^{-1}$, DW). With respect of water contents 94.4 $\mu g\,l^{-1}$ total Cd were determined in the Wupper tributary and $37.2\,\mu g\,l^{-1}$ total Pb in the Erft tributary. In case of such a pollution mostly the dissolved freight prevailed for Cd. For Pb, however, the suspended particulate matter freight predominates distinctly.

INTRODUCTION

Toxic metals have become an ecotoxicological hazard of prime importance and still-growing significance. They are, therefore, a target of high priority in ecochemical and ecotoxicological research on their level, fate, behaviour, pathways and actions in the various parts of the environment. Among the metals with high ecotoxical potential attention is to be focused in the first place on cadmium (Cd), lead (Pb) and mercury (Hg). The insidious ecotoxical hazard of these toxic metals originates from their tendency to accumulate in vital organs of man and animals. In this manner also the continuous exposure to rather low doses of toxic metals is of concern, because they cause, according to their accumulation degree, progressively increasing toxic actions over long periods of the life span of an individual.[8,17] Therefore, an important task of indispensible environmental protection is to keep the anthropogenic emissions of the toxic metals into the environment as low as possible. This refers particularly to Cd, Pb and Hg. A key prerequisite for an efficient and reliable fulfillment of this task of environmental protection is an efficient surveillance of toxic metal levels in the environment by reliable and sensitive analytical procedures providing meaningful data and intense research to deepen and complete our still rather incomprehensive knowledge on the ecochemistry and ecotoxicology of toxic metals.[18] Among the environmental chemicals of ecotoxicological concern the toxic metals are almost the only pollutant class without biodegradability. Instead they pass the environment in a biogeochemical cycle with relative accumulations and different residence times in the various types of ecosystems and with transformations to sometimes more or less toxic chemical species.

In the densely populated and highly industrialized regions of Western Europe at many locations, alarming amounts of toxic metals are emitted from anthropogenic sources on soils or into the aquatic environment. A particular burden of toxic metals is carried by the major waterways, such as the River Rhine,[6,7] which is the source of drinking-water production for millions of people.[20] The metal burden of large and busy waterways such as the River Rhine is not only of ecotoxicological concern for the river itself and its neighbouring region, but also for its estuary and the coastal waters of its discharge area where the marine environment is, with respect to the relatively shallow water depth, particularly vulnerable. Therefore, a study has been undertaken on the toxic-metal burden of the River Rhine in summer 1977 and 1978 between Lake Constance and the Dutch/German border at Emmerich operating a rather dense sequence of fifty-three sampling stations.[4] The findings for Cd, Pb and Hg will be presented in this paper. At each station samples of water and sediments have been collected. The sediments act as traps and depots of the toxic metals in an aquatic ecosystem and thus reflect the degree of the toxic-metal pollution burden. From the sediment depots toxic metals will be transferred to the dissolved state if a deviation occurs from their dissolved steady-state concentrations and if solubilizing substances, e.g. natural (humates) or synthetic chelators (e.g. NTA from detergents), interact with the sediments.

Generally the aqueous phase has important functions with regard to the fate of toxic metals in natural waters. In a river the most efficient transport of toxic metals downstream is related to the dissolved metal content and the metal amounts sorbed or bound to fine suspended particles although the 'bed roll' also moves downstream. Moreover, the dissolved state has particular significance for the transfer of metals to suspended matter, sediments and aquatic organisms while the suspended particles also play an important role as metal carriers in vertical metal transport towards the sediments and from them due to upwhirling of suspended matter. Therefore, for each station the contents of the studied toxic metals have been determined in the dissolved state and in the suspended material of the water sample as well as in sediment phase to achieve a complete information on the situation.

EXPERIMENTAL

During the last decade combined experience has clearly revealed that the attainment of reliable data on toxic-metal levels in rivers and other types of natural waters is by no means an easy task but a rather demanding application of trace analytical chemistry. Suitable analytical procedures, well adapted to the task and performed in every stage from sampling over sample pretreatment to determination, with great care, to avoid systematic errors by analytical contamination or trace-metal losses during the analytical procedure, are of fundamental importance to achieve meaningful data.

As determination methods in this study the two most versatile and common approaches in toxic-metal trace ecochemistry, differential pulse stripping voltammetry (DPSV) and atomic absorption spectrometry (AAS), have been applied.[18]

Sampling bottles and storage flasks from polyethylene were washed with tap water and triply distilled water, then soaked under intermittent shaking for 14 days in 2 M HCl, Merck suprapur, and after multiple rinsing with triply distilled water and drying, finally wrapped in dustfree polyethylene bags to be opened only at the sampling point.

Brown glass bottles for the storage of water samples for Hg determination were subjected to a similar cleaning procedure using 2 M HNO_3, Merck suprapur. The membrane filters serving to filter off suspended matter were soaked for 14 days as well in 1 M HCl, Merck suprapur, under exchange of the solution after the first week, rinsed with triply distilled water and stored in distilled water, airtight, until they were taken out with plastic covered tweezers for use. All other laboratory equipment underwent scrupulous cleaning as described in detail elsewhere.[9,10]

SAMPLING

Water samples were taken from the river bank using such facilities as groynes to reach out into the river. The collection of water samples was carried out manually by submerging the polyethylene bottle about 30 cm under the surface. The bottle was attached to an aluminium pole by a plastic device. Sediment samples were taken with a plastic bottom grab, specially fabricated by Hydrowerkstätten, Kiel. Only the top 5 cm of the sediment were taken for the sample.

SAMPLE PRETREATMENT AT THE SAMPLING SITE

Some pretreatment manipulations with the samples were carried out immediately after sampling at the station in a mobile laboratory equipped with a clean bench.[23] The water samples were filtered, dustfree, in the Sartorius apparatus, SM 16511, through membrane filters with 0.45 μm pore size under nitrogen pressure.[9] Usually 500 ml of river water were filtered using the first 250 ml for rinsing of the storage bottle and the second 250 ml portion as the actual sample for the determination of the dissolved metal concentrations. Subsequently 250 μl HCl, Merck suprapur, were added to stop biological activity of bacteria and chemical after-reactions of the dissolved trace metals. For further storage and transport the filtrate was then frozen to $-18\,^{\circ}$C. The filtered off suspended matter samples, with the filter and the sediment samples, were frozen without further pretreatment.

SAMPLE PRETREATMENT IN THE LABORATORY

The respective pretreatment procedures depend on the sample type and had to be adapted to the subsequent determination method. All reagents were of the quality grade 'Merck suprapur'. For the determination of Cd and Pb by graphite furnace AAS the acidified river-water filtrates required no further pretreatment after thawing. Only the Hg-determination by cold vapour AAS made a prior digestion of dissolved organic matter (DOM) necessary. This was achieved by heating the sample at 95 °C for 3 h after the addition of 30 ml half-concentrated H_2SO_4 and 10 ml of 3 per cent $K_2S_2O_8$ solution to 800 ml filtrate.

For the voltammetric measurements a prior decomposition of DOM is mandatory, because certain components tend to bind dissolved trace metals in rather inert complexes or chelates. The DOM decomposition was achieved without contamination risks photolytically by ultra-violet irradiation. Quartz vessels (40 ml) were conditioned with acidified filtrate (pH 2) overnight. Then 10 g of fresh acidified filtrate were put into the vessel, 10 μl of 30 per cent H_2O_2 was added to speed up decomposition and the solution subjected for 3 h to ultraviolet irradiation by a 500 W mercury vapour lamp placed 15 cm above the vessel and covered by a quartz glass with a lid. This quartz vessel was later used as a voltammetric cell.

The filtered-off suspended matter was subjected, with the membrane filter to wet digestion in 100 ml quartz flasks. After the addition of 5 ml of HNO_3 the flask was heated and the solution gently vaporized to dryness. The residue was dissolved in 5 g of acidified water (100 μl HCl per 100 ml triply distilled water) and the solution put into a 40 ml quartz flask. After rinsing the digestion flask with further 5 g of acidified water a weight of 10 g analyte was obtained in the quartz flask which was then used as voltammetric cell.

After thawing sediment samples were first freeze dried in the Beta I device from Christ, Osterode, FRG, putting the sediment sample in PTFE-covered Cr-Ni-dishes. A loose substance, made suitable for fractionation by sieving through plastic sieves, resulted. The fraction with particles below 200 μm diameter was taken for analysis in accordance with the practice of other experimenters.[1-3,16] The weight loss, due to drying at 105 °C for 2 h, was usually determined to 0.5–2 per cent. The digestion was performed with aliquots of 0.2–1 g of the sieved sediment fraction (< 200 μm particles) in PTFE vessels by adding 4–8 ml HNO_3/HCl (3:1, inverse aqua regia).[5] The PTFE vessels were put into a stainless-steel device[22] and heated to 150 °C for 3 h. After cooling to room temperature the digestion solution was filtered under the addition of triply distilled water through the filter MN 615 1/4 (Machery and Nagel, Düren, FRG). The filter had been rinsed several times with HCl and triply distilled water. The remaining silicate residue was washed with triply distilled water. As Cd, Pb and Hg of environmental pollution origin are concentrated at the surface of the sediment particles this digestion is sufficient and preferable to the usage of hydrofluoric acid which would introduce accuracy risks because of the matrix increase. The relatively small amount of acid required for the HNO_3/HCl digestion, however, causes favourable conditions for subsequent AAS determination. The satisfactory performance and accuracy of the used digestion procedure with inverse aqua regia has been proved by successful participation in interlaboratory comparisons with a River Rhine and an estuarine sediment.[15,24]

DETERMINATION PROCEDURES

With respect of the matrix and the concentration the toxic metals were determined with different modern modes of voltammetry and atomic absorption spectrometry.

The rather low dissolved concentrations of Cd and Pb in filtrated river water were determined by differential pulse anodic stripping voltammetry (DPASV) at a mercury film electrode (MFE) formed *in situ* at a glassy carbon electrode surface during the cathodic deposition stage. For this purpose the analyte was spiked with Hg^{2+} adjusting a concentration of about 2×10^{-4} M. This ultrasensitive method of which details are outlined elsewhere[11–13] has a detection limit of at least 1 ng l^{-1}, amply sufficient for the determination of these and other trace metals in natural waters. The determinations were performed under a clean bench with laminar air flow, class 100, minimizing accuracy deficiencies due to contamination from the laboratory atmosphere.

The determinations of dissolved Cd in water have also been performed by graphite furnace AAS (Perkin-Elmer 410) in order to establish with a substantial number of river-water samples that both the most common methods in toxic metal ecochemistry, voltammetry and atomic absorption spectrometry, yield coinciding results. Mercury was exclusively determined by cold vapour AAS in all matrices, i.e. water, suspended matter and sediments.

Total Hg in water was obtained by acidifying an aliquot of the unfiltered sample with HNO_3 on the sampling spot, afterwards treating the sample in the laboratory with $K_2S_2O_8$ as mentioned before, and, after cooling of the sample, reducing the Hg(II) with $SnCl_2$ to elemental Hg, which was then determined by AAS in the cold vapour mode after prior accumulation on silver wool.[15] For the determination a Perkin-Elmer 410 spectrometer in connection with a HGA 74 and a Brooks Instrument gas control unit were used.

In the analyte obtained after the digestion of suspended matter the concentrations of Cd and Pb have been determined by DPASV at the hanging mercury drop electrode (HMDE). The voltammetric parameters were: Cathodic deposition potential $E_d - 1.2$ V *vs* Ag/AgCl reference electrode, deposition time 3–5 min under stirring (600 rpm), stripping stage with pulse height 50 mV, pulse duration 57 ms, clock time of pulses 0.5 s. The polarograph was a PAR 174A. The concentrations were evaluated by two standard additions.

The concentrations of the studied toxic metals in sediments were determined by AAS, performed with the analytes obtained after the digestion of the sediment samples. Frequently the Cd and Pb concentrations in the analyte were sufficiently high (0.05–5 mg l^{-1}) to use flame AAS with air/acetylene flame (Instrument Fisher-Jarrell-Ash, Mod. 811). In case of lower concentrations graphite furnace AAS was applied. The accuracy of the AAS determinations in the sediments was established by participations in an interlaboratory comparison with a Rhine and an estuarine sediment. Details of the interlaboratory measurements and results are outlined elsewhere.[1,5]

RESULTS AND DISCUSSION

Sampling Areas and Sites

The River Rhine is the most busy waterway in Europe with respect to commercial navigation. One-third of the total population of the Federal Republic of

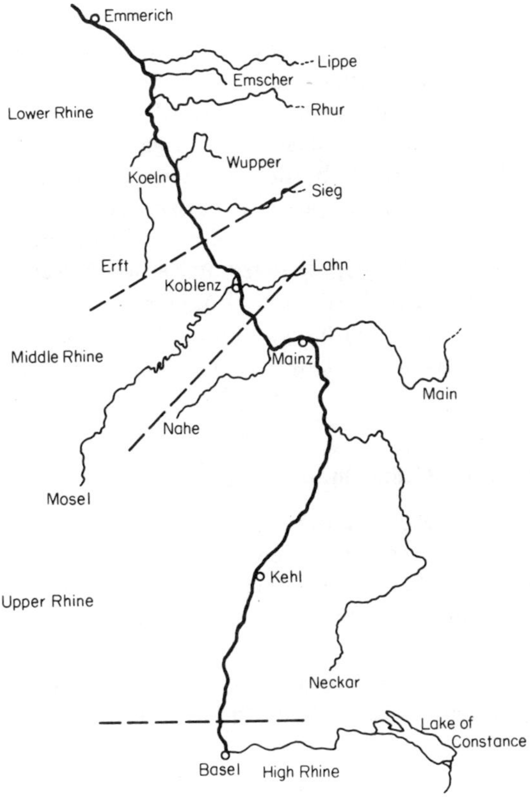

Fig. 15.1 The course of the River Rhine and of the studied tributaries

Germany (20 millions) live in the Rhine basin at a density of 500 inhabitants per square kilometre which is twice as high as the average in the country. In some regions there are remarkably high and, compared with the rest of the country, overproportional accumulations of industry, particularly iron and steel, chemical and various types of metals-producing and -processing industries. Other regions of the Rhine valley are used for agriculture, particularly wine and fruit growing with extensive use of fertilizers, and on some locations also of sewage sludge. Thus, there is a great density of many potential sources of municipal, industrial and agricultural origin for toxic metal emission into the river.[20]

Sediment samples were collected during August of 1977 and 1978. Forty sampling stations were distributed over the four main hydrological sections of River Rhine (Figure 15.1). In Lake Constance and at the High Rhine until Basle samples were collected at seven stations. In the next section of the Upper Rhine, between Basle and Mainz, eleven stations were located while the following section of the Middle Rhine until Bonn had nine sampling stations. In the last section of the German Lower Rhine until the Dutch/German border (Emmerich) were thirteen sampling stations located. Industrial and urban agglomerations are the Swiss chemical industry at Basle, the area of Karlsruhe, the regions of Mannheim/Ludwigshafen and Mainz, the urban region of Bonn/Cologne with the chemical industry belt around Cologne and the urban and heavy industrial agglomeration bordering the Rhine between Düsseldorf and the Lippe tributary. In addition samples have been collected in the mouth region of eleven tributaries to obtain information on their contribution to the toxic-metal pollution of the River Rhine. These tributaries were Neckar, Main, Nahe, Lahn, Moselle, Sieg, Wupper, Erft, Ruhr, Emscher and Lippe. Although the sampling stations were in the mouth region of the tributaries, they were chosen sufficiently upstream to prohibit mixing with Rhine water by back flow from the Rhine.

Dissolved and Particulate Cadmium in Water

The profiles of the Cd-levels in the dissolved state and at the suspended particulate matter along the course of the German part of the Rhine are given in Figure 15.2. It has to be emphasized that the concentration of Cd bound to suspended particulate matter refers also to the water volume and is given therefore in $\mu g\,l^{-1}$.

From Lake Constance to the beginning part of the Lower Rhine where, at Köln-Sürth, the southern boundary of the chemical industry belt around Cologne is reached, the dissolved Cd level remains rather low with values usually between 0.03 and 0.05 $\mu g\,l^{-1}$. It corresponds along the whole High, Upper and Middle Rhine to the Cd level in the coastal waters of the Southern Bight (Belgian and Dutch coast) and the German Bight of the North Sea and its highest average values of surface waters exceeds those of the open Atlantic by not more than a factor of 10.[11,14]

Reaching the southern part of the Cologne chemical industry belt the situation changes dramatically (see Figure 15.2). It is not only characterized by a very significant elevation of the Cd level, but also by substantial fluctuations between rather small distances. This behaviour reflects obviously the action of significant Cd discharges located in rather limited areas ('point sources'). An extremely high peak value was observed with 88.2 $\mu g\,l^{-1}$ in the mouth region of the Wupper tributary, 200 m upstream from the Rhine. In the River Rhine the Cd pollution originating from the Wupper tributary decreases downstream, but 20 km northwards it still amounts to 8 $\mu g\,l^{-1}$ and even at a distance of 40 km at Neuss it still has a level of 1.5 $\mu g\,l^{-1}$.

Further downstream the dissolved Cd level becomes again relatively low and nearly comparable with that in the High, Upper and Middle Rhine. This finding suggests that no substantial Cd-discharge sources act in this region downstream from the mouth of the Wupper tributary. It is remarkable that the River Ruhr

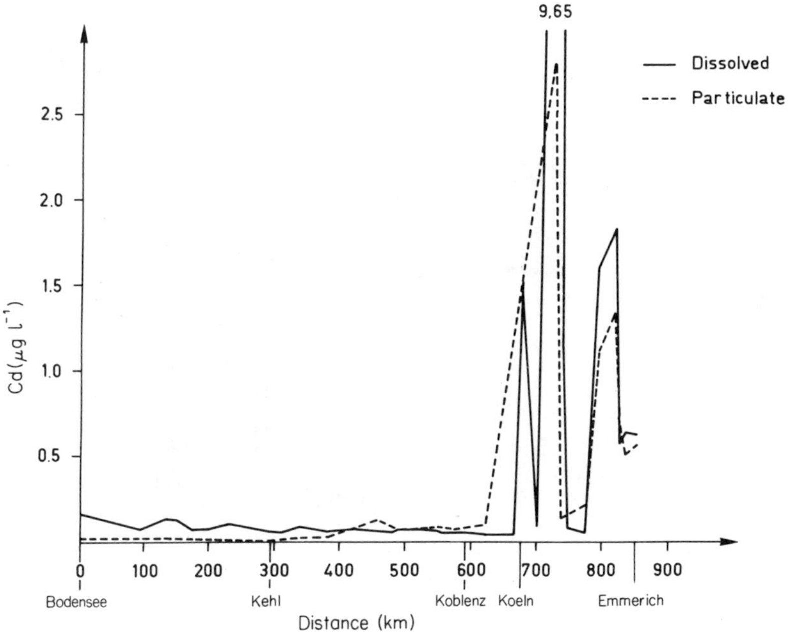

Fig. 15.2 Dissolved and particulate cadmium in River Rhine water

contributes virtually no elevation to the dissolved Cd in the River Rhine during the sampling time. A substantial elevation to about $2\,\mu g\,l^{-1}$ occurs again at Walsum and Wesel. The dissolved Cd levels remain relatively moderate in the mouth regions of the tributaries Emscher and Lippe which can be therefore excluded as significant contributors to the elevated dissolved Cd level in this section of the River Rhine. Downstream from Wesel to Emmerich at KM 848 the level of dissolved Cd remains with 0.6–$0.76\,\mu g\,l^{-1}$ rather constant and a factor 15–19 higher than the average of $0.04\,\mu g\,l^{-1}$ observed for the larger part of the Rhine between Lake Constance and Cologne.

Down to 400 km, near Speyer, the Cd bound to suspended particulate matter contributes only negligibly to the total Cd-burden per litre of water in the Rhine. Further downstream the Cd in suspended particulate matter and dissolved in water shows almost equal levels, the precise ratio is 1.12:1. After 620 km, at Brohl, the Cd content in suspended matter increases progressively until it reaches a peak value of $2.8\,\mu g$ particulate $Cd\,l^{-1}$ water downstream from the mouth of the Wupper tributary. Between Neuss and Duisberg-Rheinhausen the particulate Cd level falls rapidly to about similar levels as in the Middle Rhine region and begins to increase further downstream to a second peak value of about $1.4\,\mu g\,l^{-1}$ at Wesel falling again to about $0.6\,\mu g\,l^{-1}$ at Emmerich.

Also in the Lower Rhine region the Cd levels in the dissolved state and at

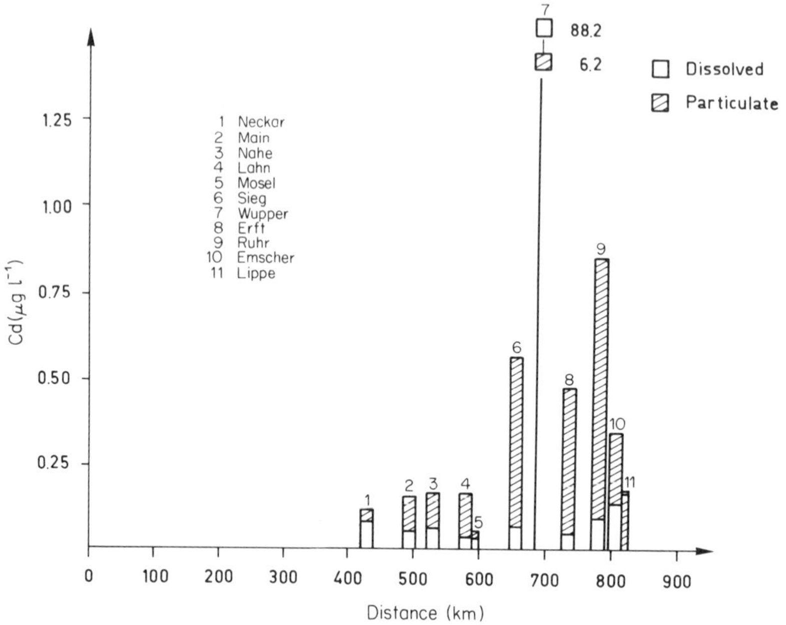

Fig. 15.3 Dissolved and particulate cadmium in the water of the tributaries

suspended particulate matter remain comparable, i.e. per litre of water nearly equal Cd freights are transported in both phases.

Figure 15.3 reports the Cd levels in the dissolved state and bound to suspended particulate matter per litre of river water in the mouth region of eleven tributaries. In general the distribution of Cd between both phases is quite different in the tributaries from that in the Rhine. With the exception of Neckar, Moselle and Lippe where the Cd levels in the dissolved state and bound to suspended particulate matter per litre of river water are comparable, in all other tributaries the Cd amount bound to suspended matter exceeds significantly the Cd level in the dissolved state. This phenomenon seems to be connected with the relatively higher suspended-matter freight of those tributaries compared with the Rhine. On the other hand the total Cd freight in all tributaries, with the exception of the River Wupper, remains below the total Cd freight of the Lower Rhine towards Emmerich. The relatively higher total Cd levels are found in the mouth region of Sieg, Erft, Ruhr and Emscher, while, as noticed before, an extraordinary peak value is observed for the Wupper. Moreover, in this tributary the major contribution is derived from the dissolved state, although a smaller Cd amount, by a factor of 13, contained in suspended particulate matter is by far the highest content found for all tributaries. The total Cd freight in the mouth region of Neckar, Main, Nahe, Lahn and Lipper is of a rather similar order as observed for the High, Upper and Middle Rhine.

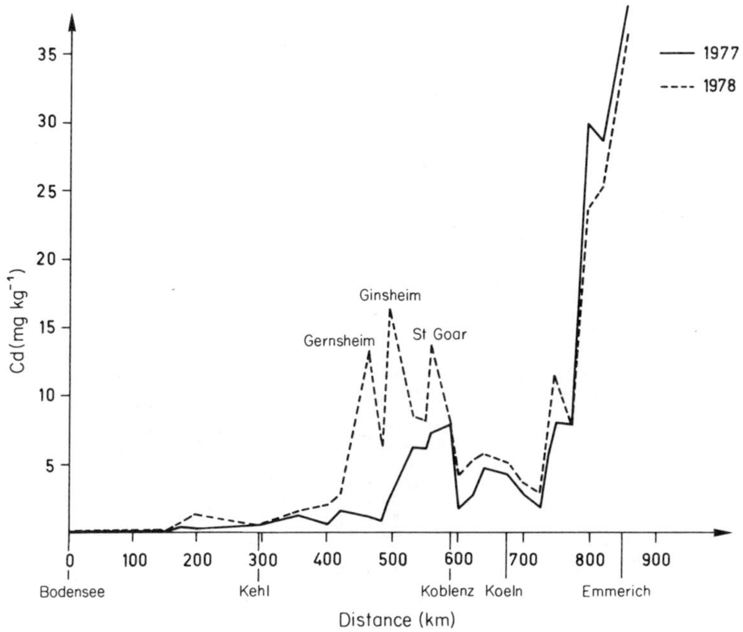

Fig. 15.4 Cadmium in River Rhine sediments

Cadmium in the Sediments

The profiles determined in 1977 and 1978 show no significant differences except in the Middle Rhine section, where, at three sampling sites, remarkably higher levels were found in 1978 with 13.3 mg kg^{-1} at Gernsheim, 16.3 mg kg^{-1} at Ginsheim and 13.7 mg kg^{-1} at St Goar (Figure 15.4). As the profile for 1977 reveals the first significant increase occurs downstream of Ginsheim between 500 and 600 km (north of Koblenz). An old bend of the Rhine River is situated at Ginsheim ('Altrhein') where a cadmium pollution up to 95 mg kg^{-1} has been observed.[6] After KM600 there follows a decrease until, at 785 km (Zons), a steep, rather monotonic increase from 1.85 mg kg^{-1} at Zons to 38.6 mg kg^{-1} at Emmerich (852 km) occurs. The latter value exceeds the Cd level typical for the Upper Rhine by a factor of about 30.

Figure 15.5 shows the Cd levels for 1977 and 1978 in sediments from the mouth region of the tributaries. The Cd levels are quite different for both years in most of the tributaries. Only for Neckar, Lahn and Lippe the differences between both years are rather small. For Moselle, Wupper, Ruhr and Emscher a higher Cd level is observed in 1977 than in 1978 while in sediments from Main, Nahe and Sieg the Cd level is more elevated in 1978 compared with 1977.

For Nahe, Lahn, Moselle, Sieg, Erft, Emscher and Lippe the Cd level in the sediments amounts for both years to about 5 mg kg^{-1} with a range of

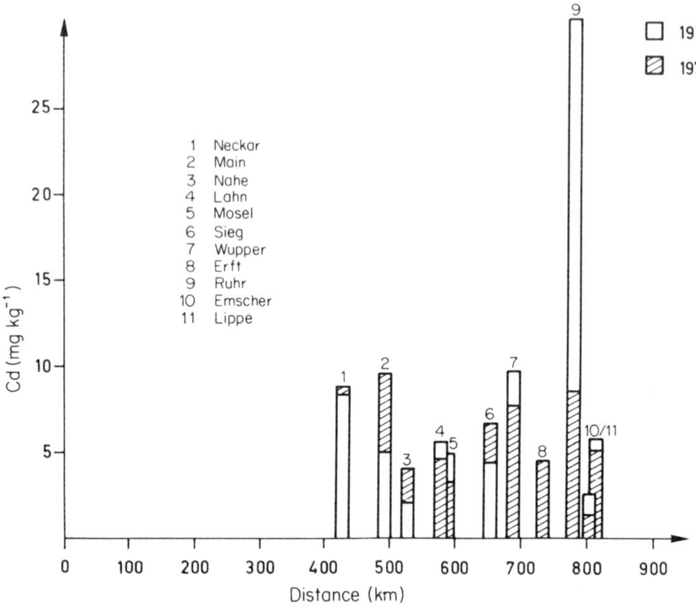

Fig. 15.5 Cadmium in tributary sediments. (*Note:* Metal levels for the different years always count from the bottom of the diagrams)

4–7 mg kg^{-1}. In the Nahe the Cd level is, however, a factor 2 smaller for 1977. Particularly low levels are observed in the Emscher amounting in 1978 to somewhat more than 1 mg kg^{-1} only. In Neckar, Main and Wupper the Cd level approaches the 10 mg kg^{-1} value. While this is the case in Neckar for both years it applies for Main only in 1978 and for Wupper only in 1977. In 1977 also the Main concentration remains about 5 mg kg^{-1}. Exceptionally high is in 1977 the level in the Ruhr sediment with 30.3 mg kg^{-1} while it falls to 8.35 mg kg^{-1} in 1978 at the same sampling site. This Ruhr value is comparable with those for Neckar, Main and Wupper in the same year. With the exception of the Ruhr value for 1977 the Cd levels in all the tributary sediments remain below that of the Rhine sediment downstream from the mouth of the Ruhr. In this context it should be noted again that, with respect to the sediment data, always the top 5 cm of the upper sediment layers have been taken.

DISSOLVED AND PARTICULATE LEAD IN WATER

As a main result of our studies with regard to Pb it was found that the highest amount occurs in the suspended-matter freight (Figures 15.6 and 15.7). Therefore an analytical separation of the suspended matter from the water body is absolutely necessary. The dominating tendency of Pb to be bound in the

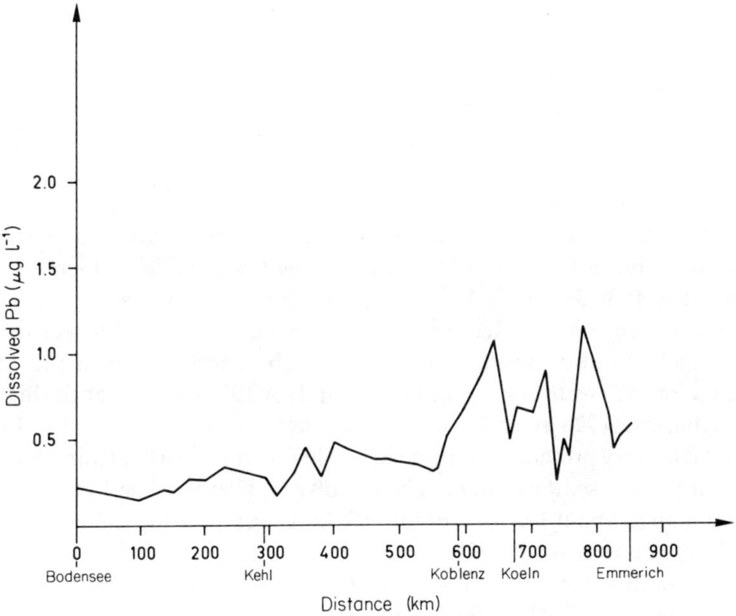

Fig. 15.6 Dissolved lead in River Rhine water

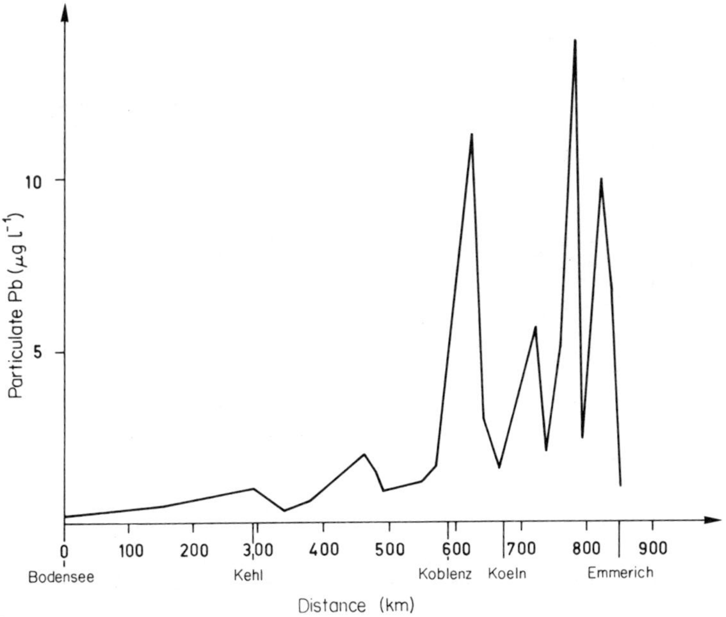

Fig. 15.7 Particulate lead in River Rhine water

suspended matter explains the rather good capacity of the river water to clean itself in the case that the suspended matter can be sedimented where the river flow is slow enough. It is to be expected that the dissolved lead levels will not vary much, because as a rule there exists sufficient suspended matter in all parts of the River Rhine to take off dissolved lead from the water body. Indeed the variation range of dissolved Pb is not as wide as observed for Cd (see Figures 15.6, 15.2 and 15.3). From Lake Constance to Koblenz the values move from 0.15–0.70 μg Pb l^{-1} (Figure 15.6). The lowest levels (0.15 $\mu g\,l^{-1}$) occurred at Rheinheim (94 km) and Freistett (0.17 $\mu g\,l^{-1}$, 312 km). The first significant increase within the flow direction seems to be connected with the entrance of the Moselle tributary. It can also be derived from the sediment data that the Moselle tributary carried much Pb into the River Rhine. The Pb concentration of the Moselle sediments for 1977 were almost the highest and for 1978 were by far the highest of all the tributaries (726 and 535 mg kg^{-1} respectively, see Figure 15.10). But also the Lahn tributary provides for a considerable Pb input into the Rhine, especially with regard to the sediment data. The dissolved Pb level in the Main before the mouth run up to about 1 $\mu g\,l^{-1}$ (Figure 15.8). About the same value was attained in the River Rhine at Brohl (620 km) and before the stream at Walsum (800 km) (Figure 15.6). Between the two maxima, however, the dissolved Pb level decreases with only one exception, i.e. the intermediate peak at Zons (0.9 $\mu g\,l^{-1}$, 718 km, industry belt of Cologne). This phenomenon proves, together with the complementary results of the suspended matter (Figure 15.7), that, with regard

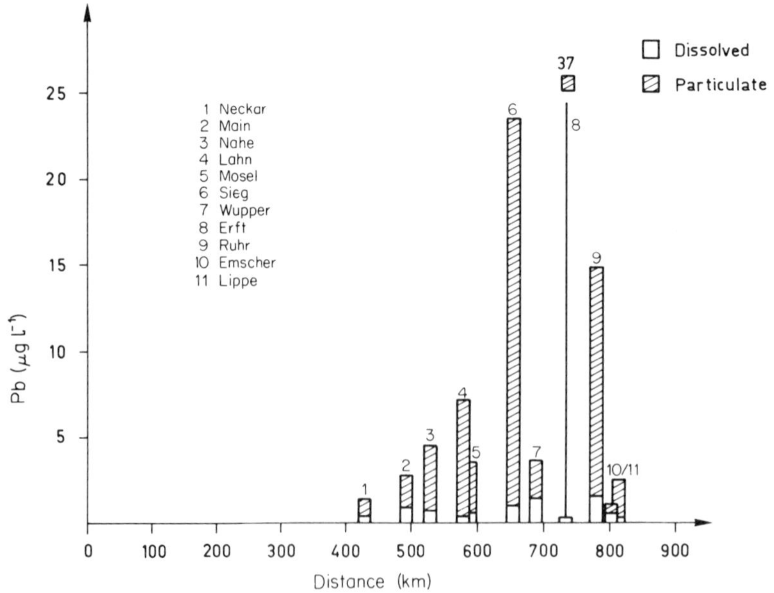

Fig. 15.8 Dissolved and particulate lead in the water of the tributaries

to Pb in water, the self purification power is not exhausted in the Lower Rhine region. Even the Pb content in the suspended matter decreased at the end of the Lower Rhine to the same levels as in the Upper Rhine. In the Upper Rhine region the relation between the Pb in the suspended matter and in the dissolved state is about 2.8:1, in the Middle Rhine 7.1:1 and in the Lower Rhine 9.5:1. Among the tributaries (see Figure 15.8) Sieg, Erft and Ruhr have especially high levels of particulate Pb. There the ratios of particulate Pb and dissolved Pb are extreme: 22.2:1 for the Sieg, 15.4:1 for the Erft and 27.9:1 for the Ruhr tributary. All the other tributaries have particulate Pb values below 10 $\mu g\,l^{-1}$.

LEAD IN THE SEDIMENTS

If the minimum levels of Pb in the Rhine sediments are combined in a diagram (Figure 15.9), then from Lake Constance to Emmerich a rather exponential increase appears. This 'baseline' of the sedimental Pb freight is superimposed by some peaks in the Middle Rhine region from Koblenz to Oberwinter (639 km). The peaks after Koblenz are caused by the extremely high Pb concentrations of the Lahn and Moselle sediments (1977: 753 $mg\,kg^{-1}$ and 726 $mg\,kg^{-1}$, Figure 15.10). In 1978 Lahn and Moselle show lower Pb levels, but in the Rhine (Figure 15.9) there is no evident difference between both years, except an increase in 1978 below the mouth of the two tributaries. This phenomenon can be derived from a

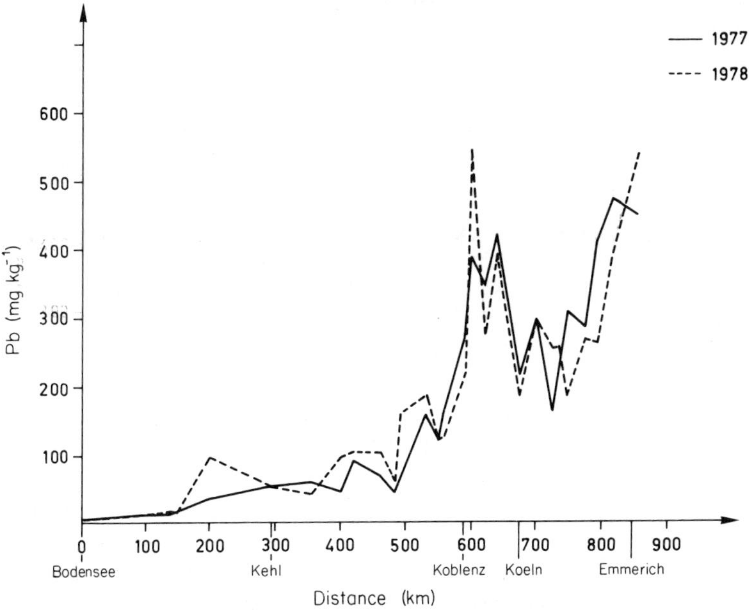

Fig. 15.9 Lead in River Rhine sediments

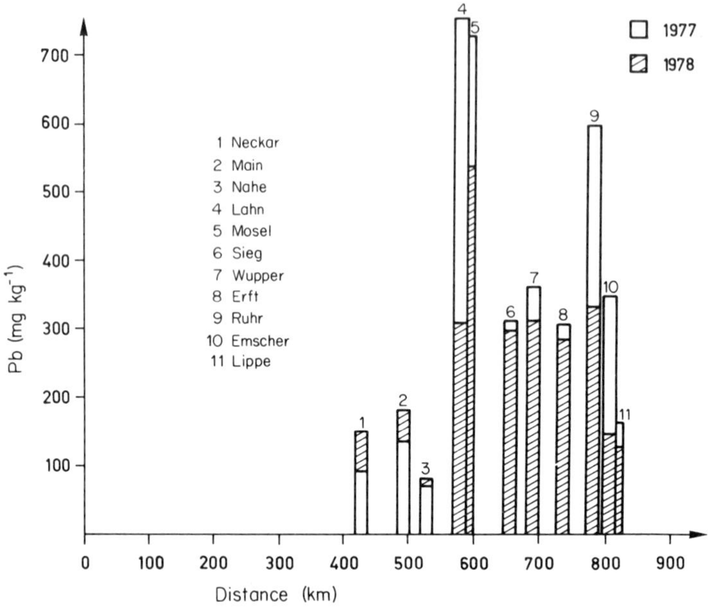

Fig. 15.10 Lead in tributary sediments. (*Note:* Metal levels for the different years always count from the bottom of the diagrams)

shifting of the sediments downstream. Also at Emmerich an increase occurred in 1978, probably for the same reason. In general there are small elevations in the Upper Rhine in 1978 and small decreases in the Lower Rhine in comparison with 1977. This corresponds very well with the respective tributaries. While the Pb concentrations in Neckar, Main and Nahe sediment increased, they decreased more or less in all the other tributaries from 1977 to 1978 (see Figure 15.10).

DISSOLVED AND PARTICULATE MERCURY IN WATER

During August 1977 samples for Hg determination were taken from Lake Constance, the Upper and the Middle Rhine and the tributaries Neckar, Main, Nahe, Lahn and Moselle. The minimal level of the unfiltered samples was about 5 $ng\,l^{-1}$ total Hg. After filtration a minimum of 4 $ng\,l^{-1}$ Hg was found. A relatively high concentration in the unfiltered samples only occurred in the Main (401 and 254 $ng\,l^{-1}$) and at the end of the middle part of the Rhine (144 and 47 $ng\,l^{-1}$).

Small elevations of about 30 $ng\,l^{-1}$ were observed near Kehl and Mannheim. Regardless of these four maximum values it follows that the average is 14.3 $ng\,l^{-1}$ total Hg in the unfiltered ($n = 23$) and 8 $ng\,l^{-1}$ Hg in the filtered samples ($n = 12$). The common ratio between the Hg connected with particles and the dissolved Hg is about 2:1.

On the 17 April 1978 five water samples were taken from a groyne in Bonn.

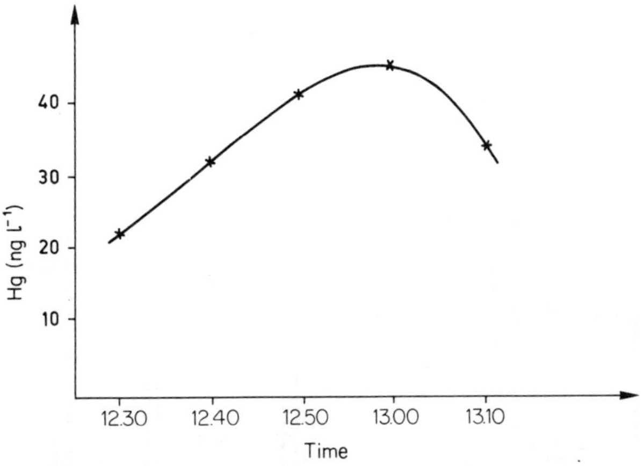

Fig. 15.11 Temporal variation of the Hg-content in the suspended particulate freight, Bonn, 17th April 1978

After 0.45 μm membrane filtration, Hg was determined in the suspended particulate matter. The results are shown in Figure 15.11.

It proves that the Hg content was changing steadily. The increase of the first three values from 12.30 to 12.50 was linear. About 13 the summit of the small Hg burden was reached. There the Hg concentration amounts to the double value of the value of the first one at 12.30. After the maximum the decrease is as steep as the prior increase.

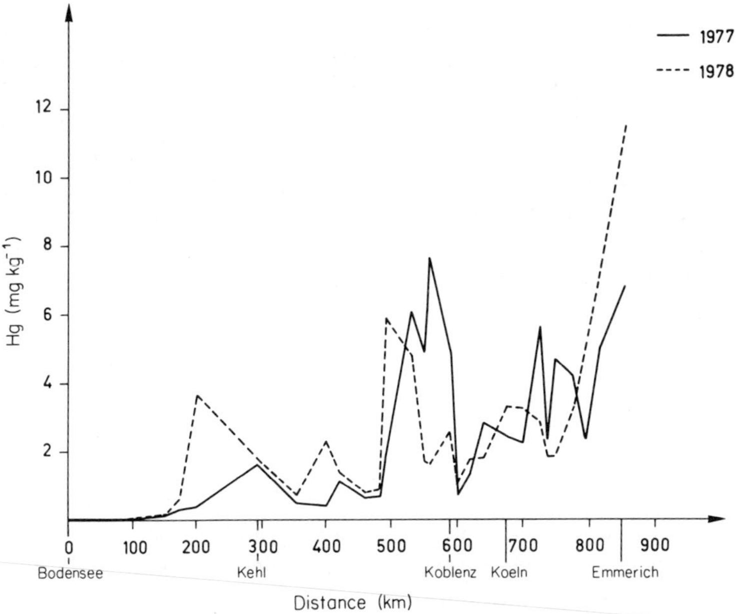

Fig. 15.12 Mercury in River Rhine sediments

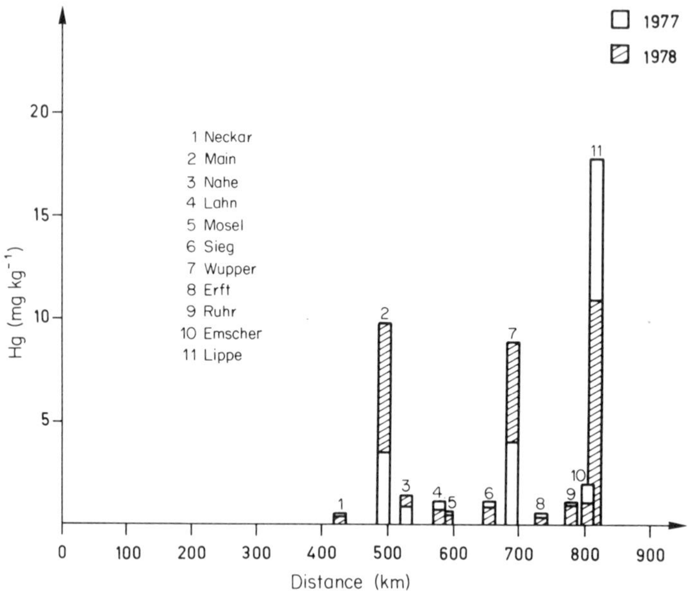

Fig. 15.13 Mercury in tributary sediments

MERCURY IN THE SEDIMENTS

Besides cadmium mercury is the heavy metal with the largest ecotoxic hazard. The enrichment factors *vs* the geochemical basic levels are, with regard to the sediment sample from Emmerich in 1978, 22.4 for Hg, 48.8 for Cd and 17.6 for Pb. Relatively high Hg concentrations occurred in 1978 at Neuenburg (200 km, 3.64 mg kg^{-1}, Figure 15.12). After maximum values at about 550 km the concentrations downstream declined to the basic geochemical data range. The high values in the Middle Rhine were probably caused by the 'Ginsheimer Altrhein' and the Main tributary. Besides the Main only two other tributaries, the Wupper and the Lippe had very elevated Hg levels (see Figure 15.13). As already noticed, the River Main had the highest Hg level in the water. The extraordinarily high Hg concentration in the sediment of the Lippe tributary (1977: 17.7 mg kg^{-1}, Figure 15.13) could partly yield the elevation at Emmerich (Figure 15.12).

CONCLUSIONS

The Rhine, with respect to ship traffic, is a busy and, in many ways, generally utilized waterway. It is therefore exposed to considerable pollution burdens. The Rhine region is a highly populated area and has in extended zones a considerable level of industrialization. The water quality is significant for several millions of people in connection with their drinking-water supply.[29]

Among the ecotoxicological hazards toxic metals, especially Cd, Pb and Hg, require particular attention, as they are rather insidious poisons tending to accumulate in vital organs of man and animals, where they display progressively increasing chronic toxic actions over long times or the total life span.

In an aquatic system the sediments constitute the depots of the toxic metals introduced into the river and are therefore an important indicator for its toxic-metal pollution degree.[6,7] However, the dissolved toxic-metal concentrations and the amounts of suspended matter are also of substantial significance, as the dissolved state and the suspended matter are the transfer states for the entrance of toxic metals into the food chain to man and for sedimentation, respectively. In this context it has to be borne in mind that toxic metals can be released from their depot at the sediments into the dissolved state, if there occur locally corresponding conditions.

As a general tendency the toxic-metal levels show a progressive but not steady increase from Lake Constance to Emmerich. Downstream from Mainz until the Dutch border several regions of particularly elevated toxic-metal levels are noticed. All major tributaries show in their mouth regions into the Rhine an enhancement of the toxic-metal burden. Within the aqueous metal freight Hg and especially Pb have a pronounced tendency to be bound to suspended particulate matter while for Cd a relatively equal distribution between the dissolved and the suspended particulate matter phase has been observed. The levels in the sediments remained rather stationary during 1977 and 1978, as a repetition of the investigation in 1978 revealed. From 1977 to 1978 only shifts of the anthropogenic sediment burden within the stream flow direction were discovered. A high Hg burden showed the Main, the Wupper and especially the Lippe tributary. Cadmium is represented most in the Wupper and Ruhr. For Pb the Lahn and Moselle sediments have extraordinary high levels.

The enrichment factors *vs* the geochemical baseline concentrations with respect to the sediment at Emmerich for 1978 were: Cd, 49; Pb, 18; Hg, 22. Thus, these three toxic metals have by far the highest anthropogenic enrichment factors in comparison with the data of, for example, Cu, Zn, As, Co, Cr, Ni and Se, which were also found,[4] but are not presented here. In future the general toxic-metal situation of the River Rhine could improve to a certain extent, because a number of sewage treatment plants have been recently built, and more are planned. These plants will remove a certain amount of the heavy metals otherwise introduced into the river. Nevertheless, the usual waste-water treatment plants are only capable of removing a part of the heavy-metal content in the waste water, predominantly from the suspended particulate-matter phase. On the other hand they discharge with the purified water, due to solubilization processes with dissolved organic matter, a certain amount of heavy metals in the dissolved state.[19] Also the heavy-metal depots in the sediments remain a long-term ecochemical hazard for the future capable of continuously releasing metals to the water. This effect would be dramatically enhanced if the river water is loaded with organics acting as efficient solubilizers, due to their strong complex potentialities. Very hazardous mobilization effects of heavy metals from their depots in the upper sediment layers and the suspended particle surfaces are to be expected in this context if plans are realized to substitute triphosphate by NTA in detergents, in order to reduce eutrophication.[21] Such a provision has to be strongly discouraged, because it would substitute one ecochemical problem (eutrophication) by another severe ecotoxical problem which is the enhancement of the levels of dissolved heavy metals in the rivers and consequently in a form most favourable for the efficient introduction of toxic heavy metals into the food chains.

REFERENCES

1. Ackermann, F., Bergmann, H., Schleichert, U., and Fresenius, Z. (1979). *Anal. Chem.*, **296**, 270–276.
2. Agemian, H., and Chau, A.S.Y. (1975). *Anal. Chem. Acta*, **80**, 61–66.
3. Aston, S.R., and Thornton, J. (1977). *Sci. Tot. Environm.*, **7**, 247–260.
4. Breder, R. (1981). Die Belastung des Rheins mit toxischen Metallen. Doctoral Thesis, University of Bonn.
5. Breder, R. (1982). *Fresenius Z. Anal. Chem.*, **313**, 395–402.
6. Förstner, U., and Müller, G. (1974). *Schwermetalle in Flüssen und Seen*, Springer, Berlin, Heidelberg, New York.
7. Förstner, U., and Wittman, G.T.W. (1981). *Metal Pollution in the Aquatic Environment*, Springer, Berlin, Heidelberg, New York.
8. Friberg, L., Nordberg, G.F., and Vouk, B. (eds) (1979). *Handbook on the Toxicology of Metals*, Elsevier/North Holland Biomedical Press, Amsterdam.
9. Mart, L. (1979). *Fresenius Z. Anal. Chem.*, **296**, 350–357.
10. Mart, L. (1982). *Talanta*, **29**, 1035–1040.
11. Mart, L. (1979). Ermittlung und Vergleich des Pegels toxischer Spurenmetalle in Nordatlantischen und Mediterranean Küstengewässern. Doctoral Thesis, RWTH Aachen.
12. Mart, L., Nürnberg, H.W., Valenta, P., and Stoeppler, M. (1978). *Thalassia Jugosl.*, **14**, 171–188.
13. Mart, L., Nürnberg, H.W., and Valenta, P. (1980). *Fresenius Z. Anal. Chem.*, **300**, 350–362.
14. Mart, L., Rützel, H., Klahre, P., Sipos, L., Platzek, U., Valenta, P., and Nürnberg, H.W. (1982). *Sci. Tot. Environm.*, **26**, 1–17.
15. Matthes, W., Flucht, R., and Stoeppler, M. (1978). *Fresenius Z. Anal. Chem.*, **291**, 20–26.
16. Merefield, J.R. (1976). *Marine Poll. Bull.*, **7**, 214–216.
17. Merian, E., Geldmacher-v. Mallinckrodt, M., Machata, G., Nürnberg, H.W., Schlipköter, H.W., and Stumm, W. (Hsrg.) (1983). *Metalle in der Umwelt*, Verlag Chemie, Weinheim.
18. Nürnberg, H.W. (1982). *Pure Appl. Chem.*, **54**, 853–878.
19. Pihlar, B., Valenta, P., Golimowski, J., and Nürnberg, H.W. (1980). *Z. Wasser-Abwasser-Forschung*, **13**, 130–138.

[20] Rat der Sachverständigen für Umweltfragen: Umweltprobleme des Rheins, 3. Sondergutachten, 1976.
[21] Salomons, W., and van Pagee, J.A. (1981). Prediction of NTA levels in river systems and their effect on metal concentrations, in, *Proceedings of the International Conference on Heavy Metals in the Environment, Amsterdam 1981*, pp. 695–704, CEP Consultants, Edinburgh.
[22] Stoeppler, M., and Backhaus, F. (1978). *Fresenius Z. Anal. Chem.*, **291**, 116–120.
[23] Stoeppler, M., and Backhaus, F. (1979). Beiträge zur Umweltforschung und Umweltüberwachung. I. Entwicklung, *Bau und Betrieb eines mobilen Spurenmeßlabors*, Jül-1571.
[24] Stoeppler, M., Valenta, P., and Nürnberg, H.W. (1979). *Fresenius Z. Anal. Chem.*, **297**, 22–34.

Pollutants and Their Ecotoxicological Significance
Edited by H. W. Nürnberg

CHAPTER 16

Estuarine Processes and Riverborne Pollutants

JAN C. DUINKER

SUMMARY

Rivers transport trace metals to the ocean in dissolved, colloidal and particulate forms. In estuaries, where river water and coastal or oceanic waters of widely different compositions are mixed, strong gradients in chemical and physicochemical properties occur. As a result, the relative contributions of the chemical species of any element to these forms can be modified considerably. A significant fraction of river borne cadmium (Cd), zinc, (Zn), copper (Cu), manganese (Mn) and iron (Fe) is deposited in bottom sediments of the Rhine estuary. This is the result of the trapping mechanism of the estuary for suspended particles as well as the removal of dissolved components into particulate components during estuarine mixing. Attempts are made to distinguish the effects of physical and physicochemical mechanisms that contribute to the observations during surveys and hourly observations at fixed stations in different salinity regimes of the estuary.

INTRODUCTION

The ocean is an ultimate sink for land-derived material, including chemical pollutants. Although atmospheric transport and deposition with precipitation can be significant mechanisms for some chemical components to the sea (e.g. lead and polychlorinated biphenyls), the supply of chemical components to the marine environment is dominated by rivers.

The erosion products of the continents will eventually reach the bottom of the oceans, thus becoming part of the earth's crust again. The residence time in sea water varies between 100 years and millions of years. The study of the large biogeochemical cycles including chemical pollutants belongs to the main subjects of marine sciences.

The influence of human activities on these cycles has been taking place over a time period that is short with respect to the time scale of the complete cycli. Generally, chemical pollutants follow the same pathways as natural components; they are thus faced with the same barriers in the transport mechanisms from rivers to the open sea. In particular, large amounts of particles may be trapped in estuaries, which are then a significant sink for pollutants.

Most rivers discharge into estuaries, being coastal water bodies with a limited connection with the sea and in which fresh water and salt water are mixed. They have been formed by sea-level rise after the last ice age.

Equilibrium between supply and removal of sediment in estuaries has not been accomplished yet in this short geological period. Roughly half of the sediment supply from land may be trapped in estuaries. This may also apply to pollutants. However, large differences occur between different types of estuaries. A Norwegian fjord may trap practically all material that is supplied; the estuaries around the North Sea are practically in equilibrium, resulting in small net gain. Sedimentation may be important locally and may be time dependent.

ESTUARINE PROCESSES

The velocity of sediment deposition in estuaries is increased by several processes that cause landward transport of suspended particles. This can be illustrated conveniently by considering estuarine circulation. As river water is less dense than seawater, a stratification occurs in many estuaries: fresh water can flow in a seaward direction on top of a more saline layer. Salt water is picked up from this layer, resulting in a bottom current in the opposite direction, preventing the escape of sediment and moreover, bringing in sediment from outside the river mouth. The highest concentrations occur in the region of maximum intrusion of seawater. This is the region of the 'maximum turbidity'.

Examples of estuaries in which the full escape of particulate matter and its associated pollutants is prevented by these processes are the rivers Scheldt, Rhine, Ems, Weser, Elb. Components transported in solution are not influenced by these processes and they may escape to the coastal zone and from there to the open sea. During estuarine mixing, the relative distribution of any particular chemical component is modified from the distribution as it occurs in the freshwater region. This is the result of the increase in the concentration of suspended matter as indicated before, but additionally to the occurrence of chemical processes causing transitions between dissolved and particulate forms during estuarine mixing. Biogenic reactions may also contribute. It is important to understand estuarine processes for predicting the geochemical behaviour of each individual element and its possible effect on organisms, as well as the role of estuaries in the mass balance between rivers and oceans.[1] We shall discuss a number of these processes in some detail.

Mixing of river and sea water is characterized by the occurrence of strong gradients in chemical and physicochemical parameters.[2,3] River and sea water have a difference in total concentrations of dissolved constituents. Chemistry in these environments may, therefore, be different as a result of changes in activity coefficients. Moreover, the relative contribution of the dominant major cations changes from Ca^{2+} and NO_3^- in river water to Na^+ and Cl^- in sea water. Dissolved oxygen concentrations in sea water are usually higher than in river and estuarine water in which complete oxygen deficiency is often observed as a result of intensive breakdown of organic matter. Dissolved oxygen concentrations are restored upon mixing with sea water; the salinity range at which this mechanism becomes effective may be different for the various seasons, values being highest in summer. pH values in river water are usually below those in sea water. Values in estuarine waters may have a maximum at low salinities.[4] This has considerable consequences for some elements.

CHEMICAL POLLUTANTS IN THE ESTUARY

The main classes of pollutants identified in the marine environment are, (a) synthetic organic chemicals, in particular organochlorines, (b) metals and (c) oil components. In the following, we shall illustrate the importance of estuarine processes for the transport of river-borne pollutants out into the sea, using information obtained from the studies on a number of trace metals.

Major Problems

The interpretation of trace-metal concentrations in estuarine water and particles in terms of relevant processes is fraught with problems. For our purposes, the major ones can be classified as follows.

The complexity of estuarine processes

An estuary is a dynamical system in which the distribution patterns of dissolved and particulate components result from the physical mixing characteristics of natural waters of widely different composition and a variety of chemical and biogenic reactions taking place independently and simultaneously. Additionally, variations in meteorological conditions, flow characteristics and biological activity are responsible for variations in the concentrations of trace metals in estuaries, with different time scales.[5] In order to understand the role of estuaries in the biogeochemical cycling of trace elements it is essential though by no means trivial, to distinguish the effects of biological and chemical from hydrodynamic processes.

Non-linearity

Several properties do not vary linearly with the parameter that characterizes the mixing series between river water and sea water, such as chlorinity or salinity.

This has been observed for several basic properties as suspended load, pH and dissolved oxygen concentration.

Analytical problems

The behaviour of a trace element during estuarine mixing depends on its chemical state, whether in solution or particulate. Although techniques are available to determine total concentrations at the levels such as occur in river and sea water solution, no general method is available to distinguish different chemical species in solution, except for a limited number of trace metals.[6,7]

The analysis of dissolved trace metals in a series of estuarine water samples is usually more difficult than for a series of river or sea water samples due to the large range in the amounts of sea salt and dissolved organic matter in the matrix. Similar problems as for dissolved species are met in the analysis of sediment samples when distinguishing different binding sites. However, general chemical methods have been proposed recently.[8,9]

Trace Metals

There has been considerable confusion in the literature on the matter of trace-metal behaviour during estuarine mixing. It has been often suggested that trace metals are desorbed from suspended matter once in contact with seawater.[10,11] The importance of the opposite process, viz. the removal of trace metals from solution has been stressed by others.[12–15]

The chemical behaviour of river-borne trace elements within the estuary is determined to a large extent by the chemical form in which they are transported by rivers, (a) in solution, (b) adsorbed on surfaces, (c) in solid organic particles, (d) in coatings on detrital particles after coprecipitation with and sorption onto mainly iron and manganese oxide, (e) in lattice positions of detrital crystalline material, and (f) precipitated as pure phases on detrital particles. The scheme allows the distinction of trace-element fractions that are readily available (dissolved and adsorbed), fractions that may become available after chemical changes (organically bound and in oxide coatings) and forms that are practically unavailable for release (in crystal structures). In the presently reported data, leachable fractions have been obtained as the amount released by the sediment upon contact with 0.1 M HCl during 18 hours.[12]

The strong salinity gradients in estuaries can be expected to cause ion-exchange reactions particularly on fine-grained particles. Adsorption and desorption reactions occur in response to a suitable free energy of sorption. Electrostatic adsorption in natural systems[16] is presently believed to give a considerably smaller contribution to the free energy of adsorption than specific interactions such as solvation, covalent and hydrogen bonding. Specific interactions may be slow and less readily reversible in the sense that equilibrium is not achieved in a reasonable time.[17] The work on the adsorption of Zn, Cu, Pb, Ni, Co, Hg, Ag, Cr, Mo, W, V on hydrated iron oxide, hydrated manganese oxide, apatite, clay, plankton and peatmoss[16] would suggest that a series of lithogenous, biogenous and hydrogenous materials adsorb trace metals from sea water. However, experimental work on the sorption of Co, Ag, Se, Mo, Sb, Cr, Cs and Rb onto and from montmorillonite, kaolinite, illite, manganese oxide, hydrated ferric oxide and peat, shows the release of substantial portions of Co, Se and Ag when the adsorbents, after contact with metal ion solutions for 1 h, were contacted with seawater for 1 h.[10] Thus, according to these results, river-borne particles might contribute additional soluble amounts of some trace metals to the ocean.

Direct information on the importance of release or removal processes of dissolved species during estuarine mixing can be obtained from dissolved metal–salinity plots,[18] keeping in mind the criteria for the identification of any non-conservative behaviour.[19] In attempts of this kind, measured concentrations of dissolved metals are plotted against salinity data in the same samples. Concentration data should be on a straight line, connecting the two end-member concentrations in river and sea water, in case of so-called conservative behaviour. Concentrations are above (or below) the ideal dilution line in case of production (or removal) of a dissolved component during estuarine mixing. Additional information can be obtained, if suspended trace metals are analysed along with dissolved species.[12,13]

Dissolved Metals

Data of this type (Figure 16.2, ref. 20) obtained during surveys in the Rhine estuary (Figure 16.1) indicate that the concentrations of dissolved Fe, Zn, Cu and Cd at low salinities in the longitudinal profile and those for Mn at higher salinity are below the ideal dilution line (Figure 16.2a). The residence time of fresh water in the Rhine estuary is in the order of 3–6 days. Relatively small variations (10–20 per cent) were observed in the concentrations of Fe, Mn, Zn, Cu and Cd in the fresh-water region during the days prior to the observations in the mixing area. The negative deviation from the ideal dilution line that is observed for these elements can, therefore, be attributed to removal processes. The importance of dissolved organic matter in this process has been demonstrated.[14]

The removal from solution does not result in a positive deviation in the plot of particulate trace-metal concentrations against salinity. In fact, a net removal of particulate matter and the associated trace metals is observed (Figure 16.2b).

The behaviour of manganese is quite different from that of the other elements. Removal from solution appears to occur also for Mn, but at a considerably higher salinity regime. The most striking feature, however, is the appearance of a maximum in the dissolved concentrations *vs* salinity plot. It has been demonstrated that the physicochemical behaviour of manganese in the estuary is

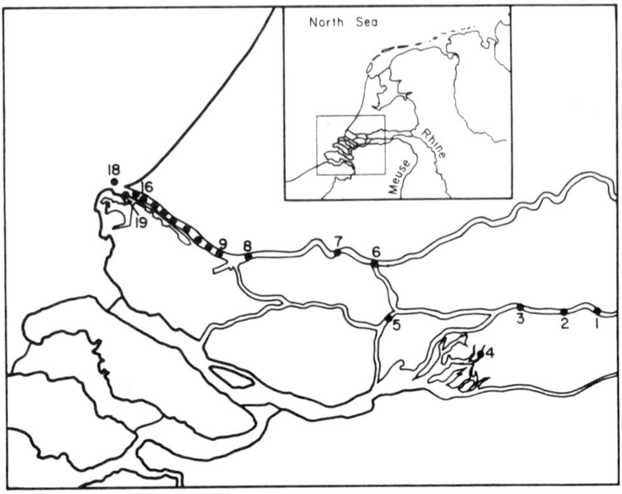

Fig. 16.1 Sampling stations in the Rhine estuary

governed by transitions between reduced (Mn^{2+}, $MnCl^{+}$, $MnCO_3$) and oxidized species ($MnOOH$, MnO_2). The variations in dissolved Mn(II) in the Scheldt and Rhine estuaries can be satisfactorily reproduced by calculations of the concentrations in thermodynamical equilibrium with the solid phases.[21] The parameters that appeared to be of primary importance are ionic strength, chlorinity, pH and Eh. Manganese (II) is removed from solution into particulate and oxidized forms in the lower estuary. Part of the particulate Mn is transported into the sea, another part is returned into the upper estuary where it can be reduced to Mn(II) as a result of the interaction between these particulates in the near-bottom layer and the overlying water.[22] This results in a positive deviation in the dissolved manganese–salinity plot at low salinities, or even a pronounced maximum like in the data of Figure 16.2.

The effects of estuarine mixing processes in the Rhine estuary have also been studied during a series of observations at fixed positions in different salinity regimes at 1-h intervals during complete tidal cycles. Station 6 is a fresh water station. The only deviations from the fresh water salinity at station 8 occur at high tide and station 17 is characterized by permanent stratification (stations are shown in Figure 16.1).

The concentrations of dissolved Zn and Cu—and less-pronounced concentrations of Fe—in the samples of station 6 are higher than in the fresh-water samples of station 8 (Figure 16.3). This may be due to removal of dissolved components in the fresh-water region, resulting from the interaction with sediment particles transported upstream from the lower estuary beyond the limit of marine salt intrusion.[23] Increased suspended sediment concentrations favour the removal of dissolved trace metals.[24] This effect is enhanced by the presence of

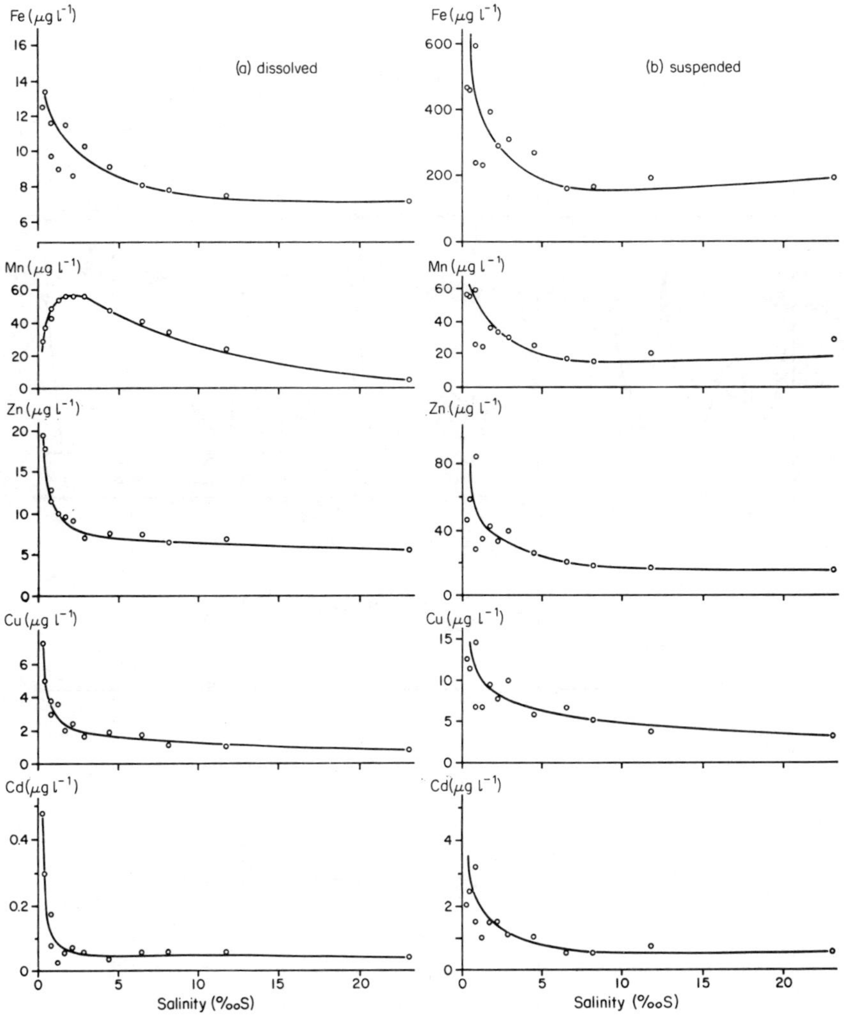

Fig. 16.2 Dissolved (a) and particulate leachable (b) metal-salinity plots for the data of the survey in the longitudinal section of the estuary. Sampling stations 6 to 18

freshly precipitated iron (and manganese) oxide and hydroxide species that are known to act as effective scavengers for trace metals.[16,25] Particles derived from the lower estuary are enriched in iron and manganese oxides.[12,13] The importance of their presence for the removal of dissolved Zn and Cu may follow from the observation of increased dissolved Mn concentrations in the samples of the fixed station 8 at high tide (Figure 16.3). This increase can only be caused by interactions involving sediment particles derived from the lower estuary.[22]

Therefore, the data of station 6 represents the fresh-water end member

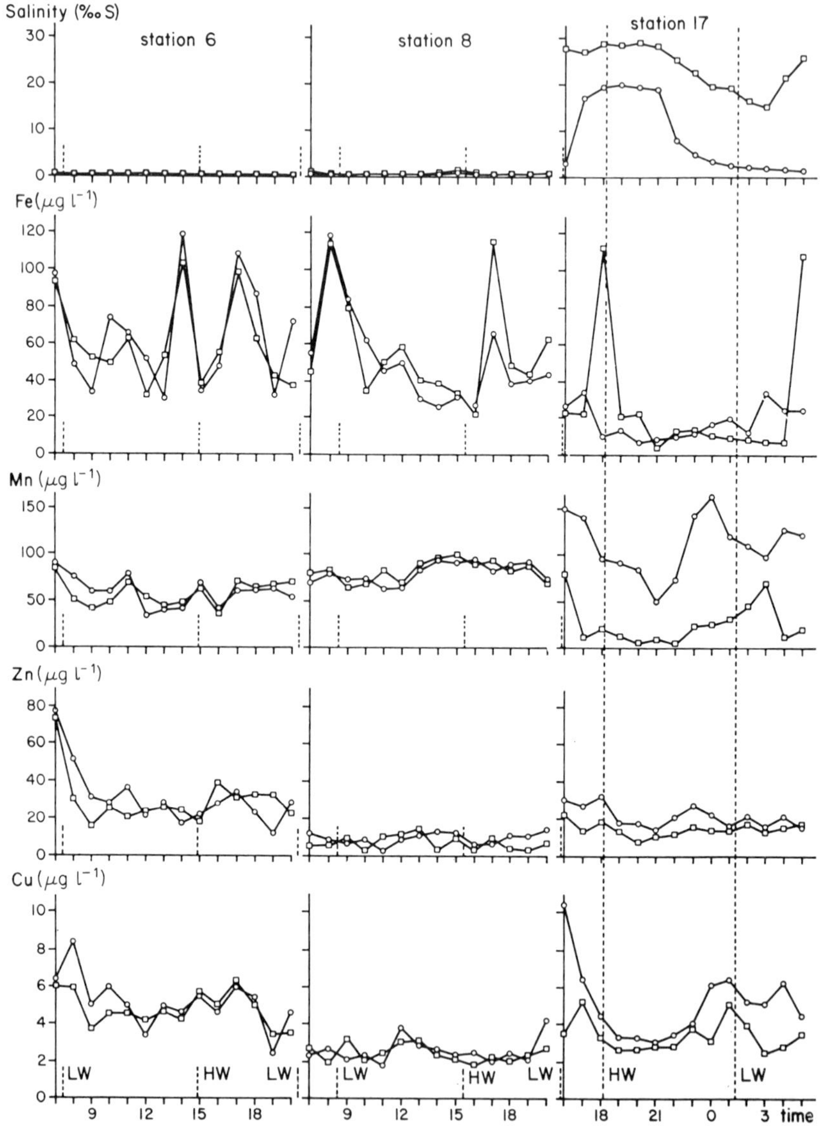

Fig. 16.3 Salinity and dissolved concentrations of Fe, Mn, Zn, and Cu at the fixed stations 6, 8 and 17 at 1-hour intervals. Surface (○) and near-bottom layer (□). Times of high and low water indicated

Table 16.1 Range of interestitial water concentrations of Fe, Mn, Zn and Cu in the five 2-cm slices of the upper 10 cm of the sediment cores taken at stations 3, 4, 5 and 7

Metal	Station			
	3	4	5	7
Fe ($mg\,l^{-1}$)	0.2–5	0.7–3.2	10–18	20–50
Mn ($mg\,l^{-1}$)	0.6–6	0.7–7	2–8	8–12
Zn ($\mu g\,l^{-1}$)	80–180	40–150	120–300	100–700
Cu ($\mu g\,l^{-1}$)	40–180	20–120	20–140	4–190

concentrations of Fe, Zn and Cu. Including the data of station 8 would tend to obscure any removal process—if present—during estuarine mixing.

A second mechanism that is important in this respect concerns a contribution of trace metals from interstitial water to be overlying water. This may be important under dynamic conditions such as during dredging operations, bioturbation or storms.[20] The importance in the present results is demonstrated by the occurrence of slightly increased dissolved concentrations in near-bottom samples at the fixed station 17 compared to surface samples. The magnitude of the interstitially dissolved concentrations in the mixing area is large enough to account for these observations (Table 16.1).

Concluding these considerations, we may state that the dissolved metal–salinity plots for Zn and Cu, also in this series of observations, suggest the occurrence of removal processes of dissolved species, or at least they indicate conservative behaviour without evidence for mobilization processes.

Suspended Metals

It is also interesting to study the relation between salinity and trace-metal concentrations in suspension. The relations between salinity and particulate leachable concentrations of Fe, Mn, Zn, Cr and Cu for the data of the fixed station 17 are almost linear (Figure 16.4) and it appears that the surface and near-bottom concentrations fit the same linear relation with a fair degree of reliability. The slope of the regression line is positive for Mn and negative for Fe, Zn, Cr and Cu. It may be concluded that suspended matter derived from near the bottom was mixed into the surface layer in nearly the same ratio as the saline water in which it was suspended. The extreme meteorological conditions during and prior to sampling may have contributed to the situation observed. A few near-bottom data for Zn, Cr and Cu at high salinities (Figure 16.4) are clearly above the mixing line. They are associated with the low suspended load samples at slack water, suggesting that the highest concentrations occur in fine grained material that

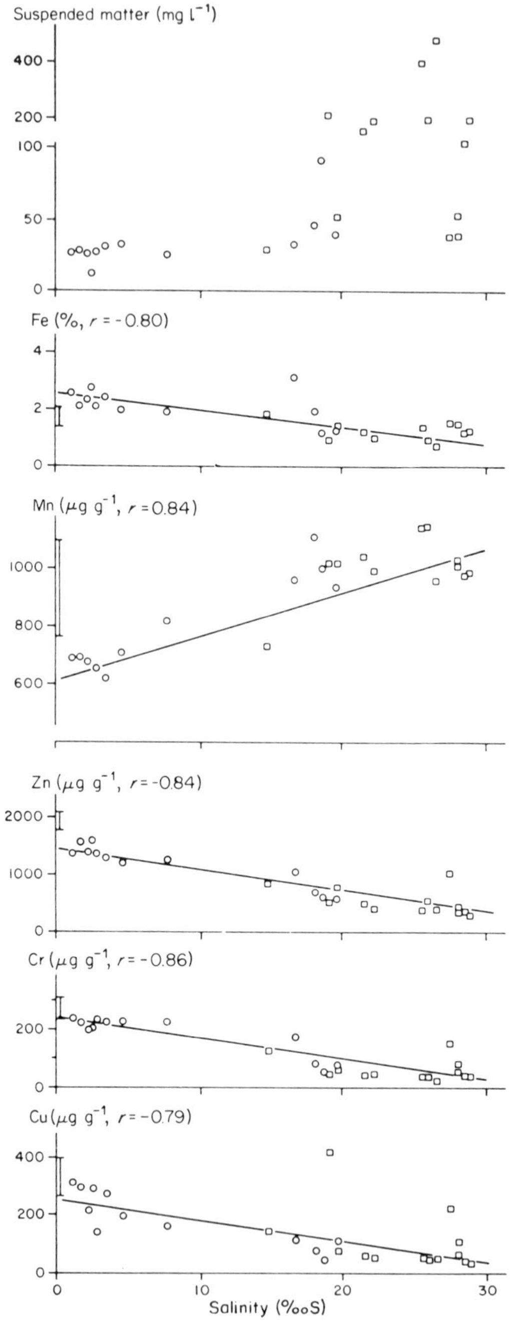

Fig. 16.4 Relationship between salinity and particulate leachable Fe, Mn, Zn, Cr and Cu concentrations ($\mu g\, g^{-1}$, per cent) in surface (○) and near-bottom (□) samples of station 17. The data of stations 6 and 8 are included as a range of values at 0.4 per thousand S

Table 16.2 Particulate trace-metal concentrations calculated from the regressions in metal-salinity plots (Figure 16.4) for a salinity of 0.4 per thousand. Sulphur ('extrapolated') and 'average' concentrations for the fresh water samples of the fixed stations 6 and 8, together with their relative standard deviations (%)

Metal	Extrapolated to 0.4‰ sulphur ($\mu g\ g^{-1}$)	Average fresh water suspended ($\mu g\ g^{-1}$)	Relative sd (%)
Fe	25000	17000	11
Mn	620	870	23
Zn	1400	2100	15
Cr	240	300	6
Cu	250	320	25

does not settle.[26] The contribution of near-bottom sediment particles to the concentrations in the surface layer is dominated by the coarser fraction. The value obtained by extrapolating the correlation line in Figure 16.4 to 0.4 per thousand S (Table 16.2) is lower than the average concentration in the fresh water suspended matter for Mn, Zn, Cr and Cu. This agrees with the model just developed. The extrapolated value for leachable Fe is higher, however. This can be attributed to the relatively large contribution of dissolved iron that is transformed into particulate forms in the early stages of mixing, increasing the leachable—and total—iron concentrations in particulate matter to values even above the characteristic fresh-water concentrations. The positive slope for particulate Mn (Figure 16.4) is due to the transition of Mn(II) into oxidized and particulate forms in the lower estuary in response to increased pH and dissolved O_2 concentrations.[21,22]

Suspended trace-metal concentrations in the surface layer can thus be accounted for in terms of mixing fluvatile and marine-derived particles, with additional contributions of originally dissolved river-borne components that are removed from solution during estuarine mixing. The same mechanism has to be taken into account to explain bottom sediment data in the Rhine estuary.[12,27–28] The removal of suspended matter and trace metals from the surface layer is indicated by Figure 16.2b.

These arguments demonstrate that in the Rhine estuary several trace metals are removed from solution during estuarine mixing and are taken up by particulate matter. Part of this load is deposited in the estuarine bottom sediments, there resulting in higher metal concentrations than can be accounted for by mixing of fluvatile and marine-derived particles. These processes may also be important in other estuaries.[30]

If removal of river-borne trace metals will appear to be a general phenomenon, only a limited part of the river transported trace metals will reach the ocean. This might assist in explaining the observed depletion in ocean water of several trace metals with respect to the solubility of any likely solid.[15]

REFERENCES

1. Mackenzie, F.T., and Garrels, R.M. (1966). *Am. J. Sci.*, **264**, 507.
2. Dyer, K.R. (1972). In, *The Estuarine Environment* (Eds R.S.K. Barnes and J. Greene), p. 11, Applied Science Publishers, London.
3. Burton, J.D. (1976). In, *Estuarine Chemistry* (Eds J.D. Burton and P.S. Liss), p. 1, Academic Press, London.
4. Mook, W.G., and Koene, B.K.S. (1976). *Est. Mar. Coast. Sci.*, **3**, 325.
5. Turekian, K.K. (1971). In, *Impingement of Man on the Oceans* (Ed. D.W. Hood), p. 9. John Wiley, Chichester.
6. Florence, T.M., and Batley, G.E. (1977). *Talanta*, **24**, 151.
7. Duinker, J.C., and Kramer, C.J.M. (1977). *Mar. Chem.*, **5**, 207.
8. Gupta, S.K., and Chen, K.Y. (1975). *Env. Letters*, **10**, 129.
9. Förstner, U. (1980). In, *Chemistry and Biogeochemistry of Estuaries* (Eds E. Olausson and I. Cato); John Wiley, New York, pp. 310–352.
10. Kharkar, D.P., Turekian, K.K., and Bertine, K.K. (1968). *Geochim. Cosmochim. Acta*, **32**, 285.
11. de Groot, A.J., and Allersma, E. (1975). In, *Heavy Metals in the Aquatic Environment* (Ed. K. Frenkel), p. 85, Pergamon Press, Oxford.
12. Duinker, J.C., and Nolting, R.F. (1976). *Neth. J. Sea Res.*, **10**, 71.
13. Duinker, J.C., and Nolting, R.F. (1977). *Mar. Pollut. Bull.*, **8**, 65.
14. Sholkovitz, E. (1976). *Geochim. Cosmochim. Acta*, **40**, 831.
15. Turekian, K.K. (1977). *Geochim. Cosmochim. Acta*, **41**, 1139.
16. Krauskopf, K.B. (1956). *Geochim. Cosmochim. Acta*, **9**, 1.
17. Parks, G.A. (1975). In, *Chemical Oceanography* (Eds J.P. Riley and G. Skirrow, 2nd ed., vol. 1, p. 241.
18. Liss, P.S. (1976). In, *Estuarine Chemistry* (Eds J.D. Burton and P.S. Liss), p. 93, Academic Press.
19. Boyle, E.A., Collier, R., Dengler, A.T., Edmond, J.M., Ng, A.C., and Stallard, R.F. (1974). *Geochim. Cosmochim. Acta*, **38**, 1719.
20. Duinker, J.C., and Noltin, R.F. (1978). *Neth. J. Sea Res.*, **12**, 205.
21. Wollast, R., Billen, G., and Duinker, J.C. *Est. Coast. Mar. Sci.* (1979) **9**, 161.
22. Duinker, J.C., Wollast, R., and Billen, G. *Est. Coast. Mar. Sci.* (1979) **9**, 727.
23. Postma, H. (1976). In, *Sediment Transport and Sedimentation in the Estuarine Environment* (Ed. G.H. Lauff), p. 158, American Association of Advanced Science Publishers, **83**, 158.
24. Aston, S.R., and Chester, R. (1973). *Est. Mar. Coast. Sci.*, **1**, 225.
25. Goldberg, E. (1954). *J. Geol.*, **62**, 249.
26. Duinker, J.C., van Eck, G.T.M., and Nolting, R.F. (1974). *Neth. J. Sea Res.*, **8**, 214.
27. Müller, G., and Förstner, U. (1975). *Env. Geol.*, **1**, 33.
28. Salomons, W. (1975). *J. Sed. Petrol.*, **45**, 440.
29. Salomons, W., and Mook, W.G. (1977). *Neth. J. Sea Res.*, **11**, 199.
30. Duinker, J. C. (1980). In, *Chemistry and Biogeochemistry of Estuaries* (Eds E. Olausson and I. Cato), John Wiley, New York, pp. 121–151.

Pollutants and Their Ecotoxicological Significance
Edited by H. W. Nürnberg

CHAPTER 17

Transport, Fate and Recycling of Heavy Metals in Sea-water Ecosystems

I. ELSKENS AND L. GOEYENS

INTRODUCTION

One can establish a relatively complete diagram of the translocation of heavy metals within the marine ecosystem of the southern part of the Southern Bight of the North Sea on the basis of the diagrams of carbon flux in the trophic chains, the determination of the degree of contamination of certain biological compartments (fish and plankton), as well as on the basis of the metal content of some physical compartments (sediments, matter in suspension and water column) if only in an *integrated form* temporally covering a complex biological and seasonal cycle, namely about a year.[1]

The figures obtained (Figure 17.1) are relatively different to those of the flux diagrams relating to carbon or nitrogen. In actual fact, the respiration of living organisms releases virtually no metals into the ambient medium, whereas the elimination of the latter is 'cumulated' to the level of excretions, dead cells and faeces; added to which there appears to be a direct interaction with the associated physical compartments (flux of suspended matter more or less rich in detrital material, etc.). And lastly, according to the flux diagram obtained for mercury and zinc in particular,[1] it is clear that the situation in the Southern Bight is governed by major flux of substances in suspension (organic and inorganic) from the Pas de Calais, the Scheldt and the considerable reservoir which sediments constitute. The high local production of organic matter *can modulate the general balance sheet* and, for example, modify the water-column content from one season to another.

Our aim, in this contribution, is to detail a few basic concepts, particularly as regards *the choice* of control parameters and the variability of the degrees of contamination observed or, in other words, to attempt to define the proposed model's sensitivity to the various working hypotheses used. In actual fact, the

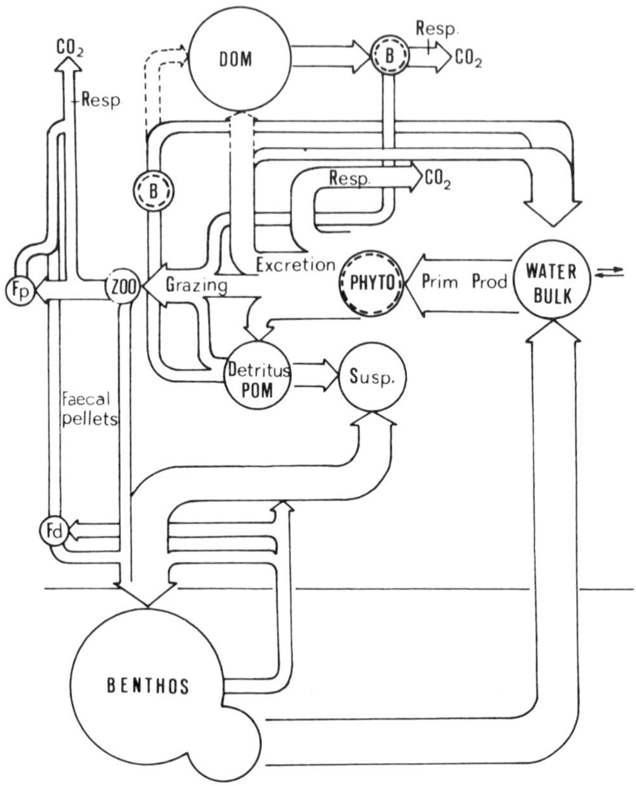

Fig. 17.1 Representation of a diagram of circulation flux of heavy metals in the marine ecosystem

advantage of a systematic approach using the mathematical model lies in considering, by definition, the elements to be monitored in the marine medium as forming part of a functional whole. By treating the complex problem in its entirety (without isolating the compartments) one can not only interpret the results in a realistic way, but also, on a scientific basis, define the most important interactions and compartments by, and only by, simplifying the complex system initially considered and reducing it to the essential elements which actually *govern* the level of contamination of the products of the sea. One cannot of

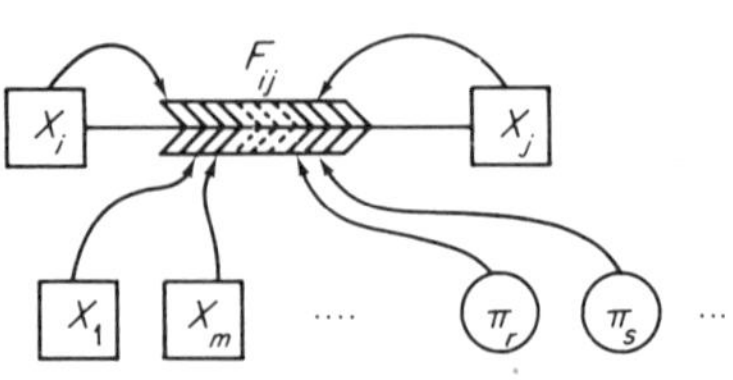

Fig. 17.2 Regulation of interactions

course consider the list of the elements listed hereunder as being exhaustive; we shall limit ourselves to a few examples which are better known than others and which are known to effectively play a prime role in the compilation of the annual balance sheet of translocations.

REGULATION OF INTERACTIONS

Principles[2]

There exists, between a donor variable, X_i, and an acceptor variable, X_j, a flux (a translocation) of matter or energy F_{ij}. This flux can be controlled (Figure 17.2) by X_i and X_j, by other variables of state X_1, X_m ... and parameters of excitation π_r, π_s, ...

$$F_{ij} = F_{ij}(X_i, X_j, X_1, X_m, \ldots, \pi_r, \pi_s). \tag{17.1}$$

In most cases,[3-5] F_{ij} can be put in the form of a product of functions according to

$$F_{ij} = X_j f_i(X_i) f_1(X_1) f_m(X_m) \cdots f_r(\pi_r) f_s(\pi_s) \cdots. \tag{17.2}$$

To simplify matters, let us assume that F_{ij} depends only on X_i and X_j

$$F_{ij} = X_j f_i(X_i). \tag{17.3}$$

Two situations can be considered.

1 When the value of variable X_i is high, the importance it represents can be found to repletion in the ecosystem and its variations have no influence on the interaction. As X_i no longer controls F_{ij} and f_i (X_i) is reduced to a constant.
2 For the low values of X_i, this variable linearly controls the interaction, f_i is then

$$f_i = \alpha_{ij} X_i \tag{17.4}$$

and F_{ij} becomes the celebrated Lotka–Volterra relation

$$F_{ij} = \alpha_{ij} X_i X_j. \tag{17.5}$$

Clearly then, these two situations are continually connected and the most common way of expressing the passage from a linear dependence corresponding to the low values of X_i to a constant for the high values of X_I is the Michaelis–Menten–Monod relation

$$f_i(X_i) = \frac{\alpha_{ij} X_i}{\beta_i + X_i}, \tag{17.6}$$

where α_{ij} is the maximum rate of consumption of X_i by X_j and β_i the constant of semi-saturation.

In point of fact, relation (17.6) is of course not the only possible one and a number of authors have proposed highly varied forms of it, some simple, some sophisticated, depending on the type of interaction studied. One can cite the threshold relations[6] and the relations which describe an inhibition of the interaction for the very high values of X_i.[7]

It is nevertheless clear that the relations which have just been represented (or suggested) for the $f_i(X_i)$ function of the simplified form (17.3) of F_{ij} can be extended to all the other functions $f_1(X_1)$, $f_m(X_m)$, ... $f_r(\pi_r)$, $f_s(\pi_s)$ ... of the general form (17.2) of F_{ij}.

If all these functions show levels of saturation and all the state variables X_i, X_1, X_m, ... (or parameters of excitation) bar one have reached saturating values, it is the latter which alone controls the interaction and plays the part of a limiting element.

More particularly, though it is possible to prove that a variable (or parameter) remains at saturating values throughout the entire period studied, it never influences the interaction considered and can, therefore, be eliminated from the expression F_{ij}.

This unquestionably constitutes a very interesting experimental means of approach to the study of translocations and interactive phenomena.

Applications

There are, in the relevant literature, a great many examples of the application of this form of theoretical approach. We have arbitrarily selected two of them, one in the field of nutrient fixation, the other in the field of pollutant fixation, in the biological compartments of the trophic chain.

Pichot *et al.*,[2] have shown that the mathematical model they presented satisfactorily simulates the circulation of nitrogen throughout the pelagic chain of the Southern Bight of the North Sea, taking into account the hypothesis selected, the relative simplicity of the system of equations and the variability of the data available. The coastal zone appears to be fairly homogeneous and can reasonably be studied by means of 'box' models (see below). On the other hand, since the coastal deposits (e.g. the Scheldt estuary) have local effects, their detailed study calls for a subdivision of the coastal zone into smaller 'boxes' and perhaps the development of a continuous spatial model.

Bouquegneau *et al.*[8] have, in a very detailed study, shown that as regards, for example, the accumulation of cadmium in higher (fish) and middle (certain invertebrates) marine organisms, the cadmium load is pursued for a very long temporal period, and the stationary state—or saturation level—is not achieved during a 180-day period. Given that, on the other hand, the cadmium fixed by the previously intoxicated organisms is released into an uncontaminated environment at an excessively low rate, a more detailed study of the fixation mechanism has led to the conclusion that the cadmium in these organisms is blocked by specific proteins synthesized as the organism's defence against the polluted environment.

These two isolated examples of application do, however, call for a few general comments from the ecotoxicological viewpoint. The entire problem of the tolerable doses would entail estimating the fauna's capacity of defence as a function of the kinetics of its production of protective proteins. There is no doubt that marine species can, without difficulty, accumulate huge, dangerous doses, doses which are indeed lethal to man. One can expect that only relatively little damage will be done to the marine world by the dumping of heavy metals. The danger of eutrophication, with its sequel of ecological alterations appears to be of much greater importance.

Both these types of pollution have an impact on man in that one reduces the quantity of food, whilst the other reduces the quantity of consumable food. On the other hand, in the eutrophized regions (coastal zones for example), the heavy metals bind readily with the dead organic matter in suspension and with the sediments. Anaerobic conditions in the water column and the sediments resulting from bacteriological activity lead to stable forms of heavy metals and their potential danger is reduced.[9] A sudden regression of the eutrophication could lead to the heavy metals being put back into circulation and result in the direct contamination of animals and plants at all levels.

ZONATION

Introduction

When it comes to describing the *evolution* of a system on the basis of chemical and biological variables, the undertaking is often revealed to be a delicate, if not simply an impossible one because of the lack of sufficient data concerning the exact description of the general terms of interaction in their spatiotemporal formulation. One can in this case use the statistical data relating to a more or less important *region* of the space studied. This region is then considered as a homogeneous system or 'box' whose overall exchanges with the outside world can then be studied. This time using a more refined approach to the study of the dynamics of the evolution and the interactions of the populations, one can also then subdivide each region or 'box' into several niches where the parameters of control appear to be identical and thus consider each niche as a homogeneous system.

It is of course essential to bear in mind, even and *especially* in this case of subdivision into groups and subgroups, the rules which always prevail in all modelling tests, namely the need for physically delimiting the ecosystem studied from the outside world and for quantifying the relations of closure which express the external influences exercised on this ecosystem and vice versa, whether this concerns the flux of advection or diffusion to or from the adjacent seas, the air–sea and sea–sediments interactions, the injection of estuary waters, the effect of the coast, etc.[2]

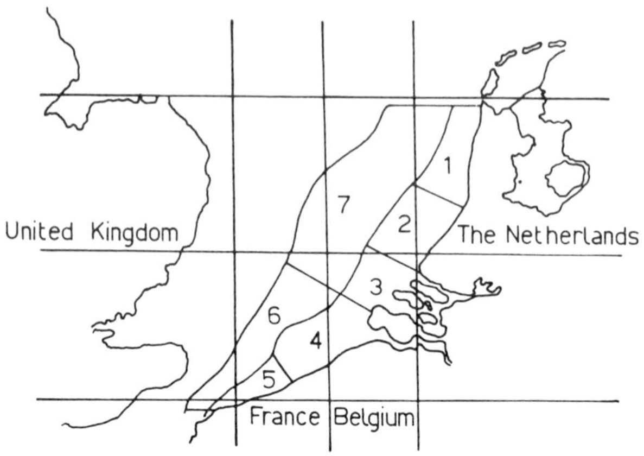

Fig. 17.3 Zonation of ecosystems of the Southern Bight of the North Sea

Geographical Repartition of the Ecosystems of the Southern Bight of the North Sea

Rather than attempting to simulate the temporal evolution of the state variables at all points of the space considered, it is preferable to study their mean values in certain characteristic zones assumed to be homogeneous.

Those conducting the experiments have been able to highlight the existence of such zones, in keeping with the results provided by the hydrodynamic models of residual circulation.[5]

They were able to distinguish (Figure 17.3) an offshore zone, characteristic of the ocean waters and comprised between the current functions of 100 and $200 \times 10^3\ m^3\ s^{-1}$, a southern coastal zone where the existence of a gyre increased the residing time of the Scheldt waters, and a northern coastal zone highly influenced by the Rhine–Meuse estuary.

For the modelling, these ecological zones were subdivided into seven boxes so as to better localize the effect of the gyre, that of the estuaries, and to highlight any dephasing from one box to another.

The temporal variation of a variable in a given box is equal to the flux of advection and diffusion between this box and the adjacent boxes, plus a sum of biochemical interactions responsible for the production or consumption of the variable within the box considered.

TRANSFER COEFFICIENTS

Introduction

On the basis of the degrees of zinc contamination of the zooplankton and fish compartments observed in the southern part of the North Sea,[1] namely 191 ppm

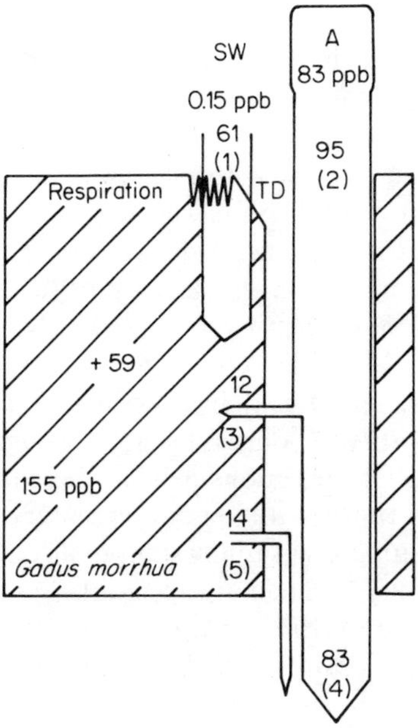

Fig. 17.4 Mercury accumulation in cod.[10] SW (sea water); F (food); (1) Hg entering via water; (2) Hg contained in food; (3) Hg assimilated via food; (4) Hg not assimilated through food; (5) Hg excreted by fish. The flux are expressed in 10^{-5} μg Hg g^{-1} of fresh weight and *per day*. (*Source:* Bouquegneau *et al.*, 1977)

and 10.3 ppm (fresh weight) respectively, a transfer coefficient of 0.054 from the zooplankton to the fish can be calculated. For the other heavy metals, the values observed for the transfer coefficients are 0.096 for Hg, 0.006 for Cd, 0.027 for Cu and 0.023 for Pb, namely values which are always lower than the unity. It may, at first glance, seem strange that at this level of the trophic chain, the predator is systematically less contaminated than the prey, but on further study, this is found to underline the importance not only of the study of excretion mechanisms, but also the detailed study of the mechanisms of contamination.

Direct and Indirect Heavy Metal Intoxication

To our knowledge, the direct study of the zooplankton–fish mechanism of transfer has not yet been undertaken in a specifically pelagic environment. It has, however, been clearly established that heavy metals are found in much greater concentrations in marine organisms than in sea water, and that they are generally more abundant still if the organisms originate from polluted environment (sea water–organisms transfer coefficients > 1.00). The accumulation of these metals in the tissues can result in direct absorption (from the water) or indirect absorption (from nutrients) or both simultaneously.

Bouquegneau *et al.*[10] have made a detailed study of the relative importance of the direct and indirect accumulation of mercury in several demersal species, particularly in cod (*Gadus morrhua*) living in the Belgian coastal zone. The results shown in Figure 17.4 illustrate that the contribution of direct accumulation is clearly greater than that of indirect accumulation (digestive system) and that a large quantity of pollutant passes through the digestive system and emerges in the form of dejections whose mercury content is highly concentrated in relation to those of the prey (alimentary ration).

One can summarize the situation as follows: input flux daily total, $(95 - 83) + 61 = 73 \times 10^{-5}\,\mu g$ Hg g^{-1} day^{-1}; net daily accumulation, $59 \times 10^{-5}\,\mu g$ Hg g^{-1} day^{-1}; direct accumulation, $61 \times 10^{-5}\,\mu g$ Hg g^{-1} day^{-1}; indirect accumulation, $12 \times 10^{-5}\,\mu g$ Hg g^{-1} day^{-1}; rate of direct accumulation, $49 \times 10^{-5}\,\mu g$ Hg g^{-1} day^{-1}.

It is therefore obvious that 85 per cent of the Hg present in the fish results from a direct absorption of the pollutant from the water. This moreover clearly accounts for the fact that one does not necessarily observe richer quantities of metal from one echelon to another of the food supply chain (transfer coefficient < 1.00): the mercury input flux (and probably that of other metals) from nutrients is lower than the output flux resulting from the excretory mechanisms of this pollutant.

HEAVY METAL CONTAMINATION OF ZOOPLANKTON

Introduction

This section on zooplankton sets out to associate to this compartment of the food-supply chain, a *representative average annual value* of its degree of contamination by each of the elements—zinc (Zn), copper (Cu), cadmium (Cd), mercury (Hg) and lead (Pb)—a value which is used in the general study of translocation.[1] The fact that Zn has been chosen here as the example is in no way prejudicial to the relative importance this metal may have in the contamination of the products of the sea. Moreover, Zn and Cu are metals essential to cellular functioning, which leads us first and foremost to believing that the mechanisms of admission of these two elements into the compartment considered could be similar, but different from those of other metals (Cd, Hg, Pb) which are not directly involved in normal cellular functioning.

Dispersion of Concentrations

The dispersion of concentrations in a given pollutant of different specimens of one and the same species of marine fish is an important factor which has already been underlined.[11] This is true also for the concentrations in the plankton considered as a whole. In the case of Zn in particular, in the fifty-six samples taken in 1974/75, spatially and temporally distributed in a relatively homogeneous way throughout the system studied, the concentration varied from 511 to 25.652 μg of Zn per gram of C.

For methodological reasons, the concentrations are expressed in grams of metal *per gram of organic carbon* and not in ppm of lyophilized dry matter. The samples analysed in fact often still contained a certain percentage of exogenic matter, NaCl, fine adsorbed particles in suspension, etc., the sole result of a more thorough rinsing prior to lyophilization being to damage the cells and modify their composition which in this case no longer corresponds to the intracellular content.

The calculation of concentration levels using units other than the aforementioned one will be based on the following coefficients, (a) the zooplankton as a whole is considered as having a total C content of 40 per cent in relation to dry matter, and (b) the moist, drained matter rid of its impurities is considered as having a degree of humidity between 80–90 per cent (namely an average 15 per cent of dry residue).

As for the fish,[11] the repartition of concentrations seems to correspond to a normal log distribution with, in addition, a certain bimodal character (see Figure 17.5), the extreme terms of which vary here in a proportion of 1–50. Without prejudice to the fact that this group (fifty-six samples) can be considered as homogeneous, the most representative value of the average level of contamination is, in a first approximation, shown by the value of the geometric mean, namely in the case of Zn: 2.591 μg Zn per gram of C.

Six particularly high values (>8700 μg Zn) seem to emerge in this group but an elementary statistical analysis shows that although a greater dispersion of the concentrations of the total fifty-six samples is significant at a level of reliability of 0.95 (F variance test) in relation to the dispersion of concentrations of a reduced sample (fifty samples), the respective mean values of both batches considered differ only at a reliability level of 0.80 (two-tailed t-test).

The distribution of the results of the 1974 and 1975 campaigns appears to be fairly homogeneous. There is no significant test to differentiate one (1974) or the other (1975) in relation to the total (1974 + 1975) except in an individual comparison: the 1974 concentrations are more dispersed than the 1975 ones at a reliability level of 0.95, but the respective averages differ significantly only at a reliability level of 0.80. One must therefore look elsewhere for the origin of this grouping and, on this basis, the most significant value of the average level of zinc contamination of the zooplankton remains the *geometric mean* cited earlier.

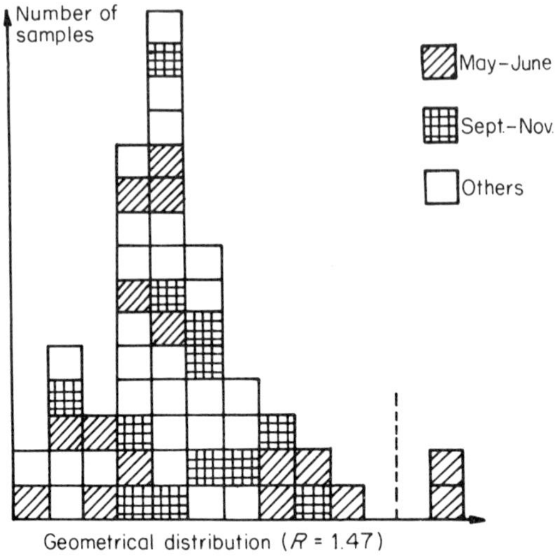

Fig. 17.5 Histogram of the zinc contamination of zooplankton, normal log distribution in 10 classes, temporal repartition

Exogenic Contamination of Samples

The hypothesis of the contamination of the zooplankton collected by exogenic matter rich in heavy metals must be envisaged (see above). Under the sampling conditions the matter could not (especially during the low season when collection is difficult and takes place over long periods) be anything other than *matter in suspension*.

Wollast[12] expresses a quantitative estimate of the composition of this matter as regards Zn in the form of an equation

$$\text{ppm Zn} = 103 + 20.6x_0$$

in which x_0 represents the mass content of organic matter as a percentage. The maximum concentration of organic matter observed (expressed in grams of carbohydrate per 100 grams of dry matter in suspension) being in the order of 30 per cent, the suspended matter richest in Zn found in the test zone gives us a value of *721 μg of Zn* per gram of dry matter. If one compares this value with the geometric mean of the concentration levels observed for the zooplankton, expressed in the same units, namely *1.036 μg Zn* per gram of dry matter, it is clear that a contamination of this kind, if it exists, *can certainly not* account for the high values recorded: on average, the zooplankton is richer in Zn than it is in suspended matter.

One cannot, however, totally exclude a contamination of this kind, particularly for the low concentrations: on the one hand, our lyophilized samples rarely contain 40 per cent (dry matter); there is, therefore, some exogenic matter and, on the other hand, the samples as a whole taken during the 1974/75 campaigns show a Zn concentration which very slightly *anticorrelates* the concentration of C, hence it would appear that the samples with a C content are the most contaminated. The contamination factor must therefore be re-examined in the light of other data such as, for example, the control of C/N ratio, the amorphous silica content, etc.

Annual Variations in the Zinc Content of the Zooplankton

If one considers, on the one hand, the overall concentration level recorded during the 1974/75 campaigns (fifty-six samples) and, on the other hand, the same concentrations grouped according to the marine zones as defined earlier (see above), namely *zone 6–7* (twenty-two samples), *zone 1–2* (thirteen samples) and *zone 3–5* (twenty-one samples) respectively, one can see that there is *no selection per zone* at a sufficient level of reliability for either the dispersion test or the comparison test of the mean values for all the possible group combinations, excluding, however, the fact that the variability in zone 6–7 is validly defined as being *inferior* to that of the 1974/75 group at a reliability level of 0.90. Consequently, from the point of view of the Zn contamination of the zooplankton, the three zones are distinguished not by a different representative mean value, but by a greater variability of the concentration levels recorded in zones 1–5.

On the other hand, a study of the histogram illustrating the difference between the high season (May–June, September–November) levels of concentration and those of the low season (see Figure 17.5) reveals that, although during the high season one simultaneously observes low and average levels of concentration, *all* the high concentration values also correspond to this same season, especially in the high-production coastal zone.

The elements of appreciation listed below obviously again bring up the problem of the definition of a *representative average annual value* of the degree of Zn contamination with regard to the zooplankton. Indeed, we know that the geometric mean 'maximalizes' the low levels of concentration and we obviously cannot—on the basis of a chance temporal distribution—ensure an equitable ponderation between the high and low season.

This is why, starting with all our observations grouped per cruise, we were prompted to reconstruct a diagram of the annual variation of Zn concentrations based on the following characteristic values (Table 17.1).

The diagram of annual variation reconstituted in this way can be compared to the diagrams illustrating the annual evolution of the zooplankton biomass.[2] The correlation per point is relatively good, taking into account a difference of approximately 21 days on the temporal scale ($r = 0.87$).

It goes without saying that the average value *most representative* of the Zn

Table 17.1 Annual variation in the degree of contamination

Date	$\bar{x}$	n	Ref. [13]
Dec.–Jan.	2591	—	MG
30.4–9.5	2960	14	P2
27.5–30.5	10162	9	P3
17.6–18.6	2596	7	B10
8.7–11.7	2050	14	B11
17.9–24.9	3929[a]	6	A2
1.10–3.10	1643	3	P5
20.11–22.11	6827[a]	3	P6

[a] These relatively 'high' values for the season are due mainly to the contribution of zones 1–5 (coastal); they are representative of the whole although the average values for zones 6 and 7 are 2122 μg Zn g^{-1} C for A2 and 3453 μg Zn g^{-1} C for P6 respectively.

contamination of the zooplankton compartment must be calculated on the basis of the integral of the curve discussed below (see Table 17.2) after smoothing.[14,15] This average value is identical for the three zones (except for the slight anomaly characteristic of zones 4–5) but, for each zone, we associated this value with a confidence interval (level of probability 0.10) which is dependent on both the number of representative samples and the characteristic variability peculiar to each zone (normal log distribution). The final values are shown in Table 17.3 below.

Lastly, taking into account the biological period of the organisms tested, we divided the total integral into portions of 1 month which made it possible for us to estimate the average value of the Zn contamination of the zooplankton. Table 17.3 consequently gives us

1 An estimated value per quarter- and per half-year of the zooplankton compartment's partial contribution (cumulated values) to the other compartments of the system (fish, return to water-column etc.); this value corresponds to the integral of the smoothed curve of Table 17.1 for the interval considered, divided by the number of months.

Table 17.2 Degree of Zn contamination of zooplankton

Zone	Value (μg Zn g^{-1} C)	Confidence interval	Value (ppm/wet)	Confidence interval
total	3181	±587	191	±35
4–5	3181	±1435	191	±86
1–3	3045	±1537	183	±92
6–7	3045	±556	183	±33

Table 17.3 Annual evolution of the degree of Zn contamination of the zooplankton

Month	Cumulated value	Uncumulated value	Month	Cumulated value	Uncumulated value
January	2201	2201	July	3297	1964
February	2302	2404	August	3080	1565
March	2404	2607	September	3011	2457
April	2505	2810	October	2987	2770
May	3273	6344	November	3232	5675
June	3519	4747	December	3181	2625

2 The estimated value of the degree of contamination which one can expect to find *at any moment* in the zooplankton compartment (uncumulated values).

If one assigns to these values a confidence interval of the same size as the one shown in Table 17.2 for each zone, one accounts perfectly for all the concentrations recorded experimentally, bar four exceptions, which shows that the smoothing technique used is correct.

CONCLUSIONS

Whilst pursuing the same objectives as those fixed for elaborating the large-scale study[11] into the contamination of other products of the sea, namely ascertaining, on the one hand, the dangers of toxicity and, on the other hand, gauging the evolution of the levels of contamination, it goes without saying that with the material currently available with regards to the zooplankton, the conclusions which emerge are limited to an enumeration of the correlative factors resulting from this *first phase* of the analytical study.

The problem relating to the *dispersion* of these concentrations has already been widely discussed in the four preceding sections. In the case of the zooplankton, it has been shown on several occasions that the distribution or dispersal of the concentrations often represents a *characteristic* of a given zone or group of zones; what in fact distinguishes zones 1–5 from the other two zones 6–7, is the variability of the concentrations recorded whilst the average levels are of the same order of magnitude. There is no doubt that the variability observed in group 4–5 is due to the influence of the gyre or, in other words, of the time of residence of the waters originating from the Scheldt (see above).

In order to study the *temporal factor*, the approach we adopted was obviously quite different from the one used for the study of the 'fish' compartment. This is because the time of residence of a given pollutant in the zooplankton compartment is determined to a great extent by the mortality or consumption (biological period and grazing), in addition to the excretion phenomena and the ingestion mechanisms. Clearly then, the annual and seasonal variations *are more pronounced* and the influence *of the medium* is also quantified with greater speed.

A close study of the evolution of the degree of contamination of the zooplankton (case of the Zn, Cu, Cd, Hg and Pb), shows us that we are dealing with two very different kinds of seasonal evolution; the first corresponds to an *increase* in the degree of contamination in relation to the annual mean concentration during the period(s) taken to be the high season; the second corresponds to a *reduction* in the concentrations in relation to the mean concentration during this same period. It is to this first group that the Zn and Cu diagrams belong, with the Hg and Cd coming under the second; Pb, for its part, appears to be a perfect intermediary between these two extremes. This is probably due to the essentially particle-like character of the Pb present in the water column, a phenomenon often referred to in the literature and one which has frequently been subject to numerous checks and re-checks.[16] It looks very much as if—for the Zn and Cu—the levels of contamination of the zooplankton increase at the same time as the biomass or with a certain temporal shift, with the phytoplankton biomass as the main food-supply. For the Cd and Hg on the other hand, it would appear that the enrichment of the water column in detrital matter following the explosion of the phytoplanktonic biomass (detrital matter is a trap for pollutants) has a *depletive* effect on the contamination levels of the zooplankton.

It is as yet too early to cavil at the intimate mechanisms of admission of the pollutants into this link of the food-supply chain, but such differences will certainly play an important part in the *choice* of options as regards the selective processes of accumulation by direct (alimentary) or indirect mean (contact contamination so to speak).

REFERENCES

1. Elskens, I., Baeyens, W., and Goeyens, L. (1980). Le transport de certains métaux lourds à l'intérieur de l'écosystème marin dans le sud de la Mer du Nord, Leçons et Séminaires de l'Ecole Européenne d'Eté d'Environnement, 4ème session 1977, *Etude et gestion du système marin*, **2**, 280.
2. Pichot, P., and Runfola, Y. (1980). Modèle mathématique de la circulation d'azote à travers l'écosystème pélagique de la Baie Sud de la Mer du Nord, Leçons et Séminaires de l'Ecole Européenne d'Eté d'Environnement, 4ème session 1977, *Etude et gestion du système marin*, **2**, 252.
3. Kelly, R.A. (1973). Conceptual ecological model of Delaware estuary, in *Systems Analysis and Simulation in Ecology*, vol. 3 (Ed. B.C. Patten), Academic Press.
4. Chen, C.W. (1970). Concepts and utilities of ecological model, *J. San. Eng. Div.*, 1085.
5. Nihoul, J.C.J. (1975). *Modelling of Marine Systems*, Elsevier, Amsterdam.
6. Steele, J.H. (1970). *The Structure of Marine Ecosystems*, Blackwell Scientific Publishers, Oxford.
7. Vollenweider, R.A. (1965). Calculation models of photosynthesis depth curves and some implications regarding day rate estimates in primary production measurements, *Mem. Ist. Ital. Idrobiol.*, **18**, 425.
8. Bouquegneau, J.M., Noel-Lambot, F., and Disteche, A. (1979). Fate of heavy metals in experimental aquatic food chains, uptake and release of Hg and Cd by some marine

organisms, Role of metallothioneins, *Rapports des Journées d'Etudes Actions Concertées Universitaires Océanologie, Belgium* (SPPS), 85.

[9] Baeyens, W., Decadt, G., and Elskens, I. (1979). A study of the distribution of mercury in the various compartments of the North Sea and Scheldt Estuary ecosystems, *Oceanologica Acta*, **2**, 447.

[10] Bouquegneau, J.M., Noel-Lambot, F., and Disteche, A. (1977). Le problème de l'intoxication directe et indirecte par les métaux lourds, *Contamination des produits de la mer, Programme National R-D Environnement—Eau, Projet Mer, Rapport final*, **9**, 255.

[11] Herman, P., and Vanderstappen, R. (1977). Analyse corrélative des facteurs liés à l'évolution de la contamination in situ des poissons, des invertébrés benthiques, de la faune et de la flore des brise-lames, *Contamination des produits de la mer*, **9**, 49.

[12] Wollast, R. (1977). Propriétés physico-chimiques des sédiments et des suspensions de la Mer du Nord, *Sédimentologie, Programme National R-D Environnement—Eau, Projet Mer, Rapport final*, **4**, 123.

[13] Elskens, I. (1977). Les niveaux de contamination de plancton par les métaux lourds, *Contamination des produits de la mer, Programme National R-D Environnement—Eau, Projet Mer, Rapport final*, **9**, 87.

[14] Kaiser, S. (1957). Numerical integration using Simpson's one-third rule, in *Numerical Analysis*, McGraw-Hill, New York.

[15] Shan Kuo, S. (1962). Interpolation (unequi-spaced data), in *Numerical Methods and Computers*, Wesley.

[16] Goeyens, L. Bijdrage tot een betere kennis van de verspreiding van zware metalen in een marien milieu, *2* lood, Licentiaatsverhandeling VUB 1973 (private communication).

Pollutants and Their Ecotoxicological Significance
Edited by H. W. Nürnberg

CHAPTER 18

Chemical Pollutants in the Marine Environment, with Particular Reference to the North Sea

J. C. DUINKER

SUMMARY

Present knowledge on the effects of metals and organochlorines derived from the Rhine/Meuse system in the adjacent coastal area is discussed. Dissolved and suspended components are distinguished as they have different transport mechanisms within the estuarine and coastal environment.

Several metals are removed from solution during estuarine mixing. However, several dissolved organochlorines behave in a conservative way during mixing in the estuarine zone. It appears that this behaviour is maintained during transport in the coastal region. High organochlorine concentrations are detected in marine mammals from the Dutch coast. This may be related to the strong decrease in the population of several species during the last decades.

INTRODUCTION

The chemical constituents that have been identified as pollutants in the marine environment can be classified as metals, oil components and synthetic organic compounds.

Metals are natural components of river and sea water, usually in very low concentrations. Serious problems as a result of their presence in the aquatic environment can occur at concentrations well above the natural values and/or when the distribution over the various chemical forms in which an element can occur is different from the natural situation. Human activities may thus cause changes in natural cyclic processes with possible long-term effects.

Oil components have been formed from natural organic matter under conditions that differ widely from those taking place at the earth's surface. The compositions of oil or oil residues are thus different from those of natural organic matter in water or sediment.

Probably most synthetic organic compounds have not been present in nature before their synthesis by man. Serious interference with natural processes in the biotic and abiotic environment result from the introduction of compounds of this type. A large group of compounds is biodegradable. However, some compounds, being stable in the environment, having detrimental effects on organisms and produced in large quantities during several decades, have been detected in practically all parts of the oceans, even in the most remote areas, as a result of their use on land. Several organochlorines belong to this group, in particular a number of pesticides such as DDT and its metabolites, dieldrin etc., and the group of polychlorinated biphenyls (PCB).

All contaminants in the sea are derived from activities on land. The transport mechanisms are rivers, pipelines and dumping activities, and atmospheric transport. The latter mechanism may be important for open ocean areas where the direct influence of the former routes is relatively small, in contrast to coastal areas such as the North Sea.

Once in the coastal marine environment, components in dissolved form will be transported from the coast out into the open sea along with the coastal water body. Components that are associated with particles behave quite differently resulting in a considerably larger residence time for particles in the coastal area than for dissolved species. River-borne pollutants that are associated with particles when leaving the estuary, will thus have a smaller opportunity to escape into the open sea than dissolved components.

Marine organisms are affected by dissolved and particulate pollutants either directly or through accumulation of chemicals in their tissues from water/sediment and from their food sources. Some chemical compounds can be metabolized by some marine species. The products can be excreted or accumulated in tissue, sometimes in a form that is more harmful than the parent compound. The distribution of pollutants over coastal and open sea areas will be illustrated for the Rhine/Meuse/Scheldt systems and the adjacent North Sea using data for metals and organochlorines. Special attention will be given to the Wadden Sea, being part of the coastal area in which particles are trapped, as a result of its topographical, hydrodynamical and biological characteristics.

TRACE METALS

Dissolved copper (Cu), zinc (Zn), cadmium (Cd) and iron (Fe) are removed from solution in the early stages of estuarine mixing, and manganese (Mn) in the lower estuary.[1,2] Further dilution with seawater from roughly 10 per thousand salinity onwards results in practically conservative behaviour, as can be deduced from dissolved metal–salinity plots and from plots showing the geographical

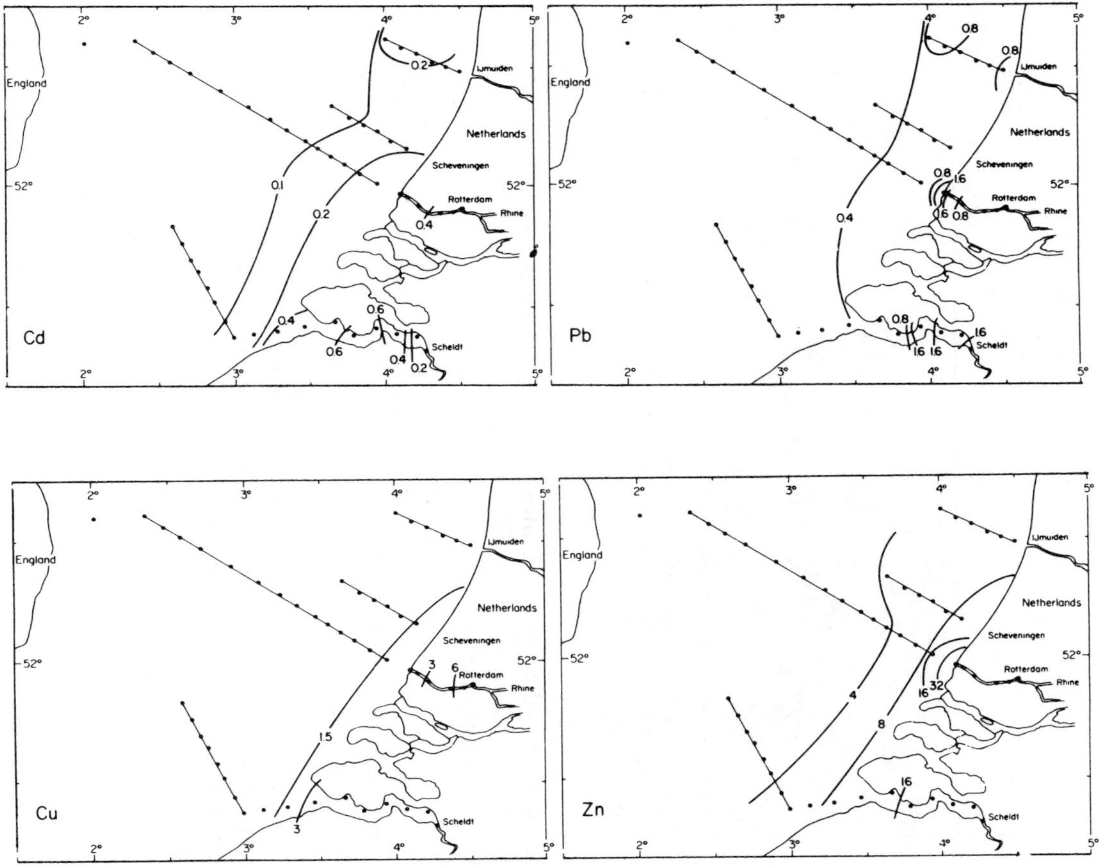

Fig. 18.1 Distribution of dissolved trace metals in $\mu g\,l^{-1}$

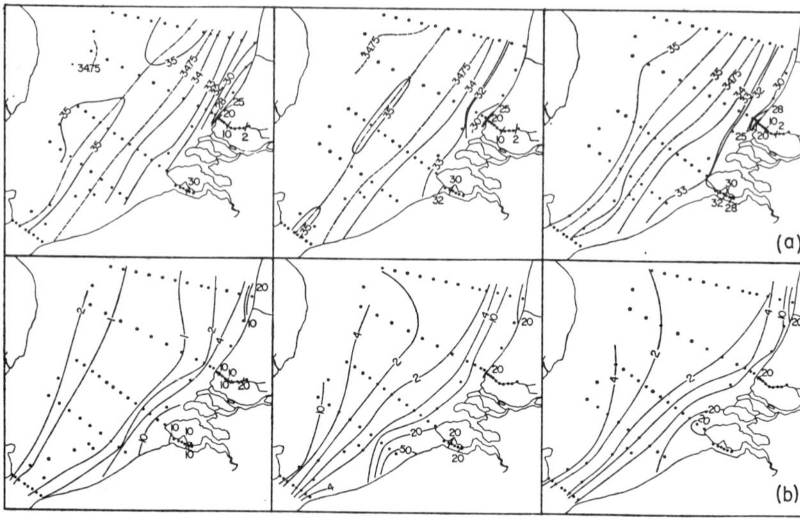

Fig. 18.2 Distribution in the Southern Bight of salinity in (a), S per thousand and, (b) suspended matter in $mg\,l^{-1}$ in September (left), October (middle) and November (right) 1974

distribution of dissolved concentrations in the coastal area near the mouth of the estuary (Figure 18.1). Dissolved concentrations in the off-shore region near the axis of maximum salinity are minimum. The direct influence of rivers appears to be absent. The more or less 'regular' geographical distribution in the coastal area may be modified considerably during storms when relatively large amounts of interstitially dissolved components may enter the water column. This effect is more important in the near coastal area than in the central part of the Southern Bight. This is a consequence of the trapping mechanism of particles that is not only operative within the estuary but also in the coastal area. For instance, several series of observations show that the axis of minimum seston concentration is shifted toward the coast with respect to the axis of maximum salinity (Figure 18.2).

The distribution of metal concentrations in suspension (in $\mu g\,l^{-1}$) from the coast to the central part of the Southern Bight follows the distribution of suspended matter closely (Figure 18.3a–c for three different surveys in transects off Rotterdam). A steady decrease in a seaward direction is observed for seston and the four metals in the data of the three surveys. However, the changes occur with different rates for the various metals. As is demonstrated in Figure 18.3a, where concentration has been expressed as the amount of metal per sediment dry weight ($\mu g\,g^{-1}$) in the River Rhine transect, values in an offshore direction decrease for Fe, Cu and Zn and increase for Mn. The same observations apply to the data of the transects more north and south (Figure 18.3c) of the 'Rhine'

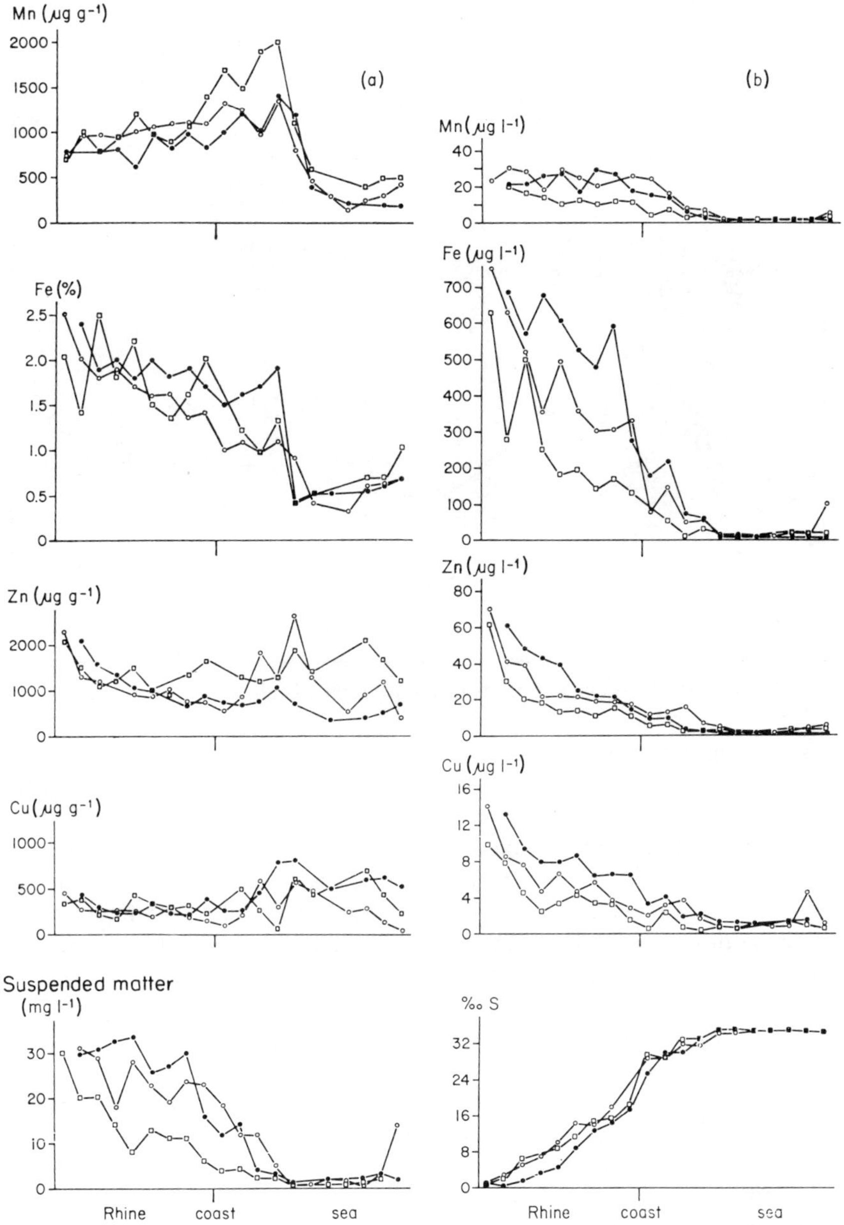

Fig. 18.3(a and b) 0.1 N HCl leachable (18 h) trace-metal concentrations in $\mu g\,l^{-1}$ (right) and in μg metal per g particulate matter dry weight (left) together with suspended matter in $mg\,l^{-1}$ and salinity in S per thousand at individual sampling stations along transect II (off Rotterdam) during cruises in September (□), October (○) and November (●) 1974 as in Figure 18.2

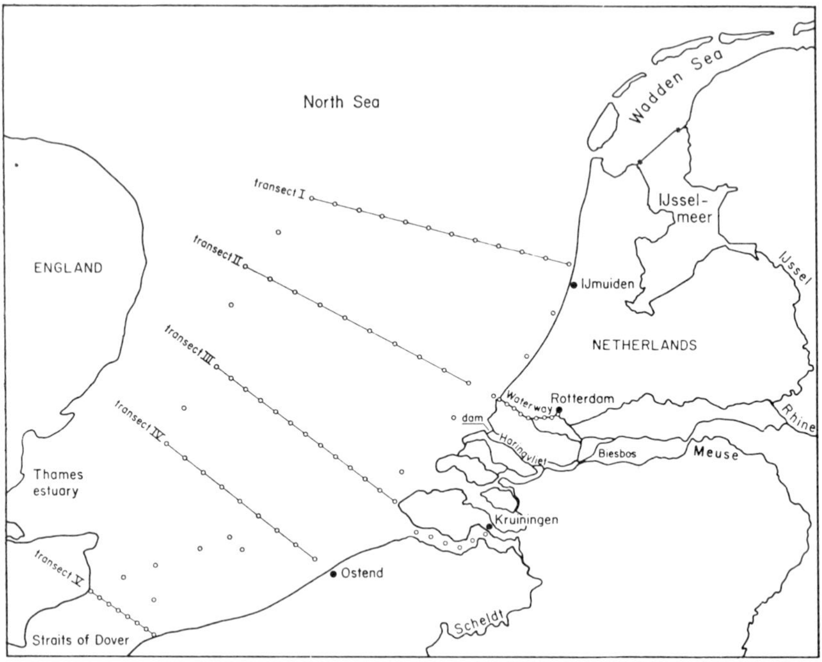

Fig. 18.3(c) Sampling stations in the Southern Bight of the North Sea

transect. A remarkable observation is the occurrence of maxima in the Cu and Zn concentrations ($\mu g\,g^{-1}$) in the central part of the Southern Bight, where seston load is only 1–2 $mg\,l^{-1}$. Manganese concentrations decrease rapidly in an offshore direction after the maximum at approximately 30 per thousand S.

These observations can be interpreted in terms of mixing in the estuary and near-coastal area of marine-derived and river-borne particles with different metal concentrations, and additionally in terms of physicochemical processes. The latter apply in particular to Mn that is removed from solution in the lower estuary and the coastal area as result of the (dissolved) Mn(II) → (particulate) Mn(IV) transitions induced by changes in pH, dissolved oxygen, chloride and bicarbonate concentrations and in activity coefficients (Figure 18.3).

A complicating factor in the interpretation is the effect of particle size. Some elements like Cu, Zn, Pb, Cd and organic C are preferentially associated with smaller, less dense particles (group I elements, Figure 18.4), dominating at low suspended matter concentrations. On the other hand, elements like Al, Fe, Ti and K (group II elements) are associated mainly with particles of larger size and/or density, dominating at larger suspended matter concentrations. We have found that Si and Ca did not show any preference for either fraction (group III). The relation between elemental contents in suspended particles and the total suspended matter concentrations is represented schematically in Figure 18.4

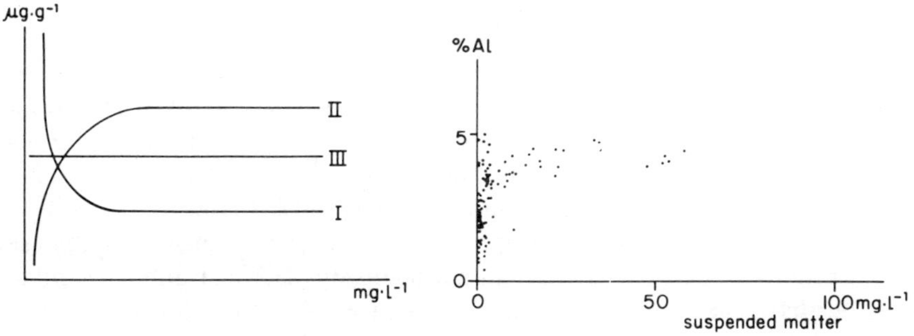

Fig. 18.4 (Left) Relation between elemental contents of group I, II and III elements ($mg\,kg^{-1}$) and seston concentrations ($mg\,l^{-1}$) and (right) actual data for Al

(left). The actual data for Al in the Southern Bight are represented in the right part of that figure. The elemental contents are practically constant at concentrations above 5–10 $mg\,l^{-1}$, depending on the geographical area. The particles giving rise to these data mainly are derived from the bottom by resuspension. The content of group I elements increases rapidly with decreasing concentrations below 1–2 $mg\,l^{-1}$. These particles may be composed of Fe and Mn oxides and detrital organic matter, being in suspension continuously.

The trapping mechanisms of particles observed in estuaries is even more pronounced in the Wadden Sea. The mechanisms of accumulation of fine-grained material against the gradient have been explained in terms of geomorphological and hydrodynamical characteristics of the area.[5–8] In parts of the Dutch Wadden Sea reducing conditions in the lower top layers exist. Under these conditions a certain part of the total metal content of bottom particles can be transformed into an exchangeable position. This process causes the metal to be more readily available to the environment. These processes are more effective in fine-grained bottom sediments.[9]

The metal content of suspended matter and water in the Dutch Wadden Sea are high; for interstitial water the observed values are even higher. The following mechanism may play an important role in the interpretation of these observations. Part of the suspended matter that is introduced from the North Sea will settle in the Wadden Sea, and lose part of its metal content under reducing conditions. This results in an enrichment of the interstitial water, from which the overlying water is enriched by water exchange and diffusion processes. Tidal and wind action will increase the efficiency of this process. This was demonstrated by observations made in the Wadden Sea at fixed positions with minor salinity variations during a complete tidal cycle. It may be a significant mechanism in the production of fine, less-dense material in suspension with high concentrations of some metals (e.g. Cu and Zn) per unit weight. Part of this material may be transported to the open ocean more or less like dissolved species.

ORGANOCHLORINES

Several organochlorine compounds derived from human activities on land are persistent in the marine environment. Their partition over the various biotic and abiotic compartments depends partly on biological and hydrodynamical factors and their relative solubilities in water and lipid material. Lipophilic organochlorines such as polychlorinated biphenyls (PCB) and members of the DDT family have high affinity for lipid tissue of marine organisms. High tissue concentrations of these compounds in marine mammals from California and the Baltic Sea have been associated with reproductive failure in sealions and seals in these areas.[10-12] Very high concentrations have also been observed in specimens from the Dutch Wadden Sea.[13-16] They are suspected to be involved in the strong reduction and the reduced pup production in the population of the harbour seal (*Phoca vitulina*) and in the disappearance of the populations of the harbour porpoise (*Phocoena phocoena*) and bottle-nosed dolphin (*Tursiops truncatus*) from Dutch coastal waters during the last decades.[17-19]

Organochlorines in river systems are transported in solution and suspension. Redistribution of any individual component over these forms may take place during estuarine mixing as result of physical, biological and chemical processes. As these forms have different transport mechanisms through the estuary out into the sea, this may have important consequences for the distribution and biological effects of organochlorines in the marine environment. Analysis of organochlorines in samples from the Rhine/Meuse estuary demonstrates the conservative behaviour during estuarine mixing of polychlorinated biphenyls, pentachlorobenzene, hexachlorobenzene, α-HCH (i.e. α-hexachlorocyclohexane), β-HCH, γ-HCH, dieldrin and endrin.[20] Data from the coastal area suggest that conservative behaviour is maintained during transport in the coastal environment. No similar data are available for other estuaries. The lowest values were found in the central part of the Bight, in the range $<0.91\ \mathrm{ng\,l^{-1}}$ for endrin to $1.0\ \mathrm{ng\,l^{-1}}$ for γ-HCH. These levels are at the lower end of the large range of values reported for open ocean areas reported by others. A striking observation for dissolved PCB in North Sea water is the relatively large contribution of lower chlorinated PCB components. This has not been reported for this area and other regions before. In all cases the PCB formulation giving the best fit on packed columns was reported to be Aroclor 1254 or Aroclor 1260, indicating the predominance of higher chlorinated components. The source of the presently reported lower chlorinated components may well be deposition from the atmosphere.

A large fraction of the organochlorines detected in water from the western Wadden Sea is derived from the Rhine estuary. Concentrations in rivers entering into the Wadden Sea directly are considerably lower than in the Rhine. Attempts to quantify the various sources are not yet complete.

Concentrations in seals from the North Sea, appear to be higher in specimens

Table 18.1 Range of concentrations of organochlorines in tissue of seals from the western and eastern Wadden Sea, in mg kg^{-1} fresh weight.[16,23] PCB are in Phenochlor DP6 equivalents

		PCB	ΣDDT	γ-HCH	Dieldrin
Blubber	W	22–600	0.51–25.4	0.03–0.39	<0.02–2.3
	E	27–564	2.2–23.3	0.04–0.98	0.06–0.9
Liver	W	1.5–36	0.1–1.3	0.001–0.02	0.01–0.07
	E	0.38–2.02	0.06–0.25	0.005–0.006	0.01–0.024
Kidney	W	1.6–31	0.05–0.76	<0.001–0.03	<0.001–0.01
	E	0.22–0.87	0.05–0.25	—	trace
Brains	W	1.4–46	0.06–3.10	<0.001–0.13	<0.003–0.02
	E	0.25–2.96	0.038–0.161	trace	trace
Spleen	W	1.1–5	0.029–0.18	0.001–0.01	0.002–0.01
Heart	W	2.1–40	0.025–0.60	0.006–0.03	0.01–0.07

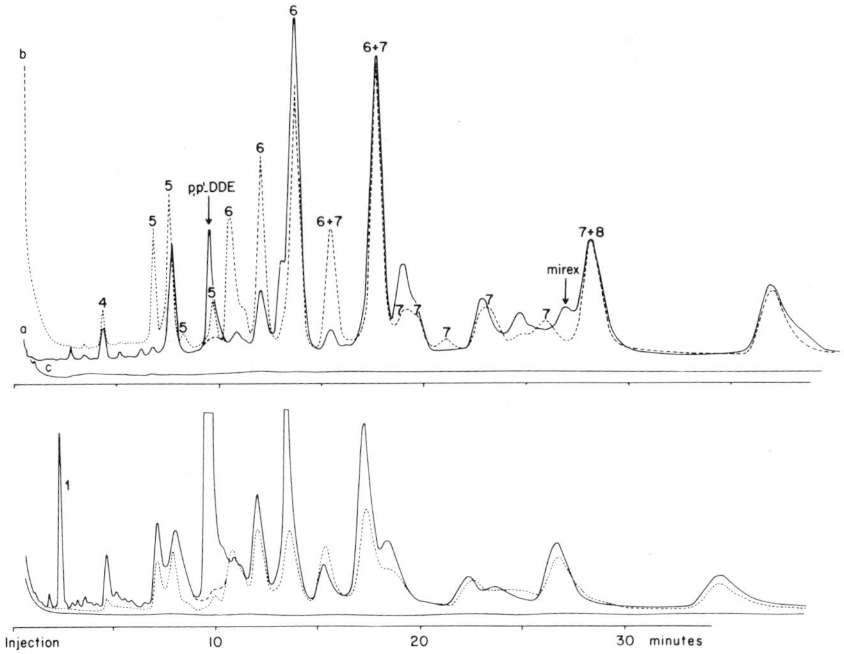

Fig. 18.5 Packed column chromatogram of the PCB fraction in seal blubber (upper) and liver of a harbour porpoise foetus (lower). Blank chromatograms are given as solid curves in the lower parts of the upper and lower part. —— sample extract, ... Phenoclor DP6, ---- sample extract after oxidation with chromic acid, leaving PCB unchanged and resulting in the disappearance of *p,p'*-DDE. The numbers in the upper part indicate the number of chlorine atoms in the eluting PCB components. (1) in the lower chromatogram indicates hexachlorobenzene. For chromatographic details see Duinker *et al.* (1979 and b)

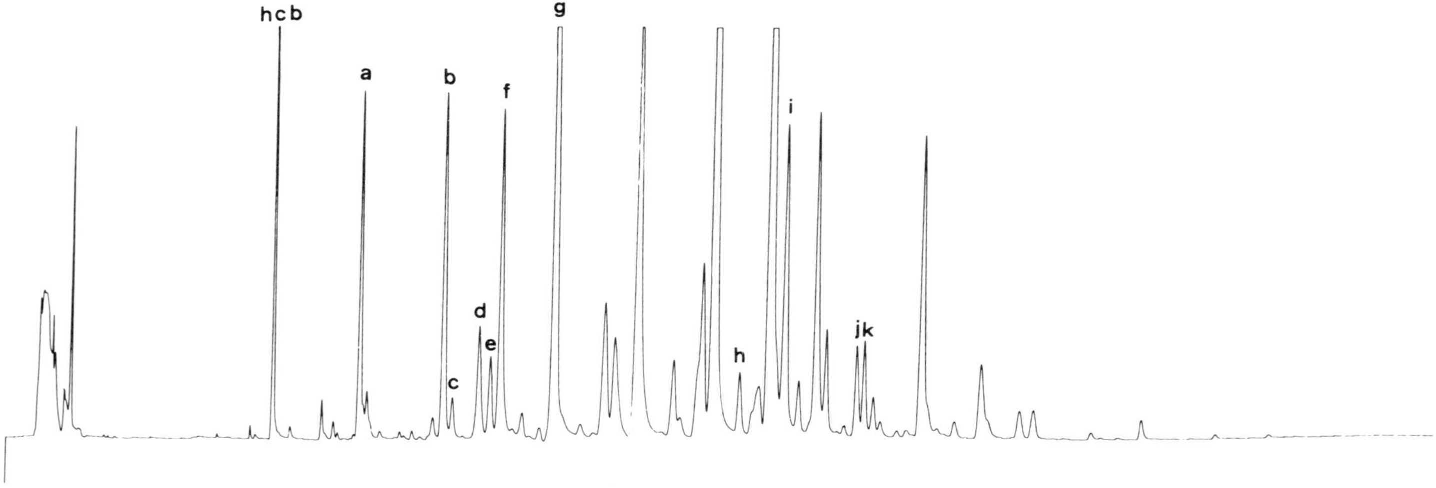
hcb
a
b
c
d
e
f
g
h
i
jk

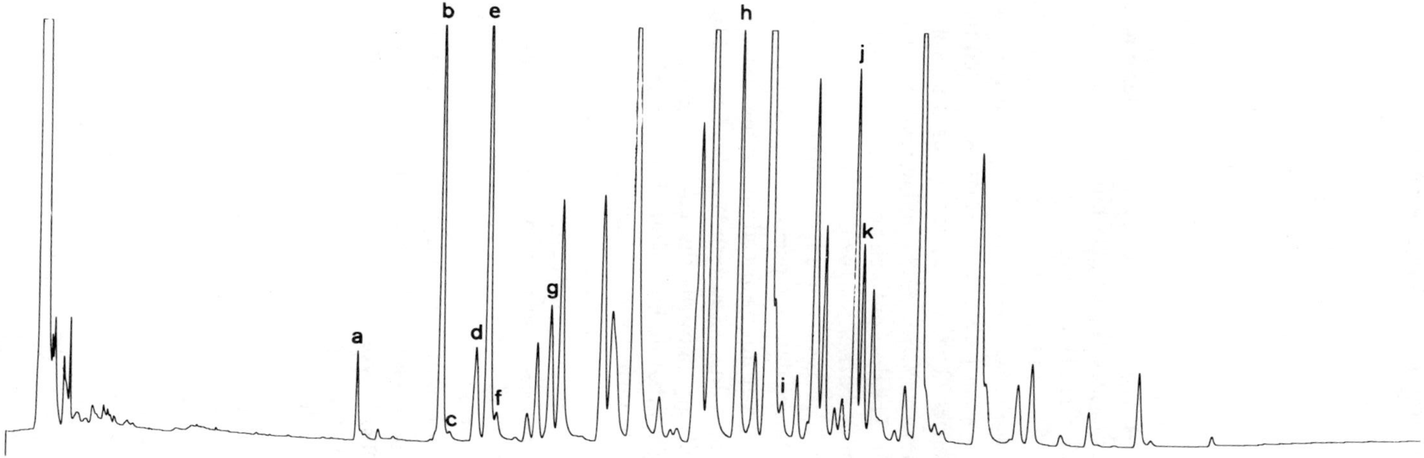

Fig. 18.6 Temperature programmed capillary gas liquid chromatograms of bubbler of two *Phocoena phocoena* (upper part) and the Clopher A60 (lower part). Peaks with same letters refer to the same PCB components (a, b, etc.). hcb = Hexachlorobenzene. For chromatographic details see Duinker *et al.* (1980)

from the Dutch coast,[13–16] than from the east coast of Britain[21–23] and the German Wadden Sea.[24] Concentrations of some organochlorines in blubber, kidney, heart, liver, spleen and brains in stranded seals from the Dutch Wadden Sea are highest for PCB, followed, although an order of magnitude or more smaller, by total *p,p'*-DDT, mirex, dieldrin, α-HCH and γ-HCH (Table 18.1).

Residue concentrations on an extractable fat basis vary among the tissues; these may thus depend on other mechanisms in addition to partition between aqueous and lipid phases. Alternatively, the lipids that determine solubility may differ between the various tissues. Thus, maximum concentration of PCB was found in heart lipids and a α-HCH and γ-HCH in brain lipids.[16] Low values of $\sum$ DDT and PCB on a fat basis in brain tissue with respect to blubber and other tissue were reported for sealions,[12] grey seals[22] and harp seals.[25] The latter authors reported ratios between 1:5 and 1:9. It was suggested that this may be due to a blood–brain barrier. The present ratios for brains and blubber are considerably larger. Values for PCB, $\sum$ DDT and mirex are typically in the order 1:1, while α-HCH and γ-HCH values are well above this ratio. These data suggest the absence of a blood–brain barrier for the latter compounds or it may be less active at the concentration levels present in the harbour seals from the Dutch Wadden Sea. Obviously, more data are necessary for a reliable conclusion.

The presence of high concentrations already in young animals has also been observed in other areas, including the eastern Wadden Sea and the east coast of England. The large similarity between the chromatograms of tissues from juveniles and adults suggests that mobilization of organochlorines from female tissue and transplacental transfer during parturition[16] and transfer to pups during lactation[26] may be important sources of organochlorines in juveniles of marine mammals, in addition to food ingested.

Figure 18.5 demonstrates the similarity in PCB patterns present in organs of a harbour seal (found dead) and in the liver of a 5-months foetus of a harbour porpoise, produced by abortion.[15,16]

It has been argued recently that PCB analyses with packed columns result in ambiguous data. The main reason is the insufficient resolution of the various individual PCB components contributing to each of the peaks in the chromatograms. The PCB components have different (physico-) chemical properties (volatility, solubility) and various physical, chemical and biological processes taking place in the environment have different effects on the distribution of the various components. Therefore, packed-column chromatograms do not represent the composition of the PCB mixture in environmental samples accurately. Thus, technical formulations such as Aroclor 1260 cannot be used for comparison and accurate quantitation even if the chromatographic patterns are very similar or even indistinguishable.[27] The ambiguity can be eliminated to a large extent by the use of capillary-column chromatography, resulting in the separation and identification of a large number of individual components.

This new type of information for environmental samples is illustrated in Figure 18.6, representing capillary chromatograms of the same samples as in Figure 18.5, obtained with packed columns. No quantitative data will be reported here. However, it can be observed by visual inspection that the relative contributions of several adjacent peaks (indicated by letters) in the sample extract differ appreciably from the corresponding ones in the technical formulation with an overall chlorine content representing the packed-column chromatogram reasonably well (Figure 18.5). The numbers for PCB contents given earlier in this paper have to be considered as approximate therefore. Nevertheless, they are still useful for comparing different areas.

It has not yet been established which are the main factors that may be responsible for the strong reduction in the population of the harbour seal and the extreme reduction in the practically and completely vanished populations of the harbour porpoise and the bottle-nosed dolphin from Dutch coastal waters. It seems important that all specimens of these populations, once found dead, are analysed in detail. The present data may assist in future analysis of the problems associated with pollution and seals, the essential aspects of which have been reviewed elegantly in a recent paper.[28]

REFERENCES

[1] Duinker, J.C., and Nolting, R.F. (1977). *Mar. Pollut. Bull.*, **8**, 65.
[2] Duinker, J.C., and Nolting, R.F. (1978). *Neth. J. Sea Res.*, **12**, 205.
[3] Wollast, R., Billen, G., and Duinker, J.C. (1979). *Est. Coast. Mar. Sci.*, **9**, 161.
[4] Duinker, J.C., and Nolting, R.F. (1976). *Neth. J. Sea Res.*, **10**, 71.
[5] Postma, H. (1954). *Archs. néerl. Zool.*, **10**, 1.
[6] Postma, H. (1961). *Neth. J. Sea Res.*, **1**, 148.
[7] van Straaten, L.M.J.U., and Kuenen, P.H. (1957). *Geologie Mijnb.*, **19**, 329.
[8] van Straaten, L.M.J.U., and Kuenen, P.H. (1958). *J. Sedim. Petrol.*, **28**, 406.
[9] Duinker, J.C., van Eck, G.T.M., and Nolting, R.F. (1974). *Neth. J. Sea Res.*, **8**, 214.
[10] Olsson, M., Johnels, A.G., and Vaz, R. (1975). Rept. SNVPM 591 Statens naturvardsverk Sweden, 43.
[11] de Long, R.L., Gilmartin, W.G., and Simpson, J.G. (1973). *Science*, **181**, 1168.
[12] Le Boeuf, B.J., and Bonnell, M.L. (1971). *Nature*, **234**, 108.
[13] Koeman, J.H., and van Genderen, H. (1966). *J. Appl. Ecol.*, **3**, (Suppl.) 99.
[14] Koeman, J.H., Peters, W.H.M., Smit, C.J., Tjioe, P.S., and de Goeij, J.J.M. (1972). *T.N.O. Nieuws*, **27**, 570.
[15] Duinker, J.C., and Hillebrand, M.Th.J. (1979). *Bull. Env. Contam. Toxicol.*, **23**, 728.
[16] Duinker, J.C., Hillebrand, M.Th.J. and Nolting, R.F. (1979). *Mar. Pollut. Bull.*, **10**, 360.
[17] Reynders, P.H. (1976). *Neth. J. Sea Res.*, **10**, 223.
[18] Reynders, P.H. (1978). *Neth. J. Sea Res.*, **12**, 164.
[19] Verwey, J. (1975). *Int. Repts. Neth. Instit. Sea Res.*, 17a and 17b.
[20] Duinker, J.C., and Hillebrand, M.Th.J. (1979). *Neth. J. Sea Res.*, **13**, 256.
[21] Heppleston, P.B. (1973). *Mar. Pollut. Bull.*, **4**, 44.
[22] Holden, A.V. (1975). *Rapp. P. -v. Réun. Cons. int. Explor. Mer.*, **169**, 353.
[23] Heppleston, P.B., and French, M.C. (1973). *Nature*, **243**, 302.

[24] Drescher, H.E., Harms, U., and Huschenbeth, E. (1977). *Mar. Biol.*, **41**, 99.
[25] Frank, R., Ronald, K., and Braun, H.E. (1973). *J. Fish Res. Bd. Can.*, **30**, 1953.
[26] Addison, R.F., and Brodie, P.F. (1977). *J. Fish. Res. Bd. Can.*, **34**, 937.
[27] Duinker, J.C., Hillebrand, M.T.J., Palmork, K.H., and Wilhelmsen, S. (1980). *Bull. Environm. Contam. Toxicol*, **25**, 956.
[28] Holden, A.V. (1978). *Mammal Rev.*, **8**, 53.

Pollutants and Their Ecotoxicological Significance
Edited by H. W. Nürnberg

CHAPTER 19

Assessing Pollution in the Mediterranean Sea

S. W. Fowler

SUMMARY

Much attention is currently being focused on assessing pollution in the Mediterranean Sea, a semi-enclosed body of water whose unique oceanographic and geographic characteristics could lead to a magnification of pollution effects. Recent studies aimed at defining the current levels and effects of heavy metals, chlorinated hydrocarbons, petroleum hydrocarbons, artificial radionuclides and sewage in environmental samples are highlighted. In general heavy metals and chlorinated hydrocarbons in coastal waters, sediments and organisms are variable with high levels found near population centres, river outflows and industrial outfalls. Although the data base is skewed toward the north-west Mediterranean, the information available suggests that this region is, in fact, under the greatest stress of pollution. Mercury concentrations in certain pelagic fish are higher than in corresponding specimens from the Atlantic. These differences have been thought to result from the large-scale anomalies present in the Mediterranean; however, this hypothesis is not in accord with the finding that mercury levels in open waters, sediments and smaller aquatic biota are similar to those in areas outside the Mediterranean. At present radionuclide levels arise entirely from atmospheric fallout, and future inputs from land-based sources are not expected to alter these concentrations significantly. The degree of oil and sewage pollution appears to be greater in the Mediterranean than elsewhere, with sewage posing health problems while oil so far is affecting only tourism. Except in the case of these latter two pollutant categories, the Mediterranean does not appear any more 'unhealthy' than other oceanic regions which have been studied to date.

INTRODUCTION

Several geographic and oceanographic features of the Mediterranean Sea make it potentially vulnerable to the effects of pollution.[1] It is a relatively shallow, semi-enclosed body of water with only limited water exchange taking place through the strait at Gibraltar and the entrance of the Black Sea. Atlantic water entering at Gibraltar slowly circulates counter clockwise around the Mediterranean where it gradually becomes warmer and more saline due to increased insolation and evaporation. Weak coastal currents and a general lack of tides combine to reduce dispersion of pollutants entering coastal waters. Furthermore, the increased construction of dams on major waterways feeding the Mediterranean may have reduced the spring run-off which tends to cleanse the continental shelf of deposited pollutants.

Besides these unique physical features of the Mediterranean Sea, there are several ecological factors that come into play when assessing its response to pollution insult.[2] The Mediterranean is oligotrophic and any perturbations of the limited productivity, especially around outfalls where productivity and pollutant levels are the highest, could have long-term consequences on the ecosystem. In addition, the Mediterranean displays a high species diversity index compared to the Atlantic and it is generally regarded that the recovery potential of a stressed ecosystem will be inversely related to its species diversity index. Finally, the wide range in surface temperatures (13 °C to ~25 °C) can pose problems for stenothermic organisms living near their tolerance limits. For example cold-loving species in summer live near their thermal tolerance threshold and any additional stress in the form of heat or pollutant input could have catastrophic consequences for the population. Because of the aforementioned physical and ecological characteristics, it is generally thought that the Mediterranean should be one of the first seas to register the effects of anthropogenic pollutant additions.

Considering the precarious nature of the Mediterranean Sea as regards its resiliency to pollution insult, there has been considerable attention focused on studies of marine pollution in this region over the last decade.[1] Most of the work has aimed at establishing baseline data on existing levels of a number of pollutants in all components of the Mediterranean ecosystem. Most information comes from coastal surveys, however, in recent years a limited number of studies have concentrated on obtaining similar data from open Mediterranean waters. In the following sections a brief summary of some of these findings is given as well as some tentative conclusions that can be drawn concerning the present health of the Mediterranean.

METALS

Most potentially toxic metals occur naturally in the marine environment at parts per billion concentrations or less, however, the anthropogenic flux of many of these metals via industrial run-off or atmospheric fallout may alter these levels in certain localized areas. Because of the concentration of heavy industrial activity and large population centres in the north-west Mediterranean, one might expect that perturbations of the natural metal levels would be more predominant in this region. Data from coastal areas do, in fact, indicate certain 'hot spots' in which high metal concentrations in water have been found. One recent survey[3] extending from Sète, France to Genoa, Italy found increased levels of copper (Cu) in waters near highly industrialized port areas such as Marseilles and Toulon. Similar studies along the Italian coast[4] have shown the influence of discharges from a chlor-alkali plant on mercury (Hg) concentration in coastal sea water. However, not all increases can be attributed to anthropogenic input; measurements made around the mouth of the Isonzo river in Trieste showed that run-off from this mining region gave rise to Hg concentrations ($\sim 0.2\,\mu g\,l^{-1}$) almost an order of magnitude greater than those normally encountered further offshore.[5]

The most comprehensive data for open Mediterranean water come from a 4-year study in which surface water, sampled from several different regions, was analysed for selected heavy metals.[6] Zinc concentrations tended to differ regionally with higher values apparent in the north-west Mediterranean and Aegean basins and lower levels noted in the Tyrrhenian and south Levantine basins (Table 19.1). In the case of Hg, comparisons are more difficult to make because of large variations in concentration in a given zone. The highest levels were found in the Aegean Sea and the lowest values in the south-west Mediterranean, nevertheless, these differences amounted to no more than a

Table 19.1 Average concentrations (± standard error) and ranges of Cu, Zn, Cd and Hg in various zones of the Mediterranean Sea[6]

Zone	Cu ($\mu g\,l^{-1}$)	Zn ($\mu g\,l^{-1}$)	Cd ($\mu g\,l^{-1}$)	Hg ($ng\,l^{-1}$)
North-west Mediterranean	<0.4 ± 0.2	2.7 ± 0.4	<0.15 ± 0.03	20 ± 3
	(<0.04–5.8)	(0.02–10)	(<0.02–0.70)	(8–32)
South-west Mediterranean	<0.10 ± 0.04	1.2 ± 0.5	<0.11 ± 0.04	14 ± 2
	(<0.04–0.60)	(0.02–6.0)	(<0.02–0.51)	(5–30)
Tyrrhenian	<0.18 ± 0.08	0.9 ± 0.3	<0.11 ± 0.04	26 ± 4
	(<0.04–0.62)	(0.02–2.3)	(<0.02–0.33)	(10–40)
Ionian-Central	<0.7 ± 0.4	1.8 ± 0.9	<0.15 ± 0.09	30 ± 10
	(<0.04–2.5)	(0.02–5.7)	(<0.02–0.57)	(5–80)
Aegean	<0.3 ± 0.1	3 ± 1	<0.07 ± 0.02	40 ± 20
	(<0.04–0.64)	(0.9–5.8)	(<0.02–0.12)	(15–80)
South Levantine	<0.04 ± 0.01	0.9 ± 0.2	<0.04 ± 0.03	16 ± 2
	(<0.04)	(0.3–1.3)	(<0.02–0.11)	(12–20)
Average	<0.33 ± 0.09	2.0 ± 0.2	<0.13 ± 0.02	22 ± 3
	(<0.04–5.8)	(0.02–10)	(<0.02–0.70)	(5–80)

Table 19.2 Heavy metals ($\mu g\ g^{-1}$ dry weight[a]) in *Mytilus* from different regions of the Mediterranean Sea. Values given are ranges

Region	Cd	Cu	Zn	Pb	Ni	Cr	Ag	Fe	Hg	Refs.
North-west Mediterranean										
(Ligurian Sea)	0.4–5.9	2.4–154	97–644	2.4–117	0.9–14.1	0.5–28.8	0.1–18.9	149–2200	0.18–0.96	8, 9
Adriatic										
(Gulf of Trieste)	1.4–1.7	6.2–9.8	87–137	3.8–15				167–219	0.28–1.3	10
Aegean										
(Saronikos Gulf)	0.06–0.08	4.5	12–87		39	0.11–7.8	0.009–0.01	17–32	0.06–0.2	11, 12
(Turkey)	6.6–12	36–64	336–452	83–110		26–55		308–356	0.89–1.1	13
South-west Mediterranean										
(Algeria)	0.3–6.5		7.2–71						0.25–0.63	14

[a] Where necessary values were converted using a wet/dry weighr ratio of 6.

factor of 3. Cadmium (Cd) and Cu were often below the detection limits of the method used; thus, those values in Table 19.1, while generally low, represent the upper limits of the average values for these metals in open Mediterranean water. The most significant finding of this study was the fact that the overall average concentrations of these metals in the Mediterranean were about the same or lower than levels reported for other oceanic areas throughout the world.

Heavy metals have also been examined in Mediterranean sediments although not in any systematic manner. Different analytical methods as well as varying sediment composition complicate regional comparisons, nevertheless, Bernhard[7] has recently summarized the data from the Mediterranean as a whole and reached several general conclusions. Concentrations in sediments vary enormously for certain metals; for example, Hg ranged from 0.01 to 47 ppm dry with the highest concentrations found near coastal regions of known Hg anomalies (i.e. Idrija and Tuscany) and the lowest levels in sediments from deep waters. Typically, the highest concentrations of most metals were found in estuarine sediments. Zinc (Zn), Cu and lead (Pb) concentrations often decreased rapidly from industrialized coastal zones towards the open sea.

The rapidity with which concentrations of these metals in sediments dropped off from known anthropogenic point sources of pollution (tens of kilometres) clearly indicates the limited extent of these local 'hot spots' at present. Many of the studies cited by Bernhard[7] showed that, as with metal concentrations in water, levels in sediments from open waters did not exceed what are considered 'typical values' and hence did not indicate an effect of pollution.

By far the most information on ambient metal levels in the Mediterranean is derived from studies on marine organisms. Bivalve mussels (*Mytilus* sp.) have been used as indicator species for metals in coastal waters and, therefore, a considerable amount of data is available on metal levels in these species (Table 19.2). These data, collected at different places and times and analysed by different techniques, are too sparse to make any detailed comparisons and should be considered only as preliminary values from these areas. Viewed as a whole, these ranges of values overlap with those reported for mussels outside the Mediterranean Sea, thus, there is no indication of wide-scale metal enhancement due to pollution.[7,8] However, many of the studies cited in Table 19.2 and others have shown the usefulness of these organisms in signalling the presence of local pollution. For example, in the north-west Mediterranean, Fowler and Oregioni[8] found the highest concentration of Zn, Pb, Mn, Cu and Fe in mussels from heavily populated and industrialized port areas such as Toulon, Marseilles, Savona and Genoa. They attributed the very high levels of Pb and Cu to a local effect arising from the large amount of boat traffic and industry around these particular ports. They also found large seasonal variations in metal levels and cautioned that increased natural loads of heavy metals in run-off can lead to difficulties in interpreting anthropogenic effects when using filter-feeding bivalves as sentinel organisms. However, in general, variations in metal levels in mussels

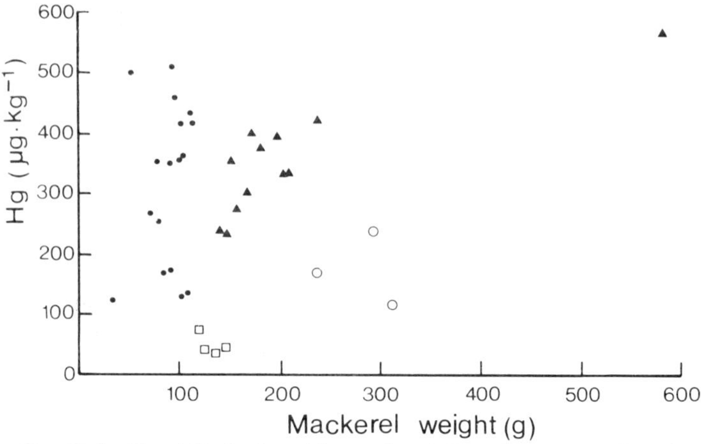

Fig. 19.1 Total Hg (ppb wet) in mackerel (*Scomber* sp.) as a function of fish weight. La Spezia, Italy (●); Elba, Italy (▲); Monaco (○); strait of Gibraltar (□). (*Source:* Baldi *et al.* 1979)

analysed in this survey correlated well with fluctuations noted in water sampled with the mussels.[3,8]

Besides mussels, several other Mediterranean species have been analysed for heavy metals.[7] Mercury has received the most attention and certain comprehensive surveys, particularly those of pelagic fish, are worth noting. Relatively high concentrations of Hg in red mullet (*Mullus barbatus*) from the Italian coast have been recorded in areas outside the influence of possible anthropogenic sources.[15] However, the areas where the fish were caught are known regions of high Hg anomalies. Anchovy, mackerel and sardines were also analysed and similarly high concentrations were found, especially in fish from coastal areas around the Monte Amiata mercury anomaly in Italy.[16] Figure 19.1 shows the relationship between Hg concentration and body weight in mackerel from several regions in the north-west Mediterranean. Fish from the Monte Amiata area clearly contain higher concentrations of Hg than similar-size individuals living in incoming Atlantic waters near the strait of Gibraltar. Comparisons of Hg levels in sardines and anchovy from both inside and outside the Mediterranean lead to the same conclusions, i.e. Mediterranean fish displayed roughly twice the Hg concentration as Atlantic specimens of similar size.[16] Tuna (*Thynnus thynnus*) ranging in size from 20 to 250 cm were also analysed for Hg.[17] Likewise, these large pelagic predators collected near the strait of Gibraltar had far lower Hg concentrations than those caught off southern Italy. Renzoni *et al.*[17] suggest that the incoming tuna from the Atlantic have less Hg because they feed on pelagic fish with low Hg concentrations where the truly Mediterranean tuna feeding on small Mediterranean pelagic fish reach much higher levels. The results of all these studies[7,16,17] have led their authors to the conclusion that the higher Hg levels noted in many pelagic fish species are not

Table 19.3 Total Hg in plankton and nekton from open Mediterranean waters.[18] Samples were collected during May–August 1977

Region	Station[a]	Microplankton net mesh size (m)	Mixed microplankton	*Euphausia* sp. (ng g^{-1} dry)	*Myctophum glaciale*
Levantine	A-1	60	36		
	A-3a	280	—		
	S-1	280	20		
	S-2	280	31		
	S-3	280	34		97
	S-4	280	18	28	116
Ionian	H-4	60	63	148	175
	H-4	280	39	100	168
	H-14	60	116	192	190
	H-14	280	48		192
Tyrrhenian	H-23	60	65	78	120
	H-23	280	36	178	279
	H-37	60	50	239	160
	H-37	280	41		
Northwestern	CS-39	500	24		
	CS-42	500	15		
	CS-44	500	21		
	CS-46	500	—	189	142
	CS-49	500	26		
	CS-51	500	78		
	CS-51	333	116		
			$\bar{x} = 46$	144	164

[a] Some stations sampled more than once or with different net sizes in the case of microplankton. These stations correspond to the following cruises: A = Atlantis II between Egypt and Malta; S = Shikmona between Israel and Crete; H = USNS Hayes between Greece and Corsica; CS = Cornide de Saavedra between Italy and Spain.

a result of anthropogenic input but rather are due to the higher than average natural environmental levels of this metal originating from the Mediterranean Hg anomaly. Although an attractive hypothesis, it clearly is at conflict with the aforementioned observations that Hg levels in both open waters and sediments are no higher than other areas in the world ocean.

This paradox has prompted other surveys to examine Hg in numerous species of open-water nekton, many of which serve as food for tuna in the Mediterranean.[18] Mercury levels in open-ocean mixed plankton and two species of nekton are given in Table 19.3. Mercury levels in mixed microplankton were generally low averaging approximately 46 ppb dry (range 15–116 ppb) for the entire Mediterranean. There was a slight tendency toward lower values ($\bar{x} = 28$ ppb) in the Levantine basin although the differences between areas were not statistically significant ($p < 0.05$). In general, overall Hg levels in mixed plankton were in the same range as those measured in similar samples collected throughout the open Mediterranean in 1975.[19] Euphausiids were usually the most prevalent nekton found at each station. Mercury levels averaged 144 ppb dry and were notably higher than the concentrations in the mixed plankton upon which euphausiids normally feed. Concentrations in myctophids averaged slightly higher than those in the euphausiids which serve as prey for these vertically migrating mesopelagic fish. Taking into account dry weight–wet weight differences, it can be seen that these open-sea pelagic fish contain considerably less Hg than has been reported for larger pelagic species[16] (Figure 19.1).

These pelagic species as well as several others have also been analysed for a suite of trace elements including As, V, Zn, Co, Cs, Ag, Se, Sb, Rb, Sc, Fe, Mn, Cu and Cd.[18,19] Broadly speaking, if one considers trace-element concentrations in all the species examined from any area within the open Mediterranean, no general regional differences in element levels were evident. On the other hand, concentrations in certain individual pelagic species did suggest regional variations in levels of some metals; however, the data are limited and more intense sampling would be needed to see if these trends hold over longer periods or whether they merely reflect short-term, local conditions at the time of sampling. Metal levels in these Mediterranean pelagic organisms have been compared with those in similar species collected from the Atlantic and Pacific Ocean.[18,19] Although direct comparisons of this nature are difficult to make (especially in the case of mixed plankton), at present there appears to be no evidence of higher metal loads in any of the Mediterranean species examined in these open-sea surveys.

The case of Hg is of particular interest in this context. Reviewing the data that are accruing for water,[6] sediments[7] and open sea biota,[18,19] there is no clear evidence that the Mediterranean Hg anomaly has led to higher ambient Hg levels outside the immediate coastal zone where in certain localities the effects of these anomalies have been well documented. In this respect the reasons for the high Hg concentrations in tuna[17] and certain other pelagic fish[16] remain elusive. The stomach contents of several tuna (10–28 kg) caught off the north-west Mediterranean coast have been found to contain large quantities of euphausiids and small myctophid fish.[18] Furthermore, none of the tuna examined with full stomachs contained fish species such as mackerel, anchovy or sardines. Since the small nektonic prey contain similar amounts of Hg as the same species living outside the Mediterranean, it is difficult to attribute these known food organisms as a source of the unusually high Hg content reported for Mediterranean tuna. Even if the tuna analysed by Renzoni *et al.*[17] were consuming mackerel, anchovy or sardines with unusually high Hg content, the question still remains concerning the origin of the high Hg levels in these latter fish which also consume small nekton like euphausiids. Until more data are available on natural Hg levels (inorganic and organic) in the Mediterranean ecosystem and on the food-chain relationships of this metal, the origin of the unusually high mercury concentrations in larger pelagic fish will remain an enigma.

Table 19.4 Heavy metals in particulate material from the Ligurian Sea[18]

Sample	Date	Mass flux ($g\,m^{-2}\,d^{-1}$)	Cd	Cu	Fe ($\mu g\,g^{-1}$ dry)	Mn	Zn
Sediment trap particulates	6/78	0.77	1.3	62	27000	812	402
	7/78	0.64	nd	21	24100	247	125
	10/78	0.77	nd	23	28700	583	99
Copepod fecal pellets			nd	950	15400	277	915
Copepods			0.9	10.1	129	5.5	71

Anthropogenic additions of heavy metals into the Mediterranean can occur by atmospheric fallout,[20] and by riverine, or direct input, of industrial wastes.[1] Direct measurements of these input rates are difficult to make. However, once in the waters it is of interest to assess the rate at which these metals sediment out of the Mediterranean. Some initial steps have been taken to measure the vertical flux of metals by collecting sinking particulate matter with the aid of sediment traps.[21] Particle collectors were placed at a depth of 100 m in waters several kilometres off the coast of Monaco and sampled at periodic intervals. Preliminary results for selected trace elements in particulates sampled during summer–autumn 1978, together with the mass flux of this material, are given in Table 19.4. It is immediately evident that trace-metal concentrations in this material are relatively high. Much of the material trapped at 100 m was in the form of zooplankton fecal pellets. Metal levels in pure copepod fecal pellets collected near the traps are also high and are similar to concentrations found in the trapped particulates (Table 19.4). In general the concentrations of metals in these particulates are much higher than those in the organisms producing them.

As expected metal fluxes were found to vary seasonally (Table 19.4). An interesting feature is that metal flux does not necessarily vary directly with mass flux, but is also strongly dependent upon the metal composition of the sinking particulates. For example in the 1978 spring–summer period the vertical flux of Zn through 100 m varied between 310 $\mu g\ m^{-2}\ day^{-1}$ in June and 80 $\mu g\ m^{-2}\ day^{-1}$ in July when the biomass and associated particulate flux was much lower. Despite a significant increase in material flux during autumn, Zn flux remained at approximately the July rate owing to a somewhat lower Zn concentration in the particulates collected in October. Data such as these, when available from all the principal regions of the Mediterranean Sea, will be extremely important in assessing rates at which heavy metals are removed from overlying waters and transported to the sediments.

CHLORINATED HYDROCARBONS

Far fewer data are available on organochlorine compound levels in components of the Mediterranean Sea. The most comprehensive coastal studies have been carried out in the north-west Mediterranean region. Raybaud[22] reported ΣDDT

Table 19.5 PCB in marine samples from the Mediterranean Sea during 1974–1976[24]

Sample type	Date	Region	No. of samples	PCB (range)	PCB ($\bar{X}$)
				$ng\,l^{-1}$	
Sea water	10.74	North-west Mediterranean	11	1.5–38	13
	2.75	Ligurian	17	1.3–8.6	3.2
	2.75	Aegean	7	<0.2–1.3	0.36
	5.75	Ionian	10	0.2–2.0	1.0
		Tyrrhenian and Algero-Provencal basin	34	0.2–5.9	2.0
	9.75	Algero/Provencal basin	8	0.6–19	4.6
			7	0.6–4.8	2.5
		Tyrrhenian	6	1.5–11.6	4.5
				$ng\,m^{-3}$	
Marine air	8.75 to 1.76	Monaco coast	13	0.1–1.0	0.4
	1.76 to 2.76	Monaco coast	12	0.03–0.08	0.06
	9.75	Algero-Provencal basin	4	0.2–0.3	0.25
		Tyrrhenian	2	0.1–0.3	0.2
				$\mu g\,kg^{-1}$	
Sediments	5.75	Ionian	3	0.8–5.1	2.8
		Algero-Provencal basin	5	0.8–9.0	4.0
		Gibraltar sill and Siculo-Tunisian sill	2	0.8	0.8
		Algerian margin	1	9.0	9.0

values of approximately 150 $\mu g\,l^{-1}$ in the surface film of the Huveaune river outfall off Marseilles. Roughly 0.5 m below the surface of the outflow, these levels dropped to 0.19 $\mu g\,l^{-1}$. However, this zone is known to be polluted and values in waters further from this source would undoubtedly by substantially less. PCBs have also been measured in the waters of this region.[23] Levels were found to range from 1.5 $ng\,l^{-1}$ near Monaco to 38 $ng\,l^{-1}$ off the mouth of the Rhone river. The consistent increase in PCB levels noted in the direction of the Rhone signalled this river as a major point source for the input of PCBs into the north-west Mediterranean. PCB concentrations in air, water and sediments are given in Table 19.5. As might be expected PCB concentrations in air and water drop off in a sea-ward direction.[24,25] Values for open waters of the Mediterranean average about 2 $ng\,l^{-1}$.[25] The data in Table 19.5 also show some regional differences; the Ionian and Aegean Seas contain somewhat lower levels than those measured in the Tyrrhenian and Algero-Provencal basins. This may result from the greater amount of industrialization in the western Mediterranean, or alternatively to the higher surface water temperature and concomitant higher evaporative codistillation rate for PCBs, or both.[24] Nevertheless, PCB levels in Mediterranean waters are of the same order of magnitude as those reported by other authors for different coastal and oceanic regions.[23–25]

A recent review[7] of chlorinated hydrocarbons in Mediterranean sediments indicates the presence of many hot spots that are correlated to anthropogenic input. For example, near the Athens outfall in the Saronikos Gulf, 1900 $\mu g\,kg^{-1}$ dry and 775 $\mu g\,kg^{-1}$ dry ΣDDT and PCB, respectively, have been measured in sediments; however, the overall extent of this input is not widespread since values drop off by one to two orders of magnitude within 5 km of the outfall.[26] For sediments outside the influence of known point sources of pollution, more typical values would appear to lie in the range 0.8–9.0 $\mu g\,kg^{-1}$ (Table 19.5).

As in the case of metals, mussels have been used to pin-point sources of chlorinated hydrocarbon pollution in the Mediterranean.[7,27,28] One survey[27] taken along the north-west Mediterranean coast showed values ranging from 0.05 to 10 ppm dry and from 0.2 to 13.5 ppm for ΣDDT and PCB, respectively. Some of the highest DDT values were found in mussels off the northern Italian coast where DDT was still being used for the flower industry. A slightly lower range (ΣDDT = 0.03–0.2 $\mu g\,g^{-1}$; PCB = 0.04–1.1 $\mu g\,g^{-1}$) has been reported for mussels from the central Adriatic coast of Yugoslavia.[28] Because of the complex interactions of mussel physiology and chlorinated hydrocarbon content,[27] it is difficult using this species to reach any conclusion about general differences in organochlorine levels inside and outside the Mediterranean. Summarizing what little information is available about chlorinated hydrocarbons in this bivalve,[7] ranges of DDT and PCB concentration in north Atlantic and Mediterranean mussels show considerable overlap.

Conclusions that can be drawn from the limited amount of data for other Mediterranean coastal and open-sea species are mixed. Bernhard[7] summarized the values for organochlorines in pelagic fish and, based on levels in sharks and sardines, concluded that Mediterranean pelagic fish were more contaminated than similar species from the Atlantic; however, he points out that levels in these fish vary considerably and depend on both season and location in the Mediterranean.

Very few chlorinated hydrocarbon data exist for truly open Mediterranean organisms; however, some recent analyses of mixed plankton and nekton can serve as a baseline for following trends in future years. PCBs and ΣDDT in seven separate mixed microplankton samples from the southern Levantine basin ranged from 15 to 230 ppb dry and 12 to 85 ppb dry, respectively. These ranges were similar to those for euphausiids from both the eastern and central Mediterranean basins (Table 19.6). It was also noted that euphausiids from the central basin had slightly higher ΣDDT/PCB ratios than the same species from the eastern region. This observation is due to a greater relative decrease in ΣDDT levels compared to PCB concentrations in going from the central to the eastern region. Furthermore, where representative plankton and nekton food chains have been sampled from a single water mass, there was no evidence for biomagnification of these compounds at higher trophic levels.[29,30]

Fowler and Elder[29] compared their data with those from similar surveys in

Table 19.6 Chlorinated hydrocarbon residues in euphausiids from different regions of the open Mediterranean. Each sample was a composite sample of several individuals collected during three cruises in 1977[29]

Region	Cruise[a]	Station[a]	PCB (DP-5)	pp′DDT	pp′DDD	pp′DDE	ΣDDT/PCB
				$\mu g\,kg^{-1}$ dry[c]			
Tyrrhenian	Hayes (6.77)	37	25	3.8	1.8	14	0.80
		37[b]	110	73	9.1	33	1.05
		23	30	6.0	3.9	24	1.11
Ionian		14	38	16	4.4	22	1.12
		14[b]	57	22	5.4	34	1.08
		4	23	8.8	2.7	11	0.97
Levantine	Atlantis II (4.77)	3a	26	3.5	0.7	6.7	0.42
		6a	65	14	3.0	18	0.53
Levantine	Shikmona (7.77)	2	26	2.5	1.7	15	0.76
		3	18	4.7	1.3	3.9	0.54
		4	9.8	nd	nd	2.5	0.26

[a] Hayes between Greece and Crete; Atlantis II between Egypt and Malta; Shikmona between Israel and Crete.
[b] Pure samples of *Meganyctiphanes norvegica*.
[c] Dry weight averages 18 per cent of wet weight.

other oceanic areas. Although the data are sparse, they concluded that PCB and DDT levels in euphausiids and other small nekton did not differ significantly from concentrations in similar species measured elsewhere. On the other hand, PCB levels tended to be slightly lower in Mediterranean mixed zooplankton. As an explanation, these authors pointed out that global PCB production had decreased since the time most of the surveys with which they compared their data were made. Since no information exists on PCB concentration in plankton or water from the open Mediterranean during the early 1970s, it is impossible to ascertain whether levels may have been higher prior to their 1977 survey. Far more data need to be collected over a longer time span in order to resolve whether these lower levels in plankton are due to a global decrease in PCB input with time or actually represent real-time geographical differences in existing PCB concentrations.

The same authors[45] also analysed a limited number of tuna (weight range = 10–28 kg) from the north-west Mediterranean to assess organochlorine levels in higher trophic level predators. Tuna muscle contained concentrations ranging from roughly 9 to 90 $\mu g\,kg^{-1}$ wet and 2.5 to 50 $\mu g\,kg^{-1}$ wet for PCB and ΣDDT, respectively. Gut contents, which were primarily composed of euphausiids, contained levels (PCB = 12–68 $\mu g\,kg^{-1}$ wet; ΣDDT = 10–35 $\mu g\,kg^{-1}$ wet) which corresponded to those in tuna muscle as well as those in euphausiids sampled elsewhere in the Mediterranean (Table 19.6). The only other Mediterranean tuna data appear to be two samples of much older individuals (145 and 165 kg, 5–7

Table 19.7 Estimates of vertical PCB flux through 100 m in the Ligurian Sea[21]

Date	Particulate PCB ($\mu g\ kg^{-1}$ dry)	Mass flux ($g\ m^{-2}\ d^{-1}$)	PCB flux ($\mu g\ m^{-2}\ y^{-1}$)
6.78	650	0.77	183
7.78	300	0.64	70
8.78	710	0.40	104
10.78	200	0.77	56
			$\bar{X} = 103$

years old) which contained 40 and 225 μg ΣDDT kg^{-1} wet and 407 and 95 μg PCB kg^{-1} wet.[31] These values are at the high end of the range or higher than those reported by Fowler and Elder[45] and the differences may be a function of fish weight such as has been shown to occur in the case of mercury in tuna.[17] Fowler and Elder[45] emphasize the lack of published information on chlorinated hydrocarbons in tuna from outside the Mediterranean; however, compared with the few available data, organochlorine concentrations in Mediterranean tuna are of the same order of magnitude as have been reported elsewhere.

The rate of removal of chlorinated hydrocarbons from the upper mixed layer via sinking particulate matter has also been studied by analysing sediment trap samples as well as freshly produced biogenic particles released from zooplankton and nekton.[21,32] Studies of this nature are particularly relevant in areas like the Mediterranean which characteristically have high sedimentation rates. Total particulate flux through 100 m depth in the Ligurian Sea was measured during a 5-month period. These values as well as PCB concentrations in the sinking particulates are given in Table 19.7. Fowler *et al.*[21] reported that the decrease in particulate flux during the summer months correlated well with the decrease of phytoplankton biomass and organic aggregates known to occur in this region. Using PCB concentrations and the mass flux values it was calculated that PCB deposition rates averaged approximately 103 μg PCB m^{-2} $year^{-1}$. Independent estimates[32] of PCB deposition rates calculated from PCB levels measured in Ligurian Sea sediments ranged from 80 to 125 $\mu g\ m^{-2}$ $year^{-1}$ and agreed very closely with the measured values. Freshly released zooplankton and nekton fecal pellets produced in the Ligurian Sea typically contained between 1300 and 38 000 μg PCB kg^{-1} dry.[21,32] These levels are normally a factor of 10 higher than either the organisms that produce the feces or the food they ingest. The general similarity between PCB concentration in fecal pellets and the trapped material was not unexpected in that fecal pellets and fecal material comprised the bulk of the particulates collected at depth. It is noteworthy that in all cases particulates collected at 100 m contained somewhat less PCB than was present in fecal pellets released from surface-dwelling organisms. This observation suggests that PCBs may be partially released from fecal matter as it sinks. The presence of relatively high levels of PCBs in both freshly produced fecal pellets and sinking fecal matter collected *in situ* indicate that zooplankton defecation contributes significantly to the downward vertical transport of these compounds in the Mediterranean Sea.[21,32]

PETROLEUM HYDROCARBONS

Oil has been a problem in the Mediterranean for a long time, but only within the last decade have quantitative data become available on the relative degree of pollution *vis-à-vis* other seas. Recent estimates suggest that between 0.5 and 1×10^6 t per year of oil enter the Mediterranean Sea,[33] a fact which, if true, would make this sea one of the most oil-polluted areas of the world. The results of several studies which have examined floating tar balls and dissolved petroleum hydrocarbons are given in Table 19.8. The wide variations in concentration are a result of both hydrocarbon 'patchiness' as well as the different sampling and

Table 19.8 Average concentrations of tar balls and dissolved petroleum hydrocarbons (PHC) in different regions of the Mediterranean Sea

Region	Year	Tar balls (mg m^{-2})	PHC ($\mu g\,l^{-1}$)	Refs.
Alboran	1969	3.3		34
	1974	1.3		35
	1974–75	9.2	25.8	36, 37
	1976	0.6	7.9	38, 39
Ligurian	1969	2.0		34
	1974	1.6		35
	1974–75	0.5		36
	1975–77	5.4	4.3	38, 39
South-west Mediterranean	1969	6		34
	1974–75	13	6.9	36, 37
	1975–77	3.9	17.5	38, 39
Tyrrhenian	1969	5		34
	1974	7.1		35
	1974–75	3	4.8	36, 37
	1975–77	0.9	7.4	38, 39
Ionian	1969	360		34
	1974–75		14.9	37
Libyan Coast	1969	125		34
	1974	6.8		35
	1974–75		24.9	37
South Levantine	1969	46		34
	1977	0.05–1.1		40
Aegean	1974	0.23		35
	1974–75		20.5	37
North Adriatic	1976–78(?)		0.1–1.0	41

analytical methods employed. Nevertheless, some trends are apparent. Generally, the highest levels of both oil aggregates and dissolved fractions occur in the Alboran Sea and along the north coast of Africa. This is undoubtedly due to both increased tanker traffic going in and out of the strait of Gibraltar and the large number of oil-loading ports on the African side of the Mediterranean. Furthermore, the relatively higher percentages of *n*-alkanes in water sampled off Libya and in the Tyrrhenian Sea indicate that fresh petroleum was introduced more directly into these regions.[37] Many investigators[1,38,39] who have compared petroleum concentrations in the Mediterranean with data from other areas generally conclude that aggregate and dissolved hydrocarbons are more abundant in the Mediterranean. At present the problem is principally an aesthetic one with the major effects causing perturbations in the tourist trade. So far no significant effects on either the marine ecosystem or man have been recorded for this region; however, in the event of a major oil spill this situation could change dramatically.

SEWAGE

Sewage is a major problem along the Mediterranean coast for both aesthetic and health reasons. Because of the large population density (particularly in summer), high ambient temperatures and often poor waste treatment procedures, health problems are more varied in the Mediterranean than those in the more northerly regions. Relatively high instances of salmonellae-type infections, cholera and numerous viral diseases are most certainly linked to the direct release of raw sewage in Mediterranean coastal waters. Morbidity rates for these diseases are much higher in the Mediterranean region than in the rest of Europe.[42] Furthermore, the problem is compounded by the relatively large consumption of raw shellfish in this area. Fresh oysters, mussels, clams, sea squirts and sea urchins form an important part of the Mediterranean diet and these species are known as vectors for transferring water-borne pathogens to man. Because of the regional demography patterns as well as the summer tourist influx, the situation is most serious in the north-western sector between Spain and Italy. Earlier figures[43] show that the annual release of the organic load in domestic sewage in terms of BOD_5 reached a level of 336 t km^{-1} of shoreline for this area, the highest figures reported in the Mediterranean basin. Besides pathogenic pollutants, sewage contains a wide variety of other contaminants ranging from heavy metals and radionuclides to pesticides and other toxic compounds. However, to date no effects of chronic, cumulative poisoning from these substances have been recorded, hence, more emphasis has been given to attacking the problem of sewage-associated pathogens. Although waste treatment and disposal practices have reportedly improved in certain areas, it is generally acknowledged that, on the whole, the situation in the Mediterranean has worsened. Furthermore, an OECD report on the environment projects that coastal usage will rise to approximately three tourists per metre of beach by the early 1980s, therefore the sewage situation can be expected to worsen unless drastic measures are taken. Rather than mount massive immunization programmes for the various diseases encountered, it is felt that the best way to combat the problem is to improve the general sanitation standards in the areas involved by widespread adoption of a 'no dumping' policy such as that featured in the sanitary improvement plans proposed by WHO.[42]

RADIOACTIVITY

The Mediterranean, like all water bodies, has its natural component of radioactivity in the form of ^{40}K, ^{87}Rb as well as a suite of naturally occurring alpha-emitting radionuclides. Gamma and beta emissions from ^{40}K and ^{87}Rb alone cause 734 dpm l^{-1} seawater.[1] Superimposed on this background of natural radioactivity are many artificially produced radionuclides originating from atmospheric fallout and nuclear energy and waste disposal processes. Among artificial radionuclides, the transuranics ($^{239,240}Pu$ and ^{241}Am) are of particular interest in the environment both because of their toxic nature and their extremely long half-lives. Recent surveys[44] of plutonium in the Mediterranean show that typical concentrations range from 0.001 to 0.019 dpm l^{-1}, 0.67 to 9.3 dpm kg^{-1} dry and 0.93 to 1.64 dpm kg^{-1} wet for water, sediment and mussels, respectively. These values are comparable to those found in similar samples from other regions in northern latitudes. Even though present levels signify no contamination origin other than fallout, it is of interest to use these values as baseline levels against which one can gauge any possible changes that may come about through increased use of nuclear energy in the Mediterranean region. It has been calculated[1] from growth projections that by the year 1987 an increase in approximately 0.004 dpm l^{-1} can be expected from all artificial radionuclides except tritium. Clearly, anthropogenic input from land-based sources is extremely small when compared to existing levels of artifical radionuclides, not to mention concentrations of natural radioactivity.

HEAT

There is much disagreement as to whether thermal input should be considered as a true form of pollution. In fact some authors,[1] have considered it a non-problem in the Mediterranean and elsewhere. It is discussed here only for the sake of completeness. No data on the effects of thermal pollution in the Mediterranean have come to my attention. Obviously in many regions of the Mediterranean where high surface temperatures are common, any additional thermal input could cause damage to marine species that are already living at or near their upper threshold of thermal tolerance.[2] On the other hand, in colder waters, thermal input may have beneficial effects by increasing the rate of biomass production. Clearly, thermal effects on the marine ecosystem will depend on local conditions and, in any event, the threat to man would be immeasurably small.

CONCLUSIONS

The levels and effects of various pollutants have been examined for different regions within the Mediterranean and compared with results from similar studies carried out in other oceanic areas. Concentrations of heavy metals and chlorinated hydrocarbons in coastal waters, sediments and organisms vary considerably but tend to be highest near centres of high population density, river outflows and industrial outfalls. Levels of these pollutants tend to be higher in the western and central Mediterranean although comparisons are limited due to a general paucity of data from the eastern and southern regions. Some data from selected coastal areas indicate that many of these pollutants are being rapidly transferred from the surface layers to the sediments via sinking particulate matter, much of which is biogenic.

In the case of heavy metals, elevated Hg levels are found in regions of known geochemical Hg anomalies, e.g. along the coast of Tuscany and the northern Adriatic region of Idrija; however, the extent of the effect from these inputs appears to be limited to the immediate area near the source. In the more open waters evidence is accruing that certain pelagic fish such as anchovy, sardines, mackerel and tuna contain significantly higher Hg concentrations than the same or similar species from the Atlantic. These differences have been explained as being due to the large-scale Hg anomalies in the Mediterranean basin, an explanation which implies higher average Hg concentrations in the aquatic environment. However, Hg levels in open Mediterranean waters, sediments, zooplankton and micronekton are similar to those in the Atlantic as well as other areas, hence, these Hg anomalies do not appear to affect all components of the Mediterranean ecosystems. Clearly, more data are needed to solve this paradox. Similar information on chlorinated hydrocarbons in open-ocean organisms is extremely sparse, but the data at hand indicate that levels are generally comparable to those in species from outside the Mediterranean.

On the other hand, dissolved petroleum hydrocarbons and tar balls tend to be higher in the Mediterranean than elsewhere. Oil and sewage would appear to pose the greatest aesthetic problems, and sewage the greatest health problem because of its action as a pathogen vector. Present radionuclide concentrations (e.g. plutonium) in the Mediterranean can be attributed entirely to atmospheric fallout and are no different from those which have been measured at similar latitudes in the northern hemisphere. Furthermore projections of future levels arising from man's activities indicate that no significant changes in total radionuclide concentrations are anticipated.

In summary, marked pollution in coastal areas is very localized and tends to be greater in the more populated north-western and central regions. Pollutant levels in open Mediterranean waters vary somewhat but on the whole are not too different from those reported for other oceanic regions, except in the case of petroleum hydrocarbons.

ACKNOWLEDGEMENTS

The International Laboratory of Marine Radioactivity operates under a tripartite agreement between the International Atomic Energy Agency, the Government of the Principality of Monaco and the Oceanographic Institute at Monaco. This work was partially supported under UNEP Contract No. FP/0503-76-09(978).

REFERENCES

[1] Osterberg, C., and Keckes, S. (1977). *Ambio*, **6**, 321.
[2] Peres, J.M. (1978). *Ocean Manag.*, **3**, 205.
[3] Fukai, R., and Huynh-Ngoc, L. (1976). *Mar. Pollut. Bull.*, **7**, 9.
[4] Renzoni, A., Bacci, E., and Falciai, L. (1973). *Rev. Int. Océanogr. Méd.*, **36/37**, 17.
[5] Kosta, L., Ravnik, V., Byrne, A.R., Stirn, J., Dermelj, M., and Stegnar, P. (1978). *J. Radioanal. Chem.*, **44**, 317.
[6] Huynh-Ngoc, L., and Fukai, R. (1979). *IVes Journées Etud. Pollutions, Antalya, C.I.E.S.M.*, 171.
[7] Bernhard, M. (1978), *Ocean Manag.*, **3**, 253.
[8] Fowler, S.W., and Oregioni, B. (1976). *Mar. Pollut. Bull.*, **7**, 26.
[9] Stoeppler, M., Bernhard, M., Schulte, E., and Backhaus, F. (1977). *Sci. Total Environ.*, **13**, 209.
[10] Majori, L., Nedoclan, G., Modonutti, G.B., and Daris, F. (1979). *IVes Journées Etud. Pollutions, Antalya, C.I.E.S.M.*, 237.
[11] Grimanis, A.P., Papadopoulou, C., Zafiropoulos, D., Vassilaki-Grimani, M., and Tsimenidis, N. (1979). *IVes Journées Etud. Pollutions Antalya, C.I.E.S.M.*, 233.
[12] Papadopoulou, C., and Kanias, G.D. (1976). *Acta Adriat.*, **18**, 367.
[13] Uysal, H. (1979). *IVes Journées Etud. Pollutions, Antalya, C.I.E.S.M.*, 213.
[14] Aissi, A. (1979). *IVes Journées Etud. Pollutions, Antalya, C.I.E.S.M.*, 207.
[15] Bernhard, M., and Renzoni, A. (1977). *Thalassia jugosl.*, **13**, 265.
[16] Baldi, F., Renzoni, A., and Bernhard, M. (1979). *IVes Journées Etud. Pollutions, Antalya, C.I.E.S.M.*, 251.
[17] Renzoni, A., Bernhard, M., Sara, R., and Stoeppler, M. (1979). *IVes Journées Etud. Pollutions, Antalya, C.I.E.S.M.*, 255.
[18] Fowler, S.W., La Rosa, J., and Oregioni, B. (in preparation).
[19] Fowler, S.W., Oregioni, B., and La Rosa, J. (1976). *Activ. Intern. Lab. Mar. Radioact.*, IAEA-187, 110.
[20] Chesselet, R., Baron, D., Benard, H., and Buat-Menard, P. (1979). *IVes Journées Etud. Pollutions, Antalya, C.I.E.S.M.*, 81.
[21] Fowler, S.W., Small, L.F., Elder, D.L., Ünlü, Y., and La Rosa, J. (1979). *IVes Journées Etud. Pollutions, Antalya, C.I.E.S.M.*, 289.
[22] Raybaud, H. (1972). Thèse Doc. Spécial., Univ. Aix-Marseille II, 64 pp.
[23] Elder, D.L. (1976). *Mar. Pollut. Bull.*, **7**, 63.
[24] Elder, D.L., and Villeneuve, J.P. (1977). *Thalassia jugosl.*, **13**, 59.
[25] Elder, D.L., and Villeneuve, J.P. (1977). *Mar. Pollut. Bull.*, **8**, 19.

[26] Dexter, R.N., and Pavlou, S.P. (1973). *Mar. Pollut. Bull.*, **4**, 188.
[27] Marchand, M., Vas, D., and Duursma, E.K. (1976). *Mar. Pollut. Bull.*, **7**, 65.
[28] Dujmov, J., Vucetic, T., Picer, M., and Picer, N. (1979). *IVes Journées Etud. Pollutions, Antalya, C.I.E.S.M.*, 137.
[29] Fowler, S.W., and Elder, D.L. (1980). *Mar. Environ. Res.*, **4**, 87.
[30] Fowler, S.W., and Elder, D.L. (1978). *Bull. Environ. Contam. Toxicol.*, **19**, 244.
[31] Alzieu, C. (1976). *Sci. Pêche*, **258**, 1.
[32] Elder, D.L., and Fowler, S.W. (1977). *Science*, **197**, 459.
[33] Le Lourd, P. (1977). *Ambio*, **6**, 317.
[34] Horn, M.H., Teal, J.M., and Backus, R.H. (1970), *Science*, **168**, 245.
[35] Benzhitsky, A.G., and Polikarpov, G.G. (1976). *Okeanologiya*, **16**, 87.
[36] Morris, B.F., Butler, J.N., and Zsolnay, A. (1975). *Environ. Conserv.*, **2**, 275.
[37] Zsolnay, A. (1979). *Mar. Chem.*, **7**, 343.
[38] Ros, J., and Faraco, F. (1979). *IVes Journées Etud. Pollutions, Antalya, C.I.E.S.M.*, 111.
[39] Faraco, F., and Ros, J. (1979). *IVes Journées Etud. Pollutions, Antalya, C.I.E.S.M.*, 117.
[40] Wahby, S.D. (1979). *IVes Journées Etud. Pollutions, Antalya, C.I.E.S.M.*, 93.
[41] Ahel, M., and Picer, M. (1979). *IVes Journées Etud. Pollutions, Antalya, C.I.E.S.M.*, 99.
[42] Brisou, J. (1977). *Ambio*, **6**, 342.
[43] FAO (1972). *GFCM Studies and Reviews No. 51*, FAO, 68 pp.
[44] Murray, C.N., and Fukai, R. (1978). *Estuar. Coast. Mar. Sci.*, **6**, 145.
[45] Fowler, S.W., and Elder, D.L. (in preparation).

SECTION III: TERRESTRIAL ENVIRONMENT AND MAN

Pollutants and Their Ecotoxicological Significance
Edited by H. W. Nürnberg

CHAPTER 20

Ecological Parameters and Their Significance for Pollution Monitoring in the Terrestrial Environment

PAUL MÜLLER

SUMMARY

Food webs are the basic structure for energy and matter transfer in ecosystems. This is why we believe analysis of such food webs is the main task we will have to work on in order to give a realistic answer to the question of the real pollution burden and endangerment for human populations. Improvement of already existing 'environmental compatibility tests' is being demanded more and more urgently. But these can only be used as relative reflections of a real risk to city life, if the metabolism of the substances is analysed in the food chains of the system in question. Parallel to this we have to evaluate areas by means of *effects* and *trend monitoring programmes* with standardized organisms, the reactions or pollution residues of which yield information rendered transferable to human populations by epidemiological examinations.

ECOLOGICAL PARAMETERS AND ECOSYSTEM RESEARCH

Information Content and 'Pollution Load Capacity' of Ecosystems

Bioindication is the breakdown of the information content of biosystems in order to evaluate spaces. Knowledge of the load capacity of ecosystems, as ecological information, is a basic pre-requisite for all planning taking the environment into consideration. However, in the same way as the 'stability' or 'elasticity' of ecosystems, the 'load capacity' of ecosystems can only be meaningfully defined with respect to certain relations among which are knowledge of the organisms' coupling type, the bonding density of the ecosystem elements and the web type (cf. Müller[29–33]). Since ecosystems are characterized both by internal as well as external states of equilibrium, load capacity and elasticity must be considered both from structural and also energetic points of view.[28] Due to the complexity of these problems, systems analytical planning models, for example, are generally much 'better' than the ecological statements which have been achieved. Interactive effects are reduced to autoecological approaches and the introduction and standardization of synecological test procedures meets with methodological and epistemological problems, although their necessity is unquestioned (cf. Ellenberg *et al.*[10]). People usually fail to realize that the methods for determining synecological effects, in the case of practical exploitability in the sense of a space evaluation, must be capable of being calibrated against natural and urban ecosystems.

The stability zone of an ecosystem can be characterized by the system states in which the disturbances generated can be compensated by limited inputs without permanent structural modifications. For this reason Webster *et al.*[45] suggested that the concept of stability should be replaced by a 'relative stability'. 'The argument that ecosystems are asymptotically stable focuses attention on the critical area of relative stability.' Their 'asymptotic stability' of an ecosystem depends on the equilibrium between all structural elements of the system, its metabolism and energy influences. 'Resistance is related to the formation and maintenance of persistent ecosystem structure. Resilience results from the tendencies inherent in ecosystems for the erosion of such structures' (cf. among others Hall and Day,[16] May,[28] Patten[38]).

In ecosystems rich in species the individual elements generally have a higher bonding density than in ecosystems poor in species. From this we can conclude that these systems can have a higher inherent internal stability. Nevertheless, they may be more changeable or unstable with respect to external influences than ecosystems poor in species. Stable ecosystems are in a position to reverse forced transformations by limited inputs, whereas changeable ecosystems diverge permanently from their original condition. The disturbance level is the quantitative expression of pollution. It indicates the measurable amount of disturbance.

Pollution is accordingly the condition in which ecosystems can still reverse forced transformations. It is inappropriate to limit this, coupling pollution to anthropogenic effects since much of the 'pollution' caused by man also occurs in the natural environment.

A system's ability to regenerate is decisively dependent on the areal effects of a disturbance factor on the bonding density and web type. There are close relations between surface area, number of individuals and genetic structure. Only extensive systems can maintain their genetic structure over fairly long periods of time. A considerable fraction of the marked population decreases in numerous animal and plant species, for example in the Federal Republic of Germany, can be attributed to 'surface reductions'.

Behaviour and Monitoring of Ecosystems

The behaviour of ecosystems is determined by endogenous element fluctuations and exogenous factors. For this reason it is therefore necessary to know more, for example, about the urban climate, particularly, however, the radiation climate and its interrelation with the living systema and the immission type.

Equally important is knowledge of substance cycles and the behaviour of chemical elements. Photochemical processes and a large number of organic carbon compounds (cf. the 1972 emission map of Cologne) indicate that we are obviously only at the top of an iceberg with limit stipulations in accordance with Technical Instructions on Air Pollution Control and propagation models. Energy transfer and mass transfer are decisive parameters in order to understand and describe the behaviour of each system. Numerous individual climatological problems (amongst others behaviour of fresh air; local wind systems) and their interconnection with the immision type (amongst other aspects, residence time of pollutants in the atmosphere, modification in precipitation and chemical composition of precipitations) have indeed already been investigated from points of view relevant to planning, however, their connection to the most important ecosystems via radiation balance and dose–effect relations have not yet been sufficiently investigated. But precisely these connections are of great significance for system stability.

We do not intend to underestimate the significance of other climatological and aerological investigations in any way. After all, they provided the first evidence

Fig. 20.1 Radiation and emission measuring station in Saarbrücken (1979). In addition to the radiation balance and important climatic factors, the gaseous immission components NO_x, CO, SO_2 and organic hydrocarbons are also measured

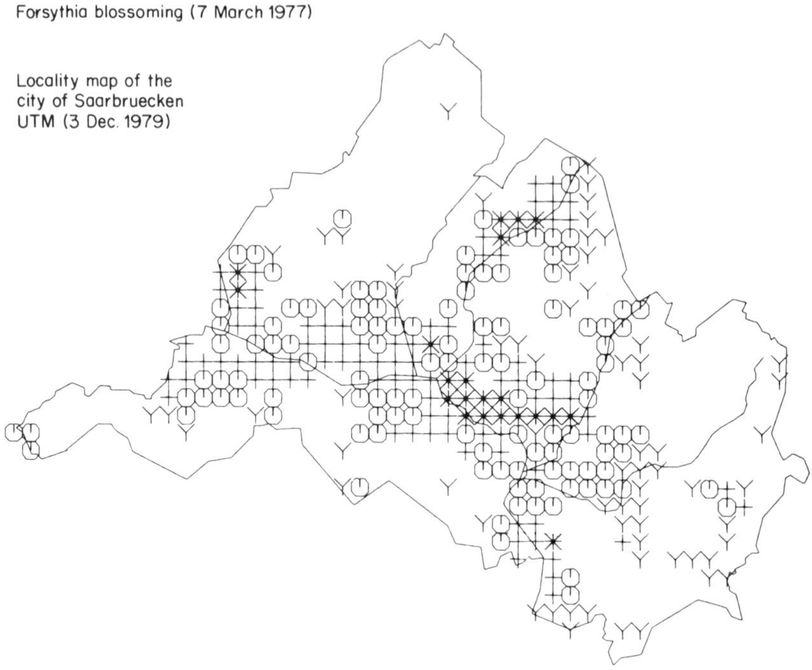

Fig. 20.2 Flowering dates of forsythia determined by the thermal and radiation balance of a city (Saarbrücken). Forsythia flowers in part 14 days earlier in the 'warmer' city centre zone (⏀) than in the suburbs (Y)

for the fact that, for example, the pollution of systems with sulphur compounds (amongst others SO_2, SO_4) can display completely different distribution patterns and thus substantiated, amongst other aspects, the low survival rate of exposed organisms at sites which, at least from the point of view of gas analysis, display uniformly low SO_2 concentrations.

All multidimensional processes in ecosystems can only be clarified by means of monitoring programs which, in addition to effect and trend maps, can lead to a realistic assessment of the pollution of a system, e.g. with 'pollutants', via specific food-chain analyses at the site. The concomitant problems become clearest in the case of urban ecosystems. Towns are our action biotope and their functional capability is determined by the quality of the decisions we make. The insights gathered with respect to urban ecosystems must, therefore, also clarify whether and why our 'form of life' makes us ill, which subsequent effects our action will have on the other systems and the highest-ranking ecosystem, the biosphere, what repercussions can be expected on human populations and whether it is possible to derive in good time the risks to our own life from the reactions of the biosystems surrounding us.

An improved breakdown of epidemiological and energetic processes, the

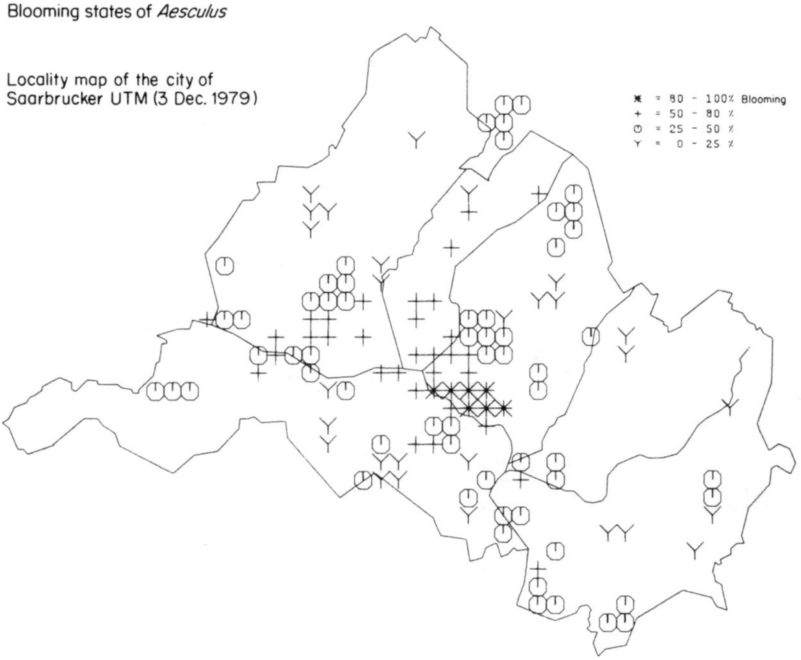

Fig. 20.3 Flowering dates of the horse chestnut (*Aesculus*) conditioned by the thermal and radiation balance of a city (Saarbrücken; study period 3–5 May, 1975)

evaluation of urban systems with bioindicators (amongst other aspects risk maps) and the definition of the pollution of urban ecosystems by means of food-chain analyses (including elucidation of mass transfer) must be analysed on a priority basis. These requirements are also reflected in the new ecosystem research programmes of the Federal Republic of Germany (cf. Ellenberg *et al.*[10], Müller[30,31]). However, they require a working group at the site adequate for the complexity of the problems.

Ecotoxicological problems are of central interest in monitoring. In the same way as our long experience with biotests for assessing carcinogenic mutagenic and/or teratogenic substances, careful basic discussions for example of the 'Limits Given by Technical Instruction on Air Pollution Control' (Federal Immission Control Act) or the 'Federal Immission Control Act' (printed paper 8/3319; 6.11.79 German Bundestag) make it clear that we are still a long way from realistically assessing the ecotoxicological behaviour of numerous organic 'environmental chemicals'.

Knowledge of for example a substance's absorption, adsorption and desorption ability, its solubility and reactivity with water, its water–octanol distribution quotient or its photochemical behaviour are only indications of its possible ecological behaviour, in the same way as knowledge gathered in the

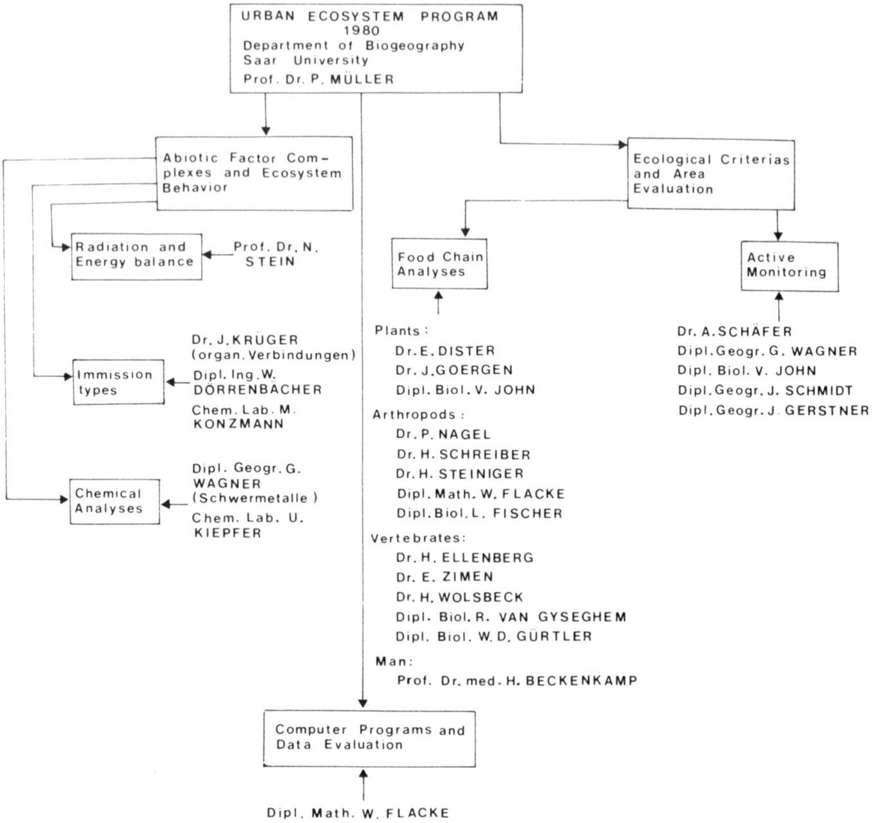

Fig. 20.4 Working group and subject areas of urban ecosystem analyses currently undertaken at the Chair of Biogeography at the University of the Saarland in 1980 (supported by the BMI—Federal Ministry of the Interior)

laboratory about its metabolization rate, persistence, bioaccumulation and/or degradability. Precise knowledge and analyses of mass and energy transfer in ecosystems' food chains with simultaneous checking of the environmental chemicals applied, with respect to the behaviour of the ecosystem, are necessary for a reliable prognosis.[29,31] At the moment this type of study is only being carried out at a few locations in the Federal Republic of Germany as part of MAB projects.[10,30,31]

It is not possible to evaluate environmental chemicals by laboratory tests alone. Further requirements are, (a) the development of test ecosystems for reproducible tests, (b) the development of standardized control procedures for biosystems in the field, (c) the development of reproducible tests on ecosystems close to nature (food controls) after application of defined substances, and (d) preparation and control of effect and trend maps (experimental monitoring).

PROBLEMS OF STANDARDIZING LABORATORY AND FIELD TESTS

Representativeness of Laboratory and Field Tests

We do not need to emphasize at this point that the transfer of laboratory results to field conditions presupposes comprehensive ecological *a priori* knowledge which is regrettably lacking for most ecosystems all over the world. The different resistance of organisms to an applied active substance leads to effects on the ecosystem such as we would not expect from knowledge, for example, of the LD_{50} values in rat tests alone.

As an example we shall use here dieldrin, which is well known from a chemical and laboratory toxicological point of view and whose effects are also familiar, at least in the non-tropics (cf. for example Brown,[4] Hill and Wright,[17] Klein,[20] Müller and Nagel[34]).

The subsequent effects of this insecticide, which has also been applied in the highlands of Adamaoua (Cameroon) in order to combat tsetse flies, on the ecosystem can only be assessed, if the system modifications attributable to the defined active substance can be reliably identified, isolated from other influencing factors and the transport and persistence as well as the transformations of the dieldrin derivatives in the food chains at the site can be reliably determined and traced. Both analytical residue studies as well as quantitative analyses of the fauna and flora have indeed already been carried out in Glossina projects in Africa (for example Koeman and Takken[21]). However, the causal connection between results of chemical and analytical and ecological population studies has been successfully tested for the first time in my working group's present study.[34] The study area (Figure 20.5), a small body of flowing water with river forest in the tropical grassland near the River Maour, was sprayed with dieldrin formulated in Shellsol R/Dutrex (Endosil EF 4220) by helicopter on 14.3.79 (dose: $750\,g\,ha^{-1}$). A qualitative and quantitative survey of the micro-, meso- and macrofauna was carried out between the end of January and the date of spraying (amongst other aspects analyses of the feed chains and diversity in the river forest, flowing water and the adjacent savannah). The investigations were continued after application of the insecticide until mid-April and repeated analogously from October to December 1979 and then until February 1980. Fifty ground traps were set in both the river forest and the adjacent savannah for the epigeal fauna, the nocturnal species were caught with luminous traps and the mortality rates determined by a modified linen-cloth method.

The quantitative results show that the total number of individuals decreases immediately after application, increases once again at the start of the rainy season and is below the initial value of the previous year after a year. On the other hand, the ecologically significant diversity values increase. However, as an analysis of the $H_{\text{diff.}}$ values indicates (species composition of the biocoenoses[34]), at the expense of far-reaching restructuring of the species.

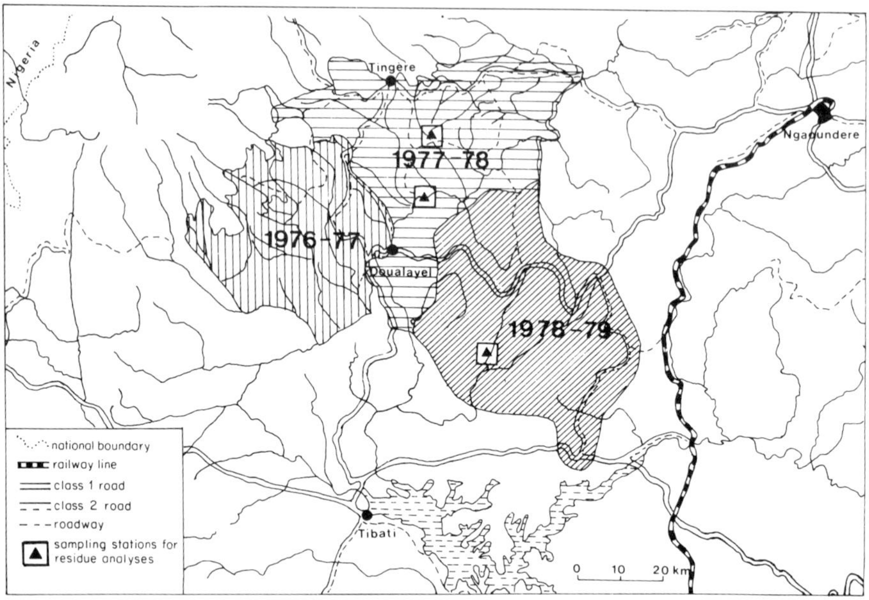

Fig. 20.5 Dieldrin spraying areas in Cameroon in the years 1976–79

Cautious computer forecasts from mortality tests at the site indicate that approximately 10 per cent of the biomass of surface arthropodes is destroyed by this application. Particular species (e.g. *Pycnoceros sulcatus*, *Coleoptera*) are especially hard hit. Analyses of the diurnal and nocturnal hypergeal fauna fit into this picture. The hypergeal biocoenosis reacts considerably more sensitively to systems disturbances by insecticide spraying than the epigeal fauna of the same savannah zone.

Additional extensive analytical residue studies were carried out before and after application in order to further define the influence of dieldrin (gas chromatograph Perkin-Elmer, Sigma 1 B; detection limits $0.001\,\mu g\,g^{-1}$ at 10 g specimen material).

Residue analyses from field populations of numerous plant and animal species are available from all parts of the world, for example OECD Programs.[2–4,7,13,17,19,20,22,24,25,27,31,32,36,39,40] However, their interpretation is frequently meaningless without any knowledge of the ecological data (for example prey, size of habitat, food chain). A further problem is that with increasing position of a species in the food chain its zone of action generally also increases. Its information content is admittedly increased by this, but the checking of significantly more complex food webs is necessary to break it down. Accumulation and degradation of pollutants in individuals, populations and food chains can take place in completely different ways and can vary according to

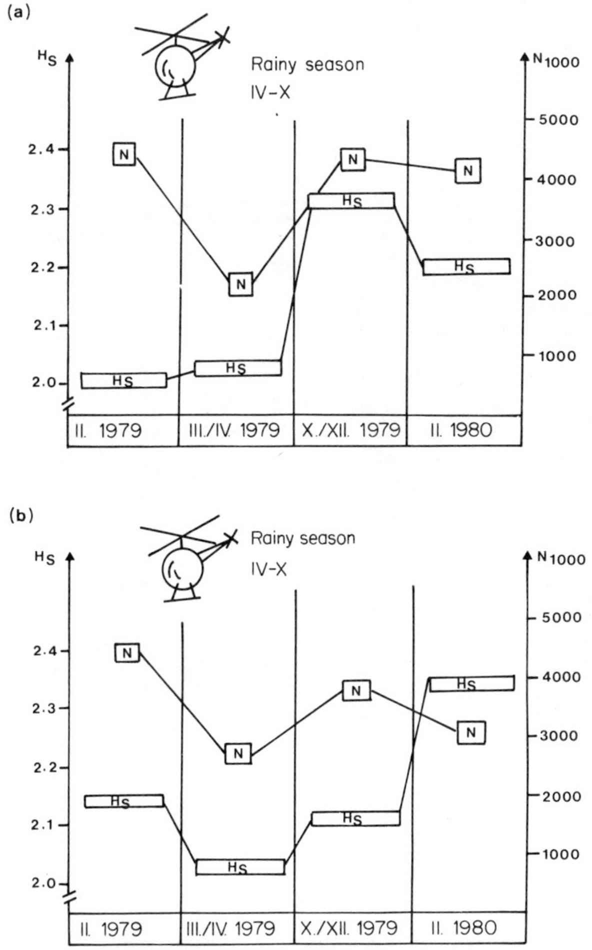

Fig. 20.6 Evaluation of Arthropoda in ground traps (without Collembola, Isoptera, Formicoidea). (a) Savannah Zone; (b) River Forest Zone.

species, sex and age. This is of considerable significance for the comparability of residue analyses in various populations.[31]

The integrators for the pollution of spaces and ecosystems are the populations and substrates woven together in the food webs. Calibration of ecological information (ecological control unit) is therefore only possible if the results of the population ecology are carried out simultaneously with a check on the pollutant persistence in the food chains. Some of the invertebrates and vertebrates captured at the site were stored in liquid nitrogen, after preparation of the livers, and taken to Saarbrücken for analysis. The analysis particularly concentrated

Table 20.1 Results of residue analyses (ppm dry weight; 1 = HCB, 2 = lindan, 3 = heptachlor, 4 = aldrin, 5 = heptachlor-epoxides, 6 = gamma-chlordan, 7 = alpha-chlordan, 8 = *p,p'*-DDE, 9 = dieldrin, 10 = endrin, 11 = DDT) from ecologically different groups of birds in the study area (collection period 15.3.79–5.4.79; directly after dieldrin spraying; supplemented according to Müller and Nagel[33])

Species	*n*	1	2	3	4	5	6	7	8	9	10	11
1. *Francolinus b.*	2	0.43			0.72			0.23		0.21 0.22		0.01
2. *Streptopelia d.*	2						0.82		1.72			1.72
3. *Streptopelia s.*	1		0.07		1.86		0.86					0.08
4. *Streptopelia v.*	1									0.12		
5. *Caprimulgus in.*	2	0.59						7.21	0.82			
6. *Ispidina picta*	8									1.28		2.35
7. *Halcyon leuc.*	4		0.23					5.35		2.70		0.94
8. *Halcyon malim.*	5							8.21		4.30		3.52
9. *Halcyon chel.*	1									2.13		
10. *Lybius bident.*	1							6.23		7.24		0.07
11. *Indicator ind.*	7	0.73						5.68		0.29		0.23
12. *Apaloderma nar.*	1									0.82		
13. *Cuculus solit.*	1									1.24		
14. Campephaga ph.	3					0.47					2.34	0.57
15. *Phyllastrephus f.*	7									0.47		3.70
16. *Phyllastrephus s.*	4									0.15		
17. *Bleda syndact.*	4									0.33		
18. *Andropadus virens*	8									0.63		
19. *Tchagra seneg.*	7									0.71		0.65
20. *Laniarius ferr.*	4									3.43		2.47
21. *Cossypha albic.*	5			0.71				0.28		0.34		
22. *Cossypha niv.*	3			0.34						0.28		
23. *Cossypha heugl.*	5			0.49						0.48		

Table 20.1—*Continued*

Species	*n*	1	2	3	4	5	6	7	8	9	10	11
24. *Cossypha pol.*	5									0.25		
25. *Turdus pelios*	7♀									0.92		3.98
	7♀									0.40		2.93
26. *Myrmecocichla n.*	1									0.42		
27. *Camaroptera b.*	1									0.08		9.27
28. *Eremomela pus.*	2	0.27						3.72				
29. *Malaenornis ed.*	3			0.62						0.82		
30. *Platysteira cy.*	5									0.05		
31. *Terpsiphone v.*	4									0.002		1.49
32. *Nectarinia v.*	2									2.44		4.93
33. *Nectarinia o.v.*	1									2.02		
34. *Lamprotornis sp.*	2									0.09		1.22
35. *Lamprotornis ch.*	1									0.21		
36. *Dicrurus adsim.*	3									3.22		2.79
37. *Cicrurus ludw.*	1									0.82		
38. *Oriolus auratus*					1.04	0.57				0.27	1.14	0.58
39. *Turdoides pleb.*	1									0.08		
40. *Hypergerus atr.*	1									0.10		
41. *Ploceus cucull.*	1									0.94		
42. *Clytospiza dyb.*	2									0.43		
43. *Clytospiza mont.*	1									0.08		

Table 20.2 Residues from *Andropadus virens* (liver; ppm dry weight) before (February 1979) and after spraying

Number of individuals	Date	Dieldrin
10	8.2.79–16.2.79	0.000–0.093 (0.018)
8	16.3.79–4.4.79	0.084–0.840 (0.630)

on the polychlorinated hydrocarbons HCB, Lindan, Heptachlor, Aldrin, Heptachloreposides, γ-Chlordan, α-Chlordan, *p,p′*-DDE, Dieldrin, Endrin and DDT derivatives.

The analytical problems (including scattering of the measured values) is depicted in detail in Egan.[9] Comparisons between the residue values from Cameroon determined by us and investigation results from other parts of the world are important for the discussion.[4–6,31] They show that for example concentrations of over 4 ppm Dieldrin or DDT seldom occur in organs of birds or mammals in Europe or North America (in contrast to fish or invertebrates). Significantly higher values can be found in the eggs of peak consumers.[7] The same is true of vertebrates from tropical areas which are agriculturally utilized and with a high application of insecticides.[21]

We were able to gather extensive material for residue analyses of numerous representative types of food web in the study period. Contamination rates were thus determined for some types of bird for the first time. The analytical results vary according to the position of the species in the food chain and its exposure (Table 20.1).

Table 20.3 Scattering of analytical residue values within the same species. The animals were gathered 8–15 days after spraying (all values in ppm/DW; whole body analyses with the exception of *Micropteropus pusillus*)

Species	*n*	Dieldrin	*m*	DDT	*m*
Aphyosemion buellanum	114	0.084–214.34	2.524	0.09–38.953	14.728
Clarias walkeri	3	0.08–3.421	1.782	0.10–3.740	2.518
Bufo regularis	17	0.071–3.420	0.932	0.731–13.411	2.438
Bufo pentoni	6	0.000–2.371	0.985	0.000–8.926	3.728
Rana occipitalis	7	0.972–1.528	1.291	8.184–22.752	14.625
Rana occipitalis Tadpoles	57	0.000–0.948	0.096	0.711–6.392	4.232
Mabuia quinquetaeniata	7	0.028–1.627	0.842	1.359–8.392	5.972
Agama benueensis	2	0.072–0.237	0.154	5.981–6.243	6.112
Micropteropus pusillus	7	133.42–174.8	142.728	11.342–18.571	14.350
Sphingidae	12	0.000–1.829	1.043	0.000–2.784	0.830
Gryllidae 25.3.79	33	2.844–62.854	14.278	0.014–27.572	14.134
Isoptera (winged) 24.3.79	28	0.000	0.000	0.000	0.000
Isoptera (winged) 25.3.79	112	0.004–1.438	0.082	0.012–0.098	0.074

Table 20.4 Dieldrin and DDT concentrations in the livers of *Turdus pelios* (ppm DW) at different times before and after spraying

	Time before start of spraying 8.2.79–16.2.79	Time after spraying	
		15.3.79–30.3.79	15.2.80
Dieldrin	0.0–0.042 ($n = 10$)	0.916 ($n = 7$♂) 0.397 ($n = 7$♀)	6.0 ($n = 1$)
DDT	—	2.9–4.0 ($n = 3$)	2.0 ($n = 1$)

Only low residues of dieldrin could be detected in the site populations before the spraying campaign. If one uses other groups of organisms for comparison, then it becomes apparent that directly after spraying a surprisingly large scattering of analytical residue values occurs. Fruit-eating pygmy flying foxes (*Micropteropus pusillus*) have dieldrin liver values above the LD_{50} values for *Rattus norvegicus*. This species was no longer detected in this area a year after application.

After spraying, Dieldrin was detectable on the leaves, fruit and on the ground in concentrations between 61 and 124 ppm. The concentrations fell to below 7 ppm 14 days after spraying (two thunderstorms with lengthy periods of rain). However, the concentrations increase at the same time in the flowing water and in the limno fauna (for example *Cyprinodontides*, *Silurides*, tadpoles, aquatic insects). In the case of some terrestrial insects the concentration increases with time.

The present findings make it clear that the uncertainty of the prediction 'modifications in ecosystems after application of only *one* active substance' increases with the diversity of the ecosystem. This means that transfer of 'limits' from the non-tropics to the tropics is not permissible without field analyses.

Genetic Comparability of Populations

A further problem which has not been sufficiently considered in ecotoxicological effect analyses in the field up to now is the different age, sex and allelotype structure. It has been known for a long time that different sex-specific accumulation rates occur in field populations. Trend analyses are, therefore, largely distorted, if these relations are not considered.

The discovery is, however, new that not only the reaction to certain substances (partial resistance) depends on the allelotype composition of a population but also that various allelotypes can display different accumulation rates without the physiological mechanism being explained in the individual case. Evolution processes are currently under way in our agricultural landscapes the knowledge

Table 20.5 Heavy-metal residues (mg/kg DS) from female and male *Carabus auratus* (Coleoptera, Carabidae) in May and June 1979 (specimens from Saarbrücken Airport)

Heavy metal	♀♀	♂♂	23.5.79	29.5.79	5.6.79	12.6.79
Airport site: centre (in front of runway)						
Zinc	♀♀		189.1	208.2	168.0	178.7
		♂♂	163.4	173.4	158.6	137.2
Lead	♀♀		1.75	1.42	1.29	1.36
		♂♂	1.73	1.34	1.14	1.27
Cadmium	♀♀		1.56	2.06	1.42	1.08
		♂♂	0.95	0.45	1.05	0.76
Airport site: west (old road)						
Zinc	♀♀		137.5	139.1	139.2	138.7
		♂♂	131.2	131.8	102.5	130.3
Lead	♀♀		1.22	1.06	1.14	1.29
		♂♂	1.05	1.03	1.12	1.26
Cadmium	♀♀		0.64	1.73	1.05	1.41
		♂♂	1.03	1.03	0.78	0.88

of which is of considerable significance for 'environmental compatibility tests'. My colleague, Dr H. Steiniger,[43] came across the existence of various allelotypes in investigating the partial SO_2 resistance of Carabidae populations from the city centre using electrophoretic separation processes (disk electrophoresis with polyacrylamide gels; Elphor varia 2 with micro-rapid-chamber, from Bender & Hobein 1537 (Figure 20.7). He discovered a correlation between allelo-polymorphism and 'pollution factors' in the case of fifty-four central European Carabidae populations (seven species) from urban and natural habitats (this was particularly well developed in the case of *Carabus violaceus*, *Carabus*

Table 20.6 Mean Pb content of female and male *Lacerta* sp. from various sites (1979)

Species	Locality	n	Sex	Pb mg kg^{-1} TS
Lacerta agilis	Bübingen	4	♂	6.07
	(Birzberg)	4	♀	3.24
Lacerta muralis	Dreisbach	6	♂	3.91
	(bend in the Saar)	4	♀	2.78
Lacerta muralis	Lemberg Iron and Steel Works	4	♂	4.45
	(Nahe)	7	♀	3.56
Lacerta muralis	Saarbrücken	8	♂	23.00
	(Central Railway Station)	4	♀	12.82
Lacerta muralis	Jägersfreude	27	♂	8.02
	(mines)	15	♀	4.61
Lacerta vivipara	University of Saarbrücken	15	♂	45.94
	(roadside)	5	♀	11.86

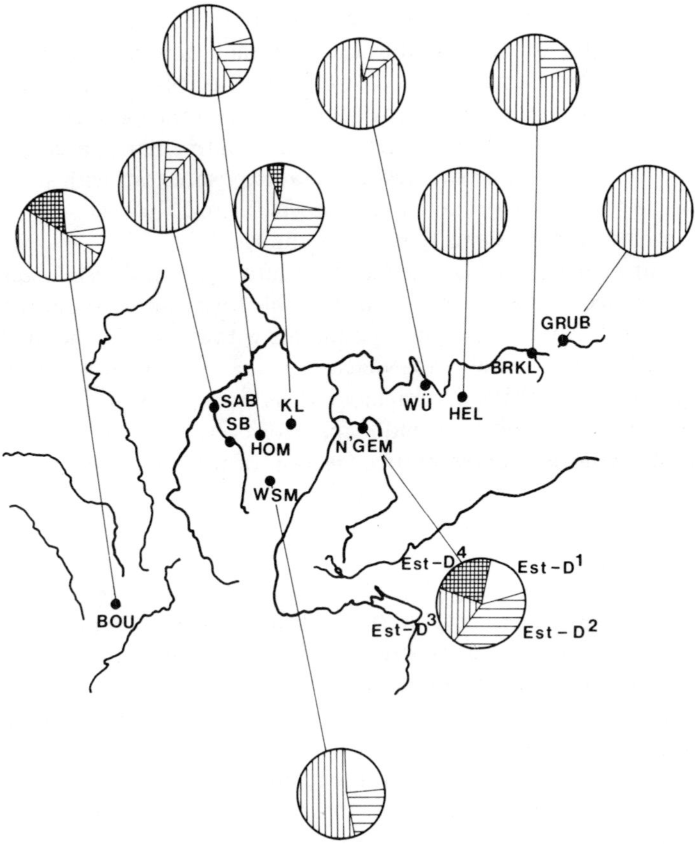

Fig. 20.7 Distribution of the allelofrequencies at the EST-D-locus of *Abax ater*. (According to Steiniger, 1978)

problematicus, *Carabus nemoralis*, *Abas ater* and *Abax parallelus*), without, however, being able to establish a causal relationship between the isoenzyme polymorphisms which occurred and their selective advantages. Zadory[46] came across a similar phenomenon in analysing the Saar populations of the snail *Physa acuta*.

The problems discussed above can be minimized, if one uses field tests with standardized organism specimens instead (experimental monitoring). However, we thus only obtain partial statements which refer to one compartment (soil, water, air etc.) of an ecosystem.

EXPERIMENTAL MONITORING

Essential information for space evaluation can best be acquired from *food chain* analyses and *ecological maps* in view of the complexity of ecosystems. Ecological maps have the task of converting the space–time behaviour of ecologically effective factors (for example modifications in radiation balance; halogenated hydrocarbons; heavy metals) and ecological elements (plant and animal species, populations, biocoenoses, ecosystems, area systems) into a usable form for a genuine space evaluation. Two test groups can be differentiated according to the degree to which they can be standardized, (a) exposure tests with standardized organisms (active or experimental monitoring), and (b) reaction and trend analyses in natural biosystems (passive monitoring).

Both groups can be further differentiated into *effect* and *trend maps*. Effect maps are intended to make conclusions possible for example about the relative toxicity of a pollutant or mixture of pollutants with respect to the site by using sensitive organisms. Trend maps determine the accumulation of pollutant concentrations in exposed or natural biosystems in time. Less sensitive organisms are therefore most suitable for trend maps. However, as 'catch substrates' these must be able to store a representative fraction of the pollutants present in the space.

Effect and Trend Maps

Only in the case of a high degree of information from the genetic structure present (plants, animals) can genetic variability as well as daily and seasonal population fluctuations enable conclusions to be made about the space quality. It is therefore necessary to draw up effects and trend maps with 'standardized' organisms.[15,22,41] Maps of this type are biotests with exposed organisms relative to area. They give information about the relative endangerment of the exposed substrates in the spaces analysed and enable us to make statements about the pollution of individual ecosystem compartments.

Effect and trend maps with exposed animals have only been developed in the last few years. Aquatic organisms (cf. Final Draft Document of the 'International Workshop on Monitoring Environmental Materials and Specimens', 23–28 October, 1978, Berlin), useful animals and terrestrial invertebrates are the main groups to be investigated.[30,31] Laboratory tests and their application under standardized conditions in the field must be possible for the species used (cf. VDI guidelines 3792 and 3795). Plants and substrates have already been used much longer in effect and trend maps covering selected areas (cf. for example the exposure of *Hypogymnia physodes*).[1,11,37,42,44]

Exposure of the lichen *Hypogymnia physodes* represents a tried and tested procedure for *effect maps*. The lichens intended for exposure, for example in a densely populated area, were taken from their natural site (outside the evaluation area) together with their bark substrate (generally oak) and applied in 6–10 parallel wood or plastic panels for one observation site. These lichen exposure panels (290 × 170 mm) with corresponding pit-shaped depressions (45 mm diameter), were applied (with an insoluble plastic paste) in such a way that the lichens just protruded above the edge of the depression. The lichen exposure panel is fixed to a mounting which ought to have a thread so that, for example, it is also possible to attach dust precipitation measuring instruments as described by Bergerhoff. Since above all immissions from one direction are acquired by means of the lichen panel, rotating exposure mounts have been developed to determine the total immission. These are driven by the wind in order to rotate about a vertial axis and thus make it possible to determine the pollutants acting upon the lichens independent of exposure.

In the conurbation around Saarbrücken four exposure boards (four points of the compass) were used at each site (1.5 m above the ground). The partly different course of damage also provided information about the different immission pollution depending on direction. At the start of exposure and at various intervals photographs are taken of the lichens on an image scale 1:2.3 and the fraction of dead lichen thalli is determined or the degree of damage is directly measured via the net photosynthesis rate. The course and extent in time of the necrotization of the thalli makes it possible to draw conclusions about the biologically effective immission dose at each site. We are able to display such correlations covering for example the Ruhr area and the Saarbrücken conurbation (Figures 20.8 and 20.9).

Effect maps, in the sense of the lichen map, only permit statements to be made about the ecotoxicological effects of, for example, one complex immission type. Elimination of a specific damage factor is in general only possible by a simultaneous combination of effect and trend maps at one site. Only

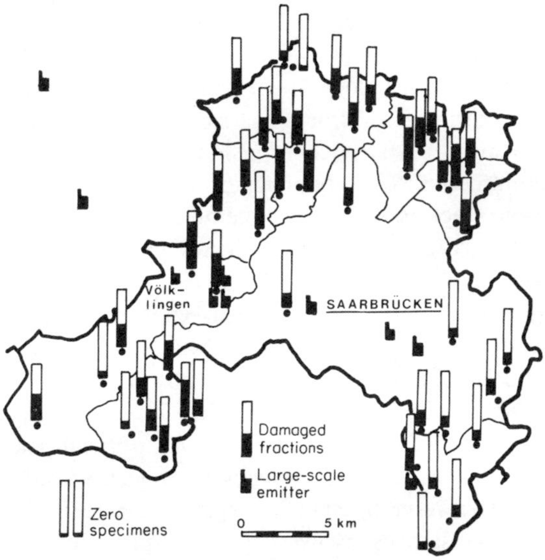

Fig. 20.8 Damage to the exposed lichens after 140 days at various sites in the Saarbrücken conurbation

Fig. 20.9 Course of damage at the exposure sites in the Saarbrücken conurbation. (Details in Erhardt and Müller, 1980)

accumulation indicators (cf. for example grass culture procedures, VDI 3792) are suitable for compiling standardized *trend maps*.

Inferences about the substances involved in a damaging effect in the case of more sensitive species (at the same site) can be made by analysing the substances contained in the exposed specimens. Transfer to human populations is also possible relative to area (particularly if we are dealing with food species) and a reference to the pollution of field populations can also be established.

At the moment experiments are being carried out with *multi-species maps* by means of which we aim to attain deeper insights into more complex immission types with the simultaneous exposure of in part ecologically different animals and plants via a multiple reaction spectrum. The different substance uptakes, which are species and, in part, sex-specific (e.g. heavy metals, DDT, PCB, PAH) is taken as an expression of the relative pollution of a site at the time of measuring. Practical experience is available for example in the case of the following plant and animal species from the Saarbrücken area: *Lolium multiflorum*, *Sambucus nigra*, *Brassica oleracea*, *Plantago major*, *Taraxacum officinale*, *Picea omorica* (Figure 20.10), *Populus* spp., *Aesculus*, *Platanus*, *Hypnum cupressiforme*, *Sphagnum* spp., *Vulpes vulpes*, *Turdus merula*, *Dreissena polymorpha*, *Physa acuta*, *Lumbricus* spp., *Lacerta agilis*, *Lacerta vivipara*, *Apis*, Carabidae. In general, an analysis of the substances contained in the exposed specimens enables us to make inferences about the substances involved in a damaging effect in the case of more sensitive species (at the same site). A trend map has been tested for Porto Alegre within the gecko *Hemidactylus mabouia*.

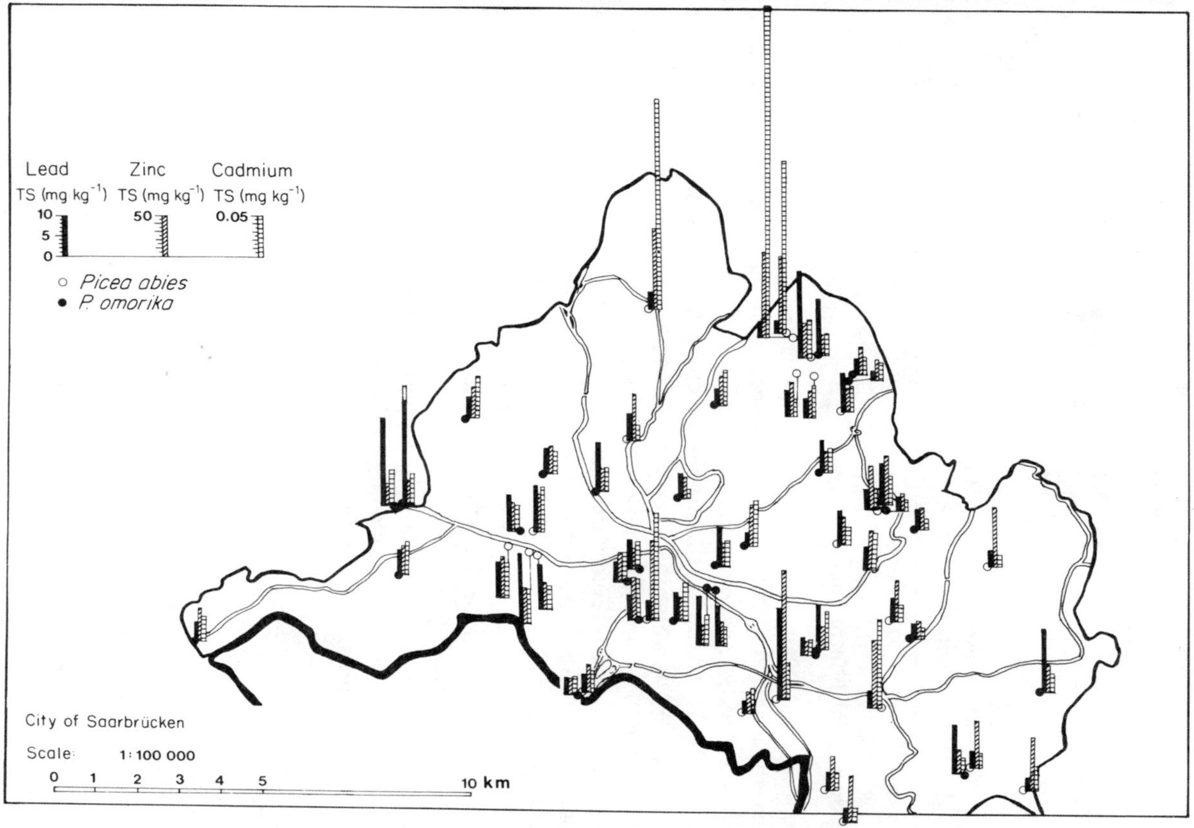

Fig. 20.10 Heavy metal content in 1-year-old spruce needles, 1978

Table 20.7 Cd, Zn and Pb residues from *Hemidactylus mabouia* from a busy road (5000 vehicles per hour; Farapos near Rodovieria) and a site in the suburbs (50 vehicles per hour; Av. Gustavo Schmidt, Porto Alegre, Brazil, 12.3.80)

Road	Cd (ppb)	Zn (ppb)	Pb (ppm)
Farapos	80–142.97 (106.48)	98.21–151.00 (128.09)	3.85–20.54 (11.81)
Av. Gustavo Schmidt	37.5–73.65 (49.55)	86.0–95.26 (92.08)	2.23–3.28 (2.74)

Ecotoxicological Information

There is no doubt that effect and trend maps can provide information relative to area about the effects of various substances relating to the biosystem used (amongst others *Hypogymnia physodes*; grass culture procedure with *Lolium multiflorum*). At the moment we do not by any means know the limits of their applicability. We do know that they are not suitable, or only to a limited extent, for checking PAHs[14] or other substances which metabolize easily when entering a biological matrix. However, if the chemicophysical properties of the substances are known they give us deeper insights into their effects in the field than pure laboratory tests.[31]

Nevertheless, neither laboratory tests nor experimental monitoring alone are sufficient to describe the actual pollution of an ecosystem. Mutagenic, carcinogenic or teratogenic long-term effects can only be determined by the simultaneous analysis of food webs with respect to the ecology of populations and chemistry, significant chemicophysical effective factors being under control (amongst others immission type, environmental chemicals, radiation and energy balance).

FOOD-WEB CONTROL AND 'ENVIRONMENTAL COMPATIBILITY TESTS'

The populations and substrates interwoven in the food webs are integrators for the pollution of spaces and ecosystems. In my opinion our dilemma is to be found in the fact that there are only a few institutes which are in a position to carry out both food-web element controls in the field (basic prerequisite: precise knowledge of the species) as well as precise chemical analyses in biological matrices.

Continuous analyses are feasible and have, in part, already been carried out covering individual towns and their surrounding area (e.g. Saarbrücken[30,31]) for 'short' food chains which have to be representative of various areas in the Federal Republic of Germany (i.e. the species involved in them must be present and

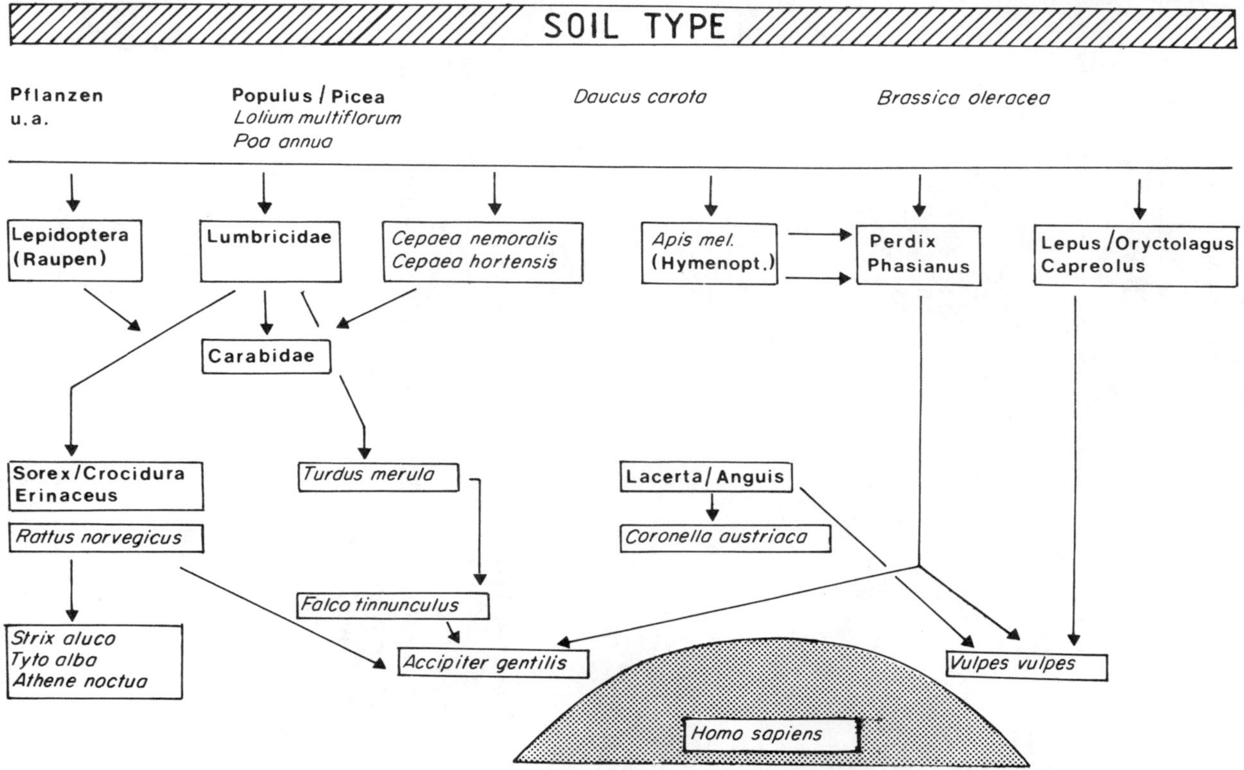

Fig. 20.11 Food-web species which are being continuously investigated in the Saarbrücken conurbation with respect to the ecology of population and residue analyses (heavy metals, PCB's, halogenated hydrocarbons) as part of a BMI research project

Table 20.8 Stomach contents of *Lacerta muralis* at various localities (2 = Jägersfreude mine; 3 = Saarbrücken Central Railway Station; 4 = Lemberg Iron and Steel Works (Nahe); 5 = bend in the Saar at Dreisbach) and quantitative results of catches from Barber traps (July 1979)

Animal group	2		3		4		5	
	Stomach	Barber trap	Stomach	Barber trap	Stomach	Barber trap	Stomach	Barber trap
Araneida	42	67	20	28	14	—	17	9
Opilionida	—	13	2	2	—	—	—	3
Acarina	—	—	—	—	—	—	—	2
Isopoda	26	33	13	14	6	—	12	7
Myriapoda	22	27	6	19	8	—	14	11
Collembola	5	16	3	2	—	—	3	—
Blattodea	—	—	—	—	—	—	2	—
Saltatoria	5	7	—	—	4	—	5	—
Dermaptera	4	8	—	—	2	—	2	2
Heteroptera	3	3	—	—	—	—	—	—
Coleoptera	19	85	4	30	7	—	6	18
Hymenoptera	6	6	3	—	—	—	3	2
Formicidae	8	17	3	7	12	—	11	8
Diptera	39	19	4	—	8	—	8	5
Lepidoptera	3	—	—	—	3	—	3	—
Oligochaeta	12	8	3	3	5	—	5	6
Slugs	8	—	—	—	—	—	4	—

available both in Kiel and also in Munich[30,31]). However, for more complex food webs they can only be meaningfully carried out via a continuous ecosystem analysis and monitoring.

After many years' experience, in part negative, and after extensive discussions with colleagues from North America (US Workshop in Washington, 1978), my working group has compiled the 'food-web species' for Germany shown in Figure 20.11 and has begun continuous population and residue analyses in the Saarland. It has become apparent that particularly analyses carried out at the same site of the soil type (various plant species), earthworms (various genera), ground beetles (Carabidae), blackbirds (*Turdus merula*), insectivores (*Sorex*), honey-bees (*Apis*) and red foxes (*Vulpes vulpes*) have led to results which can be evaluated relative to area if the ecological variables are known.

Monitoring the food-web species requires an experienced, highly qualified team which must be tightly organized and well-coordinated. Only in this way is it possible to reproduce the web variables and the pollution parameters in a continually changing environment. The basic knowledge which we still lack in this field cannot be conjured up by planning models, however cunningly thought out. Monitoring food-web species means the analysis, for example, of all prey and the substances taken up via these animals. Only in this way do we believe it is

Table 20.9 Pb and Cd residues (mg kg^{-1} DS) in *Lacerta muralis* and their prey at two sites in Saarbrücken 1979

Saarbrücken (Central Railway Station)			
Lacerta muralis	8♂	23.0 (6.39–40.44)	1.52 (0.45–2.72)
	4♀	18.82 (9.48–18.67)	0.64 (0.41–0.85)
Coleoptera		2.72	1.33
Isopoda		61.34	14.05
Araneida		17.53	5.50
Myriapoda		40.91	10.52
Slugs		19.76	5.54
Oligochaeta		77.56	26.73

possible to meaningfully interpret, for example, Pb or Cd residues in *Lacerta* populations (Tables 20.8 and 20.9).

The correctness of this approach was also underlined by the Senate Committee for Environmental Research of the DFG (1979)/Deutsche Forschungsgemeinschaft (German Research Association): 'The accumulation of pollutants in the food chains must be regarded as a highly significant problem for environmental chemicals, even, if in individual cases experimental counterevidence is available'.

Laboratory findings must, therefore, be standardized with respect to ecosystems for the ecotoxicological evaluation of environmental chemicals. In my opinion this can best be carried out via effect and trend maps as well as foodweb analyses in defined ecosystems.

REFERENCES

[1] Arndt, U. (1974). Langfristige Immissionswirkungen an ungeschütztem Nutzholz, *Staub*, **34**, 225–227.

[2] Bevenue, A. (1976). The 'bioconcentration' aspects of DDT in the environment, *Residue Rev. Rome*, **61**, 37–112.

[3] Blau, H., and Neely, W. (1976). Mathematical model building with an application to determine the distribution of DURSBAN insecticide added to a simulated ecosystem, *Adv. Ecol. Research*, **9**, 133–163.

[4] Brown, A.W. (1978). *Ecology of Pesticides*. John Wiley, New York.

[5] Brüggemann, J., Busch, L., and Drescher-Kaden, U. (1978). Changes of pesticide residues levels in different animal species of some habitats in the German Federal Republic from 1970 to 1976, *Pesticide Residues in Animals; XIII Congress of Game Biologists*. Munich.

[6] Clark, D., Laval, R.,, and Swineford, D. (1978). Dieldrin-induced mortality in an endangered species, the Gray Bat (*Myotic grisescens*), *Science*, **199**, 4335, 1357–1359.

[7] Conrad, B. (1978). Gefährdung der einheimischen Vogelwelt durch chlorierte Kohlenwasserstoffe, *Kleintier Praxis*, **23**, 87–94.

[8] Durham, W.F. (1976). Human health hazards of respiratory exposure to pesticides, in, *Air Pollution from Pesticides and Agricultural Processes* (Ed. R.E. Lee), CRC Press, Cleveland, Ohio.

[9] Egan, H. (1979). Organic and inorganic compounds—Analytical aspects, in, *Monitoring Environmental Materials and Specimen Banking*, pp. 230–246. Nijhoff, The Hague.

[10] Ellenberg, H., Fränzle, O., and Müller, P. (1978). *Ökosystemforschung im Hinblick auf Umweltpolitik und Entwicklungsplanung*. Report No. 78-10104005 from the Federal Ministry of the Interior, Bonn.

[11] Erhardt, W., and Müller, P. (1979). *Flechtenwirkungskataster im Verdichtungsraum von Saarbrücken*. Verhdl. Ges. für Ökologie, Münster.

[12] Frost, K.R., and Ware, G.W. (1970). Pesticide drift from aerial and ground applications, *Agric. Eng.*, **51**, 460.

[13] Gish, C.D., and Christensen, R.E. (1973). Cadmium, nickel, lead and zinc in earthworms from roadside soil, *Environ. Sci. Technol.*, **7**, 1060–1062.

[14] Grimmer, G. (1979). Statement of polycyclic aromatic hydrocarbons, in, *Monitoring Environmental Materials and Specimen Banking*, pp. 298–312. Nijhoff, The Hague.

[15] Guderian, R. (1977). Air pollution. Phytotoxicity of acidic gases and its significance in air pollution control, *Ecological Stud.*, **22**, 1287.

[16] Hall, Ch.A.S., and Day, J.W. (1977). *Ecosystem Modelling in Theory and Practice: An Introduction with Case Histories*. John Wiley, New York, Toronto.

[17] Hill, I.R., and Wright, S.J.L. (1978). *Pesticide Microbiology. Microbiological Aspects of Pesticide Behaviour in the Environment*. Academic Press, New York.

[18] Holdgate, M. (1979). *A Perspective of Environmental Pollution*. Cambridge University Press, Cambridge.

[19] Khan, M.A. (1977). *Pesticides in Aquatic Environments*, Plenum Press, New York.

[20] Klein, W. (1979). Organohalogenated compounds in plants and soil, in, *Monitoring Environmental Materials and Specimen Banking*, pp. 354–368, Nijhoff, The Hague.

[21] Koeman, J., and Takken, W. (1977). L'action sur l'environnement de la lutte contre la tsé-tsé état des connaissances actuelles, *FAO, Production et santé animales*, **7**.

[22] Kutz, F.W., Murphy, R., and Strassman, S.C. (1978). Survey of pesticide residues and their metabolites in urine from the general population, in, *RAO, Pentachlorophenol*, pp. 363–369, Plenum Press, New York.

[23] Kutz, F.W., Yobs, A.R., and Yang, H.S.C. (1976). National pesticide monitoring programs, in, *Air Pollution from Pesticides and Agricultural Processes* (Ed. R.E. Lee), CRC Press, Cleveland, Ohio.

[24] Matsumura, F. (1977). *Absorption, Accumulation and Elimination of Pesticides by Aquatic Organisms*, Plenum Press, New York.

[25] Metcalf, R. (1977). *Model Ecosystem Studies of Bioconcentration and Biodegradation of Pesticides*, Plenum Press, New York.

[26] Lee, R.E. (Ed.) (1976). *Air Pollution from Pesticides and Agricultural Processes*, CRC Press, Cleveland, Ohio, 264 pp.

[27] Lowman, F.G. (1979). U.S. Mussel Watch Program. The mussel watch biological monitoring research program, in, *Monitoring Environmental Materials and Specimen Banking*, pp. 393–396, Nijhoff, The Hague.

[28] May, R.M. (1977). *Thresholds and Complexity in Model Ecosystems*, Princeton University Press, Princeton.

[29] Müller, P. (1977). Belastbarkeit von Ökosystemen, in, *Energie und Umwelt*, Envitec, Düsseldorf.

[30] Müller. P. (1978). *Urbane Ökosysteme*, p. 303, BMI, Bonn.

[31] Müller, P. (1979). Basic ecological concepts and urban ecological systems. *Biogeographica*, **15**, 209–223.

[32] Müller, P. (1980). *Biogeographie*. UTB, Verl. Ulmer, Stuttgart.

[33] Müller, P. (1980). *Arealsysteme und Biogeographie*. Verl. Ulmer, Stuttgart.

[34] Müller, P., and Nagel, P. (1979). Ökosystemare Wirkungen von Insektiziden im Hochland von Adamaoua (Kamerun), *Mitteil.*, **11**, Chair of Biogeography, University of the Saarland, Saarbrücken.

[35] Müller, P., and Stein, N. (1979). Standardization of ecological criteria and their integration into the urban planning, MAB-Rep., Bonn.

[36] Murton, R.K. (1979). Considerations applicable to the monitoring of organohalogenated compounds in terrestrial wildlife, especially birds, in, Monitoring Environmental Materials and Specimen Banking, pp. 450–469. Nijhoff, The Hague.

[37] Nagel, P. and Müller, P. (1979). *Ökologische Bewertung der Saar im Verdichtungsraum von Saarbrücken.* Verhdl. Ges. für Ökologie, Münster.

[38] Patten, B.C. (1974). *The Zero State and Ecosystems Stability.* Proceedings of the First International Congress on Ecology, Wageningen.

[39] Po-Yung, L., Metcalf, R. L., Furman, R., Vogel, R., and Hasset, J. (1975). Model ecosystem studies of lead and cadmium and of urban sewage sludge containing these elements, *J. Environm. Qual.*, **4**, 505–509.

[40] Rao, K.R. (1978). *Pentachlorphenol. Chemistry, Pharmacology, and Environmental Toxicology*, Plenum Press, New York.

[41] Scholl, G. (1978). Vergleich zwischen Emissionsrate und Wirkdosis von Fluorid in standardisierten Graskulturen im Bereich eines Aluminiumwerkes, Schriftenr. Landesanst. Immissionsschutz NRW, **43**, 75–89.

[42] Schönbeck, H., and van Haut, H. (1974). *Methoden zur Erstellung eines Wirkungskatasters für Luftverunreinigungen durch pflanzliche Indikatoren*, Verhdl. Ges. für Ökologie, pp. 435–445, Saarbrücken. Verl. Junk, The Hague.

[43] Steiniger, H. (1978). Genetische Variabilität bei Carabiden-Populationen inner- and außerstädtischer Standorte (Coleoptera). Doctoral Thesis. Biogeography. Univ. Saarbrücken.

[44] Wagner, G., and Müller, P. (1979). *Fichten als 'Bioindikatoren' für die Immissionsbelastung urbaner Ökosysteme unter bes. Berücksichtigung von Schwermetallen.* Verhdl. Ges. für Ökologie, Münster.

[45] Webster, J.R., Waide, J.B., and Patten, B.C. (1975). *Nutrient Recycling and the Stability of Ecosystems. Mineral Cycling in Southeastern Ecosystems.* ERDA Symposium Series. US Energy Research and Development Administration. Springfield.

[46] Zadory, L. (1980). Allelpolymorphismus bei Physa acute DRAP. (Pulmonata, Gastropoda) in Abhängigkeit von physikalisch-chemischen Wassereigenschaften der Saar. Diploma Thesis, Biogeography, Univ. Saarbrücken.

Pollutants and Their Ecotoxicological Significance
Edited by H. W. Nürnberg

CHAPTER 21

Analytical Aspects of the Determination and Characterization of Metallic Pollutants

M. STOEPPLER

SUMMARY

A brief introduction into the aims and problems of contemporary trace and ultratrace analysis of inorganic and organometallic compounds is given. This is followed by a description of sampling, sample storage and sample pretreatment together with the necessary quality control measures to obtain reliable data. Further the current surveillance and research methods predominantly used in environmental and ecotoxicological studies are briefly reviewed, and some highlights, advantages and disadvantages are shown. Finally a comparative overview of commercially available instrumentation deals with cost–benefit estimations.

INTRODUCTION

Today there is an urgent need to elucidate the impact of harmful chemical substances, predominantly produced, used and wasted by man. Numerous interdisciplinary investigations revealed rather complex inter-relationships between different elements and compounds occurring in biological and ecological systems at relatively low concentrations (typical range: $mg\,kg^{-1}$ to $ng\,kg^{-1}$). This situation, in conjunction with similar demands from other scientific branches, stimulated a progressively intensified development of the so-called instrumental analysis, introducing new dimensions into analytical chemistry. Thus trace analytical methods became more and more accessible and indispensable for all types of scientific, medical and legal applications.

Numerous papers, presenting particular advances in the basic instrumental design and sensitivity and leading to new or improved multi-element methods, have appeared with nearly logarithmic growth rates up to the present day. A trend of efforts, focused on the highest obtainable technical performance linked with highly improved detection limits could be observed within the last decades.

Thus, relatively new methods such as different modes of activation analysis and atomic absorption spectrophotometry competed with basically older but recently improved and revised methods such as: polarography and voltammetry; mass, X-ray and emission spectroscopy; automated colorimetry; chromatographic methods.

The instrument industry increasingly designed and sold very sophisticated systems for the determination of trace elements and compounds. Therefore, an accelerated change from classical (i.e. predominantly manual), to instrumental (i.e. mechanized or even automated methods), took place. The classical analyst, a predominantly well-trained, skillful person was very often replaced by a more or less specialized technician, because modern equipment was increasingly introduced into laboratories hitherto not experienced in trace chemistry.

Making full use of the significant potentialities of advanced instrumental analysis a certain amount of outstanding analytical work was done in laboratories with appropriate analytical and general chemical background. The vast increase in analytical instruments, however, also resulted in an immense production of mediocre or even erroneous data on essential and toxic elements and compounds. That was clearly demonstrated if round-robin exercises were performed. The results of those comparative exercises very frequently showed a broad scatter of results for the concentration of a certain element or compound in a certain material (e.g. lead (Pb) and cadmium (Cd)), in whole blood and urine[1] and Hg and other trace metals at ppb (ng ml^{-1}) levels or even lower in aqueous samples.[2,3] Also at higher concentration levels remarkable deviations are observed.[4,5] While unduly broad ranges of results were obtained particularly in those large round-robin exercises which, in trace analysis, had also included less experienced laboratories, even intercomparisons between smaller numbers of more experienced institutions yielded only fairly good results at least if analysis at the ppb level is concerned.[3,6-8] Hence, analysis at ultratrace levels (i.e. below 1 ng g^{-1} or 1 ng ml^{-1}) still appears to be more a risky game with an unpredictable end than a well-defined and accurate process for laboratories not having a particular and long-lasting experience in trace analytical chemistry.[9] That brings into question the reliability of a rather large amount of analytical data being published in numerous research areas.

The reasons for these facts are many. The unsatisfactory results were often due to the application of only a single method just available in the respective laboratory without a critical check for systematic errors and also minimal knowledge of the sources of possible bias. Too many laboratories with limited experience and an insufficiently trained staff trusted too much upon a single device very often used in a 'black box' manner according to exaggerated, or even unsound but persuasive, commercial advertisements, although it could only be operated reliably by experts, and in comparison with really independent methods based on different physical principles, or applying standard reference materials of which there are, however, only for very few cases relevant types for trace analysis of environmental samples at present available.

A big influence also was the lack of knowledge of many manufacturers of analytical systems with respect to particular important side effects. Of course, they hoped, by frequent discussions with customers, to gain some knowledge on possible disadvantages of the instrumentation used, creating the necessary feedback for technical and methodological improvements. Laboratories without special expertise of the above described type, however, represent a large number of customers and are not really able to contribute with valuable ideas due to their lack of basic knowledge and experience.

On the other hand a certain number of laboratories, known for their excellent scientific standing, were frequently working apart from the practical necessities and occupied themselves by tackling problems in attaining esoteric goals. Usually they publish methods of outstanding performance, yet they can only be reproduced by experts. Regrettably those procedures were, nevertheless, often adopted by analysts without the particular experience and expertise required. Thus, attempts to apply those expert approaches in daily routine frequently failed. The above-mentioned kind of fundamental studies, however, had doubtless the decisive merit of exploring and opening new dimensions but usually only demonstrated the principal feasibility of those advanced conceptions. Therefore, the given informations should have been considered by the customers for daily use as guidelines only. There was and still is a real gap: the above-mentioned experts were not very good at the practical and logistic requirements of routine work and the routine laboratories could not reach the hitherto regarded as state-of-the-art values of the experts. Thus, results obtained during numerous scientific investigations based on trace-element data are still frequently far from the expected reliability. A big challenge for analytical chemists for the near future is, therefore, to achieve meaningful, i.e. accurate and precise, data on an extended scale, when the existing advanced instrumental procedures and the corresponding commercial instruments are used in a large number of laboratories. To reach that target some change in working philosophy by producers and users of analytical data seems indispensable. The benefits from the striking potentialities of the contemporary, highly advanced, instrumentation can be gained only if a comprehensive attitude on the analytical problems as a whole is adopted. The approach also requires the selection of the best suitable determination methods and a scrupulous performance of all working stages of an analytical procedure. In this chapter therefore the author will try to discuss the recent situation from the view of an 'intermediate' laboratory which performs fundamental development work in instrumental analysis based on experiences from applied physical chemistry and routine analysis of rather high sample amounts to be treated in its environmental and ecotoxicological research programmes.

There exist numerous excellent books, reviews and scientific papers recently published on the topics to be discussed below. The quotation of only a part of them in the frame of this condensed contribution should merely be considered as an example to given comments, not as a judgement of the mentioned publications.

SAMPLING AND SAMPLE STORAGE

The initial steps, sampling and sample storage are very frequently the source of erroneous results and hence the most important and doubtless most crucial stages of an analytical chain.

Sampling has to be performed with all possible precautions with respect to contamination and also to conservation of original sample properties. All laboratory equipment used has to be sufficiently low in the trace element content to be analysed in the respective samples.[10-12] To obtain correct data about contamination limits series of careful measurements of rinsing solutions for example have to be made to establish reliable procedures for surface cleaning of sampling vessels and equipment and to select the material best suited to the desired purpose. From our experience for trace metals and organometallic compounds quartz (Suprasil), teflon, linear polyethylene, and other polyolefine vessels are suitable materials.[11-17] If freeze drying is not possible the storage temperature for biological material should not exceed $-20\,^{\circ}C$ for short and medium periods and remain below $-50\,^{\circ}C$ for long-term storage. Water can be stored at about the same temperatures. For human and animal tissues and also plant material, if possible taken fresh, subsamples for later analysis should be divided before storage and weighed and each stored in a single vessel. Usually one refers to fresh weight, but very often this is not possible due to problems in defining fresh weight. Thus, an alternative definition of sample weight may be found, i.e. actual weight that is the weight of the material at the time of weighing which may differ considerably from the real fresh weight. If drying can be applied, this should also be done (preferably freeze drying as the usually least aggressive procedure). It leads to rather stable samples which can be stored, according to the present knowledge, at room temperature or at 4–6 °C.

However, the basic knowledge about storage conditions and disintegration rates of biological material, even at very low temperatures, is still very limited. Thus, in now ongoing projects, e.g. the pilot project on an environmental specimen bank of the Federal Republic of Germany,[18] it is intended to test carefully how far long-term storage is possible and which technical effort is needed to maintain practically unchanged materials. These studies should also yield optimal storage conditions for organometallic and organic compounds.

SAMPLE PRETREATMENT

The preparation steps prior to elemental or species determination are manifold and numerous papers have been published and reviewed in the literature covering this item from the very important purification of reagents, e.g. sub-boiling distillation of acids to errors to be attributed to release and losses of elements from and at surfaces and to accuracy deficiencies occurring in the various digestion and concentration (e.g. extraction) procedures.[11,19–26] These procedures may, in the hands of unexperienced persons, introduce such a broad variety of errors that, for example, the development of a reliable digestion procedure for a respective matrix and one or more trace elements can only be performed successfully if independent methods are applied to elucidate the best conditions for minimizing all possible errors at this stage.[11,27] For standardization and relevant application the pretreatment procedures must be carefully checked (e.g. by round-robin exercises), to obtain simple but sufficiently accurate and precise methods to be applied in routine work.

For trace-metal analysis in environmental and occupational chemistry in ecotoxicology and in medicine, differential pulse stripping voltammetry (DPSV) and atomic absorption spectrophotometry (AAS) are among the most promising methods.[11,28] Thus some procedures suitable for the preparation of various matrix types prior to determination by DPSAV or AAS are summarized later. These methods were successfully applied in the course of our extended environmental studies.[11,28,29]

Natural water samples have frequently to be treated by ultra violet irradiation or ultra violet irradiation plus H_2O_2 prior to voltammetric or AAS determination.[13] For total Hg analysis by cold vapour AAS also a digestion with $H_2S_2O_8$ is common.[16] Filter materials with particulate matter are decomposed by low temperature ashing.[13,30]

Urine usually is not accessible to immediate trace-metal analysis by DPSV. Thus, prior digestion, sometimes also for AAS, using low temperature or wet ashing is necessary.[31,32] For AAS also direct determination is feasible now applying modern instrumentation and Zeeman compensation.

Whole blood analysis of Pb, Cd, Cu, Ni, Co etc. requires prior wet ashing with for example $HNO_3/HClO_4$ for DPASV[35] a similar digestion procedure followed by solvent extraction (e.g. for nickel[32]), deproteinization/matrix modification with HNO_3 for Pb and Cd[36] and various direct methods mainly using matrix modifiers for the already mentioned or other metals for AAS.[19,34]

Tissue, plant, food and sediment samples can be decomposed by different proven methods for subsequent AAS and DPSV determinations.[18] Nitric acid pressure decomposition is used routinely with multisample units,[37] frequently followed by a second treatment with $HClO_4$ or $HClO_4/H_2SO_4$[38] or ultra violet irradiation of the resulting analyte solution.

QUALITY CONTROL

Independently from the methods applied, only rigid quality control measures can guarantee reliable results. Without discussing this topic too much in detail, the focal points of current analytical quality control shall be summarized.

Of paramount importance is the establishment of the level of accuracy. An approved aid is standard reference materials with certified concentrations.[40] Regrettably the number of materials available for this purpose is still limited.[41] Thus adding known amounts of the respective element to a homogeneous or homogenized material is a practicable approach to check or to evaluate an analytical procedure. In that case it is highly important for accurate results that the element amount added will be present after the first step, e.g. the dissolution or decomposition, in the same chemical form in which it exists in the analyte to be studied. This approach always should be accompanied by numerous careful blank and/or noise determinations and a proper statistical evaluation of the data obtained.

Unfortunately, some bias cannot be recognized in this way. For example, the base-level determination in that material can be subject to a systematic deviation. Thus, it seems more meaningful to analyse a set of samples by at least two independent methods, based also on different physical principles, as is shown in Table 21.1.

Studies of this type should be additionally combined with interlaboratory comparisons either as a current cooperation between two or more institutions or as a real round-robin exercise. This concept has been found very rewarding in the course of our long-lasting environmental research programmes.[11,29]

If DPSV is considered as the most accurate approach, for single samples the accuracy of the AAS methods lies on average within ± 15 per cent, on a group basis at ± 7 per cent. Method I on average shows the closest values compared to DPSV, method II as a direct method shows a more pronounced scatter of single values and method III probably is biased due to the more complicated procedure by a slight contamination.

During routine measurements, within-run precision, i.e. within a series of successive measurements, preferably from duplicates together with the day-to-day precision should be checked too. The most reliable approach is the aforementioned continuous application of standard materials or working standards made from excess material as similar as possible to the respective matrix. As an example, during a series of measurements of total arsenic in marine samples NBS Standard Reference Material 1571 Orchard Leaves (10 ppm As) together with a lyophilized and homogenized cod muscle working standard

Table 21.1 Accuracy check of three different AAS-procedures for total Pb in whole blood by independent DPASV analysis after a wet ($HNO_3/HClO_4$) digestion, values in $\mu g\,l^{-1}$ (ppb)

Sample	AAS-I	AAS-II	AAS-III	DPASV
1	288	307	302	266
2	112	109	90	88
3	125	137	111	105
4	78	84	99	61
5	217	223	228	203
6	72	72	74	68
7	147	144	169	142
8	106	119	121	104
9	96	114	104	82
10	82	97	99	74
11	63	66	77	57
12	92	101	104	87
13	115	118	113	104
14	92	96	82	108
15	94	101	119	94
16	62	53	52	52
17	71	98	108	105
18	294	261	286	308
19	176	140	184	178
20	178	138	173	166
21	293	300	266	260
average	136	137	141	130

Notes: AAS-I; deproteinization/matrix modification with 1 M HNO_3: AAS-II; direct, 1:10 dilution with 0.1% Triton × 100: AAS-III; pressure decomposition, followed by NaDDC-ADPC/MIBK extraction. Instruments in all cases; Perkin-Elmer 4000 with HGA 400 and AS-40 autosampler.

(approx. 20 ppm As) showed an excellent long-term reproducibility of 4 per cent.[42]

Under certain circumstances, if for example peak area measurements are possible with AAS,[34] just the running of pure, matrix-matched or matrix-adapted solutions of distinct trace-element concentration may be sufficient if the concentration is unknown to the analyst.

Since for the analysis of trace metals in biological materials there is, at present, only a small number of Standard Reference Materials available, the certification of new materials for all types of environmental materials is urgently needed. Some materials are now in the last stage of certification or just before a certification campaign,[43,44] but there is still a demand for additional materials to further improve analytical results.

ANALYTICAL METHODS

Due to their flexibility, the range of elements and detection power covered in environmental research and surveillance, all modes of atomic absorption spectrophotometry and voltammetry—the latter just at present with a remarkable increasing rate—belong to the predominantly used methods. Thus, owing to the authors long-lasting particular experience with these methods, they will be treated first, followed by a shorter description of other promising analytical methods.

Polarography and Voltammetry

Polarographic and voltammetric methods are of particular significance for numerous trace metals and metalloids (Cd, Pb, Hg, Cu, Zn, Ni, Co, As, Sb, Bi, Tl, Se, V etc.).[11,28,45–55] With respect to the required sensitivity, advanced techniques are applied and attain for various metals and metalloids (Cd, Pb, Hg, Cu, Zn, Ni, Co, Se etc.) extraordinary low determination limits linked with excellent precision. The simultaneous determination of various trace metals is performed usually by several versions of stripping (SV) and Differential Pulse Stripping Voltammetry (DPSV) employing the hanging mercury drop (HMDE), mercury film (MFE), the carbon paste and solid (e.g. gold) electrodes as working electrodes.[13,28,30,31,35,39,47–54] Polarographic and voltammetric methods combine, owing to their foundation on Faraday's Law—1 mole of substance undergoing an electrode process is equivalent to the enormous electrical charge of n 96500 Cb, with n usually between 1 and 3—extraordinary detection power with good precision and inherently high accuracy. In this context it is also to be noted that these methods are based on a very detailed and comprehensive understanding of their physicochemical fundamentals, i.e. the kinetics and thermodynamics of electrode processes. This aspect offers, after a certain training in electrochemistry and voltammetry, particular potentialities in trace analysis.[48]

Thus, advanced voltammetric approaches such as DPSV compete favourably in areas of sensitivity, detection power, and reliability with other important trace-element methods such as electrothermal AAS, inductive coupled plasma emission spectroscopy, ICP-AES and neutron activation analysis (NAA). For the simultaneous determination of Cd, Pb, Cu, Bi, Zn, Ni and Co in fresh waters, seawater, rain and drinking water. DPSV has obviously become the method of choice.[11,13,28,30,35,48–55] A significant advantage of polarography and voltammetry is that instrumentation is by far less expensive than for other analytical techniques, except colorimetry (photometry) and gas chromatography (see Table 21.3). Voltammetric instruments at present commercially available are all of multimode design and thus offer to the analyst a choice of a variety of powerful techniques including stripping voltammetry. This allows the application of the technique best suited to the respective analytical or physicochemical problem. Furthermore, the compactness of modern polarographic equipment and its low vulnerability to mechanical and electronic breakdown makes it the most favourable method for field studies in mobile laboratories[56] and on board research vessels.

An earlier disadvantage of polarography and voltammetry was the time needed for a single determination and hence the manpower required for routine analysis. The introduction of automated, computer-controlled systems will soon allow a significant speed increase. This will be possible, of course, without deficiencies in precision and specificity, as has been shown in the last few years by several authors.[57–61] The extended use of microprocessors and dedicated computers already offered by some manufacturers, promised costs even for highly sophisticated voltammetric instruments comparable to the prices of low-level AAS instruments without a need for very expensive accessories.[61] Hence for numerous research tasks and several rather important toxic trace elements, the at present still observable preference in the application of atomic absorption spectrophotometry over voltammetric methods will more and more disappear. In fact, the situation even changes, if speed is considered, because of the much lower vulnerability of voltammetry to systematic errors.[28,48]

A limitation is that only a distinct, though rather important number of metals are easily accessible to polarography and voltammetry compared with, for example, atomic spectroscopic methods. As is usually the case for other non-electrochemical trace analytical methods the direct determination of trace metals in biological materials requires prior sample pretreatment by digestion in which all organic matter is destroyed totally. This can also cause contamination and losses. The reason is that numerous organic species as well as surface active compounds often inhibit or even suppress electrode reactions. Hence, their elimination prior to the voltammetric measurement is mandatory. Sometimes dissolved organic matter can be very efficiently decomposed by prior ultra-violet irradiation of the analyte.[14,54] On the other hand, the sensitivity to a wide variety of organic, inorganic and organometallic compounds also makes polarography and voltammetry an important tool in the determination of organic compounds[45] and particularly for studies on the speciation of trace metals in sea water and fresh waters.[19,55,62,63]

Atomic Absorption Spectrophotometry (AAS)

The basic principles, practical aspects and numerous current research work have been recently extensively treated.[19,34,64–68] Thus, only a few highlights of interest for practical application in environmental research and surveillance will be presented here.

The application of AAS underwent a tremendous increase from the introduction of the method in 1955 until today with the possibility to determine

>60 elements including all important toxic metals. At first, the only available flame techniques, owing to moderate costs, versatility, sensitivity and precision despite the restriction to single-element analysis and a limited sensitivity, soon played a leading role in the analysis of minor and trace constituents. The back stroke from neglecting the influence of non-atomic absorption, however, led for years to some confusion and disappointment. The commercial availability of compensation methods could partly improve this situation. With the appearance of the first graphite cuvette for electrothermal atomization, however, a result of the highly increased sensitivity was that many problems arose from non-atomic absorption as well as from other misunderstood sources at that time; in numerous laboratories these problems were not even recognized. If the optimistic forecasts of AAS manufacturers are considered this constitutes a vast disappointment. In the meantime, and owing to the contributions of numerous experts from the theoretical as well as from the practical view, one arrived at a better understanding of possible physical and physicochemical interactions in electrothermal atomizers.[67–73] Besides non-atomic absorption, which can be overcome to some extent by improved or rather new compensation methods such as the Zeeman effect[74,75] or continuum source wavelength modulation techniques,[76] the main sources of interference are

1. Gas-phase reactions forming volatile molecules resulting in signal suppression.[77]
2. Complex and still not completely understood reactions of the matrix and the elements to be determined also with the surface of the graphite tube. The latter also include random influences of particular tubes or batches of tubes[78] changing the resistance (temperature) and life time significantly.
3. Spectral disturbances with rather sharp interfering peaks from matrix constituents,[79] also controlled in the majority by platform techniques and Zeeman correction.[80,81]

Regrettably, the AAS systems at present commercially available are far from being ideal. However, owing to the broad field of application and the existence of all possible accessories (e.g. automated sampling, fully programmable electrothermal cold vapour and hydride systems up to the possibility of advanced modes of background compensation techniques[68] and computer controlled AAS laboratories) they offer a versatile trace analytical tool if properly and critically applied. That is also the reason why AAS with the flame and electrothermal excitation (except for aqueous samples) still holds the first position in trace analytical chemistry, if sample throughput is considered.

At present, due to an often only minimal sample pretreatment and automation, this remains valid for biological materials, particularly body fluids[34] food, plants, soils etc.[18,29,82–84] But owing to frequently occurring deviations, linked with accuracy defects caused by problems with accessories and graphite tubes, the number of necessary quality control measurements is still high

Table 21.2 Typical relative detection limits for trace metals of environmental significance. Comparison of several trace analytical methods (aqueous analyte or sample solution for AAS, DPSV and ICP-AES, non-interfering matrix for NAA) values in $10^{-6}\,g\,l^{-1}$, i.e. $\mu g\,l^{-1}$

Elements	ICP-AES[a]	Flame	AAS[b] graphite furnace	Hydride/cold vapour	DPSV[d]	NAA[e]
As	100	30	<0.3	<0.03	0.1	0.1
Be	1	3	0.09		—	—
Cd	1	3	0.003		<0.0002	1.5
Co	1	15	0.15		<0.001	0.03
Cr	2	4.5	0.3		3	20
Cu	0.5	1.5	0.06		0.002	0.1
Fe	0.8	15	0.09		30	6
Hg	7.5	300	—	<0.02[c]	0.002	0.03
Mg	0.05	0.15	0.03		—	25
Mn	0.2	3	0.006		40	0.003
Mo	3	30	0.09		100	15
Ni	3	3	<0.3		<0.001	15
Pb	30	15	<0.06		0.001	300
Se	—	150	<1.5	<0.03	0.1	1
Sn	40	30	0.3	0.8	0.01	2
Ti	0.5	75	60		100	10
Tl	9	30	0.3		0.05	—
V	1	75	3		100	0.15
Zn	0.5	1.5	0.0015		0.02	2.5

Notes:

[a] ICP-AES, data from reference 85, detection limit strongly vary with the conditions applied and instrumentation used.

[b] AAS data from reference 66 and own measurements, graphite tube data are calculated for a 50 μl sample in GFAAS mode in hydride/cold vapour modes for 20 ml.

[c] For Hg a preconcentration on gold wool can be carried out and the volume increased. Thus, a detection limit of $<0.001\,\mu g\,l^{-1}$ is attainable.[34] AAS detection limits are defined as three times the standard deviation of the respective noise level owing to IUPAC recommendations.

[d] The values given for DPSV stem mainly from research work in our institute and usually refer to a PAR 174 instrument with hanging, mercury drop, mercury film and gold electrodes, respectively. Average cell volume 20 ml.

[e] Given values refer to a sample size of 0.5 g wet weight and a thermal flux of $10^{13}\,n\,cm^{-1}\,s^{-1}$. If radiochemical separation is performed much lower contents can be detected.[86]

Thus, if compared with advanced polarographic methods, the real advantage of AAS even in biological materials disappears at least for those elements for which for example DPSV is equally or more sensitive.[61] The appearance of commercially available faster DPSV systems together with optimized sample pretreatment procedures may at least break the hitherto observed dominance of AAS, even if these new electrochemical systems are comparatively expensive.

The virtues of AAS and DPSV at trace and ultratrace levels are shown in Table 21.2 for certain elements of toxicological and environmental significance in comparison to a 'classical' (neutron activation) and a rather modern (inductively coupled plasma emission spectroscopy) trace analytical method. From this table it can be seen that, despite the fact that ICP-AES can successfully compete with AAS flame, graphite furnace, hydride and cold vapour AAS, as well as DPSV, still are more sensitive at least for the most toxic elements As, Cd, Hg, Pb, Ni, Se and Tl and that NAA reaches the same detection power only for Hg and is not applicable to the analysis of Pb and Tl; for the latter NAA methods do exist in principle, but are cumbersome and time consuming, because of necessary chemical separations etc.

Other Methods for Elemental Determination

Besides the at present predominantly used AAS and DPSV methods in routine work, various other methods are also increasingly being used in all types of environmental trace-metal chemistry. These methods will be briefly reviewed in the next part of this chapter.

Multi-element Methods

The classical atomic emission spectrometry (AES), for decades frequently used in all research branches[87,88] despite poor precision and hence poor accuracy, recently underwent a development which significantly improves the signal-to-noise ratio by the introduction of plasma sources enhancing the detection limit drastically for about forty elements to the 10^{-8}–10^{-10} g range. Owing to very promising specific accessories such as very reliable electronic emission detectors and the recently introduced computer-coupled systems it is to be expected that those methods will play a more and more important role and will very soon displace flame AAS, at least where a somewhat higher detection power and/or multi-element capacity is required.[19] A disadvantage, however, as compared to flame AAS are the remarkably lower precision and the substantially higher costs for a reliable instrumentation. Nevertheless for the routine screening of materials with elevated levels of trace elements AES with ICP and other powerful sources may be the method of choice in numerous research areas.

Atomic fluorescence spectroscopy (AFS), in principle has some obvious advantages over AAS, such as broader dynamic ranges, which should together with powerful (e.g. laser) light sources, enable simultaneous multi-element analysis. Further it offers for some elements higher sensitivities.[89,90] But, despite these promising properties, a noticeable routine application has not yet been recognized. It appears that the impetuous growth of plasma emission techniques as well as the recent progress in AAS instrumentation, particularly for graphite furnaces and other accessories such as for hydride/cold vapour techniques, have prevented development and commercial introduction of AFS instruments.

Neutron activation analysis (NAA), can be successfully applied for the sensitive simultaneous analysis of up to about sixty elements.[86,91,92] The method is very versatile and thus advantageous for all kinds of single-element, fingerprint[93] and ratio studies in instrumental as well as in radiochemical versions. It offers an extraordinary approach for accurate—if properly performed practically contamination free—determinations serving to control less expensive and less accurate methods as well as for carrying out base-line studies.

Disadvantages for a broader use are the dependence on nuclear centres or reactor stations and the important additional costs for irradiation, handling and processing of the highly radioactive targets. Moreover, it is to be considered that this method cannot be used successfully for Pb, Cd and Tl analysis. Finally, the staff, applying NAA has to be very carefully selected and trained from the viewpoint of radiochemical as well as of general ultratrace work. The disadvantages concerning detection power are to some extent balanced if other activation techniques, e.g. *particle activation*, are applied. For the time being, however, there are around the world only a few laboratories devoted to this type of work.[19] Therefore for these techniques in principle the same applies as to NAA: they are important tools for basic research and quality control, but less suitable for routine analysis.

Spark source mass spectrometry (SSMS), is a very powerful tool for trace and ultratrace analysis and also has a comprehensive element coverage with a determination range to below 10^{-12} g and thus is widely used in fundamental scientific research.[94] In the last decade the method has been substantially improved by the introduction of electrical detection techniques and computer coupling. The method is extremely valuable for fingerprint and elemental ratio studies in biological and environmental materials[95] and indispensable for metabolic experiments with Pb in toxicological research.[96,97]

If only the analytical potential is considered, the disadvantages are the high cost of instruments and a relatively poor accuracy with exception of the time consuming and cumbersome isotope dilution mass spectrometry (IDMS), despite the recently improved precision (around 10 per cent under optimal conditions). The IDMS however, if all precautions for clean conditions are taken and a well-trained staff is available, is a method with extraordinary precision and accuracy.[98] If thermal ionization isotope dilution is applied and if careful operation is maintained accuracy and precision is in the order of 1 per cent. This method has been hitherto applied with excellent results in environmental research as an ultimate standardization method for analytical procedures, to measure the isotopic composition of Pb in soils, dust, and human blood,[97,98] for the accurate determination of Pb in control materials[11,18,98] and of the Pb content of sea water, snow and marine organisms.[99,100]

X-ray fluorescence spectrometry (XRF), has significantly gained in the last decade by the additional introduction of new instruments,[101,102] i.e. the multiple-crystal spectrometers with simultaneous wavelength dispersion and a capacity of up to twenty-four elements and energy dispersive multi-element instruments with Si(Li)-detectors and multichannel analysers. The advantages of the latter systems are its moderate price and relatively high sensitivity, the disadvantage against wavelength dispersive instruments the rather poor resolution (150 eV versus 3 eV). Nevertheless, these energy dispersive instruments are valuable in screening programmes with environmental materials.

As other methods, XRF needs careful standardization and—very frequently—the application of preconcentration procedures, e.g. ion exchanging materials as surface accumulators. If all these facts are considered, the method, owing to its good precision, is very useful for fingerprints and elemental ratio studios but not at ultratrace levels due to a somewhat limited detection power. Commonly if double reflecting systems[103] are not used 10^{-6}–10^{-7} g is the limit of detection.

Oligo- and Single-Element Methods

Conventional and micro colorimetric methods are still in use but mainly in the $mg\,kg^{-1}$ (ppm) range and in clinical chemistry. In environmental research the importance of these methods for metals and organometallic compounds is waning, because they frequently need separation procedures prior to measurement and thus usually cannot compete with AAS, DPSV, ICP-AES etc. However, for non-metallic compounds as anions these methods, frequently combined with catalytic amplification, sometimes automated, are still in use.[90,104–106]

Gas chromatography described again below under species separation and identification, can be used for the determination of several metals after chelation and thus under certain conditions is a rather sensitive method for the simultaneous determination of several trace metals.[107,108]

Species Separation and Identification

Doubtless studies on the fate and balance of different inorganic or organometallic species in the whole environment, particularly metabolization and accumulation are of paramount importance for the understanding of ecological implications owing to anthropogenic or naturally occurring harmful substances.

For about a decade, the determination of organomercurials in food, predominantly of aquatic and marine origin, has been performed.[39,109] There is also an increasing effort to study the principles of the biological transformation of Pb, Hg and other elements into organometallic compounds.[109,111] Promising methods for this purpose are gas chromatography with AAS, MS and ICP-AES detection as well as high performance liquid chromatography (HPLC) with for example AAS and voltammetric detection besides model studies with radiotracers.[111] With the exception of the above mentioned direct characterization and quantitation of some heavy metal species, e.g. by electrochemical and

Table 21.3 Cost–benefit estimations of devices for instrumental trace and ultratrace analysis

Method	Device	Price (approx.) 10^3 US$	Samples[a] (per 8 h)	Remarks
AAS	Basic instruments with recorder	15–30	≥200	Only flame, one element 30 s per measurement
	Complete setup with all accessories and automated sampling	45–65	20–200	Depending on mode: electrothermal, hydride, cold vapour or flame, up to 6 elements automated but sequential 12 h operation possible[b]
DPP	Basic instruments with recorder	8–12	10–20	Including DPASV up to 8 metals, species sensitive up to 5 metals simultaneously (oligo-substance method)
	Automated setup with all accessories	20–30	30–50	Computer controlled, 24-h operation possible, e.g. for drinking-water control
ICP-AES	Sequential systems, complete unit	50–100	≥200	30 s per measurement single element
	Multiple channel, complete unit	120–250	≥100	With background compensation and computer up to 50 elements, 24-h operation possible
XRF	Energy dispersive	40–60	≥100	Multi-element method, depends on sample type and program, 24-h operation usual
	Multiple crystal	≥150	≥100	
NAA	4096-Channel with computer, disk systems and detector(s)	≥100	5–50	Multi-element method access to nuclear reactor necessary. Frequently radiochemical separation unavoidable if high sensitivity required; 24-h operation usual or possible
MS	Spark source MS, complete setup	≥250	≥4	Thirty or more elements simultaneously
	Thermionic, isotope dilution	≥100	≥4	Very precise and accurate. Oligoelement mode only
Colorimetry	Complete systems with second order evaluation	5–10 10–25	<100	Single compound

(*continued*)

Table 21.3—*Continued*

Method	Device	Price (approx.) 10^3 US$	Samples[a] (per 8 h)	Remarks
GC	Basic instruments and detector	6–15	> 20	
	and automated injection	12–25	> 50	24-h operation easily possible
	Computer for GC control	20–30	> 500	Controls 14 devices
GC-MS	Complete with computer	200	5–20	24-h operation, not for routine use
HPLC	Complete with automated injection	20–35	> 20	

Notes:
[a] Samples per 8 h includes a duplicate determination either from the same solution or parallel sample including all necessary quality control measures within a normal working day, i.e. about 8 h continuous operation. For some techniques also 24-h operation is possible.
[b] If graphite tubes used, depending on element and tube properties (tube life times).

other methods in aqueous samples,[62,63] it will be necessary, after a proper sampling followed by a careful clean-up, to separate different compounds. This is then followed by the application of the instrumental combinations mentioned before.

Comments on Cost–Benefit Ratios of the discussed methods

In Tables 21.3 and 21.4 the prices of typical devices for instrumental analytical methods and their capacity for sample throughput are summarized, which is self-explanatory in principle. If great numbers of similar samples have to be analysed, partly or completely automated systems as automated AAS and DPSV, ICP-AES and multiple crystal XRF are the most suitable despite the rather high investment. For the very important continuous monitoring of toxic trace metals as Pb, Cd, Cu and Zn in tap water, rain and other water types already existing, computer-controlled DPASV systems offer a reliable and rather inexpensive approach to a definite solution of this problem.[28,60,61]

Further it is understood that very sophisticated methods such as spark source and isotope dilution mass spectrometry, neutron and particle activation analysis are ultimate and predominantly reference methods which are not normally installed in smaller research or routine laboratories. Besides the problem orientation, the use and choice of other methods often also depends frequently on local potentialities. In environmental research and surveillance the aim should at least be to have an independent method at hand to be able to check sufficiently often the main routine method(s).[11]

Table 21.4 Accessories for trace analytical chemistry

Device	Price (approx.) 10^3 US$	Samples (per 8 h)	Remarks
Programmable oven for dry ashing with quartz chamber	3–5		24-h operation usual
Computer systems for on-line control and data processing of GC, AAS	10–30		Controls up to 14 GC and up to 6 AAS systems
Low temperature asher	15–35	5–25	24-h operation usual
Nitric acid pressure digestion unit	≤3	18	24-h operation possible
Heating systems (blocks) for up to 60 samples (programmed wet digestion)	5–10	20–100	10-h operation possible
Automated wet digestion with 6 positions	10	≥18	10-h operation possible
Clean benches, laminar flow	3–10		Price depends on type and size
Liquid scintillation. Counter: up to 3 channels and computer control. 200 Sample tray	15–50	≥100	24-h operation usual for radiochemical NAA and radiotracer studies

REFERENCES

[1] Lauwerys, R., Buchet, J.P., Roels, H., Berlin, A., and Smeets, J. (1975). *Clin. Chem.*, **21**, 551.
[2] Dybczýnski, R., Tugsaval, A., and Suschny, O. (1978). *Analyst*, **103**, 733.
[3] Olafsson, J. (1978). *Marine Chemistry*, **6**, 87.
[4] Ackermann, F., Bergmann, H., and Schleichert, V. (1979). *Fresenius Z. Anal. Chem.*, **296**, 270.
[5] Kingsford, M., Stevenson, C.P., and Edgerley, W.H.L. (1973). Dept. of Scientific and Industrial Research, New Zealand, Report C.D. 2159.
[6] Carron, J.H., and Aspila, K.J. (1976). Environment Canada, Report Series No. 44, Interlab. Quality Control, Burlington, Ontario.
[7] Keppler, F.E., Maxfield, M.E., Moss, W.D., Tietgen, G., and Linch, A.L. (1970). *Am. Indust. Hgg. Ass. J.*, 412.
[8] Workshop participants, Meeting Report (1974). Interlaboratory lead analysis of standardized samples of sea water, *Marine Chem.*, **2**, 69.
[9] Tölg, G. (1976). *Naturwissenschaften*, **63**, 99.
[10] Hamilton, E.J. (1976). *Sci. Tot. Environm.*, **5**, 1.
[11] Stoeppler, M., Valenta, P., and Nürnberg, H.W. (1979). *Fresenius Z. Anal. Chem.*, **297**, 22.
[12] Stoeppler, M. (1983). In, *Trace Element-Analytical Chemistry in Medicine and Biology*, Vol. 2 (Eds P. Brätter and P. Schramel), p. 909, Walter de Gruyter, Berlin.
[13] Mart, L. (1979). Doctoral Thesis, Aachen/Juelich.
[14] Mart, L. (1979). *Fresenius Z. Anal. Chem.*, **296**, 350.
[15] Stoeppler, M., and Matthes, W. (1978). *Anal. Chim. Acta*, **98**, 389.
[16] May, K., Reisinger, K., Flucht, R., and Stoeppler, M. (1980). *Vom Wasser*, **55**, 63.

[17] Moody, T.R., and Lindstrom, R.M. (1977). *Anal. Chem.*, **49**, 2264.
[18] Stoeppler, M., Dürbeck, H.W., and Nürnberg, H.W. (1982). *Talanta*, **29**, 963.
[19] Stoeppler, M., and Nürnberg, H.W. (1983). Analytik von Metallen und ihren Verbidungen, in, *Herausg. Metalle in der Umwelt* (Eds E. Merian, M. Geldmacher-v. Mallinckrodt, G. Machata, H.W. Nürnberg, H.W. Schlipköter and W. Stumm), Verlag Chemie, Weinheim (in press).
[20] Tölg, G. (1972). *Talanta*, **19**, 1489.
[21] Tölg, G. (1974). *Talanta*, **21**, 327.
[22] Tölg, G. (1977). *Fresenius Z. Anal. Chem.*, **282** 257.
[23] Gorsuch, T.T. (1970). *The Destruction of Organic Matter*, Pergamon, New York.
[24] Bock, R. (1978). *A Handbook of Decomposition Methods in Analytical Chemistry*, translated, updated and extended by I.L. Marr, International Textbook Company, Glasgow.
[25] Flaschka, H.A., and Barnard, A.I., Jr. (1967, 1969, 1972). *Chelates in Analytical Chemistry*, Vol. 1–4. Dekker, New York.
[26] Marcus, Y., and Kertes, A.S. (1969). *Ion Exchange and Solvent Extraction of Metal Complexes*, Wiley Interscience, New York.
[27] Hamilton, E.I. (1974). *Sci. Tot. Environ.*, **3**, 3.
[28] Nürnberg, H.W. (1979). *Sci. Tot. Environ.*, **12**, 35.
[29] Stoeppler, M. *Ber. KFA Jülich*, Jül 1675–August 1980.
[30] Mart, L., Valenta, P., and Nürnberg, H.W. (1980). In, *Lead in the Marine Environment, Proceedings of the International Experts Discussion—Lead Occurrence, Fate and Pollution in the Marine Environment, Rovinj 1977* (Eds M. Branica and Z. Konrad), Pergamon, Oxford.
[31] Golimowski, J., Valenta, P., Stoeppler, M., and Nürnberg, H.W. (1979). *Talanta*, **26**, 649.
[32] Ader, D., and Stoeppler, M. (1977). *J. Anal. Toxicol.*, **1**, 252.
[33] Stoeppler, M., and Brandt, K. (1980). *Fresenius Z. Anal. Chem.*, **300**.
[34] Stoeppler, M. (1983). *Spectrochim. Acta*, 38B, 1559.
[35] Valenta, P., Ostapczuk, P., Pihlar, B., and Nürnberg, H.W. (1981). *Proceedings of the International Conference: Heavy Metals in the Environment, Amsterdam*, p. 691, CEP Consultants, Edinburgh.
[36] Stoeppler, M., Brandt, K., and Rains, T.C. (1978). *Analyst*, **103**, 714.
[37] Stoeppler, M., and Backhaus, F. (1978). *Fresenius Z. Anal. Chem.*, **291**, 116.
[38] May, K., and Stoeppler, M. (1978). *Fresenius Z. Anal. Chem.*, **293**, 127.
[39] Ahmed, R., Valenta, P., and Nürnberg, H.W. (1981). *Microchim. Acta*, 171.
[40] Uriano, G.A., and Gravatt, C.C. (1977). *Crit. Rev. Anal. Chem.*, **6**, 361.
[41] Parr, R.M. (1983). IAEA/RL/103, September.
[42] Stoeppler, M. (1980). In, *Wiss. Beiträge der Friedrich-Schiller-Universität Jena, As- und Ni-Symposium*, p. 369.
[43] van der Eijk, W. (1979). *Fresenius Z. Anal. Chem.*, **297**, 10.
[44] Cali, E., private communication.
[45] Nürnberg, H. W. (Ed.) (1974). *Electroanalytical Chemistry*, John Wiley, New York.
[46] Nürnberg, H.W., and Kastening, B. (1979). In, *Methodicum Chimicum*, Vol. 1A (Ed. F. Korte), p. 584, Academic Press, New York.
[47] Valenta, P., Rützel, H., Nürnberg, H.W., and Stoeppler, M. (1977). *Fresenius Z. Anal. Chem.*, **285**, 25.
[48] Nürnberg, H.W. (1982). *Pure Appl. Chem.*, **54**, 853.
[49] Nürnberg, H.W. (1983). In, *Analytical Techniques for Heavy Metals in Biological Fluids* (Ed. S. Facchetti), p. 209, Elsevier, Amsterdam.
[50] Sipos, L., Golimowski, I., Valenta, P., and Nürnberg, H.W. (1979). *Fresenius Z. Anal. Chem.*, **298**, 1.

[51] Sipos, L., Nürnberg, H.W., Valenta, P., and Branica, M. (1980). *Anal. Chim. Acta*, **115**, 25.
[52] Nguyen, V.D., Valenta, P., and Nürnberg, H.W. (1979). *Sci. Tot. Environ.*, **12**, 151.
[53] Valenta, P., and Nürnberg, H.W. (1980). In, *Herausgeber Gewässerschutz Wasser—Abwasser* (Ed. B. Böhnke), S. 105, Ges. Förderung d. Siedlungswasserwirtschaft, RWTH Aachen.
[54] Pihlar, B., Valenta, P., Golimowski, J., and Nürnberg, H.W. (1980). *Z. Wasser u. Abwasserforschung*, **13**, 4.
[55] Nürnberg, H.W., and Valenta, P. (1975). *In, The Nature of Sea Water* (Ed. E.D. Goldberg), p. 87, Abakon Verlagsges., Berlin.
[56] Stoeppler, M., and Backhaus, F. (1979). *Ber. KFA Jülich*, Jül-1571.
[57] Vassos, B.H., and Osteryoung, R.A. (1974). *Chemical Instrumentation*, **5**(4), 257.
[58] Kryger, L., Jagner, D., and Skov, H.I. (1975). *Anal. Chim. Acta*, **78**, 241.
[59] Bond, A.M., Grabaric, B.S. (1977). *Anal. Chim. Acta*, **88**, 227.
[60] Valenta, P., Rützel, H., Krumpen, P., Salgert, K.H., and Klahre, P. (1978). *Fresenius Z. Anal. Chem.*, **292**, 120.
[61] Ostapczuk, P., Goedde, M., Stoeppler, M., and Nürnberg, H.W. (1984). *Fresenius Z. Anal. Chem.*, **317**, 252.
[62] Nürnberg, H.W. (1977). *Electrochim. Acta*, **22**, 935.
[63] Raspor, B., Valenta, P., Nürnberg, H.W., and Branica, M. (1978). *Sci. Tot. Environ.*, **9**, 87.
[64] Dean, I.A., and Rains, T.C. (Eds) (1971). *Flame Emission and Atomic Absorption Spectrometry*, Vol. I and II, Marcel Dekker, Inc., New York.
[65] Price, W. (1979). *Spectrochemical Analysis by Atomic Absorption*, Heyden, London.
[66] Welz, B. (1983). *Atomic Absorption Spectroscopy*, Verlag Chemie, Weinheim.
[67] Sturgeon, R.E., and Chakrabarti, C.J. (1978). *Proc. Analyt. Atom. Spectr.*, **1**, 5.
[68] Slavin, W. (1982). *Anal. Chem.*, **54**, 685A.
[69] Maessen, F.J.M.J., and Posma, F.D. (1974). *Anal. Chem.*, **46**, 1439.
[70] Lundgren, G., and Johansson, G. (1974). *Talanta*, **21**, 257.
[71] C. Hendrikx-Jongerius, C., and de Galan, L. (1976). *Anal. Chim. Acta*, **87**, 259.
[72] Sturgeon, R.E. (1977). *Anal. Chem.*, **49**, 1255A.
[73] L'vov, B. (1978). *Spectrochim. Acta*, **33B**, 153.
[74] Brown, S.D. (1977). *Anal. Chem.*, **49**, 1269A.
[75] Fernandez, F.I., Myers, S.A., and Slavin, W. (1980). *Anal. Chem.*, **52**, 741.
[76] Harnly, I.M., and O'Haver, T.C. (1977). *Anal. Chem.*, **49**, 2187.
[77] Czobic, E.I., and Matousek, I.P. (1978). *Anal. Chem.*, **50**, 2.
[78] Stoeppler, M., Kampel, M., and Welz, B. (1976). *Fresenius Z. Anal. Chem.*, **282**, 369.
[79] Massmann, H., and El Gohary, Z. (1976). *Spectrochim. Acta*, **31B**, 399.
[80] Kurfürst, U. (1981). *Nachr. Chem. Tech. Lab.*, **29**, 854.
[81] Slavin, W., Carnrick, G.R., Manning, D.C., and Pruszkowska, E. (1983). *Atomic Spectroscopy*, **4**, 69.
[82] Stoeppler, M., and Nürnberg, H.W. (1979). *Ecotoxicology and Environmental Safety*, **3**, 335.
[83] Stoeppler, M., Brandt, K., Flucht, R., and Mohl, C. (1979). *Lebensmittelchemie u. gerichtl. Chemie*, **33**, 105.
[84] Breder, R. (1982). *Fresenius Z. Anal. Chem.*, **313**, 395.
[85] Boumans, P.W.J.M., and Bosveld, M. (1979). *Spectrochim Acta*, **34B**, 59.
[86] Guinn, V. P., and Hoste, J. (1980). In, *Elemental Analysis of Biological Materials*, Tech. Rep. Ser. IAEA, Vienna, **197**, 105.
[87] Bedrosian, A.I., Skogerboe, R.K., and Morrison, G.H. (1968). *Anal. Chem.*, **40**, 854.
[88] Cowgill, U.M. (1974). *Appl. Spectrosc.*, **27**, 5.
[89] Winefordner, J.D., and Elser, R.C. (1971). *Anal. Chem.*, **43**, 24A.

90 Winefordner, J.D. (Ed.) (1978). *Trace Analysis: Spectroscopic Methods for Elements, Chemical Analysis*, Vol. 46, Wiley Interscience, New York.
91 DeSoete, D., Gijbels, R., and Hoste, R. (1972). *Neutron Activation Analysis*, Wiley Interscience, New York.
92 Erdtmann, G., Nürnberg, H.W. (1974). In, *Methodicum Chimicum*, Vol. 1B (Ed. F. Korte), p. 735, Academic Press, New York.
93 Erdtmann, G., and Kayßer u. H. Petri, B. (1975). *Fresenius Z. Anal. Chem.*, **274**, 301.
94 Ahearns, A.I. (Ed.) (1972). *Trace Analysis by Mass Spectrometry*, Academic Press, New York.
95 Locke, I., Boase, D.R., and Smalldon, K.W. (1979). *Anal. Chim. Acta*, **104**, 233.
96 Rabinowitz, M., and Wetherill, G. (1976). *J. Kopple, Arch. Environ. Health*, **30**, 220.
97 Manton, W.D. (1977). *Arch. Environ. Health*, **31**, 149.
98 Facchetti, S. (1982). *Application of Mass Spectrometry to Trace Analysis*, Elsevier, Amsterdam.
99 Tatsumoto, M., and Patterson, C.C. (1963). *Nature*, **199**, 350.
100 Chow, T.J., Patterson, C.C., and Settle, D. (1974). *Nature*, 159.
101 Herglotz, K.H., and Birks, L.S. (Eds) (1978). *X-ray Spectroscopy*, Dekker, New York.
102 Campbell, W.C. (1979). *Analyst*, **104**, 177.
103 Knoth, H., and Schwenke, H. (1980). *Fresenius Z. Anal. Chem.*, **301**, 7.
104 Slanina, I., Bakker, F., Bruign-Hes, A.G.M., and Möls, J.J. (1978). *Fresenius Z. Anal. Chem.*, **289**, 38.
105 Hanocq, M., and Helson-Cambier, M. (1979). *Clin. Chim. Acta*, **91**, 177.
106 Matthes, W., Flucht, R., and Stoeppler, M. (1978). *Fresenius Z. Anal. Chem.*, **291**, 217.
107 Uden, P.C., and Henderson, D.E. (1977). *Analyst*, **102**, 889.
108 Tavlaridis, A., and Neeb, R. (1978). *Fresenius Z. Anal. Chem.*, **282**, 17.
109 Westöö, G. (1970). *Acta. Chem. Scand.*, **24**, 2299.
110 Brinckman, F.E., and Bellama, J.M. (Eds) (1978). *Organometals and Organometalloids*, ACS Symposium Series 82, American Chemical Society, Washington, DC.
111 Reisinger, K., Stoeppler, M., and Nürnberg, H.W. (1981). *Nature*, **291**, 228.

Pollutants and Their Ecotoxicological Significance
Edited by H. W. Nürnberg

CHAPTER 22

Concepts for the Ecotoxicological Evaluation of Chemicals: Ecotoxicological Profile Analysis

F. KORTE

In any discussion of environmental chemicals the scope of this term remains problematic. It is emphasized, therefore, that attention will be focused here on environmental chemicals occurring globally or supraregionally. This limitation is due to a very special reason, namely, that local environmental chemicals can be described quite well scientifically and, if enough money is used, ultimately brought and kept under control. It is only in respect of global occurrence that the situation remains apparently much less well known, even on the research side, and I should therefore like to divide this presentation into three sections.

The first section will briefly describe possible problems with environmental chemicals and whether any exist. In the second part concepts to be used to master the problems as time passes will be discussed, while the third part will give some experimental data.

We need to face up to global limitations with increasing certainty. Our ever-shrinking planet comprises 29 per cent solid land and 71 per cent ocean—thus we have ample water. The mean elevation of the land above the sea is 200–300 m, the mean depth of the sea is 3800 m, and above there is the atmosphere which becomes thinner with increasing altitude (Figure 22.1). Technologies exist for purifying water from any desired degree of contamination to any desired degree of purity. Technologies also exist today which enable the air to be cleaned to any extent we wish. In contrast, there are no methodologies for handling potentially polluted solid earth (soil) chemically or to purify it. That is very strange in view of the fact that there is legislation for keeping water and air clean. For the soil, the most limited portion of our planet, and of which around 12 per cent is used for agriculture, no antipollution laws or the like appertain. We behave as if this

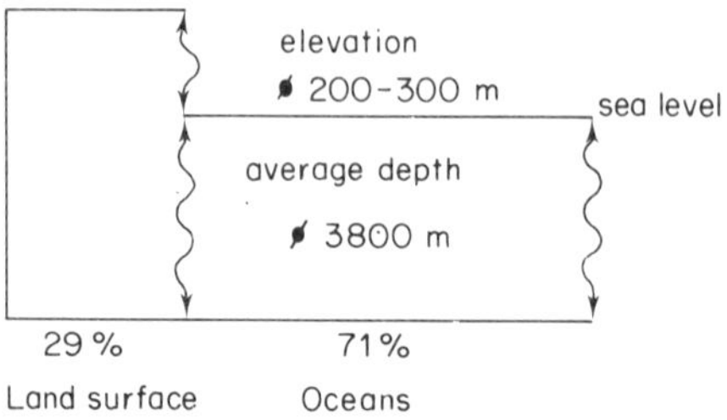

Fig. 22.1

medium solid earth is subject to no limitations whatsoever. It is surely a very remarkable situation.

The focus here is on chemicals synthesized by man because it is these which have made us conscious of our environment during the past 20 years.

Organic chemicals will be the main topic for discussion because here the difference between naturally occurring and synthetic chemicals is much more clear-cut. We have not the remotest idea what the natural level of inorganic substances is on the earth's surface.[1,2]

In 1950 7 million t of organic chemicals were produced, in 1970 the figure was 63 million t, and in 1983 it was going to be 250 million t; the belief is that it will continue at this pace. A comparison of the tonnage of synthetically produced substances with natural, emitted substances (for example from forests) shows that the two are of roughly the same order of magnitude (Table 22.1). Naturally, the production of the biomass lies some orders of magnitude higher. When one asks what is being produced, there is straightaway a profound silence. Pesticides are fairly well known because their application has been controlled by law for some time now.[3] Something is known about detergents, almost nothing about

Table 22.1 Release of organic compounds in the environment, 1970

	1950	1970	1985	Release in environment 1970	
Grand total 10^6 t	7	63	250	20	
Organic chemicals—world production					
manufactured		10^6 t		natural sources	10^6 t
solvents		10 t		methane	1600 t
detergents		1.5 t		terpene type hydrocarbons	170 t
pesticides		1 t			
gaseous base chemicals		1 t		lubricating and industrial oils	2–5 t
miscellaneous		7 t			

Source: Iliff (1971)[18]

Table 22.2 Soil input per year (no conversion, no evaporation or leaching)

	World production (million t year^{-1})	kg ha^{-1} in area of use		kg ha^{-1} when distributed on total land surface
		Locally	Globally	
Pesticides	1	2–4	0.12 (a/p/f)	0.07
Nitrogen fertilizers	15	70	3.38 (a/p) 1.76 (a/p/f)	1.12
org. Chemicals total	150			11.2

Land surface of the earth	13.392 × 10^9 ha
Agriculture	1.424 × 10^9 ha
Pastures	3.001 × 10^9 ha
Forests	4.091 × 10^9 ha

solvents. Little information is available about gaseous basic chemicals[4] and, self-evidently, there are not even clues about the many 'others'. To all intents and purposes nothing is known about lubricants and gasoline additives.[7] Recently a report in Tokyo gave the production of plastics at present as 40 million t worldwide. It is stated that these plastic materials contain from 1 to 5–6 per cent additives, that is to say, the quite large amount of 400 000 t annually. They are additives about whose environmental behaviour no one knows anything at all. The reason for including the 'old' figures by Iliff is that there has been little gain in our knowledge during the last 10 years about what we do with chemicals quantitatively. Possibly that is why legislation for an environmental chemicals law, or whatever name one applies to it, has been initiated.

Table 22.2 shows some data of how large the difference is between substances subject to regulations and uncontrolled substances. 1 million t of pesticides are being produced every year. Their importation and marketing is subject to licensing requirements. Quite a lot of information is available about them, and this probably explains why they are so often named in the media. Organic chemicals overall are 150 times 'heavier'. They are not subject to controls and little is known about their ecotoxicity and, quite often, even about their plain toxicity. It is putting the wrong emphasis in the discussion about environmental chemicals in the crudest possible way. A uniform global distribution of these manufactured substances would give 11.2 kg ha^{-1} of general chemicals but only 0.07 kg ha^{-1} of pesticides.[3,4,5,7]

A few years ago German industry meritoriously attempted to compile production figures and application patterns. The basis was that initially chemicals being produced at the rate of at least 50 000 t annually worldwide should be looked at. This procedure was continued for some years at great cost and a list of approximately 130 chemicals was drawn up. Unhappily the programme had to be discontinued because no other country, and no chemical industry of any country, wished to participate in this work. This was the only

Table 22.3 Selection of chemicals with annual global production >50000 tons

Organics	1000 tons	Inorganics	1000 tons
Acetaldehyde	2400	Aluminium oxide	17000
Acrylonitrile	2700	Ammonia	40000
Alkylbenzenes	700	Chlorine	24600
Benzene	14400	Iron sulfate	450
Carbon tetrachloride	1000	Lead oxides	363
Chloroparaffins	270	Nitric acid	23000
Cyclohexane	2800	Phosphoric acid	13000
Di-butyl phthalate	230	Sodium chlorate	600
Phthalic acid anhydride	2300	Sodium chromate and bichromate	450
Toluene	8500	Sulphuric acid	108300
Trichloro ethylene	700	Zinc oxide	420
Vinyl chloride	7730		

Note: Compiled for OECD by German ad-hoc Group

reason for ceasing this activity, which I consider to be the only useful one if the 'old' environmental chemicals are slowly to be brought under control.[14,16]

The compilation (Table 22.3) gives some data, for example 270 000 t of chloroparaffins. This is an amount which, if it enters the environment, is bound to show up in all biotic and abiotic systems, and it is no wonder that these compounds are found ubiquitously on performing an analysis. It is regrettable that the work has been stopped or put on ice for the time being.

How can predictions be made today about what will the situation be in the future? Let us start from the premise that PCB did not exist 100 years ago and that ppb ($\mu g\ kg^{-1}$) concentrations are being found today.[10] Let us assume, further that technology is going to continue to develop at past rates and that production in the chemical industry will grow at 2 per cent annually—a perfectly realistic assumption in my view, with or without an oil crisis.[8] Uninterrupted production

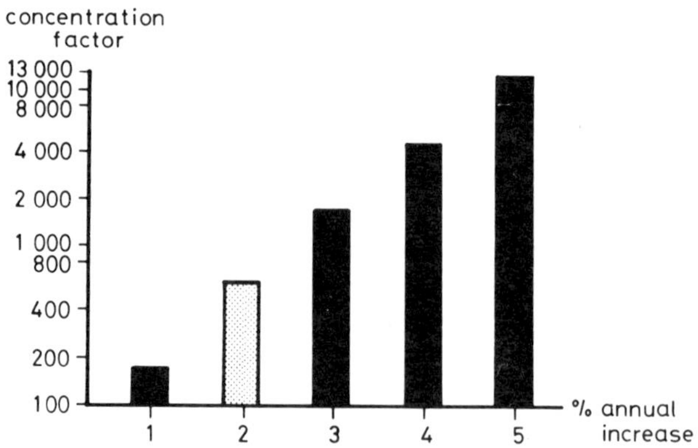

Fig. 22.2

Table 22.4

Compound	Parts per 10^{12} in Air	Rainwater	Rivers
Dieldrin	20 (London, 1966)	9–28 (England, 1964/65)	10 (Scotland, 1966) 0–118 (USA, 1965)
BHC	10 (London, 1966)	12–164 (England, 1964/65)	Detected (Scotland, 1966; USA, 1965)
DDT and analogs	10 (London, 1966)	210 (London, 1965)	5 (Scotland, 1966) 190 (USA, 1965)

will then cause the concentration to increase from ppb to ppm in another 100 years (Figure 22.2). In other words, since no noteworthy biological effects caused by ppb concentrations have been reported, but some cases are known in which human beings suffered direct harm or even died from ppm concentrations, the problem of environmental chemicals will have to be solved during the next 50–100 years. It means that there is no immediate worldwide danger, but that it is worthwhile thinking about the problem because the solution needs to follow prospectively. Globally dispersed chemicals cannot be 'recaptured'—as has been shown precisely for the example of PCB.[9]

Table 22.4 shows a classic picture as revealed in 1966. Much excitement was caused when it was found that plant protection agents never used in London suddenly appeared there in measurable amounts, and in ppb concentrations. To begin with analytical errors were postulated as the explanation, but then the basic physicochemical principles of our planet and global transport phenomena were remembered; today it is clear that all these organic chemicals disperse uniformly via the air.

Figure 22.3 is one of the first recordings in which Dr Appleby compared the distribution of radioactive material after atmospheric nuclear-bomb tests with PCB and similar types of chemicals. As expected, the concentration is revealed to be greater in the northern than in the southern hemisphere. It is also revealed that organic chemicals with the chemicophysical behaviour of, for instance, DDT—and that comprises a large portion and even the major proportion apart from plastics—are distributed in the atmosphere in exactly the same way as strontium 90. This important problem, as revealed today and even 10 years ago, really compels us finally to ascertain what it is we are introducing into the environment, and in what amount and whereabouts. Almost no activity is going on in this respect except for the work by the German industry *ad hoc* group—and this has now ceased.[7]

One of the first recorded incidents causing people to die was the 1952 fog in London in which ppm concentrations of sulphur dioxide (SO_2) plus ppm concentrations of dust produced poisoning, and within 4 days 4000 people more than normal had died.[2,7] The dust concentration was subsequently reduced and, as

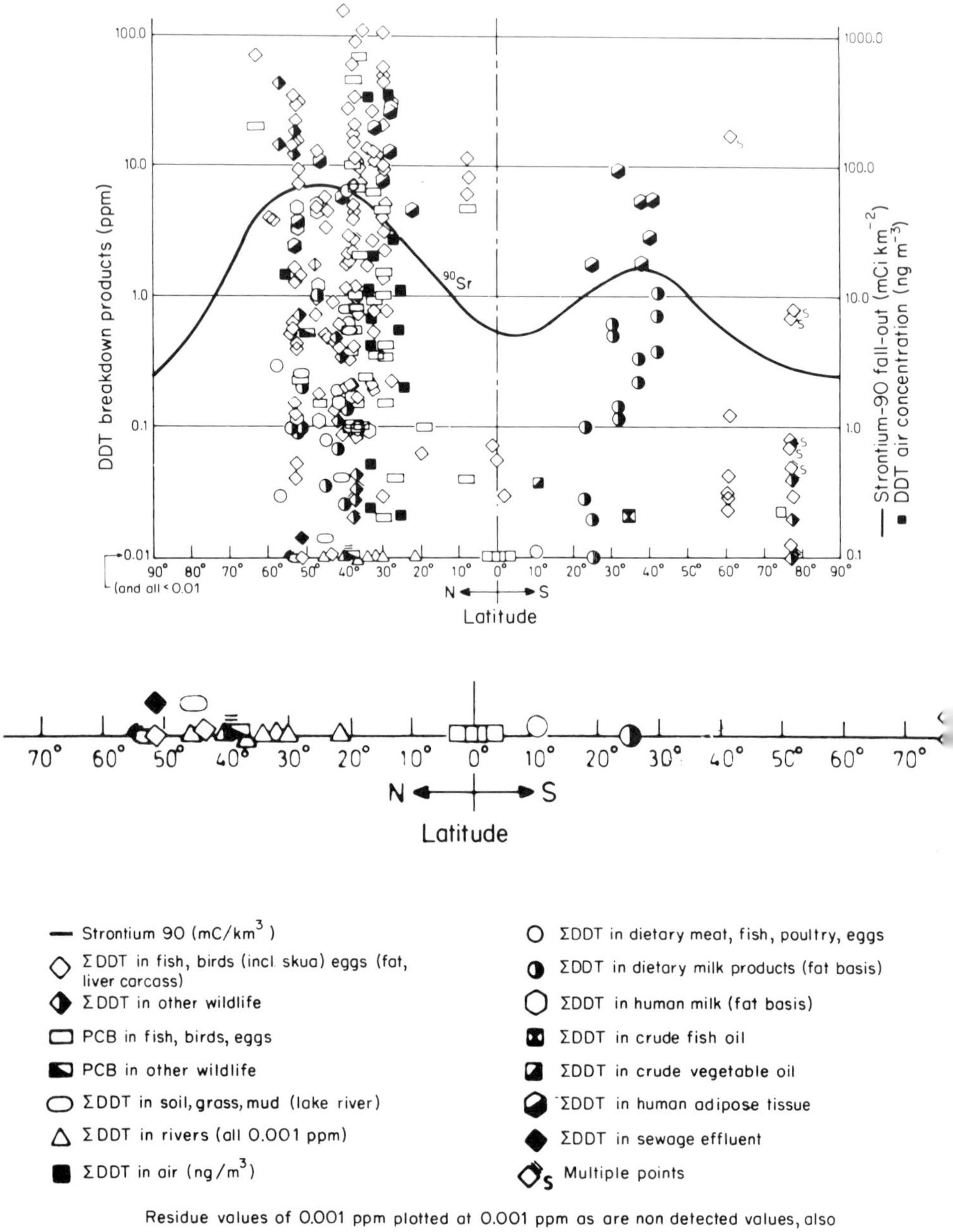

100.0
10.0
1.0
0.1
0.01
(and all < 0.01
DDT breakdown products (ppm)
1000.0
100.0
10.0
1.0
0.1
— Strontium-90 fall-out (mCi km^{-2})
■ DDT air concentration (ng m^{-3})
^{90}Sr
90° 80° 70° 60° 50° 40° 30° 20° 10° 0° 10° 20° 30° 40° 50° 60° 70° 80° 90°
N ← → S
Latitude
70° 60° 50° 40° 30° 20° 10° 0° 10° 20° 30° 40° 50° 60° 70°
N ← → S
Latitude
Strontium 90 (mC/km^3)
Σ DDT in fish, birds (incl skua) eggs (fat, liver carcass)
Σ DDT in other wildlife
PCB in fish, birds, eggs
PCB in other wildlife
Σ DDT in soil, grass, mud (lake river)
Σ DDT in rivers (all 0.001 ppm)
Σ DDT in air (ng/m^3)
ΣDDT in dietary meat, fish, poultry, eggs
ΣDDT in dietary milk products (fat basis)
ΣDDT in human milk (fat basis)
ΣDDT in crude fish oil
ΣDDT in crude vegetable oil
ΣDDT in human adipose tissue
ΣDDT in sewage effluent
Multiple points
Residue values of 0.001 ppm plotted at 0.001 ppm as are non detected values, also

Fig. 22.3

Table 22.5 Mercury in the environment

Minamata disease
First recognized 1956, 111 reported cases by 1960 including deaths.
Cause: Approx. 100 mg kg^{-1} daily in fish (as CH_3Hg^+).
CH_3Hg^+ action threshold in primates = 0.05 mg kg^{-1} body weight.

Occurrence in the diet: Hospital diet in Federal Republic of Germany

1978 = 0.2–1.0 mg kg^{-1} daily total Hg.

Sources Federal Republic of Germany 1973 (extract)

	Emission (t)	Consumption (t)
Chlor-alkali electrolysis		351
in dumps	34	
into water	42	
into air	25	
Plant protection agents (1971)	31	31
Electrotechnology	20	34
Instruments, apparatus	5	29
Catalysts		44
Chemicals, reagents	30	53
Dental		22
Others		50

is well known, SO_2 emissions have been everywhere reduced; such events have not occurred since (Table 22.5).[6]

The first two chronic intoxications are also familiar events: the Minamata and Itai-Itai diseases. Both occurred in Japan, Minamata caused by methylmercury—it is an old story and part of history today—and Itai-Itai by cadmium. In the case of mercury it is well established today that more than 40 per cent and probably as much as 60 per cent of the acceptable daily intake (ADI) is 'occupied' by natural mercury, i.e. the industrial gap is only 40 per cent. It seems now that the situation with cadmium is entirely similar if the available data are correct.[14]

The Itai-Itai illness is postulated to arise if the cadmium concentration in the food is around 10 mg kg^{-1}. In general we are below this daily absorption figure; typical soil concentrations are 0.1 mg kg^{-1} for example. In Japan a regulation has just been issued banning the use of soil containing 10 mg kg^{-1} cadmium for rice cultivation. European opinion is that the upper limit should be 5 mg kg^{-1}. If this were to be applied in Japan large rice-growing areas would have to be abandoned.

The concentration in activated sludge exceeds, or comes very close to our tolerance. Some countries take this situation seriously; thus, Sweden has now

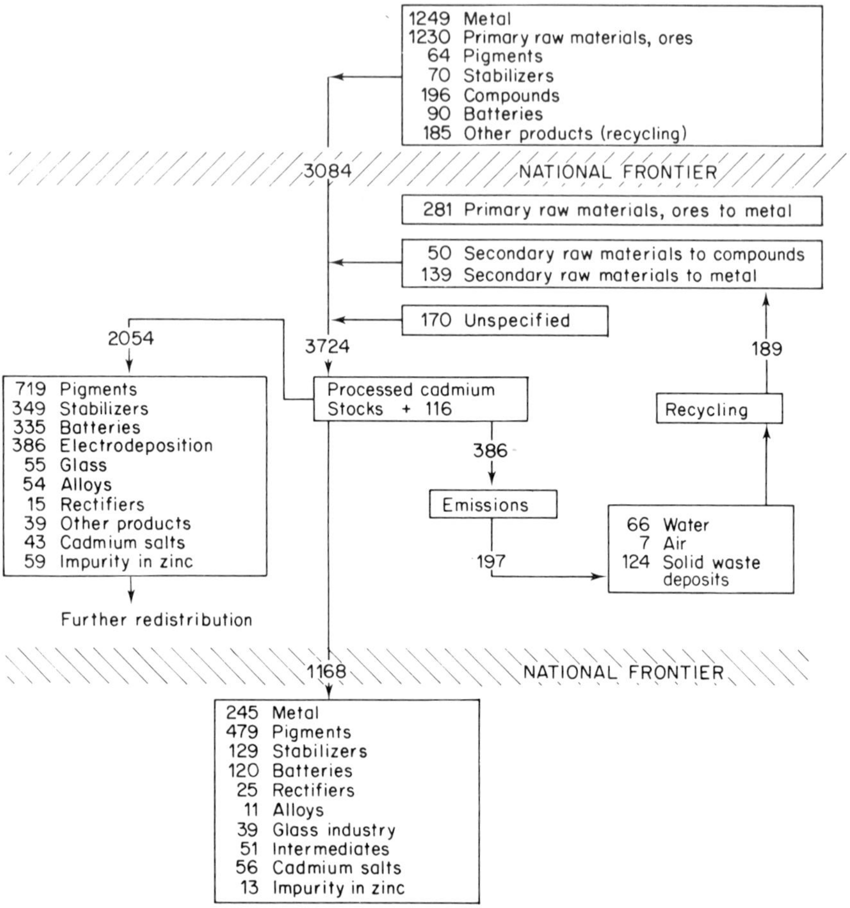

Fig. 22.4 Cadmium flow sheet in the Federal Republic of Germany, 1973 (tonnes Cd). (Source: Rauhut, 1976, private communication)

taken steps to exclude all imports of cadmium-containing products after next August.

Naturally, the question arises of how these illustrated examples of high concentrations come about. I have to ask this question despite many efforts made to find out where and what cadmium-containing products are being used in Germany. At a meeting in Tokyo very recently this phenomenon was dealt with in great detail and included a talk giving data collected in Sweden. It seems that a part of the cadmium burden arises from cadmium additives in plastic materials burnt as waste and leading to the element entering the air, spreading uniformly, and at some stage increasing the soil-cadmium concentration. If this is true, then I think it is only logical to restrict the use of cadmium where it is not absolutely essential. Sweden has gone one step further and has banned the importation of all cadmium-containing products absolutely.

Why is this example so impressive? It is because we have known about Itai-Itai disease since the 1960s. People died of it, but despite this fact it is still unknown in what form cadmium occurs in the organism and in living cells—there are no indications at all, just as there is no evidence whatever about worldwide production figures and application patterns.

Mr Rauhut, who is one of the very few who have worked with such problems using flow schemes, and deserves much merit for doing so, has shown how the fate of cadmium can be followed quantitatively. It is seen how universally cadmium is used. One ought to ask the technologists or the chemists to come along and find out what can be avoided and what is essential. No one any longer has the courage to do this.

Auto oils and lubricants contain quite a number of additives (Table 22.6). We have made an attempt to compile a list and to state the amounts contained in the different compositions, what they are used for, and the purpose to which all these additives are put. A study of their chemistry shows classes of substances but not single substances. Since self-evidently an evaluation of chemicals must always be made for individual substances and not for substance classes, this is only a beginning. The next step would be to take apart everything chemically in order to find out what passes through the engine, including what can be emitted in terms of chemical substances. Samples need to be investigated to demonstrate whether the chemical knowledge teaches what really happens. Next, the toxicologists must be called in and requested to say whether it is significant or not. However, the first step is almost entirely lacking even in countries like the Federal Republic of Germany. This is quite startling because our prosperity depends very considerably on the chemical industry. It should be remembered that there are three German companies among the five leading world chemical companies, and yet this kind of investigation is not being carried out because the personnel required are not available or for some similar reason.[15]

Available total consumption data, for instance, 550 t lead (Pb), apply mainly to the USA because it is still feasible to get some figures there, while in Europe it is almost impossible. There are 783 000 t detergent, that is to say, alkylphenates, calcium alkylates etc., but they are all classes of chemicals and not the individual chemicals required for an evaluation.

All the data in Figure 22.5 would have to be known about a substance if it is to be evaluated: How much is being produced? Where is it used, i.e., where does it enter the environment? What is its persistence? (An estimate of this is available from the chemical formula.) The dispersion can be calculated from the strontium 90 model via transport through the air. Changes under biotic conditions can be predicted qualitatively but not quantitatively. The behaviour under abiotic conditions has been overlooked so far but here, too, there are already some indications that degradation and structure can be correlated. And then there is

Table 22.6 Current and indicated future consumption of principal chemicals used as additives for lubricants, functional fluids and fuels

Chemical additives	Function	Growth (percentage/year)	US consumption (tons)	
			1975	1980
Alkenyl-succinimides	Dispersant	6.4	87400	120000
N-Substituted long-chain alkyl-salicylate	Anti-oxidant, dispersant	5.7	1185	1570
Amides and polyamides, high molecular weight	Dispersant	6.6	26500	36500
Barium phenate, normal	Detergent	−3.4	525	450
Barium phenate, overbased	Detergent, corrosion inhibitor	−2.3	3850	3425
Barium sulfonate, normal	Corrosion inhibitor, detergent, rust inhibitor	−3.8	1250	1050
Barium sulfonate, overbased	Corrosion inhibitor, detergent, rust inhibitor	−60.0	4450	50
Calcium phenate, normal	Detergent	7.6	9950	14300
Calcium phenate, overbased	Corrosion inhibitor, detergent	11.8	55932	97600
Calcium sulfonate, normal	Corrosion inhibitor, detergent, rust inhibitor	5.4	18735	24300
Calcium sulfonate, overbased	Corrosion inhibitor, detergent, rust inhibitor	7.0	113500	159200
Chlorinated paraffin wax	Extreme pressure/antiwear	−2.3	10000	9000
N,N^1-disalicylidine-1,2-diaminopropane	Metal deactivator	0.1	1000	1020
2,6-Ditertiary butyl-4-methyl phenol	Anti-oxidant	1.8	2330	2550
2,6-Ditertiary butyl phenol	Anti-oxidant	1.7	1370	1500
Ethylene dibromide	Lead scavenger	−18.1	23000	8500
Ethylene dichloride	Lead scavenger	−18.1	43000	16160
Ethylene glycol monomethyl ether	Anti-icing additive	−1.7	17600	16150
Ethylene propylene co- and terpolymers, e.g., acrylic and acrylate modified)	Cold flow improver (pour-point depressant), V.I. improver	9.4	11850	18570

Hexyl nitrate	Cetane improver (diesel fuel)	10.0	9170	14770
Hydrocarbon wax–naphthalene condensate	Pour-point depressant	4.2	1600	1970
Magnesium sulfonate, normal	Corrosion inhibitor, detergent, rust inhibitor	8.0	5850	8580
Magnesium sulfonate, overbased	Corrosion inhibitor, detergent, rust inhibitor	10.1	33000	52900
Methacrylate polymers	Pour-point depressant, storage stabilizer, V.I. improver	−2.5	375	330
Molybdenum disulfide	Extreme pressure/antiwear	0	485	485
Organophosphoric acid–fatty acid combinations	Antisquawk	3.5	5600	6650
N-Phenyl-a-naphthylamine	Anti-oxidant	2.3	3050	3400
Phosphosulfurized olefins	Extreme pressure/antiwear	3.0	14182	16.450
Polybutene amine	Detergent (gasoline)	4.7	49850	62650
Polyisobutylene	Pour-point depressant, tackiness agent, V.I. improver	−61.0	1760	16
Sodium nitrite	Corrosion inhibitor, rust inhibitor	−3.3	725	610
Styrene–butadiene polymers	V.I. improver	15.2	1620	3280
Styrene-isoprene polymers	V.I. improver	14.5	1360	2680
Sulfurized fatty olefins	Extreme pressure	3.0	2520	2910
Sulfurized lard oil	Lubricity agent	3.0	2360	2730
Sulfurized olefins	Extreme pressure/antiwear	−1.1	4820	4550
Sulfurized sperm oil replacement	Antisquawk, lubricity agent	3.4	4300	5080
Tetraethyllead	Antiknock	−17.0	166700	65800
Tetramethyllead	Antiknock	−22.5	52350	14650
Tricresyl phosphate	Extreme pressure/antiwear	4.8	7450	9400
Triethanolamine	Buffers, rust inhibitor	−3.2	1900	1600
Vinyl acetate copolymers	Pour-point depressant (cold flow improver) in residual fuel	10.0	1620	2610
Zinc-dialkyl-dithiophosphate	Anti-oxidant, extreme pressure/ antiwear, metal deactivator	5.4	74500	96700
Zinc-diaryl-dithiophosphate	Anti-oxidant, extreme pressure/ antiwear	−3.9	5130	4240
Zinc-dibutyl-dithiocarbamate	Anti-oxidant, corrosion inhibitor, metal	5.5	3920	5120

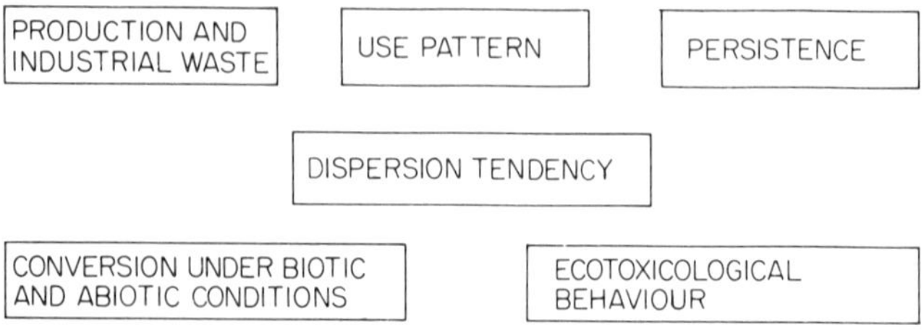

Fig. 22.5 Evaluation of the importance of environmental chemicals

the entirely new ecotoxicological behaviour. I shall present some data here. If we do not obtain these data concerning production figures and application patterns we shall continue in the same irrational vein as before. We are going to chase the polychlorinated biphenyls and hexachlorobenzene, the substances which by chance can be readily investigated analytically. We are going to study these in a big way instead of proceeding in a rational scientific manner and to begin by looking at what we are doing, and what we are introducing into the environment. The next step ought to be to reflect what the distribution of each individual substance might be; after that programmes should be developed.[12,13] The concept we use for our work today is surely scientifically quite untenable.[1,2,3]

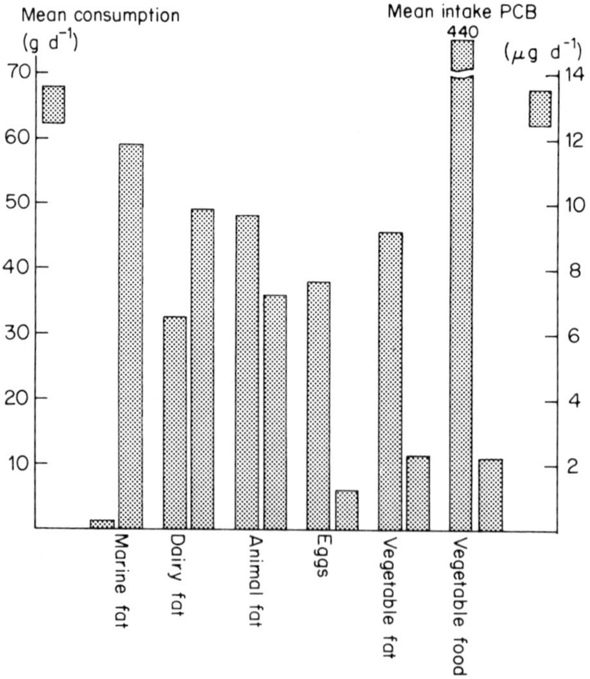

Fig. 22.6

What experimental treatment concepts are being used today or are being planned for future use?[15,16,17] Let us consider environmental monitoring to begin with. Naturally, there are great advantages associated with an investigation of familiar substances, for instance, plant protection agents in food. A start was made in the USA a long time ago. Figure 22.6 shows results obtained with this procedure; if the diet contains little marine fat the PCB burden in man is large, while dairy products help to reduce it. A vegetable diet contributes a negligible PCB burden for practical purposes. These are the data which monitoring furnishes, and it is likely very useful. It is being carried out widely in other countries, but less in Germany, and gives an 'idea' why substances accumulate in the organism. According to a second concept a predictive evaluation is built up. 'Indicator organisms' (if these exist) are chosen from human tissue, animals, microorganisms, plants, and worms and stored in a tissue bank. Such a pilot bank is being built up at present under the direction of Mr Nürnberg in Jülich. The critical issue is that, while tissues can indeed be collected, described and stored, it is also necessary to be able to analyse them reproducibly. Strictly speaking, they should be submitted to a complete chemical analysis, i.e. in respect of all substances. However, suitable techniques and detectors are not yet available to achieve this purpose; that is the concept, though. We have, therefore, first of all attempted to see whether organohalogens, the best-known compounds where analysis is concerned, can be combined together. Tissues were collected and stored and an attempt was made to analyse them using the fingerprint principle (Figure 22.7). The idea is to analyse everything and then to follow the height of the individual 'peaks' as a function of time. It is only when a peak increases considerably that an attempt is made to elucidate the constitution. The toxicology department is informed and appropriate conclusions are drawn, for example, specific investigation is indicated. This concept does not make it necessary to know the identity of the substances; all that is required is their reproducible measurement and the timing of any appreciable increase in the peak. We have performed such fingerprint analyses, and it seems to work. Relatively many substances are 'embraced'. It is naturally a very expensive concept. As is quite widely known, it is being followed up jointly by Germany and by the EPA in the USA. The analytical aspects do not look quite as hopeless as was at first feared.

Figure 22.8 illustrates the fingerprint chromatogram of a liver extract giving the molecular weights of the components. It is an investigation by combined capillary gas chromatography and mass spectrometry. All this is perfectly feasible today. It means that the tissue bank concept can apparently be realized after all and should, in my view, be continued with all possible endeavour. It has the additional advantage that reference material can be stored over several generations, thus making zero values available which we do not have today and miss so much.

The third concept uses the simulation of ecosystems. Mr Metcalf has been

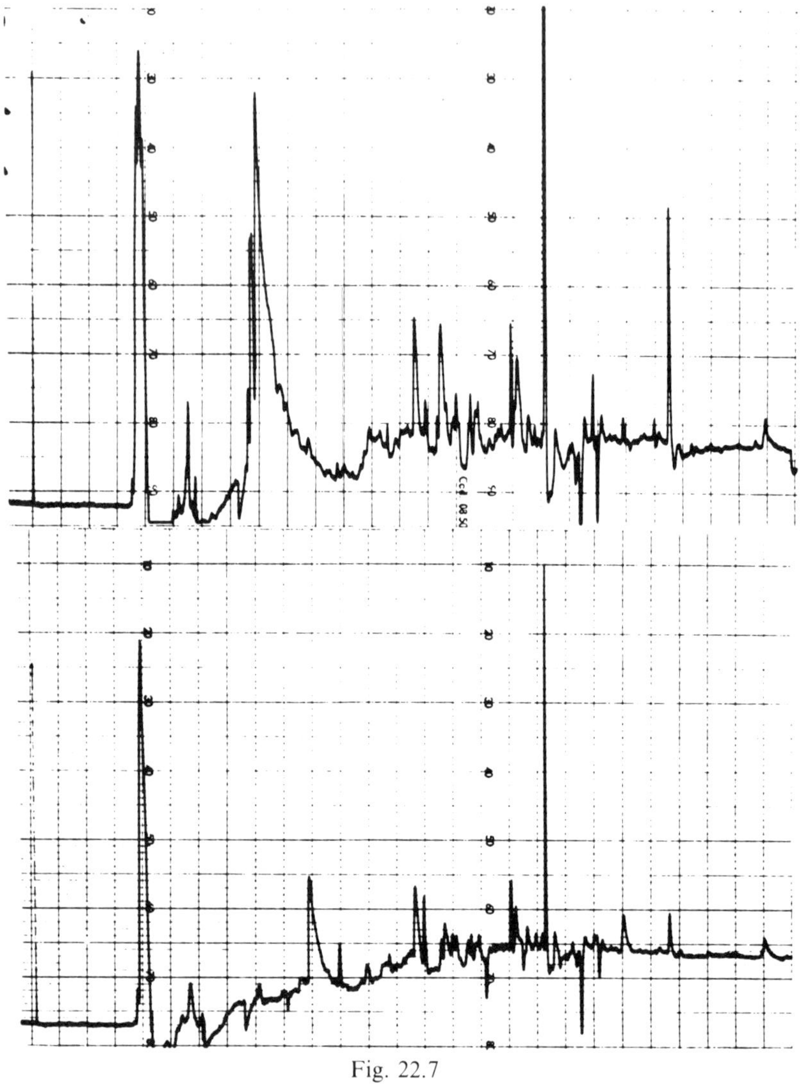

Fig. 22.7

working with this technique in Illinois in the United States. He reduces ecosystems in size in the laboratory to make them manipulatable. It has been found that this is an extraordinarily difficult matter and that it is very doubtful whether the result is relevant.

I had the opportunity to talk with Mr Metcalf about it recently. He himself admits to some doubts and concerns about the conclusions to be drawn from the results. We have chosen a different route and have conceded that we cannot model ecosystems. The reason is that no one knows what is important for

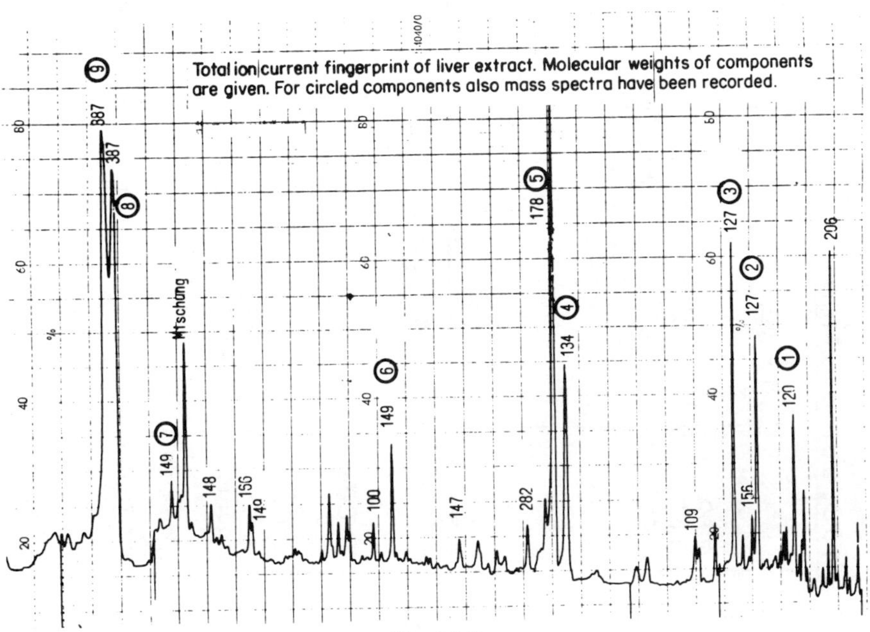

Fig. 22.8

evaluating chemicals in ecosystems. What characteristics should a chemical introduced into the environment definitely not have? The general opinion today is that substances should not remain too long in mammals, including man; anything soon eliminated is possibly better than something which accumulates. If these statements are even 80 per cent reliable it is good enough.

A high degree of accumulation in fish is undesirable (cf. Figure 22.6 showing the PCB burden) because many people and groups depend entirely on fish for food. Accumulation in algae must be measurable. Next, an interesting criterion is whether these substances disappear under photochemical, i.e. atmospheric reaction conditions; it may after all be that organic chemicals, in contrast to earlier expectations, can be degraded to carbon dioxide (CO_2) in the air by solar radiation. This is indeed the case—and in unexpectedly high measures. Experimental evidence ought to be available about the behaviour in liquid waste; the more the substance is degraded to CO_2 the better (degradation always means mineralization with formation of CO_2, H_2O etc.). If possible, the substance should be degraded to CO_2 during refuse composting, a large-scale technical process. Our tests have been designed to cover all these aspects and so that they closely match real occurrences in the ecosystem. A basic requisite has been to reduce the experimental concentration down to the ppb region. As a rule the tests are therefore carried out with 50 ppb, thus matching the practical conditions as closely as possible. Elimination from mammals (and at the same time the metabolite to non-metabolite ratio) is measured in a screening test, which is a

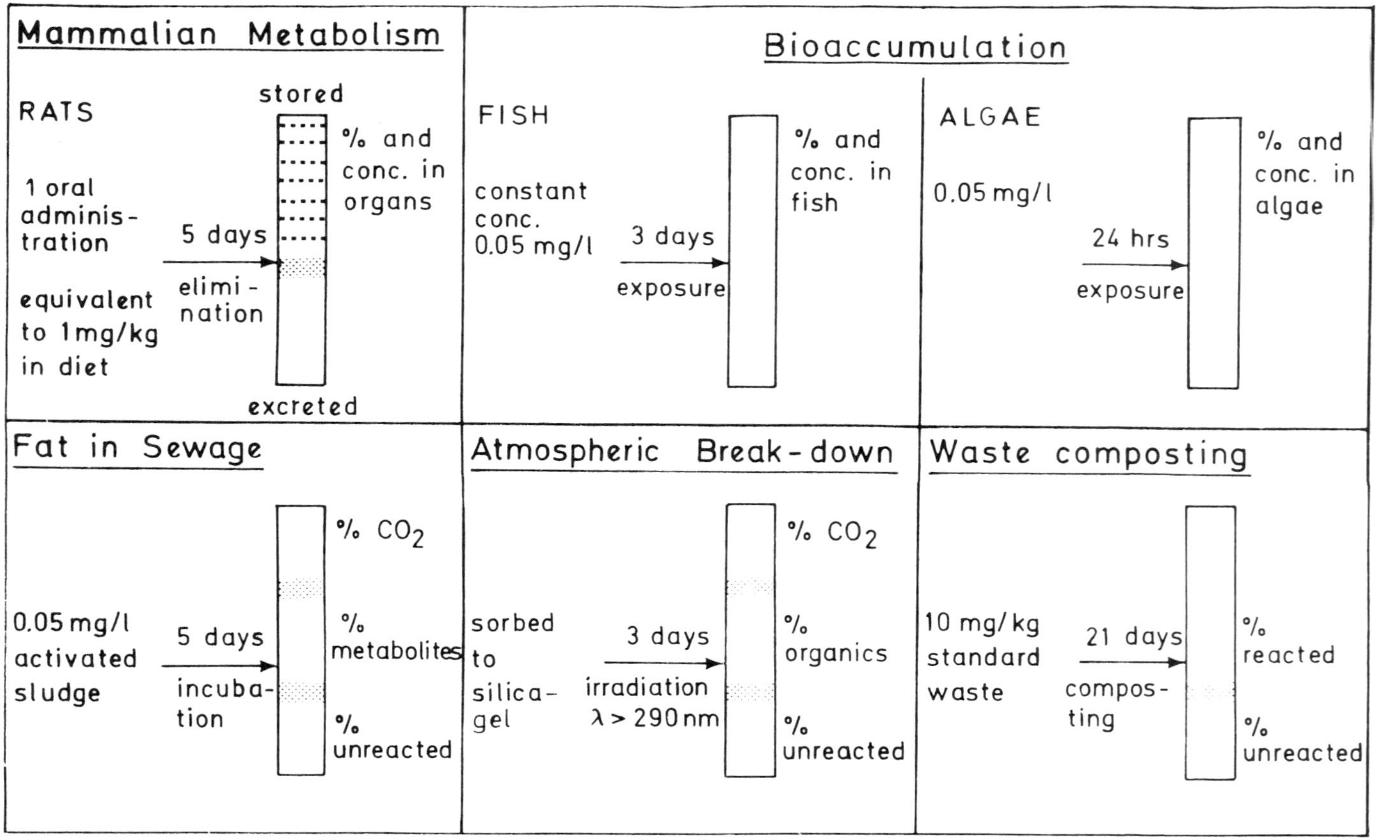

Mammalian Metabolism
RATS
1 oral administration
equivalent to 1 mg/kg in diet
5 days
elimination
stored
excreted
% and conc. in organs
Bioaccumulation
FISH
constant conc. 0.05 mg/l
3 days
exposure
% and conc. in fish
ALGAE
0.05 mg/l
24 hrs
exposure
% and conc. in algae
Fat in Sewage
0.05 mg/l activated sludge
5 days
incubation
% CO2
% metabolites
% unreacted
Atmospheric Break-down
sorbed to silica-gel
3 days
irradiation λ > 290 nm
% CO2
% organics
% unreacted
Waste composting
10 mg/kg standard waste
21 days
composting
% reacted
% unreacted

Fig. 22.9

selection test and no sensitive ecotoxicologic investigation. We have measured accumulation in fish—the test occupies 3 days—and in algae (24 h are needed), while photomineralization requires 17 hours in diffuse daylight. Activated sludge from the treatment plant in Munich has been investigated. Furthermore, we have simulated refuse composting—all testing carried out under conditions which match practical conditions, is readily accomplished and is also strictly reproducible.

Such an attempt presupposes a uniform method of analysis, and for organic substances ^{14}C analysis is the most uniform. In consequence, all substances we use are ^{14}C-labeled. It is the only way to get down to 50 ppb concentrations and yet include the metabolite to non-metabolite ratio.

In the concluding section I should like to present some data of this ecotoxicologic profile analysis. The data is that obtained with our test system. To date approximately fifty substances have been investigated. Classification is by chemistry, algae accumulation rates, fish accumulation rates, micro-organism degradation rates and elimination in the rat.

Confirmation has been obtained that the mammalian mechanism eliminates many substances much more readily than other organisms because it readily produces hydroxylation, i.e. water solubility. In consequence, it eliminates foreign substances more efficiently. These data are reproducible to within 5 per cent (Table 22.7).[11]

Table 22.7 Accumulation of phenols

	Algae (f_{24})	Fish (f_{24})	Activated sludge (f_5)	Rats (%)
Pentachlorophenol	1240	350	1100	1.6
2,6-Di-tert-butylphenol	800	230	2600	2.9
2.4.6-Trichlorophenol	116	80	40	7.8
Hydroquinone	63	13	520	0.4
p-tert-Butylphenol	35	39	240	0.1
p-Nitrophenol	11	19	30	0.4

The important feature of Table 22.8 is that a comparison procedure is obtained, for example, for hexachlorobenzene, benzene, pentachlorobiphenyl, trichlorobiphenyl, aldrin, and all the way to trichlorophenol and hexachlorocyclopentadiene. Accumulation in fish is accepted to be significant only for values of the order of 500–600; the activated-sludge test and the rat-elimination test display large variations.

It is thus possible to establish a priority list for further studies. The aim of this Ecotoxicological Profile Analysis—our name for this system—is not to utilize the results for legislative measures but to produce a selection system for further,

Table 22.8 Accumulation of chlorinated cyclic compounds

	Algae (f_{24})	Fish (f_{24})	Activated sludge (f_5)	Rats (%)
Hexachlorobenzene	24000	400	32000	57.2
2,4,6,2′,4′-Pentachlorobiphenyl	11500	770	27800	40.7
2,5,4′-Trichlorobiphenyl	7700	850	32000	3.2
Aldrin	5800	1180	18000	29.1
Pentachloronitrobenzene	3600	380	4500	0.4
2,2′-Dichlorobiphenyl	3050	830	6300	0.2
Pentachlorophenol	1240	350	1100	1.6
Hexachlorocyclopentadiene	1140	308	2350	1.9
2,4,6-Trichlorophenol	116	80	40	7.8

Table 22.9 Accumulation of nitrogen substituted aromatic compounds

	Algae (f_{24})	Fish (f_{24})	Activated sludge (f_5)	Rats (%)
Pentachloronitrobenzene	3600	380	4500	0.4
Nitrobenzene	24	0.8	40	2.3
Monolinuron	15	< 10	70	1.5
p-Nitrophenol	11	19	30	0.4
Aniline	8	1.7	1030	0.5

Table 22.10 Accumulation of dissociable compounds

	Algae (f_{24})	Fish (f_{24})	Activated sludge (f_5)	Rats (%)
HPMA	140	0.5	1260	2.6
n-Dodecylbenzenesulfonate	135	43	140	< 0.1
p-Chlorobenzoic acid	63	1.1	170	0.1
Maleic acid	14	1.4	2800	6.7
2,4-Dichlorobenzoic acid	9	0.3	9	—
2,4-Dichlorophenoxyacetic acid	6	2	17	1.1

more complicated, and thus more expensive, investigations. Basically we investigate what the newspapers report—literally just that.

Table 22.9 shows the accumulation of nitrogenous aromatic compounds. Pentachloronitrobenzene, a pesticide, is seen to accumulate much more in algae than in fish, surely an unexpected event. Accumulation in activated sludge is similar to that in algae. Warm-blooded animals excrete the substance readily because they can easily hydroxylate and thus subsequently eliminate it.

Table 22.11 Accumulation of additives and natural production

	Algae (f_{24})	Fish (f_{24})	Activated sludge (f_5)	Rats (%)
Hexadecanol	18000	18	3170	65.4
2,4,6,2′,4′-Pentachlorobiphenyl	11500	770	27800	40.7
2,5,4′-Trichlorobiphenyl	7700	850	32000	3.2
Dodecane	8000	17	1300	12.1
Di-(2-ethylhexyl)-phthalate	5400	85	3000	7.1
2,2′-Dichlorobiphenyl	3050	830	6300	0.2
2,6-Di-tert-butylphenol	800	230	2600	2.9
Biphenyl	540	110	430	0.4
Coumarin	140	<0.1	1400	0.6
p-tert-Butylphenol	35	39	240	0.1

Phenols with different substituents also show marked differences for accumulation in fish, activated sludge and in the rat. Again that in the rat is very small by comparison to the other two. Table 22.10 lists the accumulation of dissociable compounds in the form of acids, a polymaleic anhydride type of polymer etc. Considerable differences are revealed, and it is again possible to set up a priority list for further investigations and deliberation.

Accumulation of additives and natural chemicals is shown in Table 22.11. I have already mentioned that we have almost no information about additives. The figure lists a very few substances representing not even 1 per cent of those which occur in practice. Hexadecanol shows a high accumulation rate in algae due very probably to its breakdown to acetic acid and re-use; it becomes, so to speak, a quasi-autologous product. Phthalates, tertiary butylphenols and antioxidants can be included and a priority sequence established for further research work.

Table 22.12 shows accumulation data for pesticides. These compounds require to be registered but the data is none the less only incompletely available. This is

Table 22.12 Accumulation of pesticides and derivatives

	Algae (f_{24})	Fish (f_{24})	Activated sludge (f_5)	Rats (%)
Hexachlorobenzene	24000	400	32000	57.2
Aldrin	5800	1180	18000	29.1
Pentachloronitrobenzene	3600	360	4500	0.4
Pentachlorophenol	1240	350	1100	1.6
2,4,6-Trichlorophenol	116	80	40	7.8
Monolinuron	15	<10	70	1.5
p-Nitrophenol	11	19	30	0.4
2,4-Dichlorophenoxyaceticacid	6	2	17	1.1

Table 22.13 Degradation of polychlorinated biphenyls

	Activated sludge ($\% CO_2$)[a]	Photomineralization ($\% CO_2$)
2,4,6,2′,4′-Pentachlorobiphenyl	1.2	5.2
2,5,4′-Trichlorobiphenyl	0.5	5.9
2,2′-Dichlorobiphenyl	6.6	3.5
Biphenyl	9.1	9.5

[a] Including small amounts volatilized.

the first time such data have been measured under strictly comparable conditions. The differences between hexachlorobenzene, aldrin, pentachloronitrobenzene etc., show that if one were to pick up the best substance for a particular application pattern, monolinuron would be selected rather than hexachlorobenzene. If it also fitted into this same field of application, it would be a good product and better than any of the others. That is making an obvious point, but this concept has not been applied in this way so far. Strictly speaking, such a profile analysis ought to be employed right at the beginning of an industrial research project.

Table 22.13 shows degradation data for PCBs under biological conditions and by light. It is seen that light far outperforms enzymes in this respect. A little thought makes the reason clear. The biological system has 8 kcal available for a reaction while light in the form of light quanta comprises not less than 120 kcal mol^{-1}; it is able to split any type of bond. In terms of substances which are pollutants—and the argument is naturally limited to these—degradation by photomineralization is much more efficient than by biotic systems.

Table 22.14 shows more or less the same effect for other chemicals and that a

Table 22.14 Degradation of chlorinated cyclic compounds

	Activated sludge ($\% CO_2$)[a]	Photomineralization ($\% CO_2$)
Hexachlorobenzene	3.0	1.5
2,4,6,2′,4′-Pentachlorobiphenyl	1.2	5.2
2,5,4′-Trichlorobiphenyl	0.5	5.9
Aldrin	1.5	8.8
Pentachloronitrobenzene	7.0	43.0
2,2′-Dichlorobiphenyl	6.6	3.5
Pentachlorophenol	0.5	62.0
Hexachlorocyclopentadiene	2.0	46.0
2,4,6-Trichlorophenol	0.1	65.8

[a] Including small amounts volatilized.

Table 22.15 Degradation of nitrogen substituted aromatic compounds

	Activated sludge (% CO_2)[a]	Photomineralization (% CO_2)
Pentachloronitrobenzene	7.0	43.0
Nitrobenzene	1.7	6.7
Monolinuron	0.5	38.3
p-Nitrophenol	1.4	39.1
Aniline	37.0	46.5

[a] Including small amounts volatilized.

synthetic cycle may exist side by side with the natural cycle. Chemistry begins with oil; it creates something stable from it, and what is stable is not attacked by biotic systems but can penetrate into the air and is there readily degraded to CO_2 by excess solar energy. It would be possible to recreate oil from CO_2. The data available support this concept.

Table 22.15 gives some degradation figures for aromatic nitrogen compounds. It seems that in these cases where the substances are 'laden' with elements without a biochemical role photomineralization always outperforms by a long way what may be called normal mineralization. The point to be made is that one ought to try and discover as quickly as possible how high the degradation rates are. Advance extrapolations would then show the proportion of the substances to be degraded under atmospheric conditions and enable pollution of the atmosphere to be prevented. Table 22.16 once more illustrates the degradation of various compounds with photomineralization rates of up to more than 50 per cent. Tables 22.17 compares the rates of dissociable compounds including acids, sulphonates etc. Table 22.18 shows degradation figures for additives and natural products which may be compared with the accumulation data for the same compounds shown in Table 22.11.

Table 22.16 Degradation of pesticides and derivatives

	Activated sludge (% CO_2)[a]	Photomineralization (% CO^2)
Hexachlorobenzene	3.0	1.5
Aldrin	1.5	8.8
Pentachloronitrobenzene	7.0	43.0
Pentachlorophenol	0.5	62.0
2,4,6-Trichlorophenol	0.1	65.8
Monolinuron	0.5	38.3
p-Nitrophenol	1.4	39.1
2,4-Dichlorophenoxyacetic acid	0.1	26.2

[a] Including small amounts volatilized.

Table 22.17 Degradation of dissociable compounds

	Activated sludge (% CO_2)[a]	Photomineralization (% CO_2)
HPMA	3.3	12.1
n-Dodecylbenzenesulfonate	3.5	3.2
p-Chlorobenzoic acid	5.4	6.3
Maleic acid	26.9	17.0
2,4-Dichlorobenzoic acid	0.3	—
2,4-Dichlorophenoxyacetic acid	0.1	26.2

[a] Including small amounts volatilized.

Table 22.18 Degradation of additives and natural products

	Activated sludge (% CO_2)[a]	Photomineralization (% CO_2)
Hexadecanol	28.5	3.1
2,4,6,2′,4′-Pentachlorobiphenyl	1.2	5.2
2,5,4′-Trichlorobiphenyl	0.5	5.9
Dodecane	38.0	2.9
Di-(2-ethylhexyl)-phthalate	1.0	1.6
2,2′-Dichlorobiphenyl	6.6	3.5
2,6-Di-tert-butylphenol	7.7	29.5
Biphenyl	9.1	9.5
Coumarin	30.0	59.2
p-tert-Butylphenol	4.9	46.8

[a] Including small amounts volatilized.

Table 22.19 Degradation of phenols

	Activated sludge (% CO_2)[a]	Photomineralization (% CO_2)
Pentachlorophenol	0.5	62.0
2,6-Di-tert-Butylphenol	7.7	29.5
2,4,6-Trichlorophenol	0.1	65.8
Hydroquinone	7.6	57.4
p-tert-Butylphenol	4.9	46.8
p-Nitrophenol	1.4	39.1

[a] Including small amounts volatilized.

Table 22.20 Photomineralization in twin-reactor

Chemical		CO_2 (%)		$M = \frac{K_1}{K_2}$
K_1	K_2	K_1	K_2	
Lindane (1.V)	PCP	6.4	61.5	0.10
Lindane (2.V)	PCP	5.3	49.6	0.10
Nitrobenzene	PCP	6.6	65.0	0.10
HCB	PCP	2.2	67.0	0.03
PCP	PCP	65.0	68.6	0.94

Note: Ref. PCP in K_2 irradiation time 17 h.

Phenols, i.e. compounds with a hydroxyl group introduced into the aromatic nucleus, are shown in Table 22.19. Here the feature is relatively high degradation rates under atmospheric conditions in diffuse daylight. The figures shown are not values for high altitudes in the atmosphere but close to the earth in diffuse light.

Table 22.20 shows some detailed results obtained in our experiments. The photomineralization tests are performed in a photochamber on a solid surface through which air is blown; the CO_2 formed under illumination is measured (using a $[^{14}C]CO_2$ or infrared technique). Since the lamps used for illumination vary in their intensity, and it is difficult to measure light intensities, we have set up twin chambers to enable reference substances to be run simultaneously. One chamber contains the reference substance pentachlorophenol, while in the other chamber the substance under investigation is measured using the same lamp.

The results in the figure show that the method is a sound one. We can carry out photochemical degradation reactions very reproducibly in this twin chamber, which is very easy to manipulate; I express the hope that the technique will be widely used.

The pictorial compilation in Figure 22.10 presents a summary of the properties of many of these compounds. When one has collected the data, what does one do with them? It is a universal problem. The lowest common denominator needs to be picked out or else—and very much better—the substance is scrutinized in respect of use and site of application, and a possible substitute is examined. The two are then compared substance by substance. I believe that this is the test of the future. However, a prerequisite is a knowledge of what substances are being turned into environmental chemicals and where they are introduced into our surroundings—and this needs to be emphasized. It might after all be worth the effort to find out whether such a study could once more be undertaken in collaboration with the German chemical industry and interested large research establishments. It would seem that this is the one way of at long last tackling the problem of environmental chemicals more rationally in the future.

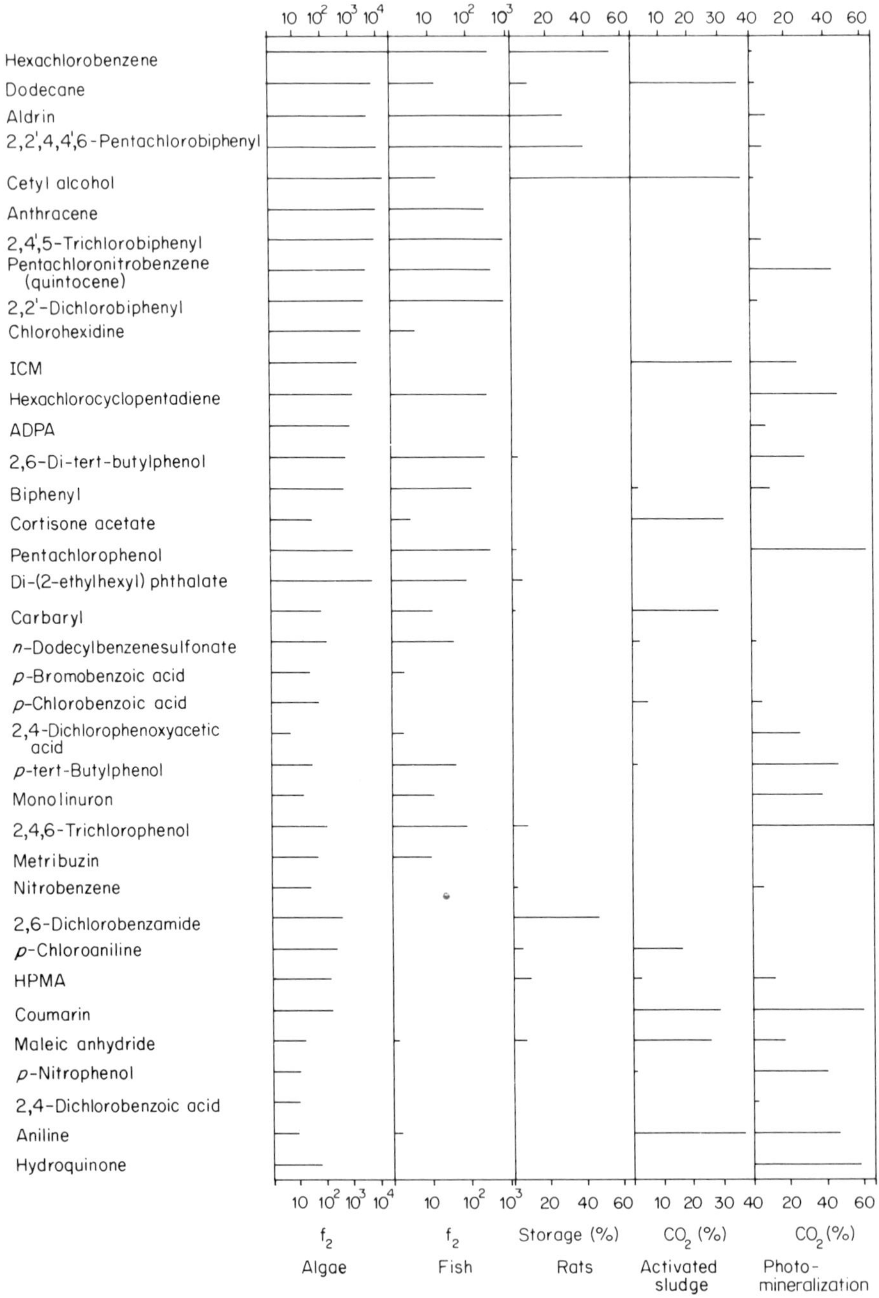

Fig. 22.10 Survey of the results of the ecotoxicologic profile analysis

ACKNOWLEDGEMENTS

The experimental work reported here was carried out in cooperation with the scientists of the Institut für Ökologische Chemie. I express my thanks to them and the technical staff of the Institute for their efforts and for the enthusiasm they have brought to the work.

REFERENCES

1 Proceedings: Chemical and Toxicological Aspects of Environmental Quality, 1st International Symposium arranged by the Institut für Ökologische Chemie der Gesellschaft für Strahlen- und Umweltforschung mbH München, Georg Thieme Verlag, Stuttgart (1969).

2 Korte, F., Klein, W., and Drefahl, B. (1970). Technical environmental chemicals, occurrence, degradation, and consequences, *Naturwissenschaftliche Rundschau*, 445–457.

3 Korte, F. (1972). Are pesticides suitable model substances for the evaluation of industrial chemicals in the environment? *OEPP/EPPO Bull.* **4**, 27–49.

4 Korte, F., and Klein, W. (1971). *Topical Problems in Environmental Chemistry*, Pharma International, Reprint, Verlag G. Braun, Karlsruhe.

5 Klein, W. (1973). Research in the Gesellschaft für Strahlen-und Umweltforschung on the evaluation of the risks involved in environmental chemicals, *Environmental Quality and Safety*, **2**, 244–247.

6 Zimmer, M. (1973). Dose versus effect relationships of air pollution in man with special reference to high-density population regions, *GSF-Bericht Ö*, **70**.

7 Korte, F. (1977). *Industrial Chemicals, Occurrence and Behavior in the Environment and Consequences*, Special publication based on data and documents referring to environmental protection, Issue no. 21 of the Environmental Meeting series, Documentation Center of the Universität Hohenheim.

8 Korte, F. (1972). What are environmental chemicals and what are their problems? *Chemosphere*, **5**, 183–195.

9 Korte, F. (1978). Effects of pesticides and related compounds in the environment, *Pure & Appl. Chem.*, **50**, 1045–1061.

10 Sümmermann, W., Rohleder, H., and Korte, F. (1978). Polychlorinated biphenyls (PCB) in foodstuffs, *Z. Lebensm. Unters.-Forsch.* **166**, 137–144.

11 Gebefügi, I., and Parlar, H. (1978). Estimating the risk posed by pentachlorophenol in the environment, *GSF-Bericht Ö*, **414**.

12 Annual Report (1969). Institut für Ökologische Chemie, *GSF-Bericht Ö*, **20**.

13 *Nachr. Chem. Techn.* (1968). **16**, 3, 44–45.

14 Korte, F., Klein, W., Parlar, H., and Scheunert, I. (1979). *Ökologische Chemie und Ökotoxikologie*, Monograph, Georg Thieme Verlag, Stuttgart.

15 Korte, F. (1969). Residue problems, Reprint from *Natur und Landschaft*, **44**, 9, 225.

16 Korte, F., Freitag, D., Geyer, H., Klein, W., Kraus, A., and Lahaniatis, E. (1978). Ecotoxicologic Profile Analysis, *Chemosphere*, **7**, 1, 79–102.

17 Freitag, D., Geyer, H., Klein, W., Kraus, A., Lahaniatis, E. and Korte, F. (1979). An approach for comparative screening of the environmental behavior of chemicals, *Ecotoxicology and Environmental Safety*, **3**.

18 Iliff, N. A. (1971). Organic chemicals (except agricultural, animal health chemicals and food additives), in Cowlston, F., Korte, F., *Environmental Quality and Safety*, **2**, 64.

Pollutants and Their Ecotoxicological Significance
Edited by H. W. Nürnberg

CHAPTER 23

Lysimeter Experiments with Selected Herbicides

F. FÜHR

SUMMARY

The radioisotope tracing technique has been a useful tool in obtaining extensive information on the fate of pesticides in the soil–plant system, including their uptake, transport and metabolism by plants; their photochemical, chemical and microbial degradation; their adsorption–desorption and translocation in soil; and their bioavailability to untreated crops. A pesticide balance study under practical field conditions using radio labelling can examine a number of factors affecting the fate of a compound at the same time and assess the magnitude of the major processes involved. On the basis of these results, more detailed studies are then formulated to be conducted under an exactly defined environment of a growth chamber or a laboratory. The use of tracer techniques in such studies is demonstrated with results from experiments with a urea derivative herbicide.

INTRODUCTION

According to data from FAO the annual losses in agriculture due to plant diseases, insect damage and weed competition are estimated to represent a market value of about $130 billion which is equivalent to the annual crop value of wheat, barley, rye, oat and potatoes.[1] In 1974 the farmers around the world have spent about $5 billion on pesticides to control these losses and the predicted yearly increase of pesticide use is about 10–15 per cent. However, even though the highest losses are observed in Asia, Africa and South America, 45 per cent of the produced pesticides are applied in the USA, 25 per cent in Western Europe, 8 per cent in Japan and only about 20 per cent in those areas where the use would be most needed to produce sufficient food from the limited agricultural lands.[2]

The application of pesticides in agricultural practice underlies specific

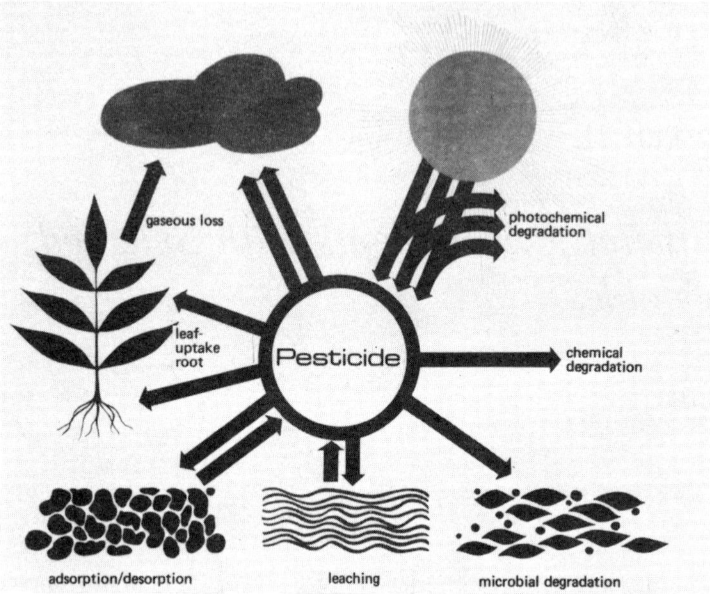

Fig. 23.1 The fate of a pesticide in the soil/plant system

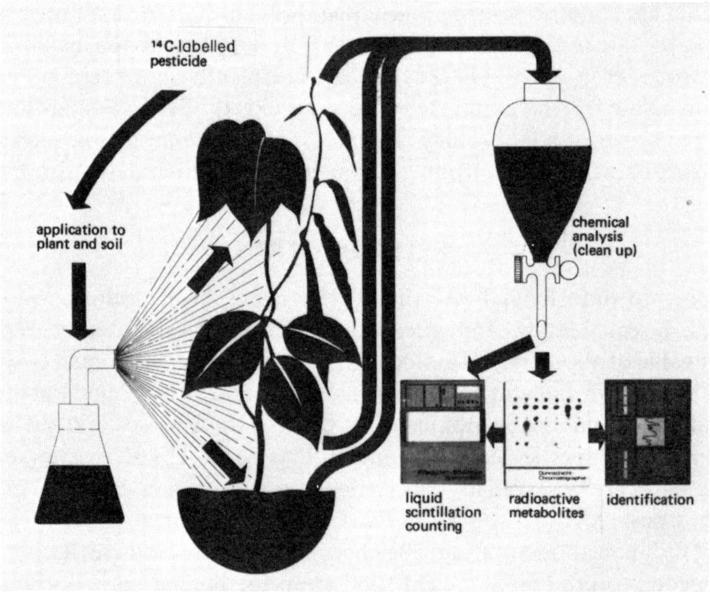

Fig. 23.2 The use of ^{14}C-labelling in pesticide studies

regulations. The pesticide itself has to meet certain standards before it will be approved for application. Already during the development of a compound the mode of action, the analytical procedures, the toxicology are investigated and certain standardized experiments on leaching, degradation and metabolism are conducted by industrial laboratories to provide the required information for the regulating agencies. But it became obvious during the discussion about the relatively stable organochlorine compounds that more detailed information is needed concerning the behaviour of pesticides in the soil–plant system. Residue analysis can only trace the original mother compound or certain metabolites of similar structure, but labelling of a pesticide molecule at a definite position became a unique technique to follow the fate of the pesticide and its metabolites formed in the soil–plant system, thus furnishing extensive information on its uptake, transport and metabolism by plants, its photochemical, chemical and microbial degradation, its adsorption–desorption and translocation in soil and its bioavailability to untreated crops (Figure 23.1).

The use of the tracer technique is demonstrated in Figure 23.2. Depending on the aim of an investigation, the compound is labelled at a definite position of the molecule and applied to plant or soil simulating spray application. The treated samples are then extracted, purified and concentrated, and labelled metabolites are separated using thin-layer chromatography, gas chromatography and high performance liquid chromatography in combination with the detection of the radioactivity by liquid scintillation counting. Hypothetical metabolites available as reference substances are cochromatographed to yield information on the possible structure of the unknown metabolites. Metabolites representing a substantial percentage of the originally applied amount of radioactivity are specially purified and concentrated for GC/MS identification.

LYSIMETER EXPERIMENTS

Many of the tracer studies with pesticides have been conducted under laboratory conditions. The data thus obtained are generally limited to evaluate individual processes. In the Jülich Atomic Research Center we are allowed to conduct small-scale, outdoor experiments with up to 30 mCi of ^{14}C per experiment. This gives us the chance to make use of a different approach which is outlined in Table 23.1.

Standardized lysimeters (1 m^2 surface area, 50 cm of packed top-soil) (Figure 23.3) and undisturbed soil cylinders (0.25 m^2 surface area, 50 cm of soil profile) have been set up in this area and the combined effect of the growth factors such as the physical, chemical and microbial processes in the soil and the climatic factors as well, on the behaviour of a ^{14}C-labelled pesticide in the soil–plant system are examined throughout the growing season including the uptake by untreated rotational crops. This balance study provides useful information on the magnitude of the uptake, distribution, and metabolism of the pesticide in plants, its displacement in the soil and leaching out of the root zone as affected by

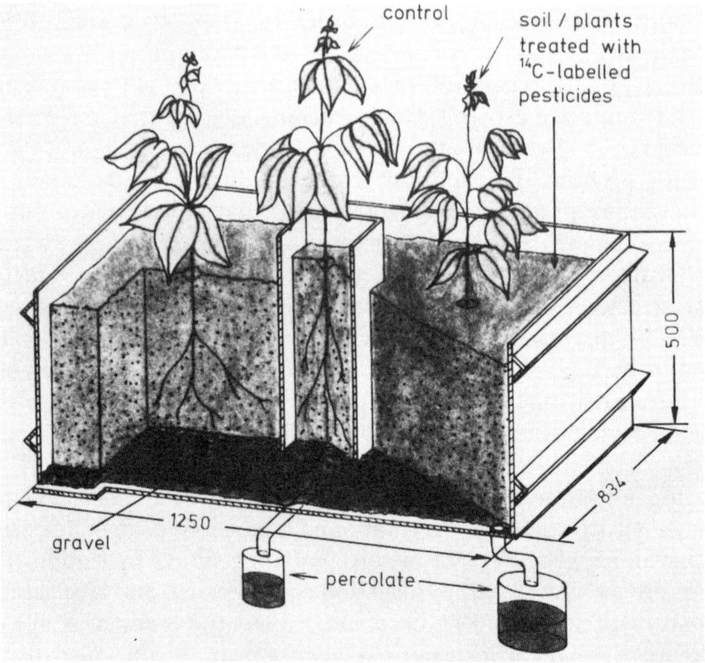

Fig. 23.3 The experimental lysimeter (measurements in mm)

precipitation and root development, as well as its metabolism and mineralization. More detailed information on this lysimeter experiment and special problems involved are given in an additional paper from our working group.[3]

As an example data are summarized from a spray application of a urea derivative herbicide (methabenzthiazuron, 1.3-dimethyl-3-(2-benzthiazolyl)-urea) to wheat at the 3–4-leaf stage of plant development.[4] The applied methabenzthiazuron concentration was equivalent to 2.5 kg Tribunil® ha^{-1} (Bayer AG, Leverkusen, FRG). The soil was a parabrown earth derived from loess. At harvest time, 111 days after spraying, 83 per cent of the ^{14}C activity was still almost exclusively in the 0–5-cm layer of the loess soil, 1.4 per cent was in the above-ground portion of the treated wheat plants, and only traces were leached from the 40 cm layer of soil (Figure 23.4). In the same lysimeter, immediately following the balance study with spring wheat, an experiment was conducted to investigate the uptake of the labelled pesticide residues or metabolites by the subsequent rotational crop rye.[4] About 83 per cent of the applied radioactivity remained mainly in the upper 10-cm soil layer at wheat harvest time (Figure 23.4). Calculated on the specific radioactivity of the applied methabenzthiazuron, the following methabenzthiazuron equivalent values were found in the treated spring wheat: straw, 16 ppm; chaff, 1.4 ppm; grain, 0.07 ppm. The

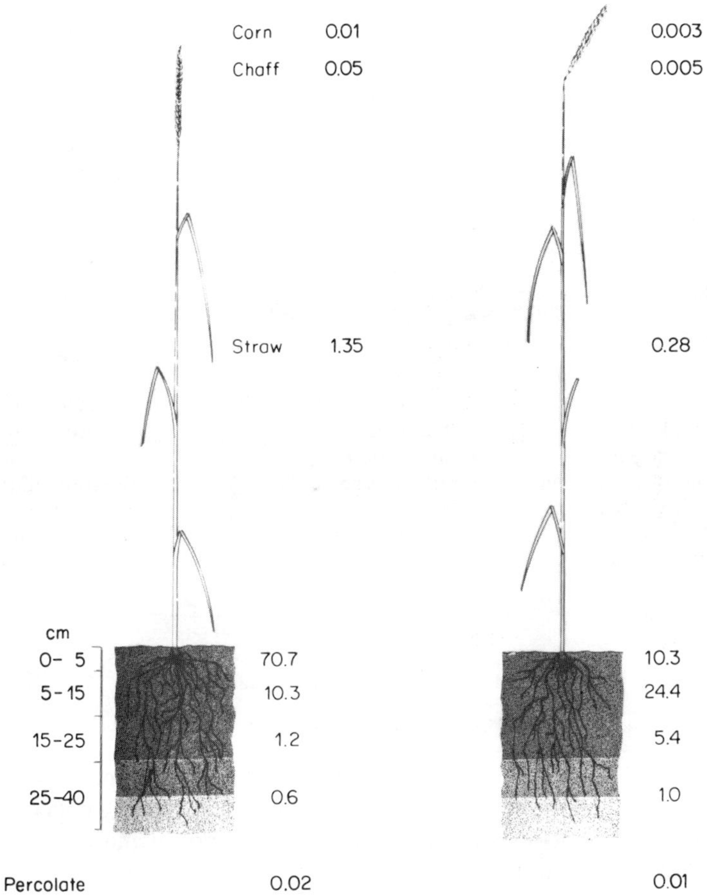

Fig. 23.4 Results from an outdoor lysimeter experiment: ^{14}C-distribution in the treated spring wheat, the rotational crop rye, and the soil after application of [benzenering-U-^{14}C]methabenzthiazuron. Radioactivity applied = 100

untreated rotational crop rye contained on the average only one-fifth to one-tenth of residual radioactivity in comparison to the treated spring wheat. About 40 per cent of the applied radioactivity remained in the 25 cm plow layer of the soil.

Based on results achieved in the lysimeter experiment, detailed studies are conducted to investigate defined steps of plant uptake, transport and metabolism (Table 23.1), and the influence of the different environmental factors on these processes as well as translocation and behaviour in soil including all processes, which may influence the bioavailability of pesticide residues to untreated rotational crops.

Table 23.1 General scheme for studies with labelled pesticides

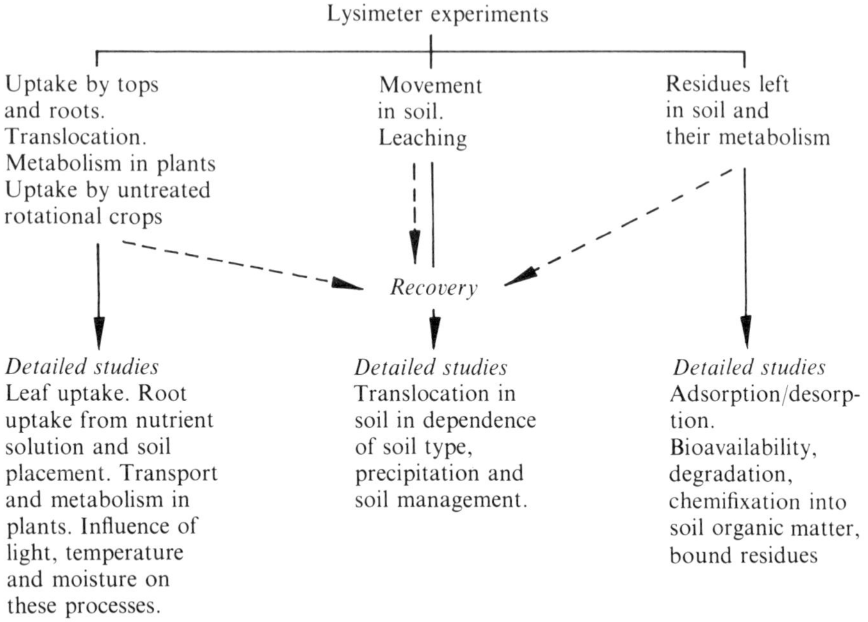

DETAILED STUDIES

So far many of the tracer studies with pesticides have been conducted under laboratory conditions. One has to be very careful in interpreting uptake results from small-scale pot experiments under laboratory conditions. Certain experimental and environmental factors are quite different from field conditions which normally favour the uptake rates extremely.

1. Distribution of the pesticide in the soil.
2. The ratio root mass to soil.
3. Maintaining of a relatively optimal soil-water content for plant roots and steady water supply.
4. Optimal conditions for plant growth and hence high transpiration rates.

Therefore, parallel to this lysimeter experiment, a pot experiment with maize was conducted in the greenhouse.[5] Four 2-week-old seedlings were planted in special experimental pots containing 700 g of a humic sandy soil with 5 or 10 ppm active ingredient thoroughly mixed into the soil The plants were grown for 6 weeks and water was supplied to the soil daily up to 65 per cent of the water holding capacity. The ^{14}C content of the dry matter accounted for an average of 11.4 per cent of the ^{14}C applied to the soil as methabenzthiazuron, almost 8 times as high as in the wheat crop from the lysimeter experiment after a full growing season.

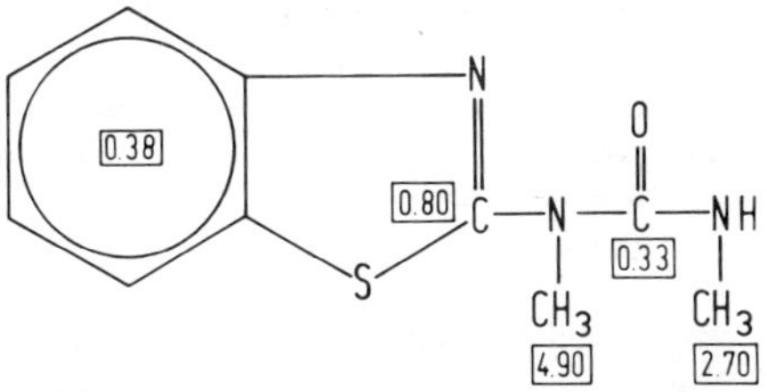

Fig. 23.5 Mineralization rates (percentage) of the different carbon atoms of methabenzthiazuron under constant conditions in the soil (1.1 per cent C, pH 6.5) at 22 °C, 65 per cent WHC, 10 ppm methabenzthiazuron, 11 weeks. Total radioactivity applied = 100 per cent

Degradation studies were conducted in 200 g of soil under standardized condition (Figure 23.5) using ^{14}C-labelled methabenzthiazuron at the different position of the molecule.[6,7]

Preferentially the methyl groups are mineralized by microbial action. Extensive analytical investigations using the soil from the lysimeter experiment were conducted to identify the major metabolite formed in the soil as 1-methyl-1-(2-benzthiazolyl)-urea.[8] Thin-layer and high-performance liquid chromatography were applied to separate and purify the major metabolite. Cochromatography of hypothetical metabolites as reference substances yielded information on the possible structure of the unknown labelled newly formed metabolite (Table 23.2). Its chromatographic properties were identical to 1-methyl-1-(2-benzthiazolyl)-urea which was confirmed by MS fragmentation yielding a parent ion at m/e 207.[8] In this identification study we started with only 70 μg of the respective metabolite.

So far the main interest is focused on the uptake of pesticides by treated crops. But several pesticides are applied to a crop either in combination with other compounds or successively. The question, therefore, remains unanswered what is the fate of the residues left in the soil after harvesting either applied for example as soil herbicides or leached into the soil from the treated plants by rain or brought into the arable soil layer via organic manuring. In the lysimeter experiments under outdoor conditions we simulate standard rotations. So far the residue uptake calculated on the basis of the specific activity of the applied compounds represented roughly one-fifth to one-tenth of that found in the treated crops. But this depends on the stability of the chemical applied, its adsorption behaviour and the chemical and physical properties of the soil. Additional detailed studies are needed to evaluate the bioavailability of pesticide residues left in the soil. Normally we deal with two fractions of residues in soil, (a) extractable pesticides or metabolites, and (b) unextractable 'bound' residues.

With time the second fraction increases, and depending on the pesticide

Table 23.1 Elution volumes of methabenzthiazuron and of different hypothetical metabolites in comparison to the major metabolite

Compound		Elution volume (ml)	
No.	Structure	μ-C 18[a]	μ Porosil[b]
1	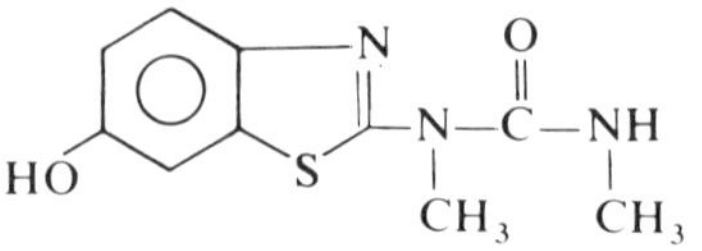	13.65	
2	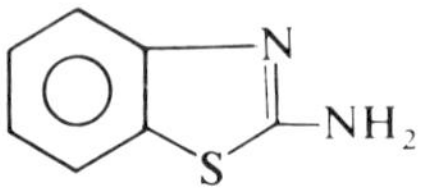	15.60	
3	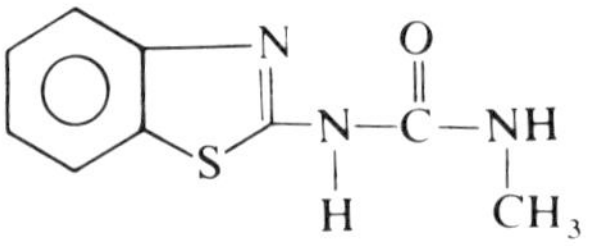	21.15	
4	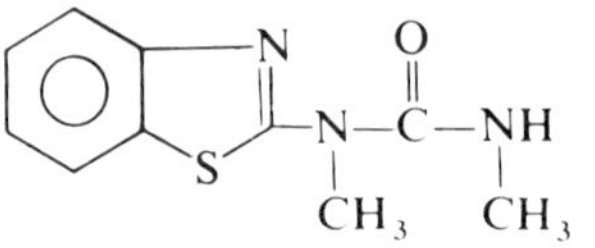	30.98	4.95
5	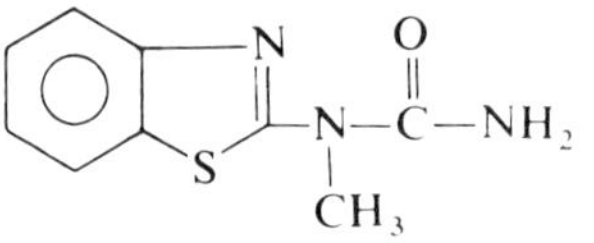	24.83	8.40
Major metabolite		24.75–14.83	8.36

[a] Solvent program no. 10. 20%–100% acetonitrile. 20 min; flowrate 15 ml min^{-1}; chart 10 mm min^{-1}.
[b] Chloroform, flowrate 15 ml min^{-1}, chart 10 mm min^{-1}.

applied anywhere a range between a few per cent to nearly three-quarters of some pesticide residues will remain in the soil in an unextractable form. So far no extraction schemes have been developed to characterize these bound residues,[9] and the question still remains whether these 'bound residues' are rather strongly adsorbed by soil organic matter binding sites or incorporated into, or assimilated by, soil organic matter.[10] With extraction procedures commonly used in soil organic matter analysis, normally quite a high percentage can be extracted, and in combination with residue analysis information on the structure and status of binding can be collected. Uptake studies with plants using the special experimental pots, will yield information on their degradation rates and bioavailability to plants.[11]

Table 23.3 Plant experiment to determine the bioavailability of the non extractable radioactivity

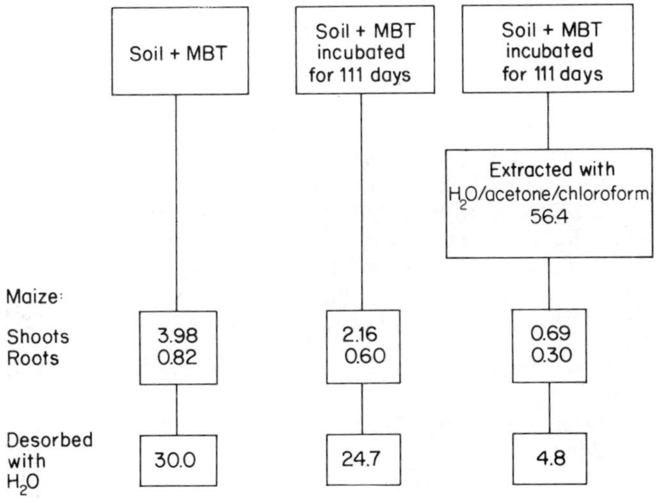

Test compound: [^{14}C] carbonyl methabenzthiazuron (MBT) Sandy soil, pH 6.4; 1.15 % C. Experimental data (Applied radioactivity = 100 %).

The extractability of methabenzthiazuron residues from soil decreases with time after applications.[12] In a special study the bioavailability of this soil bound or adsorbed fraction was investigated using maize plants.[11]

A sandy soil and [^{14}C] carbonyl methabenzthiazuron (10 ppm) was incubated at 22 °C and 65 per cent of the water holding capacity for 111 days. After exhaustive extractions with H_2O/acetone/chloroform, unextracted, bound residues remaining in this soil amounted to 41 per cent applied. Plant uptake studies were conducted in special experimental pots (1 l) which allow an airtight separation of the shoots from the root environment. After 29 days the maize shoots contained 0.7 per cent of the radioactivity originally applied to the soil before incubation in comparison to 2.2 per cent from the incubated soil not extracted (Table 23.3). From the [^{14}C] carbonyl methabenzthiazuron added to the soil immediately before the plant experiment, about 4 per cent of the radioactivity was found in the shoots. According to these results after incubation in the soil, the plant availability of the unextracted radiocarbon, as compared with the extractable portion, has decreased to about one-third, and in comparison with the availability after direct methabenzthiazuron application to about one-sixth. Desorption studies with water showed similar ratios.

This lecture should demonstrate the use of radioactive labelling in investigations on balance and metabolism of pesticides in the soil–plant system. More detailed information on this topic is given in two additional publications prepared for an Indo-German seminar on management of pesticide application in agriculture,[13,14] and more recent publications included specific descriptions of autoradiographic methods.[15,16]

REFERENCES

1 Kramer, H.H. (1974). *Pflanzenschutz und Welternte CIFAP*, **9**.
2 Metzger, H. (1972). *Chemie und Fortschritt*, **2**, 3.
3 Führ, F., Cheng, H.H., and Mittelstaedt, W. (1976). *Landw. Forsch. SH*, **32**, 272.
4 Führ, F. and Mittelstaedt, W. (1976). *Landw. Forsch. SH*, **32**, 286.
5 Cheng, H.H., Führ, F., and Mittelstaedt, W. (1975). In, *Environmental Quality and Safety: Pesticides*. (Eds F. Korte and F. Coulston) p. 271.
6 Führ, F. (1975). *Jül-Bericht*, **1234**, August.
7 Cheng, H.H., Führ, F., Jarczyk, H.J., and Mittelstaedt, W. (1978). *J. Agric. Food Chem.* **26**, 595.
8 Mittelstaedt, W., Still, G.G., Dürbeck, H., and Führ, F. (1977). *J. Agric. Food Chem.* **25**, 908.
9 Kaufman, D.D. (1976). In: *Bound and Conjugated Pesticide Residues*. ACS Symposium Series, **29**, 1.
10 Führ, F., and Haider, K. (1977). In, *Soil Organic Matter Studies*, Vol. II, p. 333, IAEA, Wien.
11 Führ, F., and Mittelstaedt, W. (1980). *J. Agric. Food Chem.*, **28**, 122.
12 Führ, F., and Mittelstaedt, W. (1979). *Z. Pflanzenernährg. + Bodenkde.* **142**, 799.
13 Führ, F. (1978). In, *Indo-German Seminar on Management of Pesticide Application in Agriculture*, Jül-1534, 39, ISSN 0366-0885.
14 Führ, F. (1978). In, *Indo-German Seminar on Management of Pesticide Application in Agriculture*, Jül-1534, 71, ISSN 0366-0885.
15 Führ, F. (1984). Rhein.-Westf. Akademie d. Wiss. Vorträge N326, Westdenberger Verlag, Opladen, 5.
16 Führ, F. (1984). In: *Isotopes and Radiation in Agricultural Sciences*, Vol. II, M. L'Annunziata and J. Ordegg (Eds), 239.

Pollutants and Their Ecotoxicological Significance
Edited by H. W. Nürnberg

CHAPTER 24

Behaviour of Organochlorine Compounds in Terrestrial Ecosystems

W. KLEIN*

INTRODUCTION

For several decades organochlorine pesticides have been heavily applied to agricultural land throughout the world. It is mainly this use which led to the identification of environmental problems with chemicals in general.

Although organochlorine pesticide use has been largely restricted throughout the world and especially in the countries of the European Communities, it must be anticipated that they are still present in large amounts in agricultural soils. Further arguments to emphasize their importance is the fact that there is still some use in Europe, e.g. hexachlorobenzene in seed-dressing. Furthermore, only a few pesticides are classified as organochlorines: DDT, cyclodienes, toxaphene and hexachlorobenzene. Nevertheless, almost one-third of all pesticides contain halogenated moieties. Therefore, scientifically they could be included in this class of chemicals, and it is not appropriate just to identify areas with previous heavy application of the classical organochlorines but to consider the whole spectrum of halogenated agricultural pesticides for appropriate scientific evaluation.

More complicated than the significance of halogenated pesticides in terrestrial ecosystems are organochlorine compounds in general. It is well established that the flora, especially the forests, have an important filter function with respect to air pollutants like sulphur dioxide (SO_2) etc. Although it is known that large amounts of organochlorine chemicals, e.g. chloroethylenes used in dry-cleaning, are emitted into the atmosphere, there is little information on the potential of terrestrial ecosystems (forests) to absorb these chemicals from the atmosphere.

One exception is some knowledge on the occurrence of PCBs in many compartments of the environment although they have never been used

* With unpublished contributions by F. Fragiadakis, P. N. Moza, D. Vockel, I. Scheunert and U. Weiß.

Table 24.1 PCB in flora

Sampling near Autobahn Pfaffenhofen or Irschenberg		Sampling remote area Alps, Tegelberg, or Hüttensee	
Plant	Dry weight (ppm)	Plant	Dry weight (ppm)
Moss	0.13	Moss	0.06
Heather	0.28	Heather	0.08
Raspberry leaves	0.15–0.27	Raspberry leaves	0.11
Larch leaves	0.95	Larch leaves	0.17
Rushes	0.12	Rushes	0.07
Hay	0.13	Fern	0.08–0.15
Grass	0.20–0.46	Juniper	0.07
Grass silage	0.40–0.42	Dwarf fir	0.09–0.10
Clover	0.33	Alder leaves	0.09
Mint	0.20		
Strawberry leaves	0.16–0.46		
Blackberry leaves	0.49		
Fir apex	0.31		
Oak foliage	0.50		
Pine leaves	0.11		
Water hemlock	0.23–0.57		

Source: Muller and Korte.[24]

intentionally and in large amounts in the open environment. The occurrence of PCBs in aquatic systems can be explained by input via sewage or landfill leachates. The occurrence in different flora species as shown in Table 24.1 can only be explained by aerial transport and absorption by plants and soil.

The fact that the same chemicals are found in polluted regions as well as in remote areas demonstrates that this type of chemicals represents not only local problems and that approaches to their evaluation should be over-regional. This is quite a difference to the agricultural chemicals where initial concentrations

Table 24.2 Waste incineration: list of selected chemicals detected in fly ash

Tetrachloroethylene	Hexachlorobenzene
Tetrachloroethane	Hexachlorocyclohexane
Tetrachloropropene	Heptachlor
Pentachloroethane	Aldrin
Hexachloroethane	Dieldrin
Pentachloropropene	Isodrin
Trichlorobenzenes	Chlordane
Tetrachlorobenzenes	Chlordene
Hexachlorobutadiene	Kepone
Pentachlorobenzene	Mirex

following intentional use might be higher by a factor of 1000 than after aerial transport. Apart from evaporation from water and land, waste incineration may be a source of many halogenated chemicals upon fall-out on terrestrial systems (Table 24.2).

IMPORTANCE OF TERRESTRIAL ECOSYSTEMS WITH RESPECT TO THE FATE OF CHEMICALS

In waste technology, the major factors to finally eliminate chemicals are microbial degradation in sewage treatment plants and burning of solid waste. This leads to the assumption of micro-organisms to have in general a high potential for the degradation of chemicals. Realizing, however, that the biomass of micro-organisms is minute as compared to green plants, their potential in the balance of the fate of chemicals needs emphasis. Global estimates of biogeochemical data are still uncertain. Nevertheless, some data are given in Table 24.3 to show the correlations of biomass of terrestrial organisms.

If, for instance, in a given ecosystem the biomass of green plants is 1000 fold as compared to the biomass of micro-organisms, the capacity for degradation would be identical for plants and microbes even if the turnover per mass unit would be 1000 fold higher in microbes. According to the present evidence, however, conversion in plants is quite efficient even for chlorinated chemicals so that higher plants really play an important role as to the degradation of chemicals in terrestrial systems.

Table 24.3 Global biomass in terrestrial ecosystems

	Biomass (10^9 t)
Forests	400–670
Woodland	100–135
Dwarf and open shrubs	6.0–13.0
Tundra	3.0–4.0
Alpine tundra	1.0
Grassland	30–34
Semi-desert	6.5
Cultivated land	7.0
Swamps and marshes	10–15
Urban area	1.0
Total	450–770
Animals	2
Men	0.02
Micro-organisms	1.5
(Soil organic matter: 2070)	

Source: SCOPE.[29]

DEFINITIONS

Some misunderstandings in the interpretation of data regarding the topic of this paper might evolve from the different understanding of the used terms. *Degradation*, for example is frequently used in the context of evaluating chemicals and represents the contrary to persistence. Many authors define 'degradation' as disappearance of an industrial chemical or the chemical under investigation. Since ecotoxicological data on a larger structural variety of chemicals are missing, it is inevitable to include conversion products in the evaluation. For organic chemicals this means that degradation should be used for mineralization or the formation of naturally abundant products only. Changes of an investigated molecule should either be named as conversion or metabolism or abiotic change as appropriate.

Furthermore, different contents might be understood for the terms *behaviour* and *fate*. Since true ecotoxicological investigations are still in their early methodological development, data on the behaviour and fate of chemicals are crucial for actual judgement and, therefore, it is most important to know what they mean. In the context of this paper both terms are used synonymous for the summarizing description of biotic and abiotic conversion as well as physical and biological transport processes. The term behaviour does not include any indication of biological effects.

It should be noticed that *ecotoxicological effects* being difficult to assess are at present described by composites of the terms 'behaviour' and 'fate', e.g. the exposure probability, or by laboratory test data on single organisms. With reference to terrestrial systems, these descriptions could be 'accumulation in birds' or 'toxicity to earth worms'.

CONCEPTS FOR INVESTIGATION

Final investigations on the ecotoxicological effects of chemicals in terrestrial ecosystems have to take into account effects on populations dynamics, and population densities of flora and fauna species. Due to the large natural variations of these parameters and the lack of exact data on these variations, at present only tests with indicator organisms are performed without definitively knowing their predictability to the real environment.

Conclusive data can be elaborated on the other hand with available methodology to assess the exposure levels of any organisms. The ideal experimental approach to this would be a large-scale monitoring for any chemical with toxicological potential. Since this is an unaffordable effort, the next best methodology would be to carry out field experiments with the respective chemicals. This is done for pesticides but cannot be done for all halogenated chemicals. Furthermore, field studies do not supply data on conversion products of the used chemical since generally the needed trace analytical methodology is not available. Therefore, an appropriate approach for a larger range of chemicals is to carry out open-air lysimeter experiments with isotopically labelled chemicals. In the future, the obtained data should be used to modify simple laboratory test systems in order to obtain predictive data with respect both to quantities and identity of chemical individuals.

In recent years, a number of laboratory models with a wide range of sophistication have been developed for terrestrial, aquatic, and mixed ecosystems. Due to the limited space and variety of species, at present these models can only give indications on accumulation and conversion but no quantitative data.

EXPERIMENTAL RESULTS

In former years, field experiments and residue analyses in crops had mainly been used to assess residues of the parent compounds and their disappearance. Today, however, it is evident that it is not sufficient to investigate only the distribution kinetics—including uptake from the soil by plants as well as excretion into soil, distribution within the plants as well as that between soil layers, evaporation into the air, and leaching out of the soil by rain—or irrigation water—of the parent compound, but also that of conversion products formed within the system.

EXPERIMENTAL MODELS FOR QUANTITATIVE RESIDUE ASSESSMENT AND LONG-TERM STUDIES

Laboratory tests on the uptake of chemicals from soil by plants are not predictive for the environment, since the uptake is much higher than in the environment when small laboratory pots are used. Under open-air conditions, the uptake of residues by plants is less than 1 per cent of residue present for the cyclodiene insecticides and other chemicals tested. Although there is strong evidence that plants actively contribute to the disappearance of residues from soil, the reduction of soil burden by harvesting is negligible.

When lysimeters of $60 \times 60 \times 60$ cm size are used under outdoor conditions with ^{14}C-labelled model substances, the quantitative data on uptake and residues in plants and soil are comparable to field data and, therefore, relevant for practical evaluation. Such a model is used since 1969 in our institute. It is shown in Figure 24.1,[28] and will be discussed in some detail.

The plants are grown in boxes $60 \times 60 \times 60$ cm constructed from water-resistant plywood. The base of the box contains holes to permit the drainage of excess water which is collected in a metal splash tray. The box is wrapped in aluminium foil on the outside to prevent temperature increases from the sunlight. The bottom 25 mm of the box is packed with stone chips of near 25 mm diameter, and the stones are covered with a 25 mm layer of well-rotted turf. The box is filled with soil to 1 cm from the top. The soil is allowed to settle for 1 month

Table 24.4 Comparison of aldrin/dieldrin residues after aldrin soil application in field trials and ^{14}C experiments

Crop	Field Trials (ppm)				^{14}C experiments ($3\,kg\,ha^{-1}$), mean value
	Application rate ($kg\,ha^{-1}$)	References	Mean value or range	Highest value	
Potatoes, whole	2–4	Shell, private communication, 1968	0.04–0.06	0.10	0.17
	2–4	Shell, private communication, 1974	0.08	0.24–0.27	
Potatoes, peeled	2–4	Shell, private communication, 1968	0.01–0.03	0.03	
	3–4	Lee, 1968/UK Ministry of Agriculture	0.01	0.14	
Potatoes, whole (treated one year prior to planting)	2.5	Shell, private communication, 1974	0.04	0.11	0.10

Sugar beet,	2–4	Shell, private communication, 1974	< 0.01–0.05	0.05	0.04
whole	2–3	Onsager *et al.*[26]	0.06	0.06	
Sugar beet, whole (treated one year	2–4	Shell, private communication, 1974	< 0.01	0.02–0.03	0.02
prior to planting)	2–3	Onsager *et al.*[26]	0.01	0.01	
Sugar beet leaves	1–6	Shell, private communication, 1974	≤ 0.01	0.01	≤ 0.01
Maize kernels	1–6	Shell, private communication, 1974	< 0.01–0.02	0.02	< 0.01
Maize leaves, dry	1–6	Shell, private communication, 1974	0.01–0.07	0.07	0.04
Wheat grains	1–6	Shell, private communication, 1974	< 0.01–0.01	0.01	< 0.01
Wheat straw	2–4	Shell, private communication, 1974	< 0.01–0.16	0.16	0.10

Fig. 24.1 Experimental set-up for studies of ^{14}C-labelled pesticides in the soil-plant system under outdoor conditions

before planting. The box is sunk into a large pit such that the upper surface of the soil is level with the surrounding ground. The soils are typical for growing of the crops used. Fertilizers are applied as in agricultural practice. Air temperature, humidity, and pressure, as well as rainfall, are recorded during the vegetation period.

COMPARISON TO FIELD DATA

For a critical evaluation of this outdoor ^{14}C experimental model, the data must be compared to those obtained from real agricultural field conditions. Again, aldrin was used as a model substance. Since analytical methods for unlabelled residues are available only for aldrin and dieldrin, not for the majority of hydrophilic or unextractable residues, comparisons can be undertaken only for these two substances. Nevertheless, these are sufficient for a rough estimate on the usefulness of these model experiments to predict to agricultural conditions. Direct comparison of soil residue data on a ppm basis after equal dosage are not significant since in agriculture the dose is applied by spraying, and a considerable portion is drifted off before reaching the soil surface. An identical application method is not possible for radiolabelled substances. However, when the residues in the samples under consideration are expressed as a percentage of the initial residue in the soil immediately after the application, data are obtained which are relatively independent on the dose, but depend on soil texture and climate. These data are comparable to those obtained from our ^{14}C-model experiments.

A series of field studies on soil residues was conducted.[3] The mean value obtained from these studies was lower than that from our ^{14}C experiments, but, since deviations between individual field data were high, the ^{14}C experiments are not outside the range of values. The slightly raised residues in the ^{14}C experiments may be due to the application method. Since field application is carried out by spraying, the insecticide may evaporate faster from the small droplets than from the normal drops in the ^{14}C experiment.

In the same study,[3] the mean percentage of dieldrin (based on the sum of aldrin and dieldrin) was found to be between 20 and 66 per cent after 1 year, dependent on the soil type. The corresponding mean value of our ^{14}C experiments was 41 per cent dieldrin.

Comparisons of crop residues between ^{14}C experiments and field experiments are shown in Table 24.4. The table shows, for the ^{14}C experiments, slightly higher data due to the higher soil residues discussed above, but they are within the range of field data. This small deviation of the results of the ^{14}C-model experiments from agricultural practice may be neglected in comparison to the great advantage of the model in the identification and assessment of conversion products for which analytical detection methods are not available.

QUANTITATIVE BALANCES

More than twenty chemicals were tested in this experimental model. For [^{14}C]aldrin, the experiments were continued for more than 10 years after soil treatment. When looking at the material balance of the system after 5 years, Table 24.5

Table 24.5 Material balance in plants and soil 5 years after application of [^{14}C]aldrin to soil

	(%)
Applied in 1969	103 mg (100)
Residue in soil 1973	34.5
Taken up by plants 1969	0.1
Taken up by plants 1970	0.2
Taken up by plants 1971	0.1
Taken up by plants 1972	0.1
Taken up by plants 1973	0.1
Leaching water 1969	1.6
Leaching water 1970	4.9
Leaching water 1971	2.4
Leaching water 1972	1.3
Leaching water 1973	0.4
Recovery after 5 years:	45.7
Volatilization within 5 years:	54.3

Table 24.6 Residues of [^{14}C]aldrin and conversion products in soil 5 years after treatment (percentage of applied radioactivity)

Sample	Aldrin	Metabolite X (unidentified)	Dieldrin	Photodieldrin	Extracted hydrophilic metabolites	Unextractable residue	Total
Soil depth 0–10 cm	0.8	<0.4	11.7	0.4	1.9	4.3	19.4
Soil depth 10–20 cm	0.4	<0.4	5.0	<0.4	1.2	1.9	8.9
Soil depth 20–30 cm	<0.4	<0.4	1.5	<0.4	0.4	1.2	3.1
Soil depth 30–40 cm	<0.4	<0.4	1.5	<0.4	0.4	1.2	3.1
Soil total	1.2	0.4	19.7	0.7	3.9	8.6	34.5

demonstrates that within 5 years half of the radioactivity applied was lost by volatilization.

The identity of the volatilized radioactivity was not investigated in this study (Scheunert *et al.*, unpublished). Thus volatilization is a major pathway for residue loss in ecosystems, and the uptake by plants and the leaching by water are of minor importance. For chemicals other than aldrin, the trend seems to be comparable. Little is known of the chemical nature and the fate of volatilized residues; they may be the unchanged compounds, conversion products, or carbon dioxide resulting from total mineralization. When looking at the

Table 24.7 Unextractable residues[a] in soil (0–10 cm depth) following soil treatment with ^{14}C chemicals (in percentage of radioacitivity present)

Chemical applied	Unextractable after 1 vegetation period	Unextractable after 2 vegetation periods
Dieldrin	<1	3
Kepone	2	2
Kelevan	8	12
Aldrin	11	18
Imugan	32	68
2,2′-Dichlorobiphenyl	42	74
3,4-Dichloroaniline	84	89
Allylalcohol	91	not analysed
p-Cl-aniline	95	92

[a] Unextractable by 48 h extraction with methanol, acetone or chloroform in a Soxhlet.

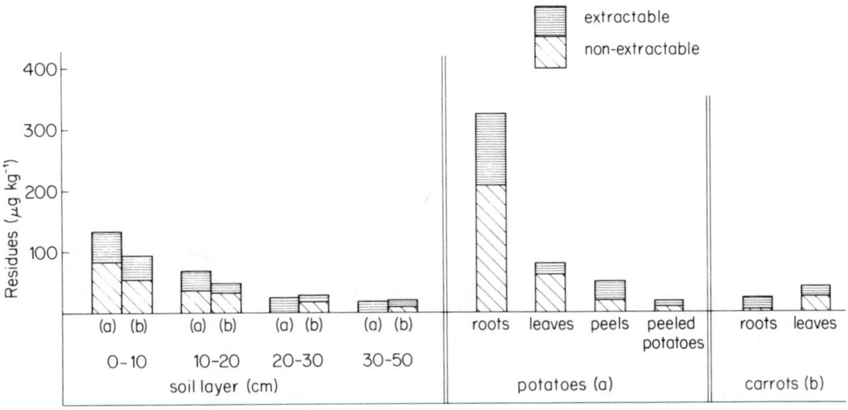

Fig. 24.2 Residues in soil and crops following soil treatment with [^{14}C]hexachlorobutadien at a dose of 6 kg ha^{-1}. (a) Residue first season, (b) residue second season

chemical composition of soil radioactivity, the significance of soluble metabolites seems to decrease. After 5 years, the only important soluble residue is dieldrin; the 'unextractable residues' have increased and are of higher importance than in the first years after application (Table 24.6).

Table 24.7 shows, for nine ^{14}C labelled chemicals, that the unextractable residues increase with time. (The reason for the slight decrease in the case of *p*-Cl-aniline is not known; probably, it is due to minor differences in experimental working-up procedures.) The table shows also that the portion of unextractable residues is negatively correlated with chemical stability and environmental persistence of at least that of the parent compound.

Hexachlorobutadiene is a chemical which has only been used in agriculture in Eastern Europe but it occurs in animal feed and human food as well as in surface waters. Therefore, its residual behaviour was investigated in the open-air system. Following its high volatilization in laboratory experiments, a dose corresponding to 6 kg ha^{-1} was used. Figure 24.2 shows a summary of the data obtained in two subsequent growing seasons. The high portion of non-extractable residues should be noticed: not only phenols and anilines but obviously also olefines lead to high non-extractable residues.

Pentachlorophenol is used in agriculture mainly to protect rice cultures, in Europe its major use is for wood preservation. Its residual behaviour was studied under controlled phytotron conditions (summary Figure 24.3). Due to the used climate, conversion ratios and total residues in soil are not far from outdoor conditions but the residues in plants are much too high. This is also reflected in the total balance sheet given for two subsequent seasons in Figure 24.4.

Similar to pentachlorophenol, a trichlorophenol, with fungicidal activity, leads to a very high portion of unextractable residues in all analysed samples. Trichlorophenol was investigated under outdoor conditions. Its residues are

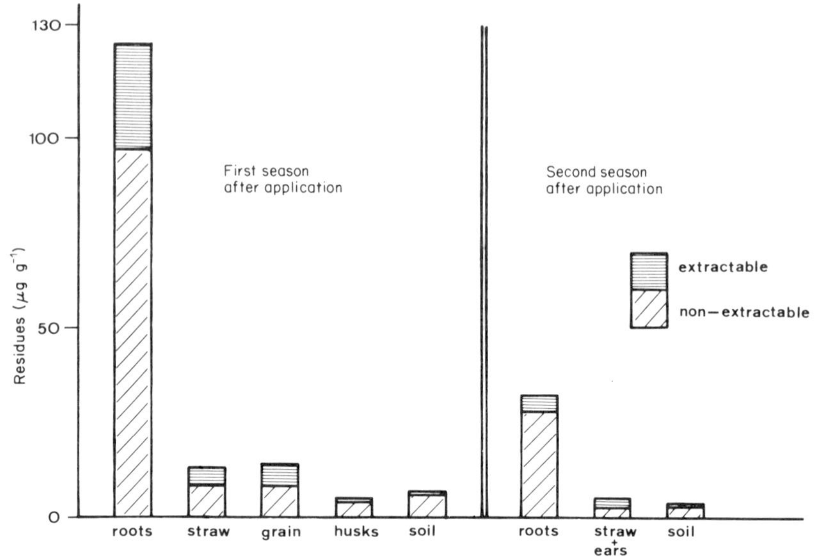

Fig. 24.3 Fate of pentachlorophenol in soil and plants (within two growing seasons, applied dose corresponding to 23 kg/ha)

given in Figure 24.5 and its 2-year balance sheet in Figure 24.6. Although from physicochemical characteristics a higher evaporation would have been expected for the trichlorophenol as compared to pentachlorophenol, it was in fact lower (despite outdoor conditions). The differences in plant uptake by a factor of 100 reflect mainly the different experimental conditions but also different crops.

Since large amounts of trichloroethylene are emitted into the atmosphere, its fate was investigated under outdoor conditions using a high dose. It was expected that it would readily evaporate due to its high vapour pressure. As shown in Figure 24.7, significant residues remain even after two seasons which is explained by adsorption and formation of non-extractable residues, respectively.

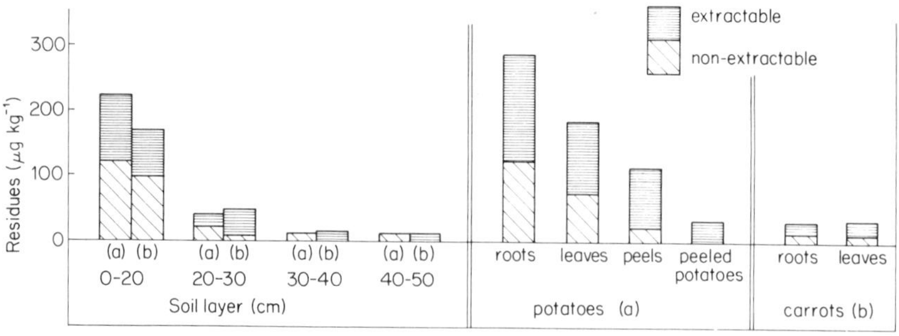

Fig. 24.4

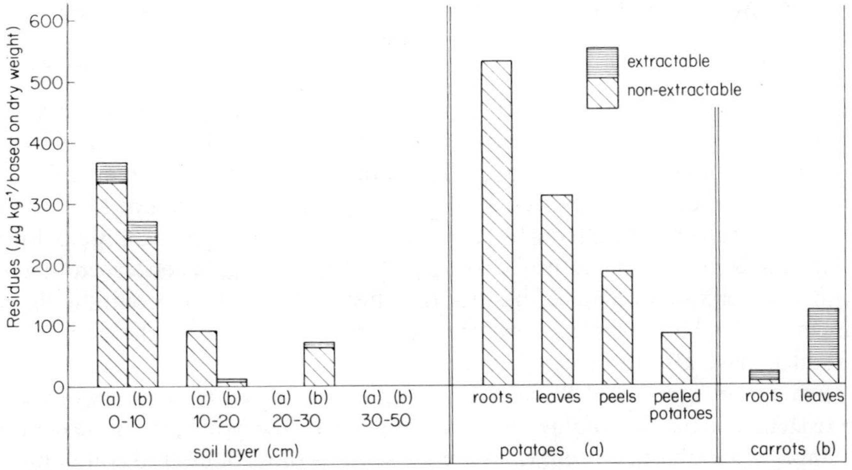

Fig. 24.5 Residues in soil and crops following soil treatment with [^{14}C]trichlorophenol at a dose of 1 kg ha^{-1}. (a) Residue first season, (b) residue second season

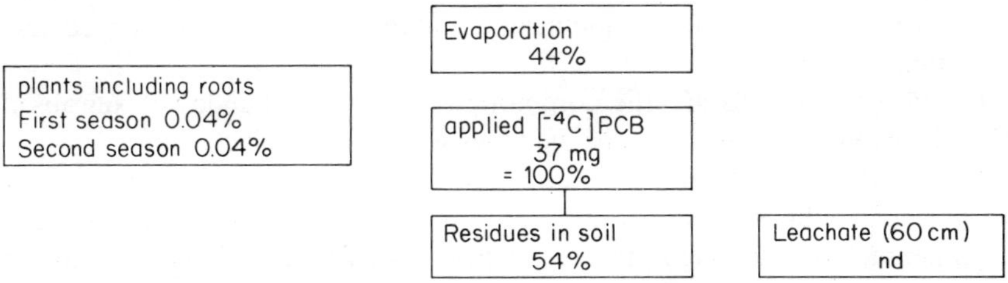

Fig. 24.6 Balance of [^{14}C]2,4,6-trichlorophenol within two growing seasons (outdoor conditions)

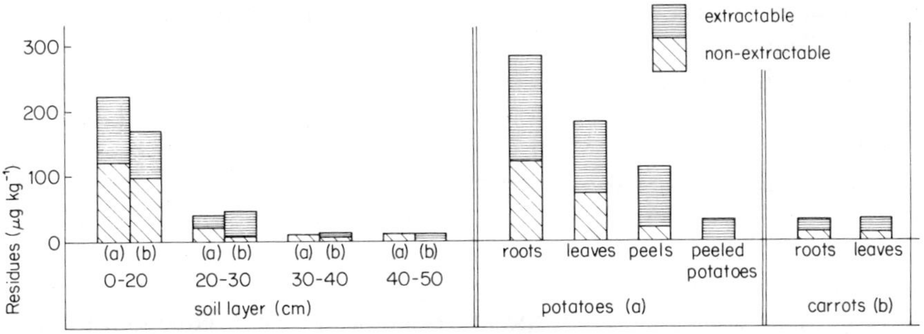

Fig. 24.7 Residues in soil and crops following soil treatment with trichloroethylene at a dose of 5.4 kg ha^{-1}. (a) Residue after first season, (b) residues after second season

METABOLIC PATHWAYS OF PESTICIDES IN THE SOIL–PLANT SYSTEM

In evaluating the enzymatic attack of organisms on environmental organic chemicals, the following situations should be considered: the chemical is easily digested and mineralized without persistent or biologically active intermediates, or it is degraded to low molecular weight fragments circulating in the carbon pool, or it is chemically altered by co-metabolism, without significant breakdown to products joining the natural carbon pool. This chapter is concerned mainly with the last situation since, besides the effects of the parent compounds on organisms and the whole ecosystem, the potential effects of the conversion products have also to be considered.

Evaluation is required for metabolic pathways as well as the amounts and persistence of the metabolites formed. In general, the principles of structural changes in xenobiotics by enzymes are the same as those of natural compounds.

For organic chemicals primary changes, mainly oxidative, hydrolytic or reductive, may be accompanied by an increase in toxicity ('activation') or by a decrease ('detoxication').[12] They are often followed by secondary changes of the primary conversion products, e.g. by alkylation, acetylation, conjugation or binding with biological molecules. Whereas in soil micro-organisms all three primary reaction types are frequent, in the higher plant oxidative processes are prevailing.

Oxidative processes are the commonest enzymatic changes of xenobiotics. Mixed function oxidases are the enzymes generally involved.[2,21]

Hydroxylation of aliphatics and aromatics, epoxidation and oxidative cleavage of double bonds, oxidation of thioethers to sulphoxides and sulphones (e.g. pentachloronitrobenzene, see below), dehydrogenations etc., are well known. Some of these processes, like the epoxidation of cyclodienes, result in a biological activation. Plant tissues are rich in peroxidase, and the abundance of phenolic products in higher plants suggests generally high oxidase activity,[2] with the result that the occurrence of metabolites with increased toxicity in plants must be taken into account; on the other hand, oxidative processes may lead to a stepwise degradation of the foreign compound. For instance, for cyclodienes (aldrin, isodrin, heptachlor), epoxidation or hydroxylation of the double bond is followed by ring cleavage and by loss of carbon atoms upon decarboxylation.

Hydrolytic conversions, e.g. by esterases and amidases which are widely distributed in nature, are also well known, as for aliphatic exters (e.g. kelevan[27]), or cyclodiene epoxides. Reductive conversions are known for xenobiotics containing nitro groups (e.g. the fungicide pentachloronitrobenzene) which are reduced to amines. The reduction of pentachloronitrobenzene to the corresponding aniline is one of the few examples where a reduction occurs in plants at relatively high rates (see below). Reductive dechlorination, like that of DDT to DDD, may also occur but only at low rates.[34] The secondary processes, in general, may be regarded as a pathway to elimination. To a limited extent, this could also be valid for the 'unextractable residues' in plants and soil. In plants, detoxication is achieved by fixation of the xenobiotic within natural macromolecular structures like cell-wall components.[10] This results in the so-called 'unextractable residues', i.e. the incorporated xenobiotic cannot be extracted from the tissue without destruction of the macromolecular bonds. In soil, even more complex reactions occur: the xenobiotic may be involved in humus formation and replace natural constituents in the humic acid macromolecule.[9] However, very little is known of the biological availability of the incorporated chemicals, i.e. whether they are taken up by animals. Even when the xenobiotic is fixed in the soil as a constituent of the polymerisate of humic acids, the possibility that it may become available again for plant uptake cannot be excluded, since the humic acids are constantly undergoing biosynthesis and degradation. Further work on this topic is urgently needed.

The photochemical influence on the conversion of pesticides in the soil–plant system is not clear. A large number of photochemical conversion products have been identified in the laboratory, but only a few have been isolated from environmental samples. One of these is photodieldrin, but its exclusively photochemical origin is still under discussion, as reported later.

The first example of metabolic conversion pathways of a pesticide which I will present here is the insecticide *aldrin*. In outdoor experiments, [^{14}C]aldrin was applied to soil at an application rate of about 3 kg ha^{-1}, and various crops were planted.[28] After the harvest, plants, soil samples from different depths, and the leaching water drained from the experimental boxes at a depth of > 60 cm were analysed, and the chemical structure of conversion products was established.

One main product in plants and soil was identical to dieldrin (Figure 24.8) which has been known to be an aldrin metabolite in various organisms and soil for many years.[4,19]

A considerable portion of the radioactivity present in plants and soil was a mixture of very hydrophilic substances, the main compound of which was found to be *dihydrochlordene dicarboxylic acid*. This substance, a product resulting from oxidative ring cleavage of aldrin (Figure 24.8), has been observed also in rabbits treated with the major dieldrin metabolite in animals, aldrintransdiol,[16,25] (Figure 24.8); later on, dihydrochlordene dicarboxylic acid was detected also directly after dieldrin application to animals.[1] In the leaching water of these outdoor experiments, which was drained from the experimental boxes at a depth of about 60 cm, dihydrochlordene dicarboxylic acid constituted more than 90 per cent of the radioactivity present.[23]

The other substances from the hydrophilic group could not be identified in soils and plants. In the leaching water, however, a hydrophilic byproduct was identified as *aldrin-trans-diol* (Figure 24.8), the major animal metabolite of aldrin[18] and dieldrin[17] mentioned above. It cannot be excluded that this substance is also a byproduct in the hydrophilic metabolite group of plants and

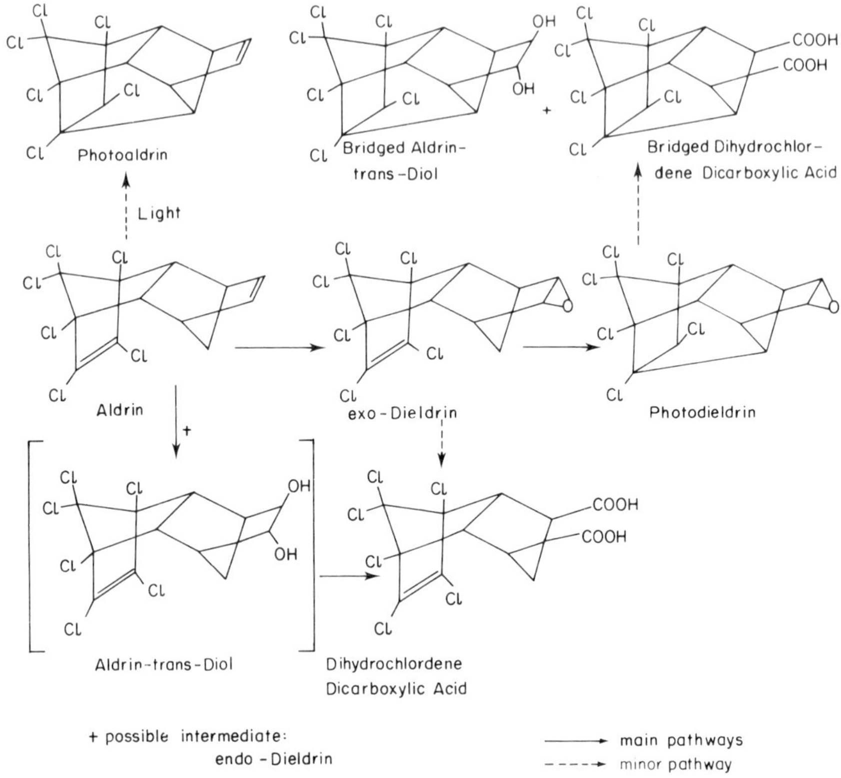

Fig. 24.8 Degradation pathways of aldrin in the soil–plant system

soils; the attempts to detect it in these extracts, however, failed due to high amounts of biological byproducts which occurred much less in water extracts (i.e. detection limit in water is better than in soils or plants).

The known photoisomerization product of dieldrin, *photodieldrin* (Figure 24.8), was found preferably in leaves or straw (5–10 per cent of the total residues); smaller amounts were detected in soils, especially in soil with wheat or maize. The dependence of photodieldrin formation in soil upon the plants may be explained by the different capabilities of the plants to shade the soil or by metabolic reactions; the question cannot be decided in experiments under normal outdoor conditions. The theory of metabolic photodieldrin formation is supported by the following observation. Since photodieldrin is not a photoproduct of aldrin, but of its conversion product dieldrin, the photoproduct of aldrin itself, photoaldrin (Figure 24.8), should be found in higher concentrations than photodieldrin when aldrin is applied to soil; but this did not occur. Photoaldrin was not at all detected in plants. In soils, it was observed only in traces (<0.5 per cent of the total residues); higher concentrations (up to 5 per cent) were found only in one soil sample. It should be mentioned also that metabolic formation of photodieldrin has been reported for micro-organisms.[22]

Besides these metabolites, plant and soil extracts contained, in varying concentrations, radioactive products which were *not extractable with organic solvents*. Their percentage of total radioactivity was dependent on the kind of soil and on depth, that means, it was higher in deeper soil layers. It was low in most of the plant samples. From a combined soil sample, 55 per cent of these residues could be solubilized with dilute ammonia or sodium hydroxide, and this solubilized portion was identified as dihydrochlordene dicarboxylic acid. It may be concluded that the unextractable radioactivity is neither aldrin nor dieldrin, but hydrophilic metabolites which are bound to soil complexes (e.g. humic acids), cell-wall constituents (e.g. lignin) or other macromolecules, as discussed above. Furthermore, it cannot be excluded that part of the unextractable radioactivity are normal biological plant constituents resulting from assimilation from $^{14}CO_2$ or other low-molecular degradation products.

The metabolites identified thus far from outdoor experiments indicate at least two independent major metabolism pathways (besides the abiotic minor pathway to photoaldrin); these are shown in Figure 24.8. Both pathways are started by an attack on the most reactive molecule site, i.e. the non-chlorinated double bond. The first metabolism pathway is the epoxidation by mixed function oxidases resulting in dieldrin which may be rearranged by light and/or metabolism to form photodieldrin. Both epoxy groups—that of dieldrin and of photodieldrin—are chemically stable and react only very slowly. The hydrolysis of dieldrin to form aldrin-trans-diol—a normal metabolic pathway for mammals[1,5,17]—has not been demonstrated in plants or normal soils thus far; the hydrolysis of photodieldrin to form the corresponding bridged diol in soil is only 1 per cent within 1 year.[31] Similarly, the dihydrochlordene dicarboxylic acid is formed from dieldrin very slowly,[14] as is the formation of the corresponding bridged acid from photodieldrin.[31] These data suggest that the first metabolism pathway of aldrin in soil and plants—the epoxidation—has a 'dead end' in respect of skeleton degradation.

In contrast to dieldrin, aldrin is decomposed to dihydrochlordene dicarboxylic acid in relatively high rates by the second metabolism pathway—without intermediacy of dieldrin. The intermediacy of aldrin-trans-diol in this pathway was confirmed experimentally. After application of the [^{14}C]diol to plants, 8 per cent of the recovered radioactivity had been converted to dihydrochlordene dicarboxylic acid within 4 weeks, and after application to soil, more than 80 per cent.[11] The high conversion rate of the trans-diol could be the explanation by the fact that it is not found in soils or plants. Thus, the second major metabolism pathway of aldrin implies the oxidation of the double bond to form the diol, followed by oxidative ring cleavage to form dihydrochlordene dicarboxylic acid. Of course, there could exist additional possibilities for the formation of the acid.

The oxidative ring cleavage was also observed for other cyclodiene insecticides,

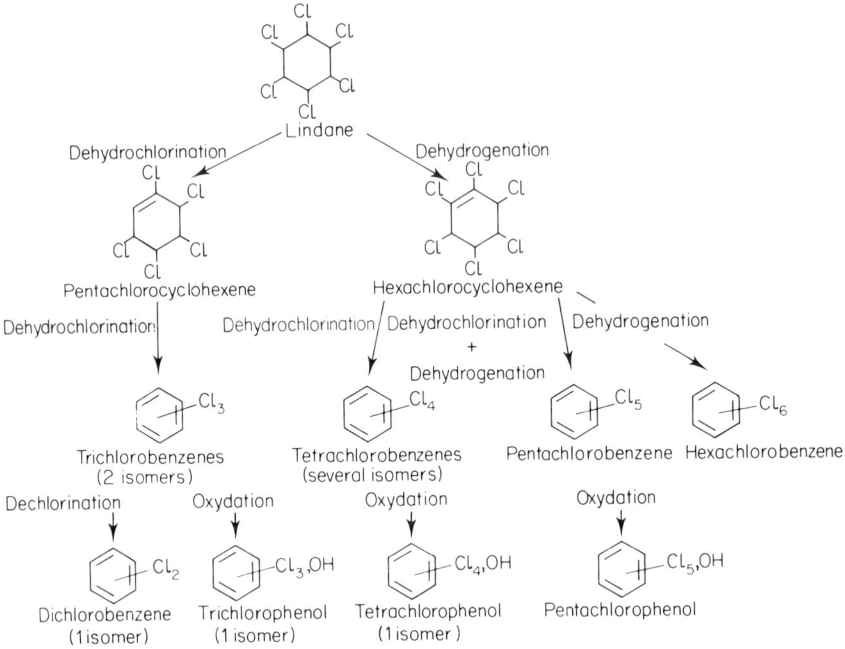

Fig. 24.9 Conversion pathways of lindane in the soil–plant system

e.g. for isodrin[33] or heptachlor.[32] In these cases, however, the resulting dicarboxylic acids were unstable and were degraded further by losses of carbon atoms. A stable final product for both insecticides was—among others—Prill's acid.

The next pesticide to be discussed will be *lindane*. Following its agricultural use, [^{14}C]lindane was applied to the leaves of lettuce or endives in the open air in two successive years; a laboratory hydroponic culture experiment was also carried out. Figure 24.9 shows the conversion products identified and their formation pathways.[13,15]

The abundance of metabolites, most of which have also been found in other organisms, shows that lindane metabolism is more complex than was assumed in former years. Some years before, it was assumed that the dehydrochlorination via pentachlorocyclohexene was the only pathway of lindane metabolism. However, the tetra- and pentachlorobenzenes and -phenols isolated by several authors were hardly explainable by this pathway. In the experiments discussed here, another intermediate has been found—hexachlorocyclohexene—which is formed from lindane by dehydrogenation without chlorine elimination. Via this intermediate, the formation of tetra-, penta- and hexachlorobenzene and their respective phenols is easily explainable by dehydrochlorination and/or further dehydrogenation. Dechlorination also takes place—dichlorobenzene is formed from trichlorobenzene.

The occurrence of hexachlorobenzene as a lindane metabolite needs a short discussion. Hexachlorobenzene was found to be radioactive, as were all other metabolites and, therefore, is indeed a lindane metabolite and no random impurity from general environmental contamination. The occurrence of hexachlorobenzene as a metabolite of lindane is interesting with respect to environmental hexachlorobenzene contamination. It is known that hexachlorobenzene occurs in food, especially when it is of animal origin, in human fat tissue, and in milk. The amounts are higher than are expected from its agricultural use as a fungicide. It is also of interest that, contrary to all other environmental chemicals tested in this laboratory, [^{14}C]hexachlorobenzene was found to undergo no detectable conversion in plants or soil within one vegetation period.[7]

Comprehensive world production figures for hexachlorobenzene are not known. However, it is anticipated that its annual release into the environment from use as a fungicide, as a waste product from industrial syntheses like chlorine manufacture, benzene- and hydrocarbon-chlorination, as a minor by-product in pentachlorophenol and pentachloronitrobenzene, and as a reaction product from sewage disinfection by chlorination, altogether is similar to the total annual use of lindane. Under this assumption, the data given here would indicate that the actual contribution of lindane use of the environmental load with hexachlorobenzene is well below 1 per cent.

Pentachloronitrobenzene is a fungicide used for a number of vegetables, for soil and seed treatment. Figure 24.10 shows its metabolism in onions.[35] The primary enzymatic attack is that on the nitro group which is either reduced or eliminated and replaced by the hydroxyl or —SR group. Secondary reactions like methylation, formylation, acetylation or oxydations follow. Reductive dechlorination seems to be a minor pathway.

Thus, the *total* residues of non-persistent compounds may be similar to those of persistent ones. Table 24.8 compares the urea herbicide buturon[8] to the cyclodiene aldrin, both in outdoor experiments in the soil–plant system. The high portion of unextractable degradation products of buturon seemed to be due to its primary metabolite *p*-chloroaniline. Therefore, [^{14}C]*p*-chloroaniline was also studied in the soil–plant system under outdoor conditions. As expected, the

Table 24.8 Comparison of disappearance and residual behaviour of aldrin and buturon within one vegetation period after use

	Aldrin	Buturon
Percentage of residues in top 10 cm soil layer based on applied ^{14}C	55.5	42.4
Percentage of parent compound in total residue, top 10 cm soil layer	40.6	31.8
Percentage of non-extractable portion of residues in top 10 cm soil layer (solvent: methanol)	7.6	53.5
Percentage of residues in leaching water based on applied ^{14}C	1.9	0.3

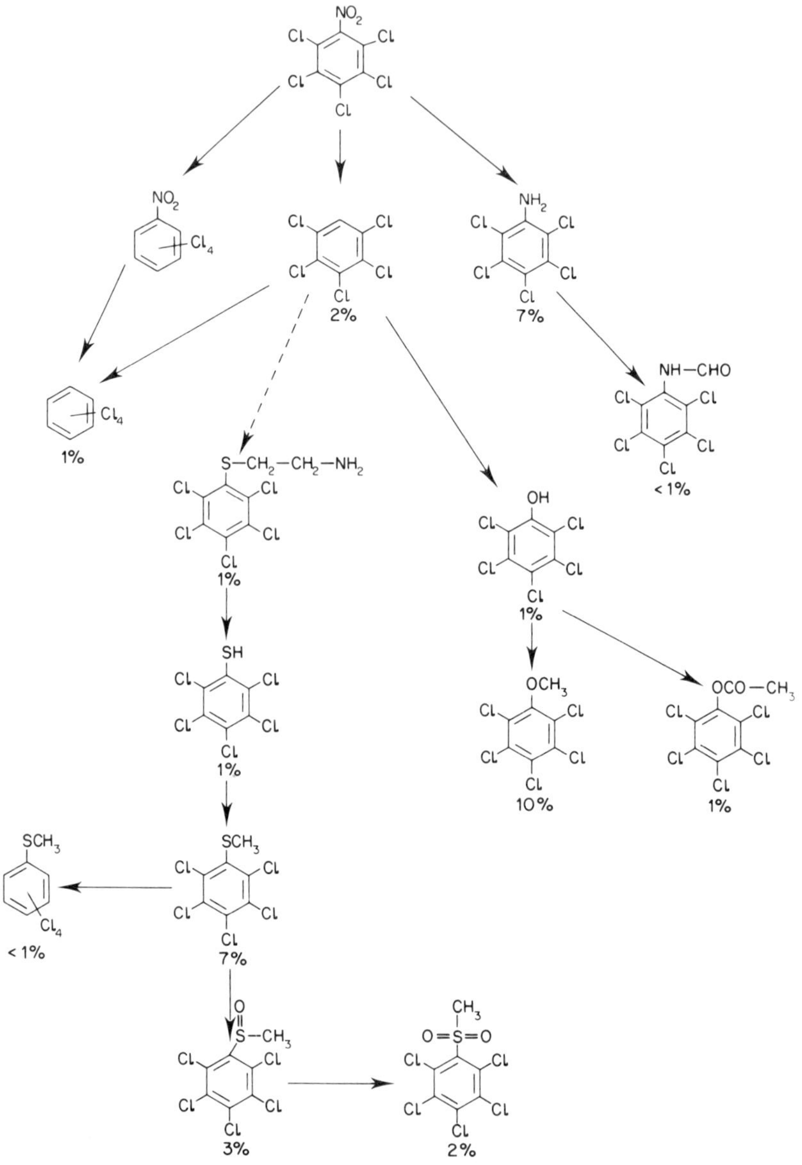

Fig. 24.10 Conversion pathways of pentachloronitrobenzene in onions

Table 24.9

Metabolites number	Substance	Formula	M^+ in mass spectrometry	Rf-values[a] (1) Hexane/benzene/acetone (7:3:1)	Rf-values[a] (2) Benzene/ethyl-acetate (3:1)	Occurrence: Soil	Occurrence: Barley
I	3,3',4,4'-Tetrachloro-azobenzene	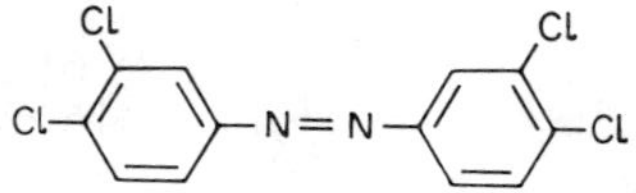	318	0.86	0.92	+	(+)
II	3,4-Dichloro-formanilide	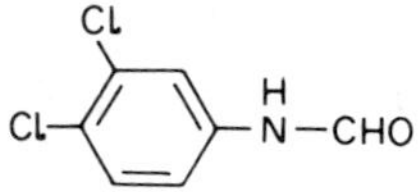	189	0.09	0.25	+	–
III	3,4-Dichloro-acetanilide	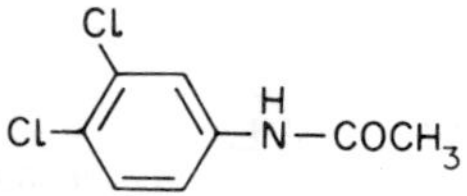	203	0.09	0.22	+	(+)
IV	Unidentified		249 (2Cl)	0.09	0.25	–	+
V	Unidentified		224 (2Cl)	0.09	0.22	+	–
VI	6-Hydroxy-3,4-dichloro-acetanilide	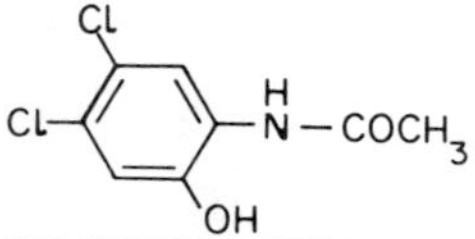	219	0	0	(+)[b]	–

a = TLC on Merck silicagel G-60 plates, 0.25 mm thickness
b = Analysed as methylated derivative, Rf 0.13 in solvent (1) and 0.31 in solvent (2)
+ = Present
– = Not detected
(+) = Identified by comparison with authentic sample only on TLC

Table 24.10 Dynamics of organic xenobiotics in cultivated soil—radioactive residues in agricultural soil, 1–10 years after application of ^{14}C-labelled chemicals (in mg kg^{-1}, equivalent to the parent compound)

Chemical	Experimental time (years)	Application rate (kg ha^{-1})	Residues (0–10 cm depth), after one season				Residues (0–10 cm depth), after the total exposure time			
			Total	Parent compound	Soluble conversion products	Unextractable residues	Total	Parent compound	Soluble conversion products	Unextractable residues
Aldrin	10	2.9	1.88	0.87	0.80	0.21	0.35	<0.01	0.24	0.11
Aldrin	10	2.9 1st year +3.5 2nd year	1.58	0.78	0.64	0.16	0.65	0.01	0.47	0.17
Dieldrin	8	0.8	2.63	2.58	0.03	0.02	0.15	0.10	0.02	0.03
p-Cl-aniline	7	1.4	0.40	<0.01	0.02	0.38	0.22	<0.01	0.02	0.20
Metribuzin	7	1.0 1st year +1.0 2nd year	0.32	0.03	0.06	0.23	0.20	<0.01	0.01	0.19
Kepone	6	1.2	0.87	0.85	0.01	0.01	0.51	0.49	<0.01	0.02
Monolinuron	6	2.5 1st year +1.9 2nd year	0.58	0.09	0.06	0.43	0.22	0.01	0.01	0.20
Kelevan	5	1.8	1.87	0.14	1.58	0.15	1.19	0.03	0.64	0.52
3,4-Dichloro-aniline	5	1.4	0.83	0.01	0.12	0.70	0.55	0.01	0.04	0.50

Hexachloro-benzene	4	0.7[a]	0.68	0.67	0.01	0.01	0.22	0.18	0.02	0.02
2,5,4′-Tri-chlorobiphenyl	4	1.3	0.31	0.25	<0.01	0.06	0.18	0.17	0.01	<0.01
2,4,6,2′,4′Penta-chlorobiphenyl	4	1.1	0.49	0.45	<0.01	0.04	0.32	0.26	0.02	0.04
Chloroalkylene-9	4	1.3	0.19	0.08	0.03	0.08	0.10	0.02	0.01	0.07
Pentachloro-nitrobenzene	4	36.1	9.3	3.67	4.57	1.06	5.61	0.16	2.91	2.54
Pentachlorophenol	4	1.1	0.36	<0.01	0.04	0.32	0.08	<0.01	<0.01	0.08
Allylalcohol	3	170	7.5	1.1		6.4	5.7	0.5		5.2
2,2′-Dichloro-biphenyl	2	1.1	0.24	0.12	0.02	0.10	0.15	0.03	0.01	0.11
Trichloroethylene	2	5.4	0.22	010		0.12	0.17	0.07		0.10
Imugan	2	0.3	0.21	0.12	0.02	0.07	0.11	0.02	0.01	0.08
Hexachloro-butadiene	2	6.2	0.13	0.05		0.08	0.09	0.04		0.05
Trichlorophenol	2	1.0	0.37	0.02	0.02	0.33	0.27	0.01	0.01	0.25

[a] As seed application.

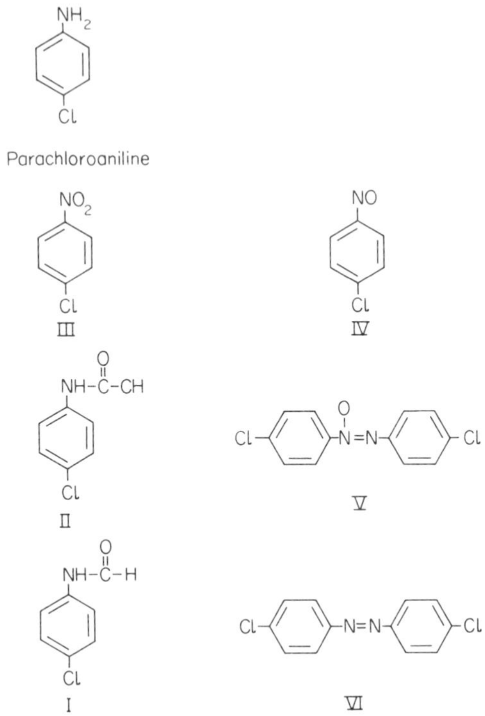

Fig. 24.11 Conversion products of *p*-Cl-Aniline in the soil–plant system

largest portion of radioactivity was bound as unextractable residue in soil and plants (in soil more than 90 per cent already after one vegetation period). The extracted identified conversion products, as shown in Figure 24.11, represent only a minor fraction.

Free *p*-chloroaniline was not detected in any sample. Upon alkaline hydrolysis of the unextractable residue in soil, however, *p*-chloroaniline could be identified. Similar conversions and residual behaviour were found for 3,4-dichloroaniline which is a precursor of several pesticides.[30]

CONCLUSIONS AND RECOMMENDATIONS

The data presented underline the introductory statements and show the diversity of chemical and physical processes in terrestrial systems even if considering plants and soil only. For further conclusions, Table 24.10 shows a summary of validated data of the discussed open-air experiments. The list of chemicals includes those generally classified as organochlorines, other substances containing chlorine in their structure, and one non-chlorinated pesticide. When

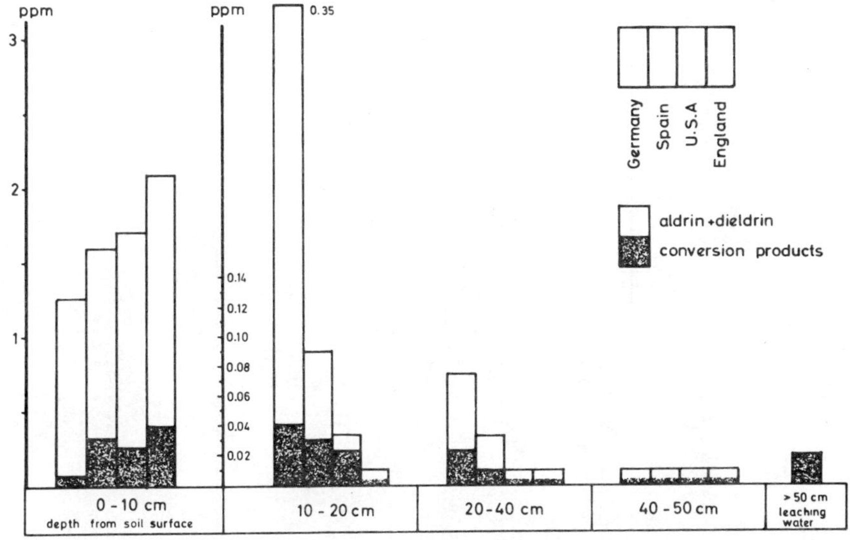

Fig. 24.12 Fate of [^{14}C]aldrin in treated soil

comparing the short-term (one season) and long-term (up to 10 years) fate of these chemicals some unexpected conclusions can be drawn.

A lipophilic chemical which is persistent against conversion into hydrophilic derivatives results in high residues within one season as compared to less persistent ones (compare dieldrin with aldrin and others). Over several years, however, the long-term data show a more efficient elimination of the persistent substance than for the less-persistent ones. This can be explained by the combination of 'bound residue' formation (which show long-term persistence) of the hydrophilic substances and by their leaching into deeper soil layers and reduced evaporation. This even applies for allyl-alcohol, The non-chlorinated chemical. For allyl-alcohol, the reduction of total radioactive residues after the first to the end of the third season is only very small due to the high portion of 'sedimented' bound residues which in this case could be natural products. According to Table 24.9, all investigated chemicals show no large differences with respect to long-term total residue formation despite a variety in bioconvertability, volatility etc. Therefore, it is most urgent to establish a methodology for a feasible and reliable structure elucidation of bound residues. If their evaluation would reveal no long-term impact on soil quality for the whole of chemical input (not only selected chemicals but the sum of all chemicals reaching soil), the assessment of the potential for effects in soils would be much improved.

Since open-air experiments are rather time consuming, costly, and cannot be completely standardized, recent research efforts aim at the quantitative prediction of the fate of chemicals from feasible laboratory experiments. The data presented here may serve as reference for the adoption of laboratory tests. In the real

Table 24.11 Volatilization of organochlorine compounds

Chemical	p (20–25 °C)	Water solubility ($\mu g\,l^{-1}$ 20–25 °C)	Volatilization from water (test, % ml^{-1} in 1 h)	Volatilization from loam (test, % ml^{-1} in 1 h)	Volatilization from soil (outdoors, % of application, in one season)
2,4,6-Trichlorophenol	4×10^{-2}	8×10^{5}	1.41	0.73	36
p-Chloroaniline	3.3×10^{-2}	3.9×10^{6}	0.62	0.70	67
2,5,4′-Trichlorobiphenyl	5×10^{-4}	1.1×10^{2}	30.2	0.41	68
Pentachlorophenol	1.5×10^{-4}	2×10^{4}	2.57	0.31	40
2,4,6,2′,4′-Pentachlorobiphenyl[a]		31	26.0	0.24	42
Aldrin[a]	8×10^{-6}	17	16.3	0.21	12
Kepone	3×10^{-7}	3×10^{3}	0.025	0.04	16

[a] In volatilization test: applied concentration above solubility data from literature.

environment, the overall fate of a chemical varies from season to season and under different climates by about ±20–50 per cent. Distinct data may vary by more than one order of magnitude as shown in Figure 24.12 for aldrin investigated under identical experimental but different climatic conditions (see soil layer 10–20 cm). As a consequence of these variations in nature, laboratory tests need to supply only the correct order of magnitude for the factor requested.

For some organochlorine compounds physicochemical, laboratory test, and outdoor data are compiled with respect to volatilization. Vapour pressure of these chemicals varies by about five orders of magnitude, volatilization in a laboratory test with loam soil varies only by a factor of about 20 but is well parallel to vapour pressure. Volatilization in the environment, however, varies only by a factor below 6 and—more important—is not parallel to vapour pressure and test data. This is due to structure dependent formation of less volatile conversion products as well as 'binding' to soil. Thus, the final conclusion is that prediction of the environmental behaviour of chemicals in terrestrial ecosystems from tests has to be based on test data for all relevant factors like adsorption/desorption, volatilization, degradation, leaching etc., and needs careful scientific evaluation of the significance of these factors for each chemical separately.

REFERENCES

1. Baldwin, M.K., Robinson, J., and Parke, D.V. (1972). A comparison of the metabolism of HEOD (Dieldrin) in the CF_1 mouse with that in the CFE rat, *Fd. Cosmet. Toxicol.* **10**, 111–351.
2. Brooks, G.T. (1972). In *Environmental Quality and Safety*, Vol. I, p. 106, Georg Thieme Publishers Stuttgart, Academic Press, New York.
3. Elgar, K.E. (1975). In *Pesticides; Environmental Quality and Safety*, Suppl. Vol. III, p. 250, George Thieme Publishers, Stuttgart, Academic Press, New York.
4. FAO/WHO (1968). In *1967 Evaluations of Some Pesticide Residues in Food, The Monographs*, p. 9, Rome.
5. Feil, V.J., Hedde, R.D., Zaylskie, R.G., and Zachrison, D.H. (1970). Dieldrin-^{14}C metabolism in sheep. Identification of trans-6,7-dihydroxydihydroaldrin and 9-(syn-epoxy)hydroxy-1,2,3,4,10,10-hexachloro-6,7-epoxy-1,4,4a,5,6,7,8,8a-octahydro-1,4-endo-5,8-exo-dimethanonaphthalene, *J. Agric. Food Chem.* **18**, 120–124.
6. Freitag, D. (1976). Dissertation Universität Bonn.
7. Freitag, D., Weisgerber, I., Klein, W., and Korte, F. (1974). Beiträge zur ökologischen Chemie LXXVI. Schicksal von Hexachlorbenzol-^{14}C in Sommerweizen und Boden nach Saatgutbehandlung, *Chemosphere*, **3**, 139–142.
8. Haque, A., Weisgerber, I., Kotzias, D., and Klein, W. (1977). Conversion of ^{14}C buturon in soil and leaching water under outdoor conditions, *Pest. Biochem. Physiol.* **7**, 321–331.
9. Hsu, T.-S., and Bartha, R. (1976). Hydrolyzable and nonhydrolyzable 3,4-dichloroaniline humus complexes and their respective rates of biodegradation. *J. Agric. Food Chem.* **24**, (1), 118–122.
10. Kaufman, D.D., Still, G.G., Paulston, G.D., and Bandal, S.K. (1976). Bound and conjugated pesticide residues. ACS Symposium Series 29, American Chemical Society, Washington, DC.

[11] Kilzer, L., Detera, S., Weisgerber, I., and Klein, W. (1974). Beiträge zur ökologischen Chemie LXXVII. Verteilung und Metabolismus des Aldrin-Dieldrin-Metaboliten trans-4,5-Dihydroxy-4,5-dihydroaldrin-^{14}C in Salatpflanzen und Boden. *Chemosphere*, **3**, 143–148.
[12] Klein, W., and Korte, F. (1970). In Chemie der Pflanzenschutz- und Schädlingsbekämpfungsmittel, Vol. 1, p. 199, Springer Verlag, Berlin, Heidelberg, New York.
[13] Kohli, J., Weisgerber, I., and Klein, W. (1976a). Balance of conversion of ^{14}C lindane in lettuce in hydroponic culture, *Pest. Biochem. Physiol.* **6**, 91–97.
[14] Kohli, J., Weisgerber, I., Klein, W., and Korte, F. (1973). Beiträge zur ökologischen Chemie LIX. Rückstandsverhalten und Umwandlung von Dieldrin-^{14}C in Kulturpflanzen, Boden und Sickerwasser nach Bodenapplikation, *Chemosphere*, **2**, 153–156.
[15] Kohli, J., Weisgerber, I., Klein, W., and Korte, F. (1976b). Contributions to ecological chemistry CVII. Fate of lindane-^{14}C in lettuce, endives, and soil under outdoor conditions, *J. Environ. Sci. Health B*, **11**, (1), 23–32.
[16] Korte, F. (1972). In, Radiotracer studies of chemical residues in food and agriculture, Panel Vienna, October 1971, IAEA STI/PUB 332, Vienna, p. 3.
[17] Korte, F., and Arent, H. (1965). Metabolism of insecticides, IX. Isolation and identification of dieldrin metabolites from urine of rabbits after oral administration of dieldrin-^{14}C, *Life Sci.* **4**, 2017–2026.
[18] Korte, F., and Kochen, W. (1966). Insektizide im Stoffwechsel XI. Ausscheidung, Verteilung und Umwandlung von Aldrin-^{14}C und Dieldrin-^{14}C in der Ratte, *Med. Pharmacol. exp.* **15**, 404–408.
[19] Korte, F., Ludwig, G., and Vogel, J. (1962). Insektizide im Stoffwechsel II. Umwandlung von Aldrin-(^{14}C) und Dieldrin-(^{14}C) durch Mikroorganismen, Leberhomogenate und Moskito-Larven, *Liebigs Ann. Chem.* **656**, 135–140.
[20] Lahaniatis, E.S., Parlar, H., and Korte, F. (1977). Über Das Vorkommen chlorierter Kohlenwasserstoffe in Flugaschen von Müllverbrennungsanlagen. Beiträge zur ökologischen Chemie CXXXII, *Chemosphere*, **1**, 11–16.
[21] Mason, H.S. (1957). Mechanisms of oxygen metabolism, *Science*, **125**, 1185–1188.
[22] Matsumura, F. (1972). In *Environmental Quality and Safety*, Vol. I, p. 96, Georg Thieme Publishers, Stuttgart, Academic Press, New York.
[23] Moza, P., Weisgerber, I., and Klein, W. (1972). Beiträge zur ökologischen Chemie L. Auswaschen eines wasserlöslichen Aldrin-^{14}C-Abbauprodukts aus Boden, *Chemo*sphere, **1**, 191–195.
[24] Müller, W., and Korte, F. (1976). Beiträge zur Ökologischen Chemie CXVII. Polychlorierte Biphenyle und Hexachlorbenzol in Umweltproben aus dem südduetschen Raum, *Chemosphere*, **2**, 95–100.
[25] Oda, J., and Müller, W. (1972). In *Environmental Quality and Safety*, Vol. I, p. 248, Georg Thieme Publishers, Stuttgart, Academic Press, New York.
[26] Onsager, J.A., Rusk, H.W., and Butler, L.I. (1970). Residues of aldrin, dieldrin, chlordane, and DDT in soil and sugarbeets, *J. Econ. Entomol.* **63**, 1143–1146.
[27] Sandrock, K., Bieniek, D., Klein, W., and Korte, F., (1974). Beiträge zur ökologischen Chemie LXXXVI. Isolierung und Strukturaufklärung von Kelevan-^{14}C-Metaboliten und Bilanz in Kartoffeln und Boden, *Chemosphere*, **3**, 199–204.
[28] Scheunert, I., Kohli, J., Kaul, R., and Klein, W., (1977). Fate of [^{14}C]aldrin in crop rotation under outdoor conditions. *Ecotox. Environ. Safety*, **1**.
[29] SCOPE (1977). *Workshop on Biogeochemical Cycling of Carbon*, Hamburg/Ratzeburg.
[30] Viswanathan, R., Scheunert, I., Kohli, J., Klein, W., and Korte, F. (1978). Long-term

studies on the fate of 3,4-dichloroaniline-^{14}C in a plant-soil-system under outdoor conditions, *J. Environ. Sci. Health*, B, **13**, (3), 243–259.

31 Weisgerber, I., Bieniek, D., Kohli, J., and Klein, W. (1975a). Isolation and identification of three unreported photodieldrin-^{14}C metabolites in soil, *J. Agric. Food Chem.*, **23**, 873–877.

32 Weisgerber, I., Detera, S., and Klein, W. (1974). Beiträge zur ökologischen Chemie LXXXVIII. Isolierung und Identifizierung einiger Heptachlor-^{14}C-Metaboliten aus Pflanzen und Boden, *Chemosphere*, **3**, 221–226.

33 Weisgerber, I., Tomberg, W., Klein, W. and Korte, F. (1975b). Beiträge zur ökologischen Chemie CXV. Isolierung und Strukturaufklärung einiger hydrophiler Isodrin-^{14}C-Metaboliten aus Weißkohl, *Chemosphere*, **4**, 99–104.

34 Zimmer, M. and Klein, W. (1972). Rückstandsverhalten und Umwandlung von p,p′-DDT^{14}C und seiner Analogen p,p-DDE-^{14}C und p,p′-DDD-^{14}C in höheren Pflanzen. *Chemosphere*, **1**, 3.

35 Begum, S., Scheunert, I., Haque, A., Klein, W. and Korte, F. (1979). Conversion of (^{14}C)pentachloronitrobenzene in onions. *Pest. Biochem. Physiol.* **11**, 189.

Pollutants and Their Ecotoxicological Significance
Edited by H. W. Nürnberg

CHAPTER 25

Contamination by PCBs and Organochlorine Pesticides of Belgian Birds of Prey, Their Eggs and Their Food, 1969–1982

C. Joiris and K. Delbeke

SUMMARY

Three complementary methods were used in order to describe and to understand the contamination of Belgian birds of prey: the analysis of unhatched eggs ($n = 350$), the analysis of raptors found dead ($n = 189$) and the determination of changes in eggshell thickness ($n = 1432$). The results obtained by the three methods are coherent.

As an example, the contamination by residues of the DDT group is discussed in some details. The level of the contamination is clearly species dependent: the birds-eating species are more contaminated than the mammals-eating raptors.

For some species (owls), a difference of contamination is found between two zones in Belgium: owls coming from the northern, agricultural zone are more contaminated than birds of the same species coming from the southern, forest zone. Their eggs also show a higher DDT level and a greater eggshell thinning.

Year-to-year variations in the level of DDT contamination are also detected for the owls and their eggs.

The analysis of some prey items (House Sparrow and small mammals) show that the main factor allowing to understand all results lies in the diet characteristics of the raptors: the more birds they eat, the higher their contamination. This is valid for the different species, and for the differences in contamination noted for a given species in the northern and southern zones, as well as from one year to the other.

The main conclusions are: first, that it is very useful to utilize different methods in order to determine the contamination of biological material, and second, that an integrated study is needed in order to understand the levels of contamination and their variations. This study must include data on the population dynamics of the raptors, and mainly on their local feeding habits. In such conditions only can birds of prey be utilized as 'bioindicators' of terrestrial ecosystems, and can the differences of their contamination, as a function of space and of time, be understood.

INTRODUCTION

In terrestrial ecosystems, birds of prey can be considered as good 'bioindicators' ('targets') for stable pollutants: they are at the top of food pyramids and thus have high levels of persistent pollutants accumulating along the food web, such as organochlorine residues. They also represent an integration of small-scale variations of the contamination, because of their range and of their life-span. It must be kept in mind that the situation is completely different in aquatic ecosystems, where the carnivores of the highest level do not necessarily have the highest contamination. First, because the ecological structure of aquatic ecosystems cannot be represented as pyramids of biomass, as the productivity (the ratio production to biomass, or the turn-over time) of the different levels can be completely different (from a few days for phytoplankton to a few years for fish). Second, because the animals do not necessarily get their contaminants through their food only: fish, for instance, can be directly contaminated from the water and only partially through polluted food.

When starting a study on the contamination of Belgian terrestrial ecosystems in 1969, we were severely limited in the possibilities of analyses and in the financial support, and chose birds of prey as the best biological material. For ethical reasons, we decided not to destroy any raptor, but to use biologically lost material: unhatched eggs and raptors found dead. This leads of course to somewhat biased results, as the samples include sterile eggs and those birds having died because of a high pesticides level. But the main cause of mortality in raptors being still of human origin—shooting, accidents with cars and trains—the bias seems rather limited and acceptable in the frame of such a study.

RESULTS AND DISCUSSION

A summary of the obtained results is presented in this section; as an example, the contamination by residues of the DDT-group (that is mainly DDE) is used in the discussion.

The contamination of unhatched raptor's eggs is clearly species dependent. As far as DDT is concerned, the levels are low in mammal-eating raptors, and high in bird-eating raptors. When the species are classified as a function of their diet, the contamination is increasing with increasing percentage of birds in their food (Table 25.1).

Table 25.1 Organochlorine pesticides and PCB contamination of unhatched eggs of birds of prey, collected in Belgium from 1969 to 1979 (ppm/wet weight, median values).[1–3,12] The species are classified in function of increasing percentage birds in the diet

Species	Diet % birds	(ref.)	*n*	PCB	Lin.	HCB	H.E.	Diel.	Total 1	ΣDDT
Pernis apivorus	0	4	5	0.6	0.02	0.002	0.06	0.01	0.09	0.08
Aegolius funereus	2	4	11	1.2	0.01	0	0.01	0.01	0.05	0.04
Buteo buteo	4.5	4	17	1.8	0.03	0	0.08	0.03	0.17	0.02
Falco tinnunculus	4.5	4	23	0.3	0.02	0	0.04	0	0.06	0.04
Athene noctua	5	5	13	1.4	0.03	0	0.13	0	0.17	0.14
Tyto alba	5	6–8	171	1.8	0.01	0	0.49	0.18	0.68	0.53
Asio otus	12.5	10	2	0.7	0.10	1.11	0.12	0.30	1.62	5.36
Strix aluco	13.5	9	77	0.4	0.01	0	0.04	0.01	0.06	0.12
Circus pygargus	—	—	6	1.4	0.01	0	0.06	0.01	0.08	0.10
Accipiter gentilis	92	4	5	4.5	0.03	0	0.13	0.49	0.67	1.12
Accipiter nisus	98	4	20	3.4	0.02	0	0.30	0.25	0.57	2.43

Lin., Lindane; H.E., Heptachlore epoxide; Diel., Dieldrin; Total 1, Lindane + Heptachlore epoxide + Dieldrin; ΣDDT, DDT + DDE + DDD.
n, Number of eggs analysed.

For the Barn Owl (*Tyto alba*) and the Tawny Owl (*Strix aluco*), a difference in contamination is to be noted as a function of the geographical origin of the samples: the eggs coming from the northern zone (i.e. north of the rivers Sambre, Meuse and Vesdre) and more contaminated than the eggs coming from the southern zone (Table 25.2).

Yearly variations are also noted in the contamination of Barn Owl's eggs (the limited amount and distribution does not allow determination whether the same phenomenon exists for other owl species or not) (Table 25.3).

The study of the contamination of the dead raptors leads to the same type of

Table 25.2 Geographical differences in the DDT contamination of three owls.[1–3,11,12,15] DDT (ppm/wet weight, median values)

Species	Zone	Diet % birds	(ref.)	DDT Eggs	(*n*)	Muscle	(*n*)	Liver	(*n*)
Tyto alba	S	0	6–9	0.04	(22)	0.31	(8)	0.49	(8)
Strix aluco	S	2	9	0.12	(36)	0.02	(2)	0.19	(2)
Asio otus	S	3	—	—		0.08	(8)	0.01	(8)
Tyto alba	N	10	6–9	1.14	(101)	2.52	(24)	1.69	(24)
Strix aluco	N	20	9	0.18	(9)	0.91	(7)	0.39	(7)
Asio otus	N	30		—		3.86	(20)	9.49	(20)

S, southern zone; N, northern zone (see text).
See legend Table 25.1.

Table 25.3 DDT contamination of the Barn Owl (*Tyto alba*) in different years[1,3,11,12,15] (ppm/wet weight, median values)

Date	ΣDDT					
	Eggs	(*n*)	Muscle	(*n*)	Liver	(*n*)
1972	0.03	(17)	—		—	
1973	0.87	(21)	0.40	(3)	0.72	(3)
1974	0.11	(46)	0.24	(3)	0.17	(3)
1975	0.29	(22)	1.36	(9)	0	(9)
1976	0.95	(6)	1.21	(3)	0.93	(5)
1977	1.28	(26)	3.62	(2)	3.43	(2)
1978	1.60	(27)	0.52	(5)	0.39	(5)
1979	0.81	(5)	1.50	(11)	7.97	(9)

conclusions: the DDT level is species dependent, and the contamination shows a general tendency to increase with increasing percentage of birds in the diet (Table 25.4). For the Tawny Owl, the Barn Owl and the Long-eared Owl (*Asio otus*), a difference in the northern and the southern birds is also found (Table 25.2). A yearly variation in the contamination is also detected (Table 25.3).

Another method to be used in order to follow the secondary effects of the organochlorine pesticides consists in measuring the eggshell thickness, eggshell thinning being mainly due to a DDT contamination.[13] We used two methods: the direct determination of the eggshell thickness and the determination of the 'Ratcliffe's index', defined as the ratio between the product of the outer

Table 25.4 Organochlorine pesticides and PCB contamination of birds of prey found dead between 1972 and 1982 (breast muscle and liver) (ppm/wet weight, median values)[1,11,12,15]

Species	Diet (% birds)	*n*	PCB		Total 1		DDT	
			Muscle	Liver	Muscle	Liver	Muscle	Liver
Circus aeruginosus	—	1	0.33	0.29	—	0.05	—	0.04
Pernis apivorus	0	1	—	0.07	—	0.35	—	0.13
Milvus milvus	—	1	0.50	2.30	0.12	0.63	0.10	0.42
Buteo buteo	4.5	28	0.88	0.67	0.32	0.45	0.06	0.05
Falco tinnunculus	4.5	18	0.91	0.91	0.75	1.53	0.07	0.10
Athene noctua	5	31	0.43	0.76	0.76	0.64	0.38	0.15
Tyto alba	5	44	7.60	10.00	1.25	2.47	1.26	0.97
Asio otus	12.5	32	1.82	1.52	0.64	1.95	0.79	2.06
Strix aluco	13.5	13	4.31	0.90	1.48	3.95	0.33	0.62
Falco columbarius	90	2	—	2.69	—	5.16	51.3	55.2
Accipiter nisus	98	18	5.17	3.84	1.21	0.81	2.31	2.15

n, Number of birds analysed; Total 1, mainly Heptachlore epoxide. See legend Table 25.1.

measurements of the egg (both in mm) to the weight of the eggshell (in mg).[14] The results of both methods are strongly correlated (for example see Figure 25.1).

The obtained results are coherent with the levels of DDT detected in the eggs and in the birds of prey: a strong decrease of eggshell thickness can be noted for the bird-eating raptors only, while the species having a lower DDT contamination keep normal eggshells (Table 25.5). For some species, the

Table 25.5 Changes of eggshell thickness in Belgian birds of prey, during the periods 1950–1982. The results are expressed in function of the pre-1950 values taken as 100 percent.[12,16]

Species	Diet (% birds)	Period (years)	*n*	Percentage decrease[a] Thickness	Index
Pernis apivorus	0	50–59	8	(−2.9)	(−4.8)
		60–69	10	(6.2)	(−0.6)
		70–79	4	(−12)	(6.4)
Buteo buteo	4.5	50–59	36	(2.5)	3.5
		60–69	41	5.3	4.6
		70–79	16	0	(−2.4)
Falco tinnunculus	4.5	50–59	26	8.3	(−1.5)
		60–69	23	0	(−0.8)
		70–79	48	0	(−0.8)
Athene noctua	5	50–59	22	9.1	(4.7)
		60–69	29	9.1	6.7
		70–79	13	0	(−2.7)
Tyto alba	5	50–59	20	0	(−3.2)
		60–69	19	(8.7)	16.0
		70–79	153	4.2	9.8
Asio otus	12.5	50–59	25	4.2	6.8
		60–69	11	(8.7)	14.6
		70–79	24	25	30.6
Strix aluco	13.5	50–59	16	0	−5.6
		60–69	25	(7.1)	0
		70–79	32	0	(1.4)
Falco subbuteo	50	50–59	3	(12.5)	(14.9)
		60–69	12	(12.5)	18.0
Accipiter gentilis	92	50–59	15	12.8	13.8
		60–69	23	10.0	14.9
		70–79	7	(−4.5)	(−6.7)
Accipiter nisus	98	50–59	21	27.3	25.2
		60–69	63	27.3	26.4
		70–79	5	(16.7)	(9.4)
Falco peregrinus	100	50–59	6	(9.1)	18.2
		60–69	6	(28.6)	29.1

See legend Table 25.1.

[a] The values are presented between parentheses when not significant at the 0.02 level in a Student T-test.

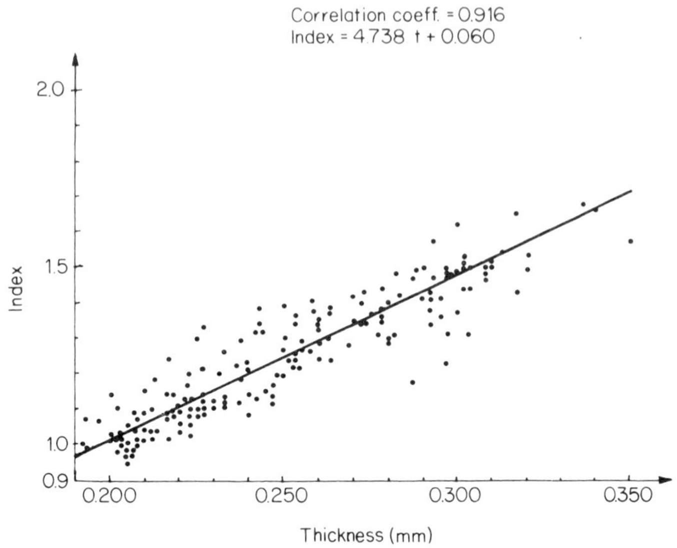

Fig. 25.1 Relation between index and thickness: the example of the Sparrowhawk, *Accipiter nisus*

decrease is progressive and the highest thinning is detected for the period 1960–69 (for example see Figure 25.2).

The comparison of eggshells coming from the two zones shows again a clear geographical difference for the Barn Owl and the Tawny Owl: no decrease in the southern zone, significant decrease in the northern zone (Table 25.6).

A first important aspect to be pointed out is that the results obtained with three different methods: contamination of the eggs, contamination of the raptors and changes of eggshell thickness, are coherent and lead to the same conclusions. The coherency of the results can be quantitatively expressed as a series of

Table 25.6 Geographical differences in the eggshell thinning for three owls[12,16]

Species	Zone	Diet (% birds)	*n*	Percentage decrease Thickness	Index
Tyto alba	S	0	21	(−4.0)	(−9.8)
Strix aluco	S	2	39	(−3.3)	−9.8
Asio otus	S	3	23	4.2	(−6.8)
Tyto alba	N	10	167	4.2	9.9
Strix aluco	N	20	37	19.0	20.2
Asio otus	N	25	20	7.1	(3.6)

See legend Tables 25.3 and 25.5.

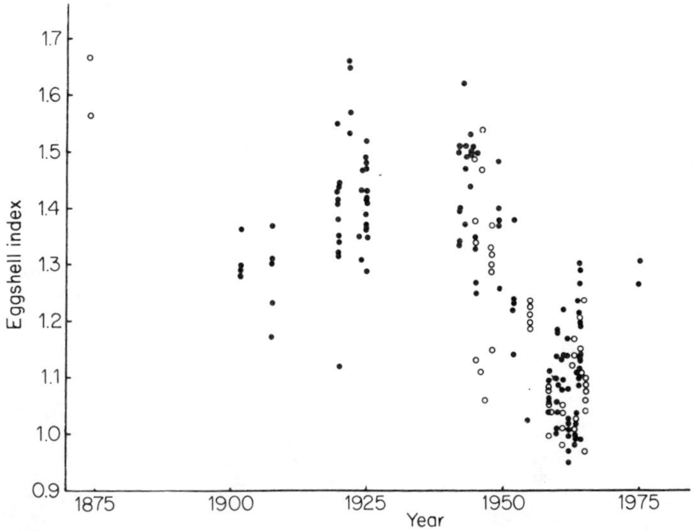

Fig. 25.2 Evolution of the value of the Ratcliffe index for the eggshell of the Sparrowhawk, *Accipiter nisus*. ○ Northern zone; ● southern zone

relationships between the different parameters: contamination of liver versus muscle, eggs versus muscle, eggshell thinning as a function of DDT contamination of the muscle, of the eggs and of the diet of the raptors (Figures 25.3 to 25.7). This provides a precious confirmation for the validity of these conclusions.

Another point to be discussed is the geographical variations: it is interesting to understand why two populations of the same species (owls), living not far from each other, can show such important differences in their contamination. Two main factors could explain this difference. On the one hand, the northern region is an agricultural and industrial one: it seems logical to suppose that the utilization of pesticides is high and that the ecosystem is heavily contaminated; the southern region consists mainly of forests and meadows: the use of pesticides is probably limited and the contamination of the ecosystem low. But on the other hand, the diet of the owls is not identical in both regions: they eat more birds in the northern zone than in the southern[6-10] and this can also explain a higher DDT level in owls from the northern zone.

In order to determine the relative importance of those two mechanisms, a series of prey items were taken in the two regions and analysed. The results show a clear difference in the DDT contamination of sparrows and small mammals, but no significant difference between the zones (Table 25.7). The differences in feeding habits are thus the main cause of the different DDT contamination of owls in the two zones.

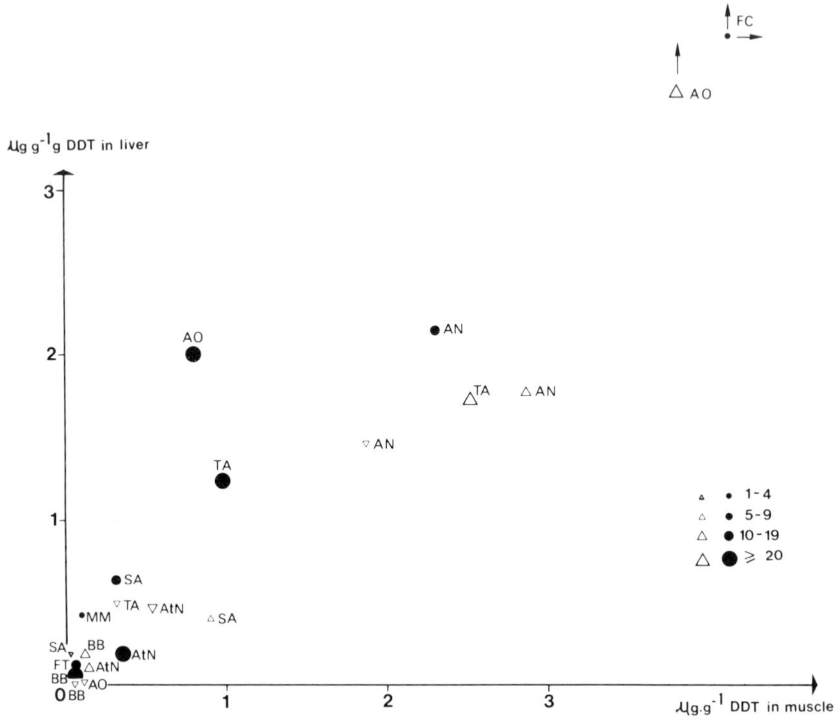

Fig. 25.3 Relation between the DDT contamination of liver and breastmuscle of raptors found dead in Belgium (median values) (1, 11, 12, 15). △ Northern region; ▽ Southern region; ● both regions; BB: *Buteo buteo*; AN: *Accipiter nisus*; FT: *Falco tinnunculus*; FC: *Falco colombarius*; At N: *Athene noctua*; AO: *Asio otus*; SA: *Strix aluco*; TA: *Tyto alba.* The number of the samples are reflected by the size of the symbols. *Asio otus* (North) has a median liver DDT level of 9.49 $\mu g\, g^{-1}$ (ppm); the *Falco columbarius* has a muscle and liver DDT contamination of respectively 51.3 and 55.2 $\mu g\, g^{-1}$.

This means that populations of a given owl species can be contaminated at different levels, even if they live in equally contaminated ecosystems. This also implies that an integrated study is needed in order to understand and compare values of contamination by organochlorine pesticides in birds of prey: the population dynamics, and mainly the diet characteristics are to be known in order to interpret such results. As these feeding habits can show important local and yearly variations, a detailed local study is necessary together with each study of the contamination.

A third aspect is the ecological stability of the residues. As the utilization of organochlorine pesticides was forbidden in our regions from the late sixties on, it is normal to expect a progressive decrease in the contamination of our ecosystems. But 10 years after the ban of, for instance, DDT, its concentration

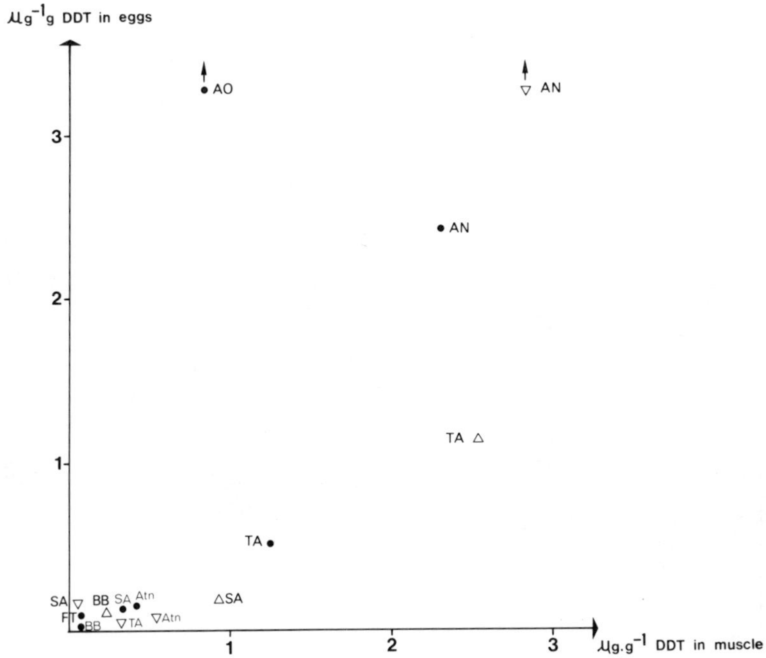

Fig. 25.4 Relation between the DDT contamination of unhatched eggs and breastmuscle of raptors found dead in Belgium (median values) (1–3, 11, 12, 15). △ Northern region; ▽ Southern region; ● both regions. BB: *Buteo buteo*, North (10 birds, 6 eggs), both regions (28, 17); AN: *Acipiter nisus*, South (6, 2), both (18, 20); FT: *Falco tinnunculus* (18, 23); At N: *Athene noctua*, South (10, 7), both (31, 13); AO: *Asio otus* (32, 2); SA: *Strix aluco*, North (7, 9), South (2, 36), both regions (13, 77); TA: *Tyto alba*, North (24, 101), South (8, 22), both (44, 17). *Accipiter nisus* (South) and *Asio otus* had DDT levels in their eggs of respectively 10.4 and 5.4 $\mu g\, g^{-1}$ (ppm).

Table 25.7 Organochlorine pesticides of some prey-items taken in the two zones of Belgium[12,17] (ppm/wet weight, mean values)

Species	Zone	*n*	Lin.	HCB	H.	H.E.	Diel.	ΣDDT
House sparrow	S 1973	10	0	nd	nd	0.025	0.001	0.146
Passer domesticus	S 1978	14	0.026	0.002	0.089	0.024	0.0003	0.013
	N 1977	10	0.029	0.001	0.154	0.052	0.001	0.021
	N 1978	11	0.054	0.006	0.160	0.049	0.014	0.034
Total		45	0.028	0.003	0.130	0.037	0.004	0.049
Small mammals	S 1973	12	0.184	nd	nd	0.073	0.008	0
	S 1978	11	0.018	0.002	0.082	0.013	0.008	0.001
	N 1978	10	0.070	0.002	0.170	0.108	0.113	0.001
Total		33	0.094	0.002	0.123	0.064	0.040	0.001

See legend Table 25.1; nd, not determined.

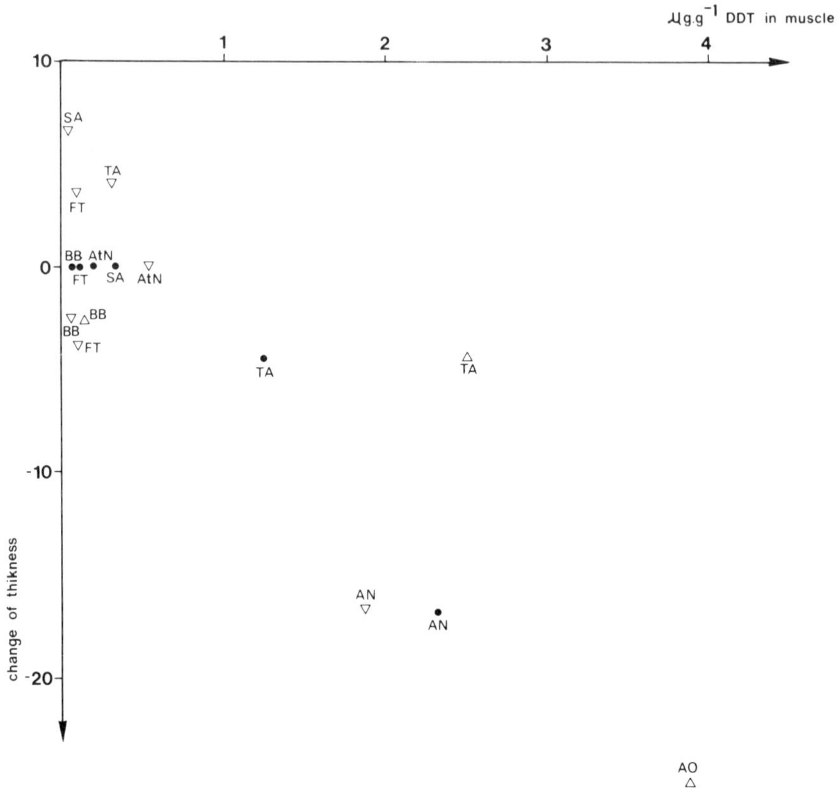

Fig. 25.5 Relation between the eggshell thickness change and the DDT contamination of the birds' breastmuscle. Only eggshells collected since 1970 are considered (median values) (1, 11, 12, 15, 16). △ North region; ▽ Southern region; ● both regions; BB: *Buteo buteo*, North (10 birds, 9 eggs), South (9, 4), both (18, 16); AN: *Accipiter nisus*, South (6, 5), both (18, 5); FT: *Falco tinnunculus*, North (9, 14), South (3, 5), both (18, 48); At N: *Athene noctua*, South (10, 12), both (31, 13); AO: *Asio otus*, North (18, 24); SA: *Strix aluco*, South (2, 18), both (13, 32); TA: *Tyto alba*, North (24, 101), South (8, 7), both (44, 153).

only starts to slowly decrease in nature: this shows a much higher stability than is expected from the biological stability in individual animals.[12,18] Another argument is to be found in the yearly variations of the contamination: the reappearance of high levels after years of low contamination does not seem to be due to the use of DDT (the ban is considered as being effective). This means that residues excreted and eliminated from the raptors are recycled within the ecosystem and re-accumulated along the food web: the ecological stability is thus the main parameter influencing the decrease of the DDT contamination in natural ecosystems, not the biological stability as measured in isolated animals.

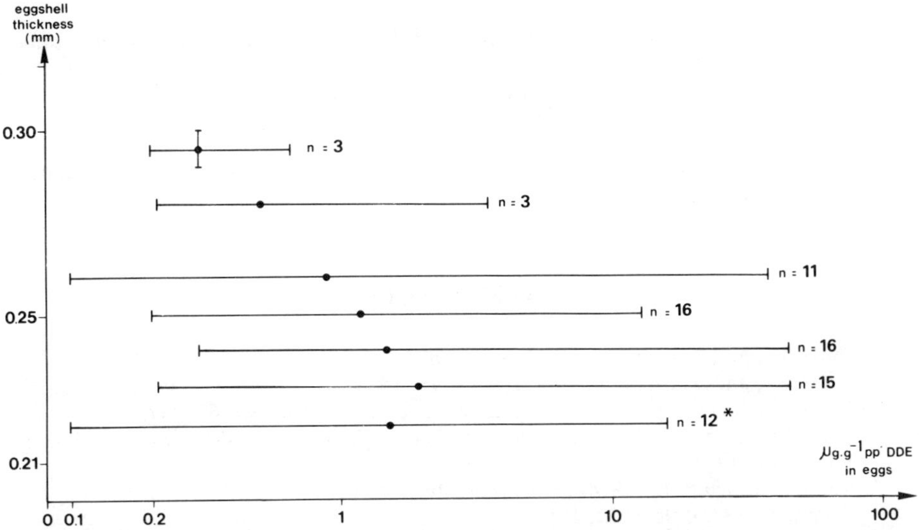

Fig. 25.6 Relation between the shell thickness and the DDE contamination of unhatched eggs of the Barn Owl (*Tyto alba*) (median values, minimum – maximum; n = number of samples; * including 2 eggs with a shell thickness of 0.21 mm).

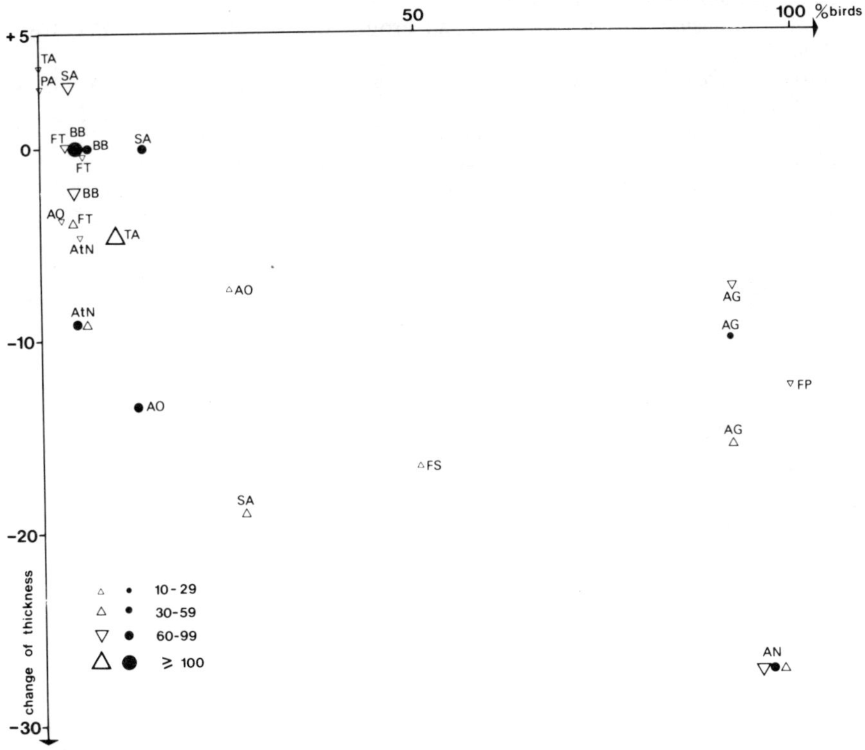

Fig. 25.7 Change of eggshell thickness in the period 1950–1970 as a function of the percentage of birds in the diet of the raptors. See legend Figure 25.3. PA: *Pernis apivorus*; AG: *Accipiter gentilis*; FP: *Falco peregrinus*; FS: *Falco subbuteo*.

ACKNOWLEDGEMENTS

A few results incorporated in this study were obtained by last year students: A. Willems and C. Crabbe.

REFERENCES

1 Joiris, C., and Martens, P. (1971). *Aves*, **8**, 1.
2 Joiris, C., and Martens, P. (1973). *Aves*, **10**, 158.
3 Joiris, C., Lauwereys, M., and Vercruysse, A. (1977). *Gerfaut*, **67**, 447.
4 Uttendörfer, O. (1952). Neue Ergebnisse über die Enährung der Greifvögel und Eulen.
5 Libois, R. (1977). *Aves*, **14**, 165.
6 Van der Straeten, E. (1974). *Wielewaal*, **40**, 145.
7 Godin, J. (1975). *Aves*, **12**, 105.
8 Leurquin, J. (1975). *Aves*, **12**, 127.
9 Delmée, E., Dachy, P., and Simon, P. (1979). *Gerfaut*, **69**, 45.
10 Delmée, E., pers. comm.
11 Joiris, C., Lauwereys, M., and Vercruysse, A. (1973). *Aves*, **10**, 161.
12 Joiris, C., and Delbeke, K. (unpublished).
13 Blus, L.J., Gish, C.D., Belisle, A.A., and Prouty, R.M. (1972). *Nature*, **235**, 376.
14 Ratcliffe, D.A. (1967). *Nature*, **215**, 208.
15 Joiris, C., Delbeke, K., Martens, E., Lauwereys, M., and Vercruysse, A. (1979). *Gerfaut*, **69**, 319.
16 Joiris, C., Dejaegher, J., and Delbeke, K. (1979). *Gerfaut*, **69**, 195.
17 Joiris, C., Lauwereys, J., and Vercruysse, A. (1973). *Aves*, **10**, 171.
18 Ludwig, G., Arent, H., Kochen, W., Poonawalla, N., Rechmeier, G., Staasni, M., Vogel, J., and Korte, F. (1966). *Scientif. Plant. Protect. Conf.*, Budapest, Feb. 22nd, 1.

Pollutants and Their Ecotoxicological Significance
Edited by H. W. Nürnberg

CHAPTER 26

*Man—a Target of Ecotoxicological Influences**

H. P. BERTRAM, F. H. KEMPER AND CHRISTA ZENZEN

SUMMARY

Amongst ecotoxicological influences on human beings halogenated organic compounds, e.g. chlorinated polycyclic hydrocarbons, as well as inorganic substances, e.g. toxicological significant heavy metals, are of importance.

Beside the features of hazards of exposure target organs of toxicity and depot compartments, the following aspects are mentioned.

1. Organochlorine pesticides: results of analytical determination in human organs (blood, liver, fatty tissue) of several collectives and organs distribution of DDT and HCB are given (Real Time Monitoring and autopsy material). Correlation to age and pathophysiological processes (arteriosclerotic disease, tumorchachexia) are discussed. Some analytical results are compared with regard to different reference systems ('extractable lipids' and 'whole tissue').
2. Toxic heavy metals: mechanism and symptoms of intoxication with lead, mercury and cadmium are described. Examples for epidemic environmental load with organomercurials and cadmium are reviewed. The extended halftime of cadmium in human organism leads to a rapid load of most tissues with cadmium. Some results of Cd-determinations in human liver and kidney specimens are discussed in relation to age. Interactions between toxic and essential trace elements may play an important role by toxification or detoxification of the pollutants. Therefore, the determination of a 'metal fingerprint' containing the analytical results of essential and non-essential trace elements and bulk electrolytes is proposed.

* The respective investigations were sponsored in part by BMFT/UBA (Federal Ministry for Research and Technology/Federal Environmental Agency).

Numerous environmental pollutants must be classified as health hazards for man. In Figure 26.1 sources of possible xenobiotic exposure are listed: e.g. domestic poisons (detergents, pigments, propellants), naturally occurring substances (plant or animal toxins), food components (additives, bacterial toxins, nitrosamines), stimulants (nicotin, ethanol, drugs).

From the toxicologists and pharmacologists point of view it is of interest in connection with xenobiotics to look for the target organs, the possibility of accumulation and moreover hazards of exposure. The results of analytical work in 'Real Time Monitoring' as well as long-time-trend analysis are the basis for medical estimation of the ecotoxicological risk concerning man's life.

Moreover possibilities of conversions of pollutants in different ecosystems must be taken in consideration. The conversions may result in substances with increased or lowered toxic qualities, as shown in Figure 26.2 for mercury and organohalogenated hydrocarbons.

In the following some points of interest concerning the toxicity of some non-essential trace metals and human exposure to organochlorine pesticides are discussed.

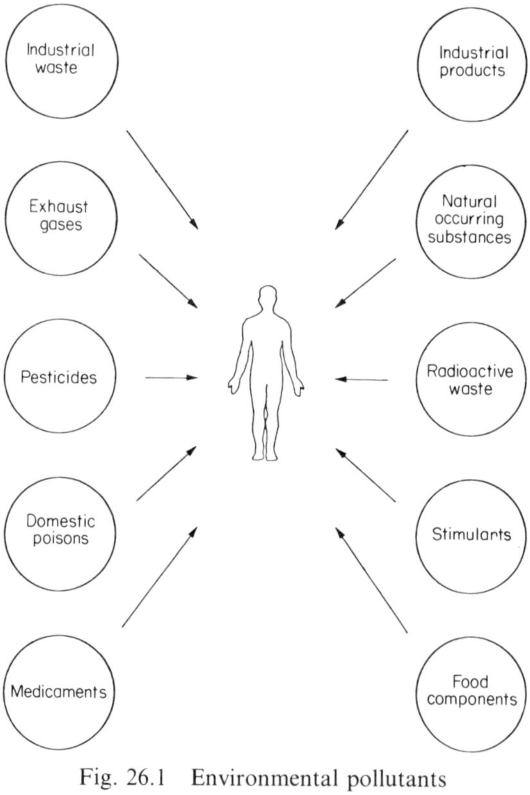

Fig. 26.1 Environmental pollutants

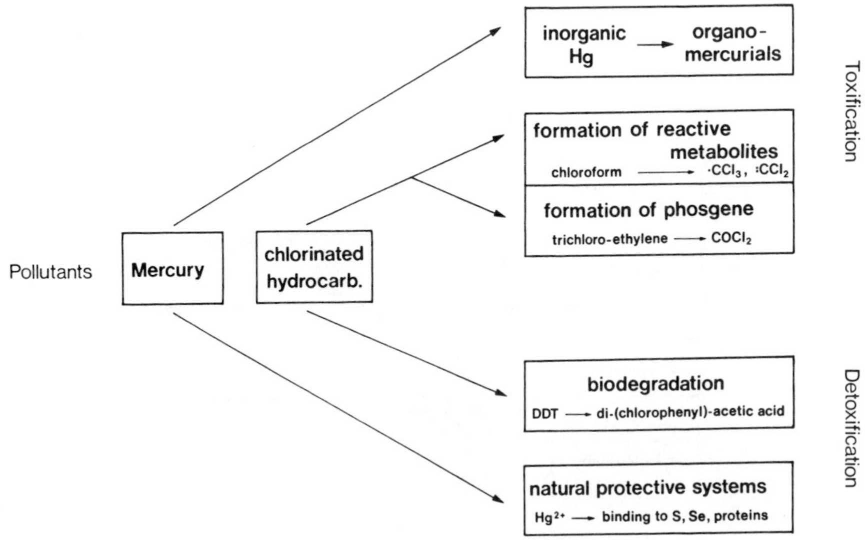

Fig. 26.2 Conversion of pollutants

ORGANOHALOGENATED COMPOUNDS

Organohalogenated compounds are of important ecological and toxicological interest because of their presence in various parts of human environment. Figure 26.3 shows the hazards of exposure, the target organs of toxicity and the depot compartments of organohalogenated compounds. Among the possibilities of exposure, insecticides with prolonged halftime (DDT, hexachlorobenzene etc.) and substances with possible cancerogenic risk, e.g. solvents (chloroform, trichloroethylene, binylchloride), should have the highest priority in scientific evaluation. For that reason the observation of their ecotoxicological behaviour and not at least the respective environmental exposure of man is needed. Our institute is engaged in a programme testing the human exposure to organochlorine pesticides (OCP) as DDT, HCB, BHC etc. This programme is divided in two parts, part 1, the 'Real Time Monitoring' where the OCP content in blood of living people is measured, and part 2, the 'Environmental Specimen Bank Program', where we measure the OCP content in various organs of deceased people. The results of the 'Real Time Monitoring' are presented in Table 26.1.

The age of the persons analysed was within 20–30 years. Most of them were students. The concentrations are given in $\mu g\,l^{-1}$, once the total average and once separated in female and male. The highest amount was found for HCB, the lowest for *op*-DDT. As usual the main part of total DDT is DDE. The maximum value measured is 13.5 $\mu g\,l^{-1}$ for hexachlorobenzene, found in the blood of a woman,

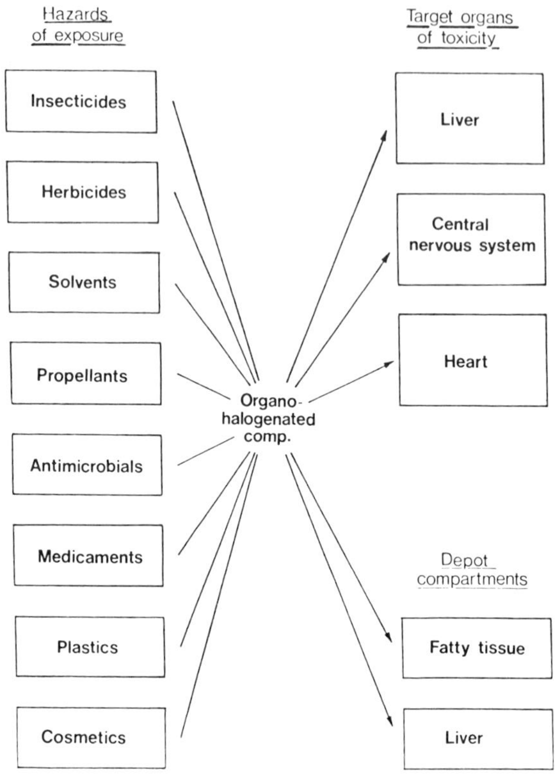

Fig. 26.3

engaged in agriculture, where she had to dip young plants in a HCB solution, an evidence of the influence of the occupation to body burden.

In the 'Environmental Specimen Bank Program' we analysed human adipose tissue and human liver. In Table 26.2 and 26.3 the OCP concentrations in both tissues are presented. The sequence is nearly the same as in whole blood, but the concentrations are about a thousand-fold higher. The concentrations listed here represent an average of a group of seventy-nine deceased human beings from the age of 1 h to 91 years. That is the reason for the wide range of OCP content.

Figure 26.4 gives further explanation, while it presents the coherence between the age and total DDT and HCB concentration in human adipose tissue. Although all extremely high values were eliminated, the coherence is obvious, especially for the HCB content. Figure 26.5 shows the same criteria compared, but shown as a histogram. The extremely high values are included here. The age-concentration dependence is evident, so that the OCP concentration in human adipose tissue represents a pollutant–exposure–(time) gradient, not present in the histogram of human liver content (Figure 26.6).

Table 26.1 Organochlorine pesticides in human blood: decreasing concentrations ($\mu g\, l^{-1}$)

	Average			Maximum	Minimum
	Total	Female/ Male			
HCB	3.89	F	4.11	13.52	0.05
		M	3.69		
DDT total	2.26	F	2.36	8.61	0.06
		M	2.15		
pp'-DDE	1.79	F	1.88	6.46	0.06
		M	1.70		
pp'-DDT	0.37	F	0.39	1.51	0.02
		M	0.34		
γ-BHC	0.29	F	0.30	1.38	0.02
		M	0.27		
α-BHC	0.20	F	0.22	0.92	0.01
		M	0.19		
DDD	0.17	F	0.20	1.81	0.003
		M	0.13		
Heptachloro-epoxide	0.13	F	0.13	0.71	0.02
		M	0.13		
Dieldrin	0.12	F	0.13	0.38	0.01
		M	0.11		
op-DDT	0.04	F	0.04	0.28	nd
		M	0.03		

Table 26.2 Organochlorine pesticides in human adipose tissue: decreasing concentrations ($\mu g\, g^{-1}$ extractable lipids)

	Average	Maximum	Minimum
HCB	6.11	45.38	0.14
DDT total	4.21	30.19	0.09
pp'-DDE	3.58	28.92	0.09
pp'-DDT	0.33	2.97	nd
Dieldrin	0.17	0.98	nd
γ-BHC	0.12	0.66	nd
DDD	0.09	0.69	nd
Heptachloro-epoxide	0.08	0.79	nd
α-BHC	0.07	0.35	nd
op-DDT	0.02	0.16	nd

Table 26.3 Organochlorine pesticides in human liver: decreasing concentrations ($\mu g\,g^{-1}$ extractable lipids)

	Average	Maximum	Minimum
HCB	4.34	37.54	0.63
DDT total	2.51	16.53	0.10
pp′-DDE	2.06	10.04	0.08
DDD	0.34	1.20	0.05
pp′-DDT	0.27	2.72	0.05
Dieldrin	0.26	1.51	0.06
Heptachloro-epoxide	0.14	0.46	0.02
γ-BHC	0.09	0.41	nd
α-BHC	0.025	0.36	nd
op-DDT	0.02	0.41	nd

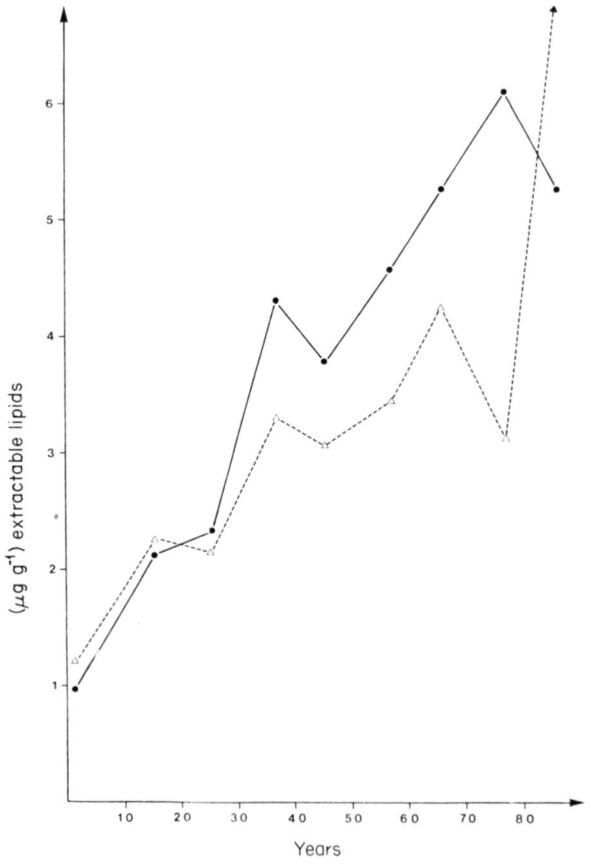

Fig. 26.4 HCB (●) and total DDT (△) content in human adipose tissue

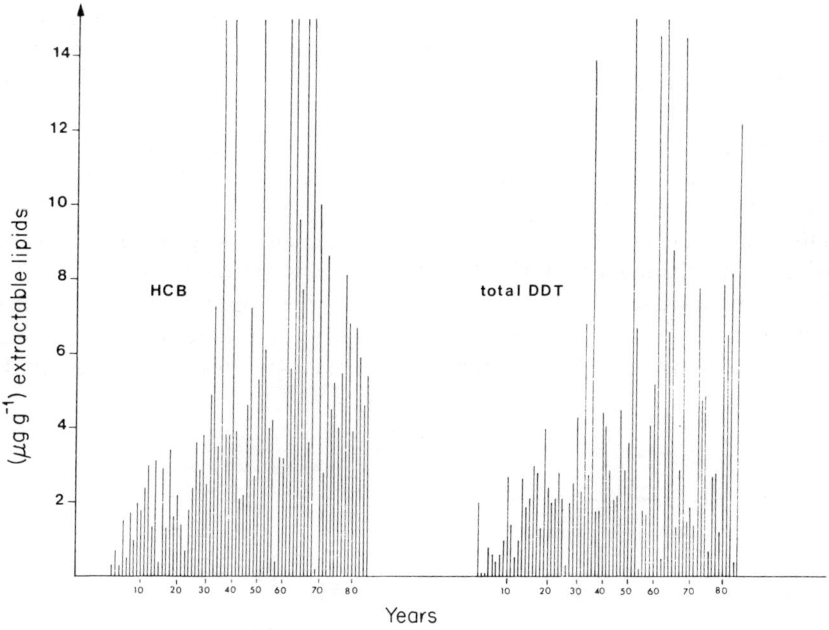

Fig. 26.5 HCB and total DDT content in human adipose tissue (relation to age)

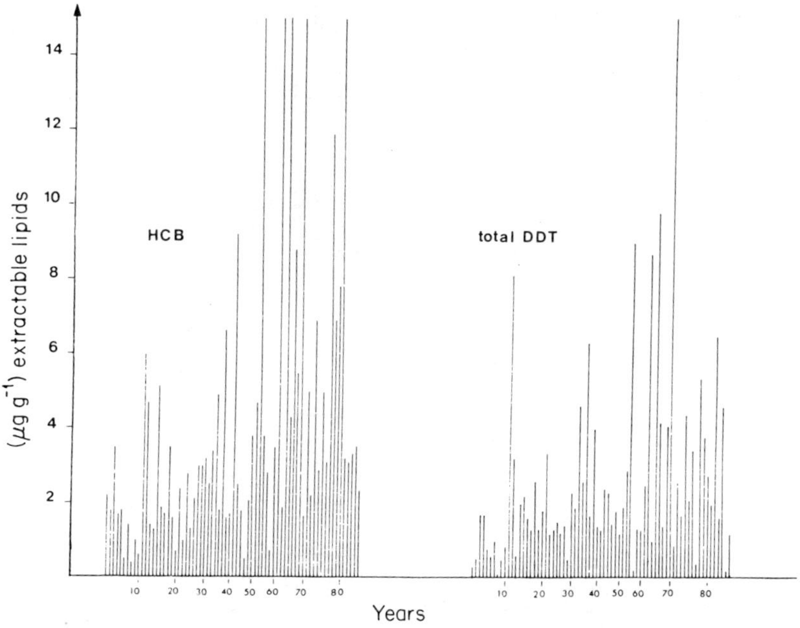

Fig. 26.6 HCB and total DDT content in human liver (relation to age)

Probably there is a slight ascent in the HCB-concentration but not at all in the total DDT-content. It is evident, that age may influence the analytical results at least for the HCB-concentration. Therefore comparisons of the results of other collectives may become difficult.

Another criteria, influencing the comparability of results received by different laboratories, is the refernce basis. On all the figures shown before, the concentration was noted as $\mu g\,g^{-1}$ *extractable lipids*. Organochlorine pesticides accumulate in the lipid fraction of a tissue or organ and, therefore, the higher the lipid content, the higher the OCP content. Comparing the same organ of two different persons, or different organs of the *same* person—expressing their concentrations in $\mu g\,g^{-1}$ *tissue* would probably cause false low or false high results, according as the lipid content in these tissues is low or high, respectively.

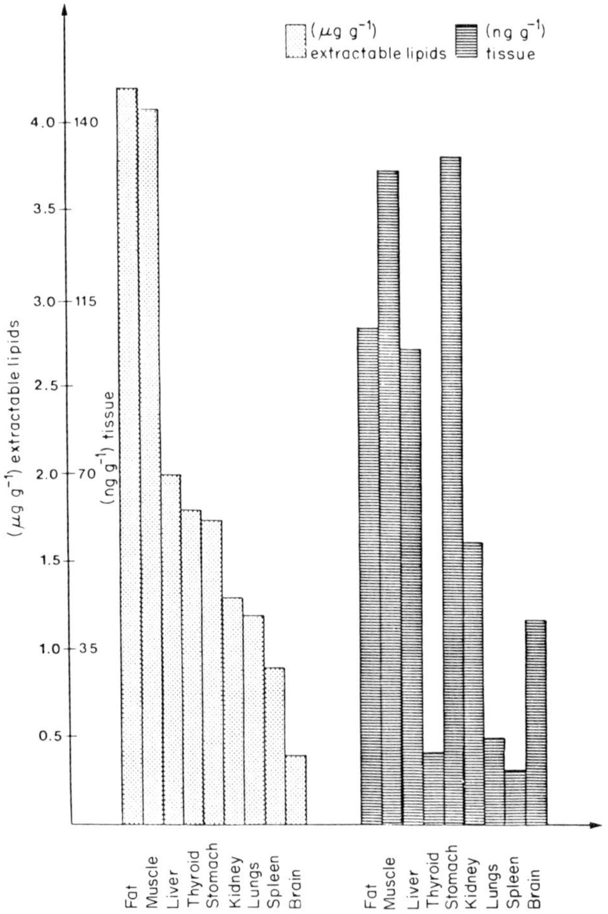

Fig. 26.7 Total DDT distribution in human organs

Therefore, noting results in $\mu g\,g^{-1}$ extractable lipids is the only possible way to compare results, but on the other hand, expression of results on this relation base *only* could never give an absolute OCP content, although knowing the organ weight.

In Figure 26.7 this influence of the relation basis on the distribution pattern in nine different human organs is presented. The columns on the left side show the concentration in $\mu g\,g^{-1}$ extractable lipids in a decreasing grade. Fat and muscle reach highest values, spleen and brain the lowest.

As would be expected, the DDT concentrations in relation to fat basis are not

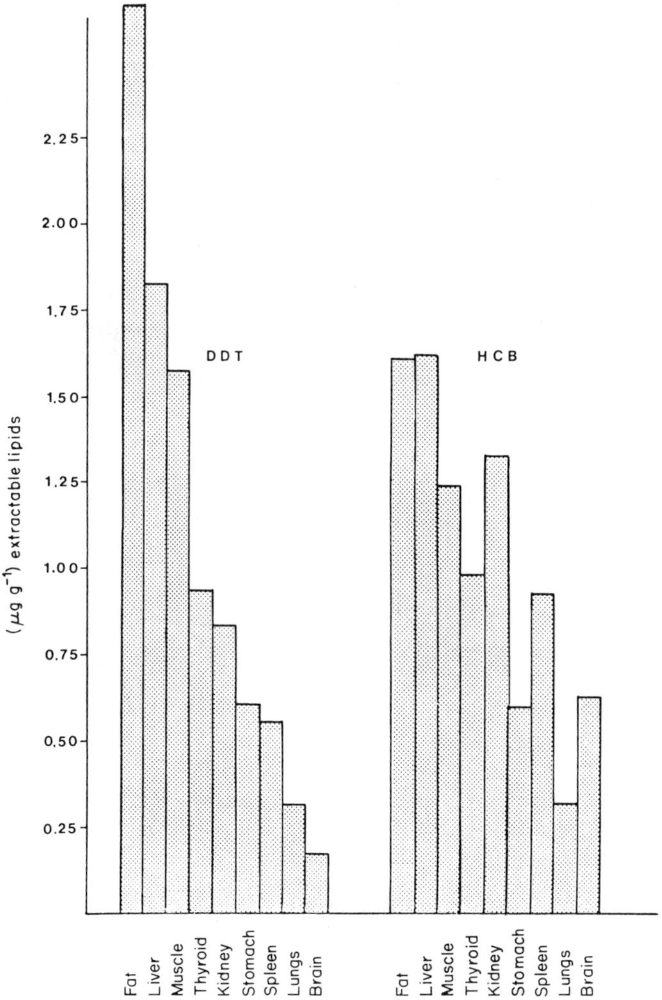

Fig. 26.8 Total DDT and HCB distribution in human organs

quite the same in each organ, but more or less different. A possible explanation may be the different extent of blood circulation in these organs. The columns on the right side are arranged in the same sequence: fat, muscle, liver, thyroid, stomach, kidney, lungs, spleen and brain, but the highest value is now in the stomach, followed by muscle, fat and liver, the lowest in thyroid and spleen. Fat, muscle, stomach and liver of this man are the tissues with the highest lipid content, the explanation for the high values received on this reference basis.

With regard to these differences it is advisable that the analytical results have to be given together with data, allowing recalculation to a wet-weight basis, and allowing comparisons and statistical evaluations under different aspects.

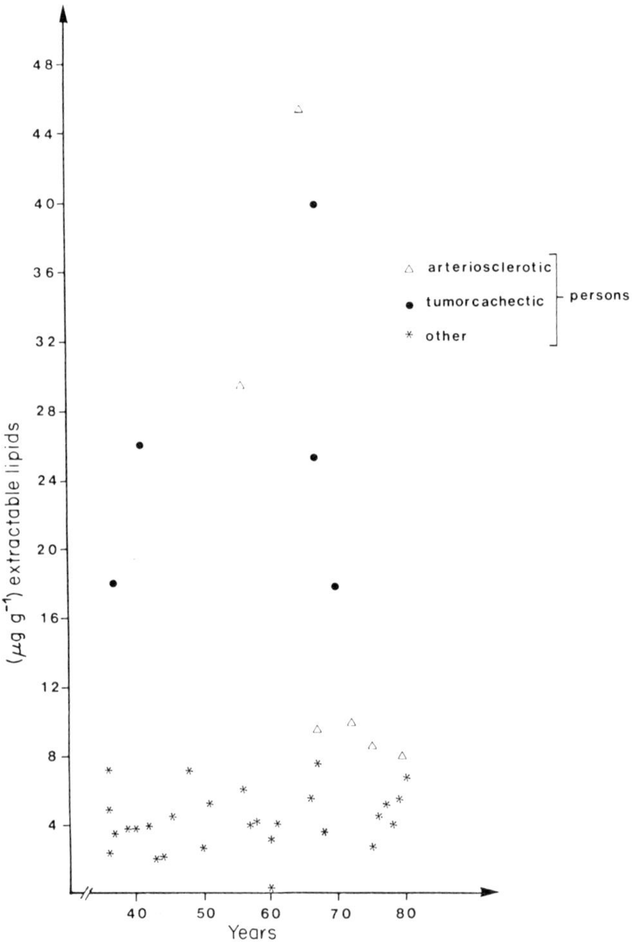

Fig. 26.9 HCB concentration in human adipose tissue. Persons: △ arteriosclerotic; ● tumorcachectic; * other

The DDT content in individual tissues is not necessarily representative for the distribution of *other* pesticides in these tissues.

In Figure 26.8 the distribution of DDT is shown in comparison with that of HCB. It is obvious that the relative distribution of DDT and HCB is different with regard to these compounds. With regard to other pesticides analysed, there are similar changes in distribution pattern.

An impressing example for the influence of some diseases on the distribution and concentration of pesticides in human organs is given in Figure 26.9. Shown is the HCB concentration in human adipose tissue in tumorcachectic, in arteriosclerotic, and in people who died by another disease. The reference is $\mu g\,g^{-1}$ extractable lipids. The HCB content ranges from about 0 to 46 ppm. None of the cachectic patients—only the extremely emaciated persons are

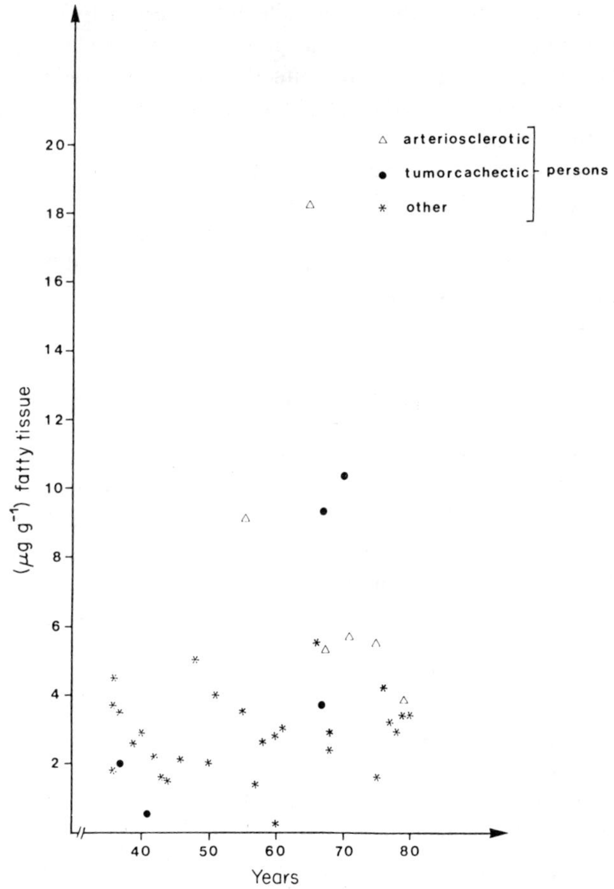

Fig. 26.10 HCB concentration in human adipose tissue. Persons: △ arteriosclerotic; ● tumorcachectic; ∗ other

considered—has a HCB concentration within the so-called normal range. The difference within the arteriosclerotic people—only those with serious sclerotic signs are considerd—is much larger, but here none of the values is within the normal range, too.

The explanation for the high content in cachectic people may be found in the shifted lipid content of the adipose tissue. The effect of starvation results in an *increase* of pesticide content in fatty tissue. The normal ratio adipose tissue to lipids is about 1:0.6. In these patients, however, the ratio increased to 1:0.2. The conclusion of this condition may be that the pesticide level has diminished during the time of emaciation, but not with the same speed as the quota of lipids, so that the persisting amount of OCPs, referred to a smaller portion of lipids, simulate higher values.

Figure 26.10 may be the proof for the former conclusion. It presents the HCB content in the fatty tissue of the same collective referred to *tissue* basis. The difference between cachectic and 'normal people' is compensated although the conten in *arteriosclerotic* persons is not effected by the changed reference basis. Probably the explanation for the latter group may be the restricted blood circulation in these people.

Analytical measurements of persisting pesticides in the environment and especially in human beings are important and necessary for toxicological and legal evaluations. But it has to be considered that the comparability of results gained by different laboratories, of the different collectives or different countries is difficult or even impossible because of the varying conditions and numerous other parameters influencing the results in a very high manner.

HEAVY METALS

From the view of the toxicologists there are two forms of poisoning, in many cases differing in symptoms: the acute and the chronic intoxication. With regard to environmental exposure the symptoms of chronic poisoning due to continuous or intermittent incorporation of low doses, are important and will be reviewed below for lead, cadmium and mercury.[1–4]

Lead

In Figure 26.11 some base points of toxicology of lead are listed. The possibilities of exposure may be ranked: most important as environmental hazards are the working area, paint pigments and car exhaust gases. The importance of organo-lead compounds as antiknock agents in environmental toxicity is based upon their precursor role for inorganic lead produced in the burning process.

Target organs of toxicity of lead are: bone marrow, muscles of stomach and gut and (as almost all heavy metals) the central nervous system. The toxic mechanism of lead in these organs is not yet clear in all points. In bone marrow

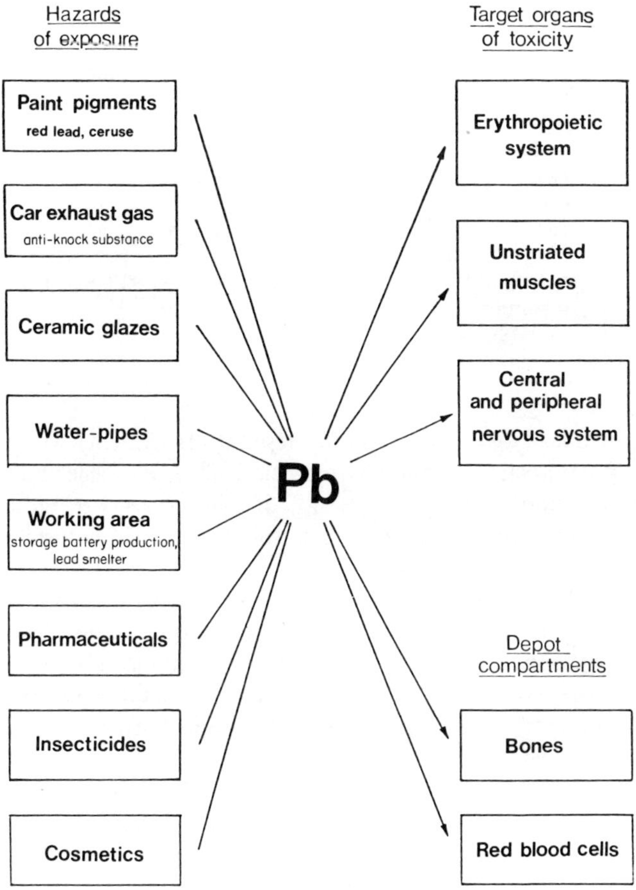

Fig. 26.11 Lead

lead involves a suppression of haem synthesis by interference with several enzymes. This results in an increased urinary excretion of various precursors.

Most important for diagnostic purposes is the depression of δ-aminolevulinic acid dehydratase, causing an elevation of the urinary ALA concentration. Moreover, enhanced ALA can pass through the blood–brain barrier and possibly cause some of the neurological effects of lead toxicity.

Table 26.4 shows the typical symptoms of lead poisoning in workers working in metal cutting (scrap metal work etc.) with lead paint pigments.[5] It can be seen, that anaemia and dysfunction of the gastrointestinal tract are common; serious effects on the brain are rare.

The gingival 'blue line' (similar in other heavy-metal poisoning) is found only when there is poor dental hygiene: sulphur bacteria produce hydrogen sulphide, which combines with lead to form lead sulphide at the the gingival margin.

Table 26.4 Symptoms of lead poisoning in workers

	Common
Fatigue	
Anaemia	
Abdominal pain	
Blue line	
Gastrointestinal disturbances	
Headache	
Metallic taste	
Irritability	
Insomnia	
Convulsions	
	rare

Deposition of lead takes place in bone (depot compartment). More than 90 per cent of the total body lead is found here as relatively insoluble lead phosphates. It is particularly found in the growing ends of long bones. For monitoring lead exposure, measurement of blood lead is the most obvious test, but must be combined with other parameters.

The resulting limits of exposure degrees are not defined exactly in all countries. The upper limits of *normal* levels are considered to be: blood lead $400\,\mu g\,l^{-1}$, urinary coproporphyrin $150\,\mu g\,l^{-1}$, urinary δ-aminolevulinic acid $4\,mg\,l^{-1}$. Values above these may be acceptable; the dangerous range of lead absorption yields: blood lead, $>1000\,\mu g\,l^{-1}$, urinary coproporphyrin, $>1500\,\mu g\,l^{-1}$, urinary δ-aminolevulinic acid, $>40\,mg\,l^{-1}$.

As mentioned earlier, children are more sensitive to lead exposure than adults due to their greater absorption rate. In a recent study,[6] concerning the imission of a lead-zinc smelter in Nordenham (northern Germany), blood and hair lead of children were analysed. Both were significantly elevated compared with a control group from the isle of Helgoland. Symptoms of lead intoxication were not seen.

In another study,[7] the relation between blood lead in children and intoxication signs are given. Nearly 60 per cent of children with blood-lead levels of $600\,\mu g\,l^{-1}$, when suppression of ALA-D becomes evident, were without symptoms. With $1000\,\mu g$ Pb l^{-1} 30 per cent showed no signs of intoxication.

Table 26.5 Tests for monitoring workers exposed to lead

Lead in blood
Lead in urine
Coproporphyrin in urine
δ-aminolaevulinic acid in urine
δ-aminolaevulinic acid dehydratase of red blood cells
Photoporphyrin in erythrocytes

Mercury

All mercury compounds are poisonous, particularly the fat-soluble methyl mercurials and the water-soluble mercury salts. From the toxicological and environmental point of view this classification into fat- and water-soluble compounds gives important characteristics for the difference in effects of these two groups on human organism.

Organocompounds of metals may more easily pass the various membranes in the body, resulting especially in an increased toxicity to the central nervous system.

The toxicity of organomercurials must be mentioned separately because there is a complicated cycle of mercury in nature. Micro-organisms are able to transform mercury salts to metallic mercury, or methyl mercury compounds.

Target organs of chronic mercury toxicity are shown in Figures 26.12 and 26.13.

1. Changes and symptoms in the mouth are observed only in severe intoxications.
2. Proteinuria at continued exposure to mercury with massive losses of albumin and alpha-2-globulin is found only in exceptional cases of chronic mercury poisoning. In microscopic examination the renal cells (glomeruli) do not show any abnormality. The mechanism of the proteinuria is unknown.
3. Mercurialentis is a greyish-brown or yellow reflex from the lens, caused by a deposit of mercury. This reflex is not combined with ocular disturbances.
4. The main target organ of toxicity of mercury compounds is the central nervous system. The first symptoms often falsely diagnosed as neurasthenic syndrome or hysteria are called micro-mercurialism, because the clinical picture is based mainly on a minor intensity of classic symptoms of mercurial erethism.

Details concerning this syndrome may be used in the diagnosis of mercury intoxication. The massive form of central nervous system symptoms is seen rarely and only in workers with excessive exposure. One of the most characteristic features of mercurialism is the presence of mercurial tremor in clinical symptomatology. The tremor develops gradually in the form of fine trembling of the muscles beginning in the fingers.

Some examples of environmental intoxication due to methyl mercury compounds will now be reviewed. Seed grain is normally treated with organomercurials as fungicides, then highly toxic and marked as poison. In 1956, in northern Iraq, seed grain was imported, but people would not wait until the grain has grown. They saw the poison symbols on the bags and decided to make a biological test: they fed the mercury-treated seed to chickens for several days and after observing no ill effects they ate it themselves; some days later the chickens

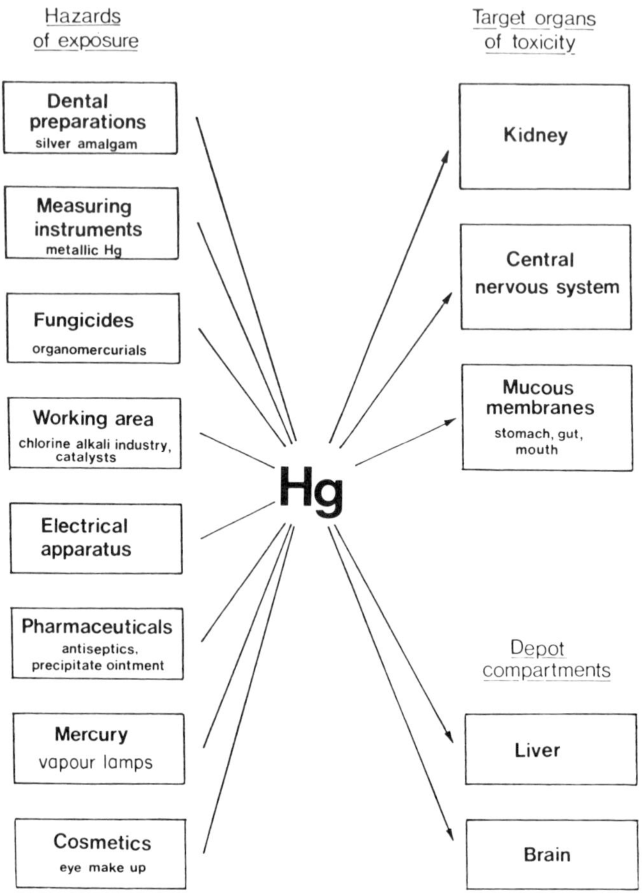

Fig. 26.12 Mercury

died. In the following weeks several hundred severe intoxications in man appeared with typical signs of central nervous system toxicity.

The most serious environmental mercury intoxication took place in the years from 1956 onwards in southern Japan. The so-called Minamata disease[8] was caused by ingestion of fish and shellfish contaminated by mercury discharged from a chemical factory of the Chisso Corporation, a major plastic manufacturing plant.

In 1959 mercury concentration in the sludge near the drainage channel reached 2010 $\mu g\ g^{-1}$ (wet weight).[9] At that time the Chisso Corporation was using a large amount of inorganic mercury in a synthesis process and continued to discharge mercury-polluted effluents until 1968. Total mercury dumped into the Minamata Bay was calculated to be as much as 200–600 tons.

The effects on the ecosystem were dramatic: dead fish floated on the surface of

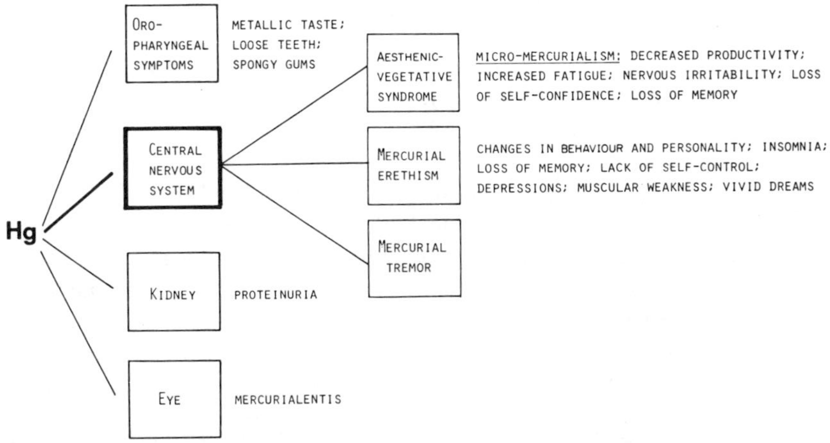

Fig. 26.13 Chronic mercury poisoning (inorganic Hg)

the sea. In 1952 birds began to drop into the sea, apparently unable to fly. On the sea surface octopus floated, so weakened that children could catch them with their bare hands. First intoxications of mammals appeared in 1953 in cats and dogs. The symptoms lead to the description of the malady as 'Dancing cat disease'. Up to now 3000 humans were poisoned by this methyl mercury, especially persons with high consumption of fish and shellfish. Symptoms of chronic Minamata disease appear for the first time possibly caused by former brain damage surfacing now as a result of age. Originally it was believed, that the cause of poisoning was biomethylation of inorganic mercury. Later it became known, that the factory was discharging methyl mercury in its effluents. The symptoms of methyl mercury intoxication are caused by damage of central nervous system tissue.

In pregnant women with only mild symptoms methyl mercury passes from the mother to the foetus, causing intoxication called congenital Minamata disease. Massive mercury incorporation may be seen in the analytical data obtained from autopsy cases showing several times higher mercury concentrations than in other diseases (Table 26.6).

Table 26.6 Mercury content in human organs (μg/g wet weight)

	Autopsy cases of Minamata disease	Autopsy cases of other diseases
Liver	22.0–70.5	0.18–1.06
Kidney	12.8–144.0	0.25–3.02
Brain	2.6–24.8	0.05–1.54

Source: Harada.[9]

Cadmium (Figure 26.14)

Cadmium is surely the most dangerous heavy metal in the field of health aspects because of its extremely extended half time in human organism. For renal cortex a half time of about 30 years has been calculated. Investigations on the content of cadmium in different parts of all ecosystems are therefore necessary with high priority to estimate the health risk and draw up balances as shown in Figure 26.15. A continuous load of most tissues with cadmium can be seen. Figure 26.16 shows some of our own results of autopsy material. Cadmium content of kidney and liver from birth to age 20 years are shown. A rapid load from the first days or month of life is obvious.

Maximum kidney cadmium content is reached at age 50–60 years (Figure 26.17), then a decreasing concentration of cadmium is observed. This decrease is

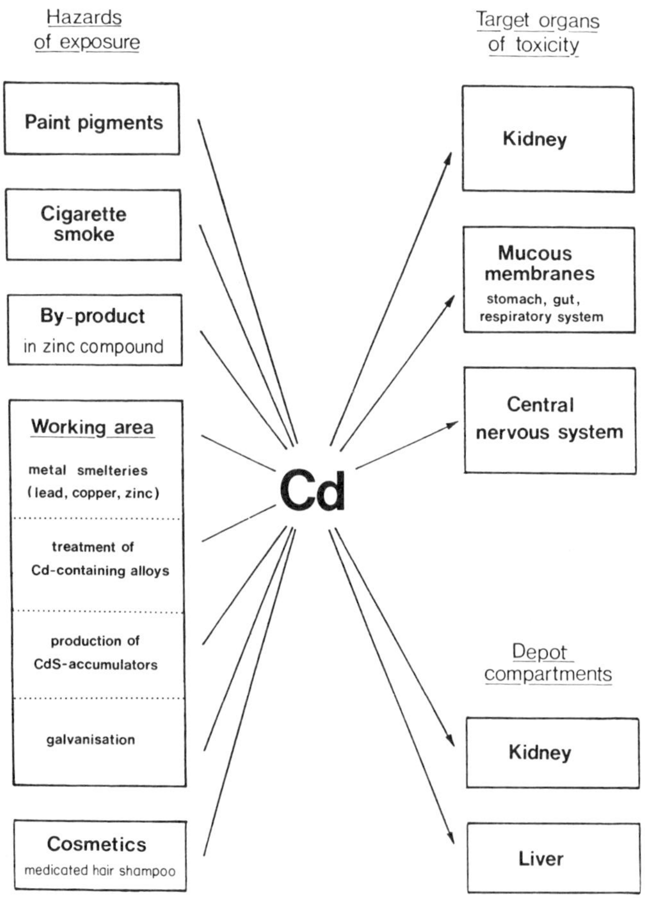

Fig. 26.14 Cadmium

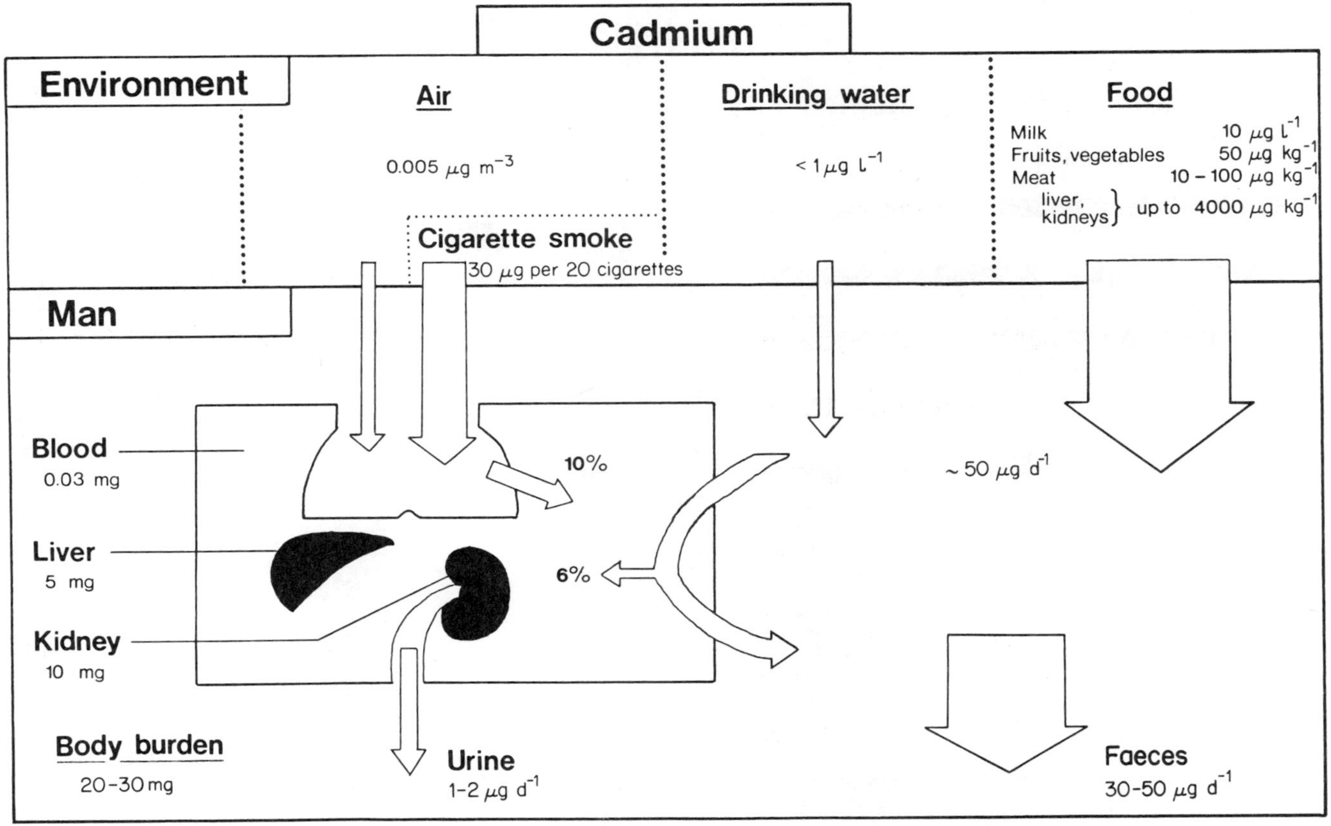

Fig. 26.15 Cadmium

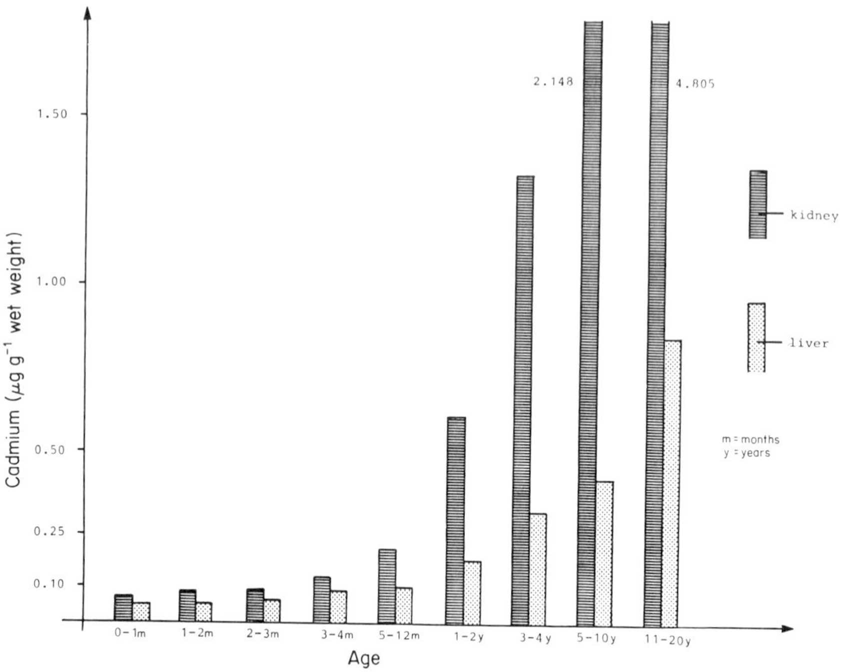

Fig. 26.16 Cd content in liver and kidneys of children

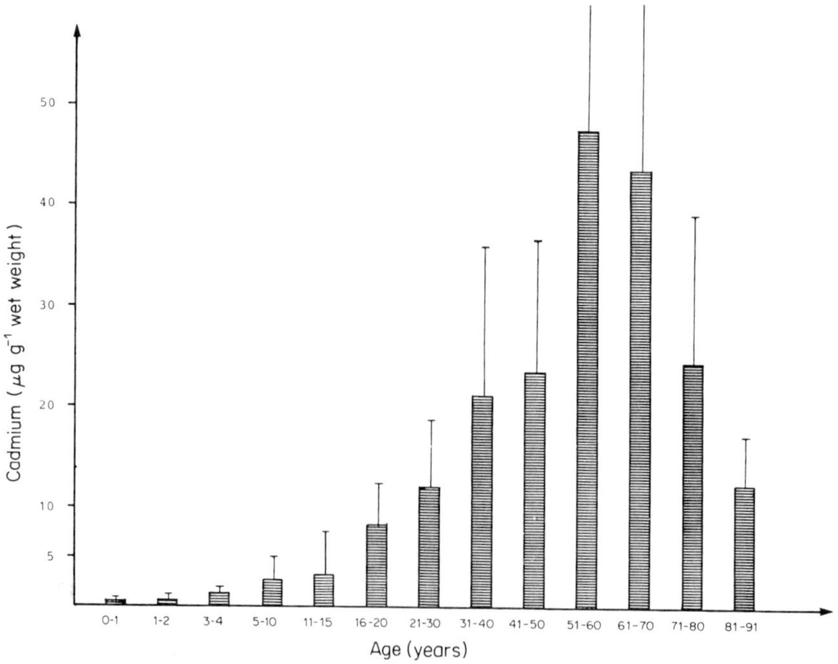

Fig. 26.17 Cd content kidney relation to age

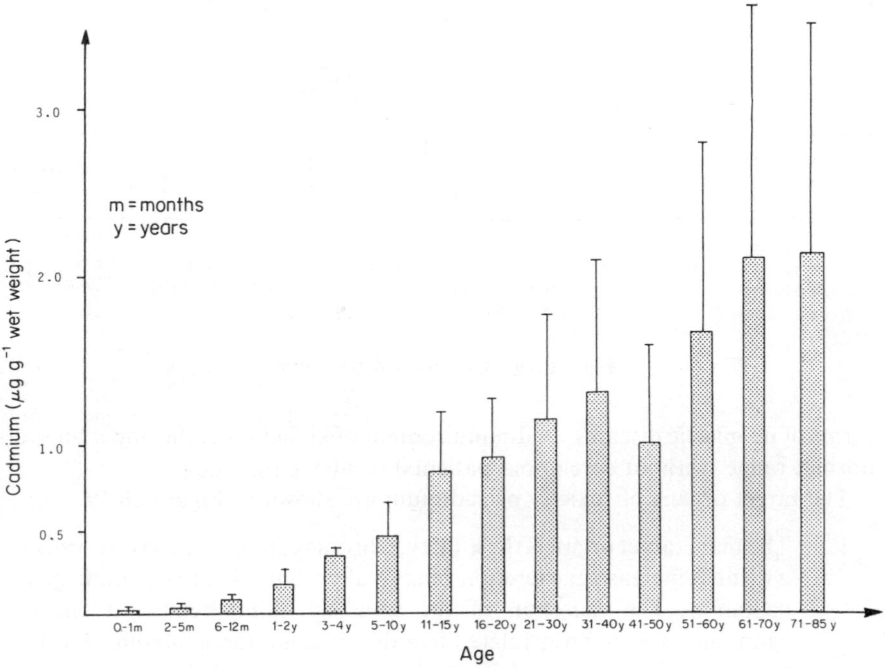

Fig. 26.18 Cd content liver relation to age

described by other authors too.[10] At present it is very difficult to explain these changes. They might be connected with factors related to a low cadmium intake a long time ago and with different food habits in older age groups. But these explanations are not perfectly satisfactory. Other tissues, such as liver from the same persons, show no decreasing trend (Figure 26.18).

Perhaps metabolic changes in kidneys in old age are of great importance. Decreased lipid content of organs or variation of organ size may explain another cadmium kidney effect. Table 26.7 shows four cases of patients with different

Table 26.7 Correlation cadmium content kidney carcinoma

Age	Sex	Form of cancer	Cd kidney $\mu g\ g^{-1}$ (ww)	Range of groups without Ca patients	
				Age	Cd $\mu g\ g^{-1}$
16	F	pharyngeal	22.22	15–20	4.53–11.22
37	F	mamma, metastasising	21.66	31–40	7.75–17.47
70	M	pancreatic	34.77	71–80	9.05–35.52
76	M	esophageal	42.53		

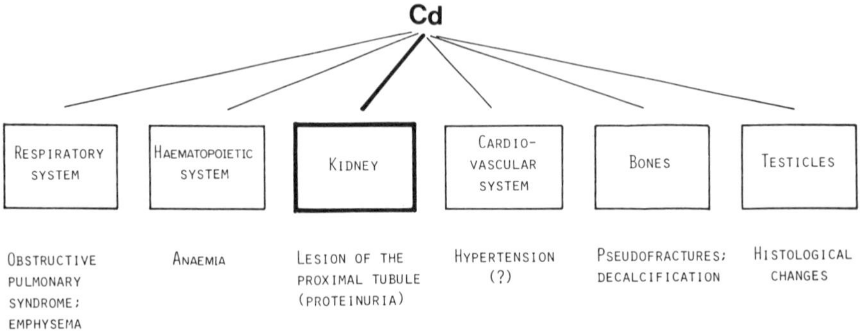

Fig. 26.19 Chronic Cd poisoning

forms of neoplastic diseases. Cadmium content of kidneys is at the upper limit of normal range (without carcinoma patients) or above this range.

The target organs of toxicity of cadmium are shown in Figure 26.19.

1. The main target organ is the kidney. Chronic cadmium exposure leads to a distinct increase in excretion of several proteins due to cell damage by cadmium. The excretion of the so-called 'minialbumin' in chronic cadmium exposure was related to a defect in immunoglobulins, but this theory was not yet proved. The other target organs of chronic toxicity of cadmium are of lesser importance.
2. In the respiratory system cadmium oxide dust or fumes may cause emphysema.
3. Chronic cadmium exposure as a cause for hypertension in humans is until now a controversial theory. There are only some animal experiments possibly indicating this effect.
4. Testicles are damaged mainly in acute poisoning.
5. Decalcification of bones may lead to serious clinical pictures, as seen in a severe environmental cadmium poisoning in Japan, called 'Itai-itai disease':

The cadmium contaminated waste waters of the biggest zinc mine in Japan were used for the irrigation of rice fields; great damage to rice plants appeared and people living there incorporated large amounts of cadmium over years by contaminated rice and river water. Since the Second World War hundreds of intoxications have appeared, until it was officially declared in 1971 that this disease is mine poisoning. The intoxication is characterized by extreme loss of minerals from the bones in combination with intense pain.

Accumulation of toxic metals in tissues of poisoned persons are unbelievably high: e.g. the femur from the body of an extremely severe case contained 15000 μg cadmium g^{-1} which is 1.5 per cent![11]

Some animal, and other studies, showed later that there could be, in addition to the increase of cadmium uptake, a nutritional deficiency of calcium and/or vitamin D in the poisoned patients. This leads to the important question of metal interferences. There are innumerous interrelations between trace elements, not only between essential metals, also in non-essential and in bulk elements. These interactions may be kinetic or metabolic and are particularly powerful among the environmental contaminants. An important aspect is, for example, the detoxifying role of selenium for mercury and other heavy metals.

Figure 26.20 shows our own results: increasing cadmium content of kidneys is related to an increase of zinc in the same tissue specimens, indicating the induction of certain detoxifying proteins, which may bind cadmium, zinc and other metals in the same way. To reveal possible interactions, we developed a 'metal fingerprint', containing the results of analysis of several elements in one picture. Figure 26.21 shows an example. This is a fingerprint of a female proband for blood and hair. The arrows indicate the concentrations within the normal

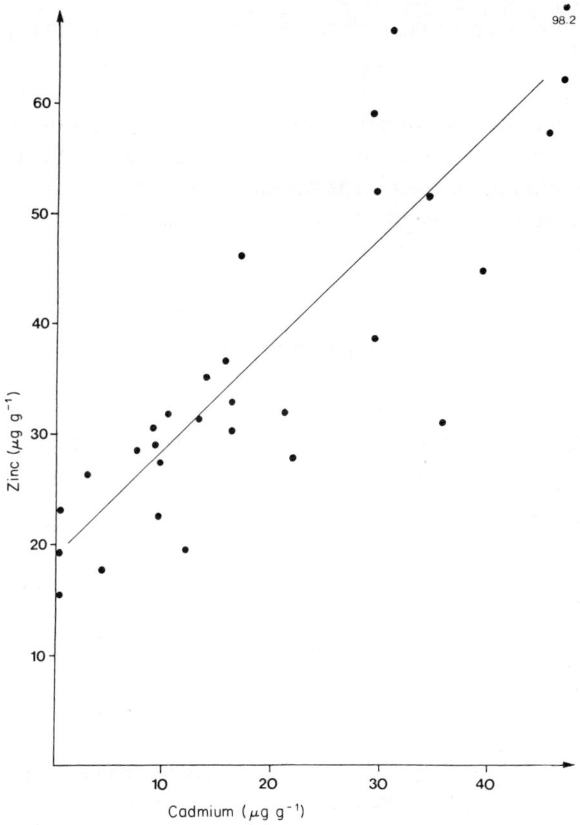

Fig. 26.20 Cd/Zn relationship human kidney

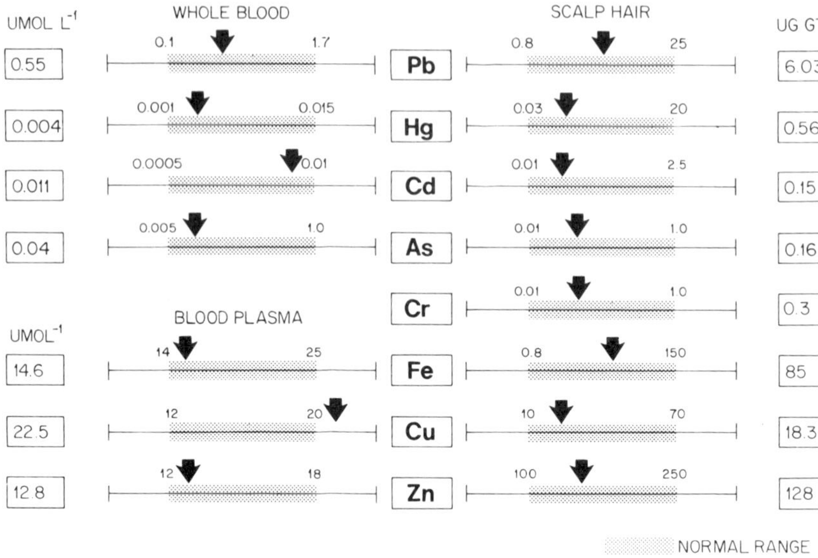

Fig. 26.21 'Metal fingerprint' RTM 2055

range. For example the increase in copper and decrease in zinc blood concentrations may be related to the intake of an oral contraceptive.

Interactions among the trace elements may be so powerful, that environmental studies involving single elements can lead to dangerously erroneous conclusions.[12] Moreover the pattern of metals may give more detailed information on possible pathogenetic processes and causal statements.

REFERENCES

1. Prasad, A.S., and Oberleas, D. (Eds). (1976). *Trace Elements in Human Health and Disease*, Vol. II, *Essential and Toxic Elements*, Academic Press, New York, San Francisco, London.
2. Friberg, L., Piscator, M., Nordberg, G., and Kjellström, T. (1974). *Cadmium in the Environment*, 2nd edn, CRC Press, Cleveland, Ohio.
3. Friberg, L., and Vostal, J. (1972). *Mercury in the Environment*, CRC Press, Cleveland, Ohio.
4. Goldwater, L.J., and Stopford, W. (1977). *Mercury*, in, *The Chemical Environment* (Eds J. Lenihan and W. W. Fletcher), p. 38, Blackie, Glasgow, London.
5. Moore, M.R., Campbell, B.C., and Goldberg, A. *Lead*, in: *The Chemical Environment* (Eds J. Lenihan and W. W. Fletcher), p. 64, Blackie, Glasgow, London.
6. Thron, H.L., Engbert, N., Krause, Ch., Laskus, L., and Sonneborn, M. (1978). *Bleibelastung von Bevölkerungsgruppen, Bericht des Instituts für Wasser-, Boden- und Lufthygiene des Bundesgesundheitsamtes*, Berlin.
7. Guinee, V.F., Davidow, B., and Tytun, A. (1975). Clinical and environmental correlations with blood lead levels of children in New York City, in, *Proceedings of the International Symposium 'Recent Advances in the Assessment of the Health Effects of Environmental Pollution'*, Vol. II, p. 545, Luxembourg.

[8] Tsubaki, T., and Irukayama, K. (Eds). (1977). *Minamata Disease*, Kodansha, Tokyo, Elsevier, Amsterdam, Oxford, New York.

[9] Harada, M. (1978). Methyl mercury poisoning due to environmental contamination, in, *Toxicity of Heavy Metals in the Environment*, Part 1 (Ed. F. W. Oehme), p. 261, Marcel Dekker, New York, Basel.

[10] Curry, A.S., and Knott, A.R. (1970). 'Normal' levels of cadmium in human liver and kidney in England, *Clin. chim. Acta*, **30**, 115.

[11] Kobayashi, J. (1978). Pollution by cadmium and the Itai-Itai disease in Japan, in, *Toxicity of Heavy Metals in the Environment*, Part 2 (Ed. F. W. Oehme), p. 199, Marcel Dekker, New York, Basel.

[12] Underwood, E.F. (1978). Interactions of trace elements, in *Toxicity of Heavy Metals in the Environment*, Part 2 (Ed F. W. Oehme), p. 199, Marcel Dekker, New York, Basel.

Pollutants and Their Ecotoxicological Significance
Edited by H. W. Nürnberg

CHAPTER 27

Ecogenetics

M. GELDMACHER-V. MALLINCKRODT

Ecogenetics denotes a hereditary predisposition for an individual reaction to environmental influences. We may interpret it as an interaction of genetic inheritance and environment. Forms of ecogenetic reaction are evident in all parts of animated nature: in bacteria, plants, insects, fish, mammals and also in man.

Quite naturally this phenomenon is particularly interesting to man. It provides an important framework for the consideration of problems and questions dealing especially with pharmaceuticals, industrial substances, pesticides and insecticides, radiation, exhaust gases, poisons found in the environment, but also natural substances and even some types of foods.

The effect of the substances mentioned can be most diversified. In some individuals poisoning may be marked, while the health of other individuals will not be affected. Such problems are, therefore, very important from the standpoint of preventive medical care and occupational and industrial health. These problems, however, also influence the development and testing of pharmaceutical products, planning of food exports, breeding of suitable strains of plants etc.

The earliest observation of an ecogenetical reaction may date back to the year 1852.[33] When rabbits were fed leaves of the belladonna plant, an unexpected phenomenon was noticed. While some of the animals died, no adverse effects were observed in others. At the time of this discovery, no explanation of the discrepancy was possible. Fifty years later, at the beginning of this century, the answer was found. The sera of the rabbits surviving after eating the leaves of the belladonna plant contained an enzyme, atropine esterase, which metabolizes atropine into tropic acid and tropine.

Cross-breeding rabbits with high and with no atropine esterase activity in the

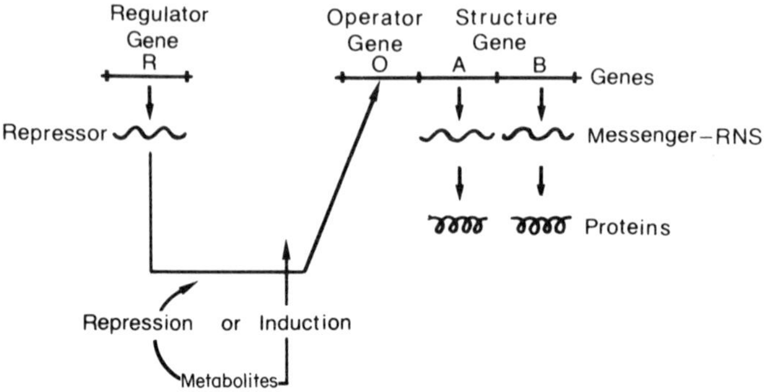

Fig. 27.1 Jacob and Monod's hypothesis of proteinsynthesis regulation

serum yielded animals with a median activity in the F-1 generation. In the subsequent generation, in accordance with Mendel's laws, rabbits with no median and high activity in the ratio 1:2:1 were found. This result allows the interpretation that the presence of atropine esterase is linked to a single gene with two alleles.

The clue to understanding the basic life processes of ecogenetics may be derived from the fact that many of the processes we are interested in within this context involve enzymes. Enzymes are proteins, and the synthesis of proteins in organisms is genetically controlled and determined. To explain the control of the synthesis of proteins, Jacob and Monod (cited in reference 14) formulated the simplified hypothesis depicted in Figures 27.1 and 27.2.

While we are most interested in the problems associated with man, ecogenetic responses in other living organisms are also important for us since these

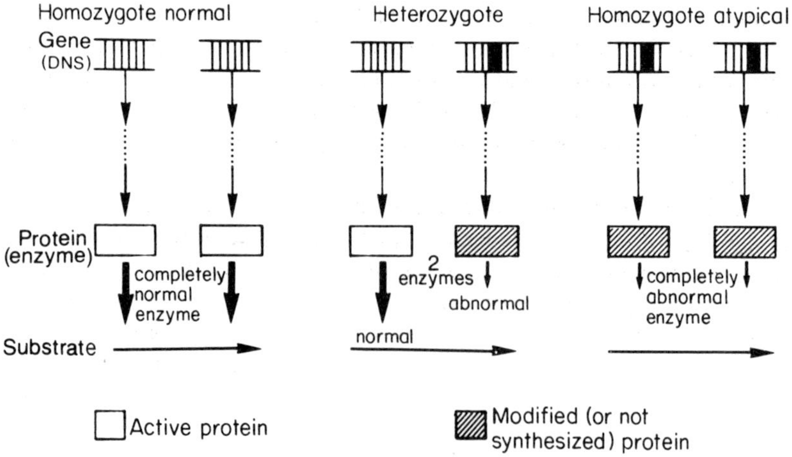

Fig. 27.2 Gene–dose effect in the synthesis of enzyme variants

organisms either represent sources of human food, or are enabled by genetical disposition to accept food with a modified or changed composition. For these reasons several ecogenetic processes in simple organisms should be reviewed.

BACTERIA

In a certain strain of common *Escherichia coli* two different enzymes were detected[35] which are responsible for a genetically determined resistance to inorganic and organic mercury compounds. Both enzymes were isolated.

1 An organo-mercury hydrolase, decomposing the mercury–carbon link in compounds such as methyl-mercury-chloride, phenyl-mercury-acetate and ethyl-mercury-chloride.
2 A mercury reductase, which reduces bivalent mercury to metallic mercury. Subsequently mercury escapes in the form of vapours from the culture medium, allowing renewed growth of the bacteria.

Corresponding resistance phenomena were observed in a strain of *Pseudomonas*.

PLANTS

It is well known that in soils contaminated with heavy metals only those plant populations which are genetically tolerant to these metals will grow. Accordingly plants exist which are resistant to lead (Pb), zinc (Zn), copper (Cu) and also to cadmium (Cd),[26,37] while plants not tolerant in this respect would perish in the soils containing high concentrations of these metals.

WORMS

Even in some species of earthworm varied sensitivity to heavy metals can be found.[19] Considerable differences in the concentrations of Zn and Pb in different species were detected.

INSECTS

Reports on variations of the ADH enzyme of the fruit fly, *Drosophila melanogaster*, which metabolizes ethanol into acetaldehyde, have been published.[21] Well known in insects are the genetically determined resistances to pesticides which respectively manifest themselves.[20]

MAMMALS

Certain inbred strains of mice are resistant to Cd necroses of the testes.[42] This property is governed by a single autosomal recessive gene. The testes of these mice assimilate less radioactively labelled Cd than the testes of other strains of mice sensitive to Cd. Although the protective mechanism is not known, it may well be assumed that a protein exists which binds Cd and is specific for these organs. Such proteins are known for the transport of metals within the organism, particularly metallo-thioneines. In the strain of mice resistant to Cd, the gene responsible for this protective action is probably located in chromosome 12.

The resistance to atropine in rabbits has already been mentioned. A hereditary Zn deficiency in mice milk has also been observed.[29] Suckling newborn mice, drinking exclusively the milk of mice homozygous for this lethal mutation, cannot survive. If the young mice were additionally offered Zn, the mortality was drastically reduced. It is assumed that the gene responsible for this Zn deficiency modifies the transfer of Zn from the maternal blood into the milk. Another animal model is the lethal trait A46 in cattle. Symptoms are paraceratosis, thymic hypoplasia, hereditary Zn deficiency.[37]

MAN

It is highly probably that genetically determined disorders of Zn metabolism are also responsible for the human disease acrodermatitis enteropathica which occurs comparatively seldom and is also treated by the administration of supplementary Zn.[28,32] In this section only ecogenetical phenomena concerning man will be reported.

Firstly we are able to describe genetic factors associated with different responses to substances introduced into the body, such as, drugs, alcohol, poisonous industrial products and foods.

Drugs

Besides the previously mentioned study on rabbits, the first observations of genetically determined differences in metabolism or in the response to substances foreign to the body involved drugs.[31] The term 'pharmacogenetics' was coined in 1953 for this branch of ecogenetics. Using drugs as an example, the technique of solving ecogenetical problems can be demonstrated as follows.[14]

Evaluating the effects of a drug on a sizeable number of participating patients, a uniform response distribution with respect to a mean value is generally observed (Figure 27.3a). Variations in the response can be caused by differences in age, sex, food intake, endocrine status etc. Sometimes, however, considerable individual differences in response in the form of a bimodal or multimodal distribution are found (Figure 27.3b). Such discontinuous distributions are highly interesting for the geneticist, because often such occurrances indicate the existence of a specific gene, which may only be recognizable after the administration of a particular drug.

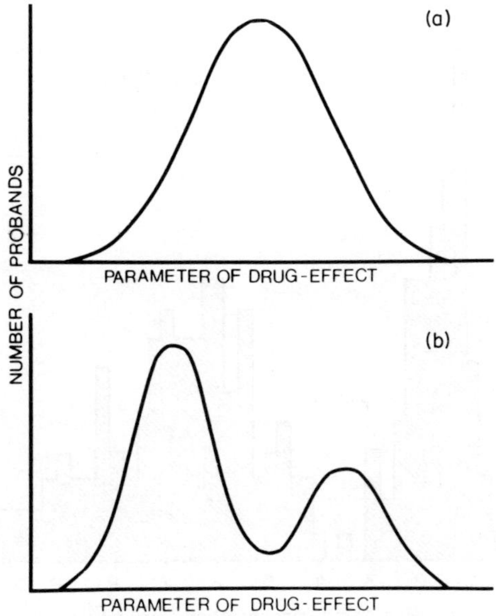

Fig. 27.3 Unimodal and bimodal distribution. (a) Uniform distribution; (b) non-uniform distribution

A very well investigated pharmaceutical example is the difference in the acetylation of INH, a drug used in the treatment of tuberculosis.[34] The plasma levels of INH, determined 6 h after the administration of the drug, exhibit a bimodal distribution in a non-selected sample of the European population (see Figure 27.4). Low plasma levels of INH are associated with a high acetylation rate. In individuals showing a high plasma level of INH, little pharmacologically inactive acetylINH is found.

Investigations involving families and twins proved a gene with alleles for high and low *N*-acetyl-transferase activity to be responsible for the bimodally distributed INH levels in the plasma. In different populations the incidence of individuals acetylating INH rapidly or slowly varies (Figure 27.5).

In Japan approximately 90 per cent rapidly inactivating individuals, in the population of Europe approximately 50 per cent rapidly acetylating individuals are found. Apparently the success of the therapy with INH remains unaffected by the different rates of acetylation. Polyneuropathia as a side effect is, however, observed less frequently in individuals acetylating INH rapidly, damage to the liver more frequently.[34]

Investigations on the substrate specificity of *N*-acetyl-transferase revealed that very likely substances such as INH, sulphadimidine, phenelzine, hydralazine, serotonine and perhaps other pharmaceuticals feature a similar genetically determined polymorphism of the *N*-acetyl-transferase.

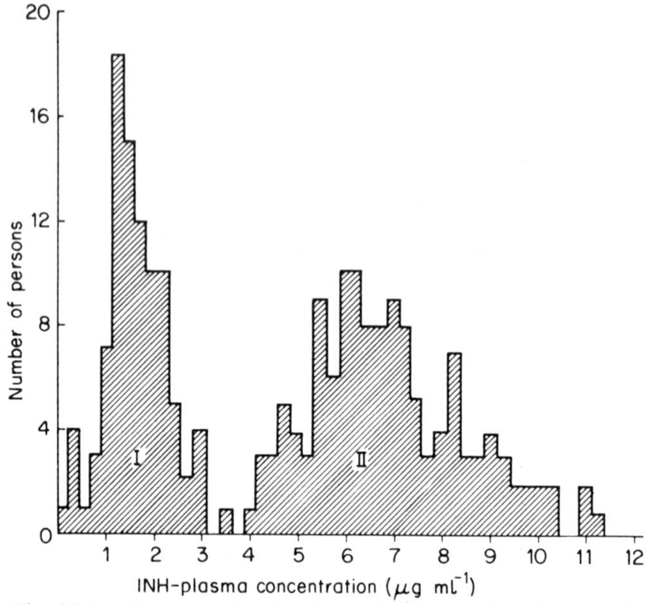

Fig. 27.4 Bimodal distribution of measured values (INH excretion = abscissa) examined in a sampling of 267 probands (ordinate). I = rapid acetylation probands, II = slow acetylating probands

Another group of substances including *p*-amino-benzoic acid, *p*-amino-salicylic acid, sulphanilamide, aniline and others do not exhibit this polymorphism, whereby aniline is important in industrial toxicology.[17]

These findings generally allow the conclusion of the existence of different acetylating systems. There is at least one polymorphous and one monomorphous acetylating system. The enzyme *N*-acetyl-transferase is also interesting because it is involved in the metabolism of certain psychopharmaca, which are alleged to have a role in the serotonine metabolism of the brain.[34] This is a further reason for closer investigations of the biochemical and genetical causes of the various psychoses.

The example just presented can be explained more or less clearly by the differences in the *metabolism* of the administered drug.

A different mechanism is found in another genetically determined disorder, the lack of glucose-6-phosphate dehydrogenase (G-6-PD) in the erythrocytes which manifests itself in a particular reaction of the individuals to drugs, poisonous industrial products and certain types of food.[2]

In Table 27.1 drugs and poisonous industrial products are listed which initiate a haemolytic effect in patients lacking G-6-PD. This disorder represents a lack of enzyme in the red blood cells, resulting in a too low concentration of reduced glutathione. This phenomenon can cause the red blood cells to be more

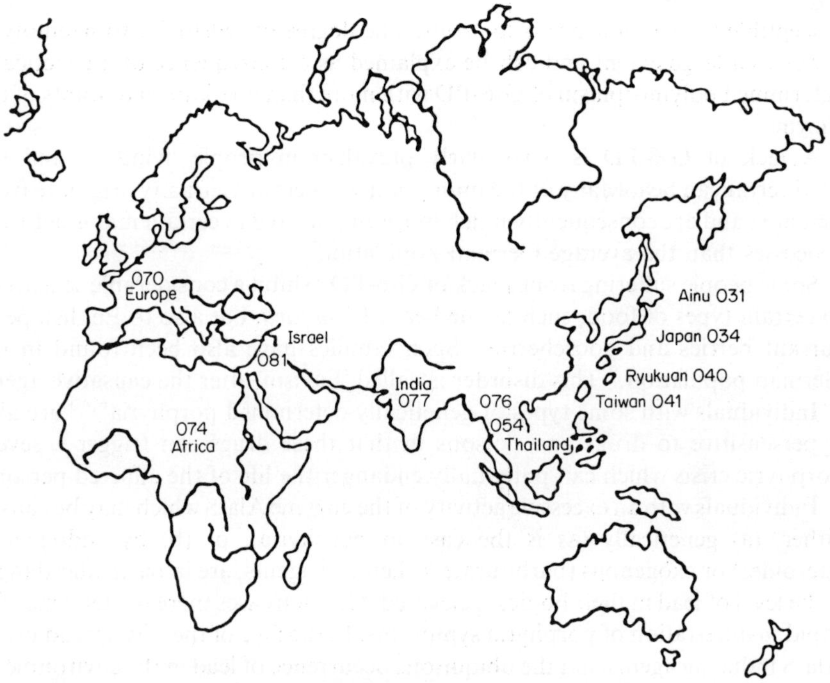

Fig. 27.5 Polymorphism of isoniacide excretion/acetylation in different populations. (Gene frequencies for the characteristic 'slow acetylation'; 0.81 = 81 per cent)

Table 27.1 Selected agents which cause haemolysis in glucose-6-phosphate dehydrogenase-deficient subjects

1	Aniline derivatives
	a *N*-Acetylarylamines: acetanilide, phenacetin (acetophenetidin), [?]paracetamol (acetominophen)
	b Phenylhydrazine
	c Sulphonamides: N^4-acetylsulphanilamide, sulphanilamide, sulphamethoxypyridazine, sulphapyridine, salicylazosulphapyridine (Azulfidine®)
	d Sulphones: diphenylsulphone, thiazolesulphone
2	Chloramphenicol
3	[?]Menadione derivatives
4	Naphthalene
5	Nitrofuran derivatives
	a Furazolidine
	b Nitrofurantoin
	c Nitrofurazone
6	Quinoline derivatives
	a 8-Aminoquinoline antimalarials: Pentaquine, Primaquine, Quinocide
	b [?]Quinine, [?]Quinidine
7	[?]Salicylates (acetylsalicyclic acid, *p*-aminosalicyclic acid)

susceptible to the influence of poisons. The degree of sensitivity to haemolysis varies to a large extent and can be explained as a consequence of a genetically determined polymorphism of G-6-PD with more than fifty known mutants of the enzyme.

A lack of G-6-PD is particularly prevalent in people living around the Mediterranean Sea. Many of the immigrant workers in Germany originate from that area and are consequently much more endangered in certain manufacturing processes than the average German population.

Some people suffering from a lack of G-6-PD exhibit a considerable sensitivity to certain types of food, such as the bean, *Vicia fava*, but also to English peas, currant berries and gooseberries. Such families have also been found in the German population.[9] This disorder is called 'Favism' after the causative agent.

Individuals with some types of genetically determined porphyria[6,40] are also hypersensitive to drugs and poisons in that these drugs can trigger a severe porphyric crisis which can potentially endanger the life of the afflicted persons.

Individuals with an excessive activity of the enzyme Ala-S which may be caused either, (a) genetically (as is the case in porphyria), or (b) by endogenous (steroides) or exogenous (barbiturates) chemical agents, are in particular danger if the level of lead in their bodies is elevated.[4] In such cases there is a tendency for rapid manifestation of porphyria symptoms. In the face of the widespread use of Ala-S inducing agents and the ubiquitous occurrence of lead in the environment, a more precise specification of this health risk would be very important.

Other genetically determined disorders caused by hereditary tendencies to accumulate, or to disregard, biologically important metals such as Cu, Fe[40] or Mg[25] are known in man.

The copper-thioneine of the liver may have a major role in the occurrence of hereditary Wilson's disease.[1] In the course of this disorder, a considerable accumulation of Cu occurs in the liver, since the elimination of excessive Cu via the bile fluid is impaired. Severe disease of the liver or forms of behaviour resembling the symptoms of schizophrenia are the medical manifestations.

Ethanol

Biochemical differences in ethanol metabolism were first proved in mice, but have since also been demonstrated in man.[44,45]

Different enzyme systems metabolize ethanol in man. Liver-alcohol dehydrogenase (ADH) is one of the most important of these enzymes. Here a genetically determined polymorphism is present: (a) the variant found in Europe with a low metabolic rate is termed 'normal' enzyme (approximately 90 per cent of the European population), while (b) the variants with comparatively high metabolic rates are classified as 'atypical' enzymes. The classification as 'normal' applies, however, only to Europeans. In other populations up to 90 per cent of the individuals have the fast metabolizing enzyme.[7]

Table 27.2 Differences in alcohol tolerance and symptoms in Caucasoids and Asiatics. Subjective symptoms after alcohol imbibition in Caucasoid and Asiatic populations

Symptom	Caucasoids ($n = 34$)		Asiatics ($n = 78$)	
	%	Number	%	Number
Heartburn	5.8	2	52.5[a]	41
Tachycardia	2.9	1	43.5[a]	34
Muscle weakness	2.9	1	25.7[b]	20
Dizziness	8.6	3	37.2[b]	24
Fatigue	5.8	2	33.4[b]	26
Sleep induction	0		18.0	4

[a] $p < 0.001$.
[b] $p < 0.01$.

Table 27.2 shows the differences in the tolerance for alcohol in Europeans and Asians, which will probably be manifest in such investigations.[15]

Industrial Products

In this field genetically determined differences in metabolism may result in potential danger to individuals carrying corresponding genes.[41] It has already been mentioned that individuals lacking G-6-PD, who can be found predominantly in populations native to the area around the Mediterranean Sea, are sensitive to numerous chemicals. It is also known that in cases of too low activity of the enzyme α_1-antitrypsine,[15] which is usually genetically determined, lung emphysema occurs at a comparatively early age, and that in children disorders of the liver may even develop.[22]

The enzyme α_1-antitrypsine (PI), a protease inhibitor, is a protein which migrates in the α_1-globuline fraction during serum electrophoresis. The exact function of α_1-antitrypsine is not yet known. The genetic polymorphism of α_1-antitrypsine was first detected in 1963 in patients exhibiting only 10 per cent of the normal activity. Since then many different alleles of this system have been observed (see Table 27.3). Differences in the activity of the enzyme have been found between Europeans, Lapps, the population of the Middle East and African negroes.[39]

Individuals with this genetic trait should not be employed in positions involving additional stress on the lungs, since, according to data published in the current literature, repeated prolonged inhalation of certain industrial gases may trigger disease manifestations in cases of corresponding genetic disposition. Smoking has the same adverse effect.[24]

Table 27.3 Gene frequencies and frequencies of α_1-antitrypsin-variants in European populations

Allele	Frequency	Genotype = Phenotype		Frequency (%)	Serum concentration (%)	Susceptibility to emphysemic bronchitis
Pi^M	0.9025	Pi^M	Pi^M	81.45	100	—
Pi^S	0.0312	Pi^S	Pi^S	0.97	=60	(++)
		Pi^M	Pi^S	5.63	=80	(+)
Pi^Z	0.0111	Pi^Z	Pi^Z	0.12	=10	++++
		Pi^M	Pi^Z	2	=50–60	++

The values do not total 100 per cent because further alleles occur, e.g. Pi^F.

The lack of sulphite oxidase, which has been observed in several instances in the past years,[5] may also be important in occupational toxicology. Persons afflicted with this deficiency are particularly sensitive to SO_2. Sulphur dioxide, however, constitutes a very predominant pollutant.

Our team was successful in proving another genetically determined difference in the metabolism for an occupationally used substance, the pesticide parathion.[11–13]

Parathion is metabolized in the organism as shown in Figure 27.6. An important reaction is the oxidation to paraoxon, which is very toxic to warm blooded and many other animals. Parathion and paraoxon are hydrolyzed in body fluids and organs. Of special interest is the hydrolysis of paraoxon in

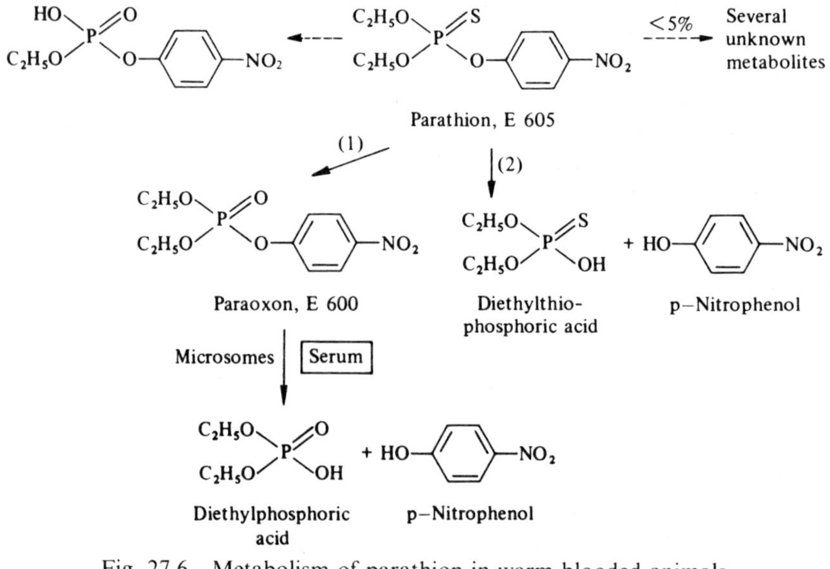

Fig. 27.6 Metabolism of parathion in warm-blooded animals

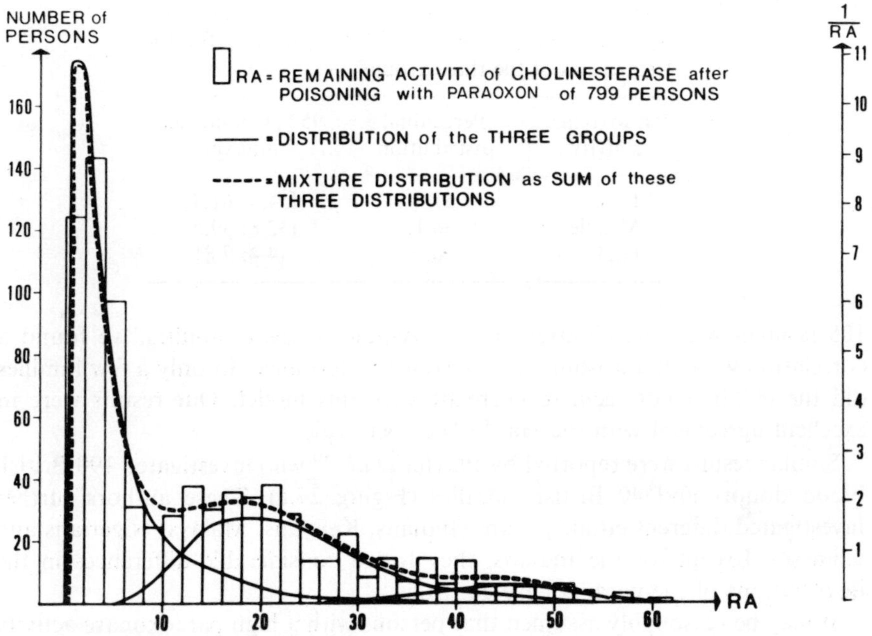

Fig. 27.7 Frequency distribution of the cholinesterase residual activity in the sera of 799 nonrelated persons

human serum. After the incubation of human sera with paraoxon in a suitable concentration, we found different remaining activities of the serum cholinesterase between 0 and 67 per cent of the original cholinerterase activities. This effect is caused by differences in the activity of the human serum paraoxonase, an enzyme that hydrolyses paraoxon. The paraoxon concentration is varied during the incubation by these different activities of the serum paraoxonase, so that different degrees of inhibition of the cholinesterase result. A high remaining activity of the cholinesterase is correlated with a high paraoxonase activity ($r = +0.82$). We were able to show this by using both methods to determine the paraoxonase activity in the sera of test persons.[10,12,13]

Other authors[23,30,46] determined the activity of human serum paraoxonase by spectrophotometry via the liberated *p*-nitrophenol. These authors also found polymorphism in human serum paraoxonase. In further investigations, we determined the paraoxonase activity of 799 unrelated test persons by using the anticholinesterase technique (Figure 27.7) and found the following percentage distribution of the three activity groups (Table 27.4). The limits of the three groups overlap, but by applying the method of Hommel[18] we have established the following limits with the highest probability: between groups 1 and 2, 11.2 per cent residual activity; between groups 2 and 3, 39.0 per cent residual activity of the serum cholinesterase after incubation with paraoxon. We further investigated

Table 27.4 Estimations of frequencies, means, variances and 95 per cent confidence intervals for the frequencies of the three groups

Paraoxonase activity	Percentual distribution	95% Confidence interval
Low	57.9	[54.4; 61.4]
Middle	36.1	[32.8; 39.5]
High	6.0	[4.4; 7.8]

105 families with 190 children. In ninety-nine of these families, we found a correlation with an autosomal intermediary inheritance. In only a few families did the children not seem to correlate with this model. Our results were in excellent agreement with the Hardy-Weinberg rule.

Similar results were reported by Playfer *et al.*,[30] who investigated 190 British blood donors and 40 British families (Figure 27.8). These authors further investigated different ethnic groups (Indians, Kenyans, Malays, Nigerians and Chinese). Except for the Indians, they found considerable differences in the distributions of the paraoxonase activities.

It may be reasonably assumed that persons with a high paraoxonase activity might be better protected against the effects of parathion. This could be important for individuals employed in the production of the pesticide and for the

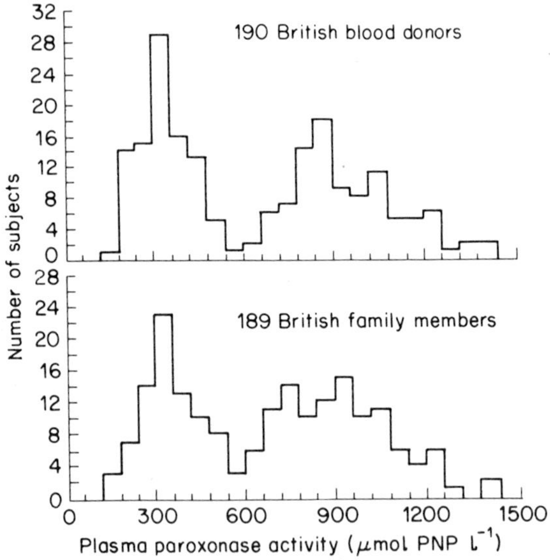

Fig. 27.8 Frequency distribution histograms of British blood donors and British families

agricultural use of E605. Individuals with a low serum paraoxonase activity would consequently be more endangered.

Food

There are a large number of hereditary enzyme deficiency conditions associated with intolerance to certain types of foods. Fortunately these deficiencies are comparatively rare.

Fructose intolerance has been known for a relatively long time.[27,40] In cases of hereditary fructose intolerance the fructose in food, for example as a component of cane sugar or in fruit juices, cannot be metabolized and used. Individuals afflicted with fructose intolerance experience profuse perspiring, tremour, drowsiness, and eventually mild jaundice. If administration of fructose is nevertheless continued, a diffuse enlargement of the liver, poor nutritional status and proteinuria will develop in children. Retarded growth and intelligence deficiencies have also been observed.

It is very important to keep in mind that the alcohol sorbite, which is contained in many infusion solutions for therapy of patients, is potentially lethal in cases of fructose intolerance since it is completely converted to fructose *in vivo*. In addition to deaths resulting from the administration of fructose, cases with fatal results after infusion of sorbite have been reported. It has been suggested that patients with fructose intolerance be issued a special health passport.[36]

Due to hereditary enzyme deficiencies some persons cannot tolerate galactose, a component of lactose. Intake of this type of sugar can also result in damage to the liver, mental impairment and clouding of the eye lens. Children suffering from this condition must be protected by a special diet.[40]

Most common and perhaps of the greatest consequence is lactose intolerance.[38,40] The genetically determined total absence of the enzyme lactase in the mucosa of the intestine which decomposes lactose is very rare. Most mammals including man have this enzyme during infancy. The enzyme activity may, however, decrease substantially with advancing age. If lactose is nevertheless consumed (by drinking milk), lactose malabsorption characterized by abdominal pain, flatulence and severe diarrohea will result. Relating this phenomenon to the total human population it becomes evident that in most populations the lactase activity in the intestinal mucosa decreases drastically after infancy. Only in those cultural groups traditionally engaged in dairy activities and consequently consuming large amounts of milk is lactose intolerance infrequent.

Lactose is relatively well tolerated in northern Europe, where milk products have been a staple part of the diet for several thousand years, and in certain regions of Africa. The phenomenon of lactose intolerance usually develops during adolescence in large populations in Africa, southern Europe, the Near East and the Far East.

More than 80 per cent of the adult Europeans maintain a high activity of the enzyme lactase in the intestinal mucosa, but 95 per cent of orientals as well as 70 per cent of Indians and 75 per cent of the Negroes in the USA do not show lactase activity as adults. Lactose intolerance is transmitted autosomal recessively.

This intolerance is of considerable consequence for the export of foodstuffs for example into starving areas of the Third World. Powdered milk cannot always be considered a nutrition support for these people.

Genetically determined defects in the enzymatic decomposition have also been reported in the metabolism for a considerable number of amino acids contained in animal proteins.[40] These deficiencies resulted in severe organic damage in children afflicted with this disorder.

CARCINOGENOUS DISPOSITION

Induction of microsomal enzymes causing an increase of the metabolic rate following exposition to certain carcinogens (benzanthracene), steroids (methylcholanthrene) and pharmaceuticals due to hereditary disposition may be of importance[16,31] among others to, (a) workers in oil refineries or automobile service stations, (b) individuals exposed to exhaust gases, and (c) heavy smokers. Aryl-hydrocarbon-monoxigenases, which catalyse hydroxylating reactions in metabolism, may become retarded or accelerated in their activity by inductions with the substances mentioned above. Experiments with cultures of human cells proved that individuals, in which a higher rate of metabolism is induced by one of the substances mentioned above, may in a special way be predisposed to develop lung carcinomas.

MUTAGENOUS EFFECTS (GENE DAMAGE)

Conclusive proof for a direct relationship between the activity of chemicals and ensuing actual damage to human genes is difficult to obtain.[8] Under special favourable circumstances it may be possible to microscopically demonstrate certain effects (changes in the structures of the chromosomes) in individuals exposed to carcinogenic chemicals (see Table 27.5). The consequences of these and other changes in the genes (mutations) are not immediately manifest in the affected individuals, but in the case of certain somatic mutations many years later (malign degeneration, cancer), or in the case of mutations in the gonads in the descendants of the carrier of the mutations, recessive mutations are only detectable several generations after their initiation.

The relationship between the initiating noxa, and the induced mutational events, is additionally obscured by the fact that the consequences of the expected mutation are extremely diversified. In many cases these consequences may also be manifest only under specific or special conditions.

It is, therefore, not admissible to assume that all mutations will manifest

Table 27.5 Mutagenic effects of drugs and chemicals

Classification substance	Chromosome aberration		
	In human cells		In other test systems (including mutations)
	In vitro	*In vivo*	
Alkylation agents			
Trenimon, TEM	+	+	+
N-lost-derivative	+	+	+
Psychotropic agents			
Psilocybin	+	+	?
Phenothiazine	(±)	(±)	(±)
LSD 25	±	±	+
Nucleic acid antimetabolites			
Propylthiouracil	+	+	?
6-Mercaptopurin	+	+	+
Antibiotics			
Mitomycin C	+	+	+
Daunomycin	+	+	?
Pesticides			
Organophosphorus compounds	+	+	+

themselves as severe deformation syndromes, debility, hereditary metabolic disorders or other classical hereditary disorders. Parallel to this well-known spectrum of manifestation, less-significant deviations from the regular pattern and standard can be expected. If it has no lethal consequence, the resultant gene damage will be copied again together with the other hereditary characteristics and be transmitted to future cell and individual generations, and hence be multiplied and disseminated.

It is well known that experiments with animals do not allow absolute conclusions with respect to man.

The phenomena outlined so far are most important. As a consequence, action concerning the development and clinical testing of new pharmaceutical products, as well as the general acquisition of data on the biochemical–genetical constitution of individuals and of populations with respect to medical care, industrial development and the food supply appears to be necessary.

Recent scientific findings sometimes allow us to predict if a predisposition is present. Even though their number is still insufficient, preoccupational screening tests are available to determine whether a certain individual may work in a specific working location without being endangered.

A number of years ago, Goedde recommended the introduction of a so-called 'pharmacogenetic passport', which now should be extended in the direction of an 'ecogenetic passport', similar to the 'allergy passport' and the 'accident prevention cards' of serogenetics to register the individual reaction of frequently used metabolism and detoxifying reactions such as hydroxylation, acetylation and esterization.

Additional investigations on man are particularly important since the results of experiments with animals are only to a limited extent directly applicable to man. This field constitutes a new direction of research which will become very important in the future. Teams of geneticists, biochemists and industrial chemists should cooperate with clinical pharmacogeneticists in the exploration of these problems.

During the last international meeting in Titisee in 1977,[43] entitled 'Human genetic variation in response to medical and environmental agents: pharmacogenetics and ecogenetics', training programmes, interdisciplinary workshops and summer courses were recommended.

As mentioned at that meeting by Dr Gilbert S. Ommen of the Office of Science and Technology Policy, Executive Office of the President, Washington, DC, President Carter was very interested in these problems and initiated the creation of a corresponding institute.

REFERENCES

1 Altmann, H-J. (1978). Toxikologische Bedeutung der Metallothioneine, *Bundesgesundheitsbl*, **21**, 317.

2 Atlas, S.A., and Nebert, D.W. (1976). Pharmacogenetics and human disease, in *Drug Metabolism—From Microbe to Man*, Taylor & Francis, London.

3 Brummerstedt, E., Basse, E., Flagstadt, T., and Andresen, E. (1977). Animal model of human disease, *Amer. J. Pathology*, **87**, 725.

4 Calabrese, E.J. (1978). Will elevated levels of lead exposure precipitate clinical symptoms of porphyria in individuals with the latent condition? *Med. Hypotheses*, **4**, 282.

5 Drew, R.T., Johnson, J.L., and Rajagopalan, K.V. (1977). Sulfite oxidase deficiency, *Arch. Environ. Health*, **32**, 141.

6 Druschky, K-F. (1978). *Die akute intermittierende Porphyrie*, Thieme, Stuttgart.

7 Fenna, P., Mix, P., Schäfer, O., and Gibert, J.A.L. (1971). Ethanol metabolism in various racial groups, *Canad. Med. Assoc. J.*, **105**, 472.

8 Gebhart, E. (1977). *Chemische Mutagenese*, Fischer, Stuttgart, New York.

9 Gehrmann, G., Sturm, A., and Amelung, D. (1963). Favismus in Deutschland, *Dtsch. Med. Wschr.*, **88**, 1865.

10 Geldmacher-v. Mallinckrodt, M., Baumgartner, W., Petenyi, M., Burgis, H., Lindorf, H.H., and Metzner, H. (1972). Korrelation zwischen der unterschiedlichen Vergiftbarkeit der Serum-Cholinesterase durch E 600 und der Aktivität des E 600-spaltenden Enzym-Systems in menschlichen Seren, *Hoppe Seylers Z. Physiol. Chem.*, **353**, 217.

11 Geldmacher-v. Mallinckrodt, M., Hommel, G. and Dumbach, J. (1979). On the genetics of the human serum paraoxonase, *Hum. Genet.*, **50**, 313.

[12] Geldmacher-v. Mallinckrodt, M., Lindorf, H.H., Petenyi, M., Flügel, M., Fischer, Th., and Hiller, Th. (1973a). Genetisch determinierter Polymorphismus der menschlichen Serum-Paraoxonase (EC 3.1.1.2), *Humangenetik*, **17**, 331.

[13] Geldmacher-v. Mallinckrodt, M., Lindorf, H.H., Petenyi, M., Flügel, M., Fischer, Th., and Hiller, Th. (1973b). Genetisch determinierter Polymorphismus der menschlichen Serum-Paraoxonase (EC 3.1.1.2); Diskussionsbemerkung, Schlußwort, *Humangenetik*, **19**, 353.

[14] Goedde, H-W. (1972). Genetically determined variability in response to drugs, *Pharm. Weekblad*, **107**, 437.

[15] Goedde, H-W. (1977). Genetische Aspekte der Pharmakologie, *Öff. Gesundheitswesen*, **39**, 485.

[16] Goedde, H-W. (1979). Ökogenetik, *Fortschr. Med.*, **97**, 127, 165.

[17] Goedde, H-W., and Altland, K. (1970). Genetisch bedingte Variabilität der Arzneimittelwirkung, *Med. Klin.*, **65**, 1507.

[18] Hommel, G. (1978). Estimation of the distribution of a genetically determined enzyme polymorphism by a stochastical outside criterion, *Biomet. J.*, **20**, 371.

[19] Ireland, M.P., and Richards, K.S. (1977). The occurrence and localisation of heavy metals and glycogen in the earthworms *Lumbricus rubellus* and *Dendrobaena rubida* from a heavy metal site, *Histochemistry*, **51**, 153.

[20] Kalow, W. (1962). *Pharmacogenetics*, W.B. Saunders, Philadelphia, London.

[21] Kamping, A., and van Delden, W. (1978). Alcohol dehydrogenase polymorphism in populations of *Drosophila melanogaster*. II. Relation between ADH activity and adult mortality, *Biochem. Genetics*, **16**, 541.

[22] Karitzky, D., Lesch, R., Goedde, H-W., Witt, I., Boehm, N., Beckmann, R., Jobke, A., and Künzer, W. (1978). Hepatopathie bei homozygotem α_1-Antitrypsin-Mangel, *Dtsch. Med. Wschr.*, **103**, 161.

[23] Krisch, K. (1968). Enzymatische Hydrolyse von Diäthyl-*p*-nitrophenylphosphat durch menschliches Serum, *Z. Klin. Chem. Klin. Biochem.*, **6**, 41.

[24] Larsson, C., Eriksson, S., and Dirksen, H. (1977). Smoking and intermediate α_1-antitrypsin deficiency and lung function in middle aged men, *Brit. Med. J.*, **2**, 922.

[25] Lombeck, I., and Bremer, H.J. (1977). Primary and secondary disturbances in trace element metabolism connected with genetic metabolic disorders, *Nutr. Metab.*, **21**, 49.

[26] Macnair, M.R., Major genes for copper tolerance in *Mimulus guttatus*, *Nature*, **268**, 428.

[27] Odièvre, M., Gentil, C., Gautier, M., and Alagille, D. (1978). Hereditary fructose intolerance in childhood, *Am. J. Dis. Child.*, **132**, 605.

[28] Overhoff, H., and Forth, W. (1978). Biologisch essentielle Elemente (Spurenelemente), *Dtsch. Ärzteblatt*, **75**, 301.

[29] Piletz, J.P., and Ganschow, R.E. (1978). Zinc deficiency in murine milk underlies expression of the *lethal milk* (lm) mutation, *Science*, **199**, 181.

[30] Playfer, J.R., Eze, L.C., Bullen, M.F., and Evans, D.A.P. (1976). Genetic polymorphism and interethnic variability of plasma paraoxonase activity, *J. Med. Genet.*, **13**, 337.

[31] Propping, P. (1978). Pharmacogenetics, *Rev. Physiol. Biochem. Pharmacol.*, **83**, 123.

[32] Reich, H., and Opitz, K. (1976). Zink als Therapie der Wahl bei Acrodermatitis enteropathica, *Med. Klin.*, **71**, 1.

[33] Sawin, P.B., and Glick, D. (1943). Atropinesterase, a genetically determined enzyme in the rabbit, *Proc. Nat. Acad. Sci. Wash.*, **29**, 55.

[34] Schloot, W., and Goedde, H-W. (1976). Psychopharmakogenetik, *Med. Klin.*, **71**, 481.

[35] Schottel, J.L. (1978). The mercuric and organomercurial detoxifying enzymes from a plasmid-bearing strain of *Escherichia coli*, *J. Biol. Chem.*, **253**, 4341.

[36] Schulte, M., and Lenz, W. (1977). Todesfälle durch Sorbit-Infusionen, *Dtsch. Ärzteblatt*, **74**, 1947.

[37] Simon, E. (1977). Cadmium tolerance in populations of *Agrostis tenuis* and *Festuca ovina*, *Nature*, **265**, 328.

[38] Simoons, F.J. (1978). The geographic hypothesis and lactose malabsorption, *Am. J. Digest. Dis.*, **23**, 963.

[39] Srivastava, L.M., Benkmann, H.G., and Goedde, H-W. (1977). Review on genetic traits in Europeans, Middle East orientals and negroes: serum proteins, *Indian J. Physic. Anthropol. Hum. Genet.*, **3**, 85.

[40] Stanbury, J.B., Wyngaarden, J.B., and Fredrickson, D.S. (1972). *The Metabolic Basis of Inherited Disease*, McGraw-Hill, New York.

[41] Stokinger, H.E., Mountain, J.T., and Scheel, L.D. (1968). Pharmacogenetics in the detection of the hypersusceptible worker, *Ann. New York Acad. Sci.*, **151**, 968.

[42] Taylor, B.A. (1976). Linkage of the cadmium resistance locus to loci on mouse Chromosome 12, *J. Hered.*, **67**, 389.

[43] Vogel, F., Buselmaier, W., Reichert, W., Kellermann, G., and Berg, P. (1978). Human genetic variation in response to medical and environmental agents: Pharmacogenetics and ecogenetics. International Titisee Conference Schwarzwald-Hotel, Titisee, Black Forest, Federal Republic of Germany, October 13th–15th, 1977, *Human Genetics*, Suppl. 1.

[44] v. Wartburg, J.P., Papenberg, J., and Aebi, H. (1965). An atypical human alcohol dehydrogenase, *J. Biochem.*, **43**, 889.

[45] v. Wartburg, J.P., and Schurch, P.M. (1968). Atypical human liver alcohol dehydrogenase, *Ann. N.Y. Acad. Sci.*, **151**, 936.

[46] Zech, R., and Zürcher, K. (1974). Organophosphate splitting serum enzymes in different mammals, *Comp. Biochem. Physiol.* (B), **48**, 427.

SECTION IV
REGULATORY AND GENERAL ASPECTS

Pollutants and Their Ecotoxicological Significance
Edited by H. W. Nürnberg

CHAPTER 28

German Regulations for Environmental Chemicals

F. SCHMIDT-BLEEK

Like most other highly industrialized countries, the Federal Republic of Germany has, during the last 10 years, passed legislation for the protection of man and his environment from the hazards of potentially dangerous substances. In particular, in the areas of emissions from stationary sources progress has been made. In concert with recent international developments to control the quality of chemical products—as far as their potential danger to man and his environment is concerned—the Government of the Federal Republic of Germany is preparing basic legislation now. This 'Chemikaliengesetz' will reflect decisions reached among member states of the European Community, and ensuring regulations will take fully into account the present intensive efforts by OECD member states to harmonize technical and scientific details for the evaluation of chemicals, such as test guidelines, good laboratory practices as well as data transfer issues.

At the present time, some 45 000 substances are traded worldwide, as such, in mixtures and in a vast variety of articles. More than 150 chemicals are produced in excess of 50 000 tons per annum. The Federal Republic of Germany exports some DM 40 billion worth of chemicals each year, corresponding to better than 15 per cent of its export volume.

In spite of the huge number of different commercial substances and their economic significance, very little is known about their environmental mobility, persistence, and the dangers arising from their presence in the environment, leave alone about their decomposition products, synergistic or indirect effects. Known problems, however, give rise for very real and general concern: methylmercury, DDT, PCB, PCT, FHC, 2,4,5-T and many others.

There exists in the Federal Republic of Germany today a considerable body of laws and regulations designed to control environmental chemicals emitted from production plants, other stationary and mobile sources, as well as toxic substances from waste disposals. At this time, however, no general product safety legislation is in force in the Federal Republic.[1]

In concert with the other member states of the European Community, the Federal Republic is presently engaged in negotiations to precipitate common legislation. Since common control legislation has as a consequence trade among the member states which is free from any quality control barriers, testing and evaluation procedures must be specified and harmonized to a very substantial degree before the law can be adopted. (The EEC regulation was passed on June 19, 1979 (6. Amendment of Directive 67/548/EEC).)

Parallel to the negotiations in Brussels, an administrative draft for a chemicals control legislation is being readied in the Federal Republic of Germany. (This draft (Entwurf zum Chemikaliengesetz) was approved by the Federal Cabinet on June 20, 1979 and is now before the legislative bodies for consideration and approval. An English copy can be made available by the author.) Its main features are:

1. All *new* substances must be notified with the national authorities prior to marketing. The notification includes submittal of a considerable number of information and test results designed to permit first estimates of the potential hazard to man and the environment (base set) and marketing can commence 45 days after completion of notification.
2. Suspicious *existing* (old) substances can be subjected to notification and testing on an individual basis.
3. Exceptions from notification will include: new substances marketed for the purpose of testing according to the law; new substances marketed as reagents up to a certain quantity; new substances marketed in quantities of less than 1 t $year^{-1}$ (limited notification may be required); new substances requiring specific further development outside the premises of the producer for up to 1 year (limited notification and permission by authority may be required).
4. The notifier is required to submit data beyond the base set under the following conditions (Stufenplan, step sequence): certain marketing quantities are surpassed (e.g. 50 t $year^{-1}$ and 1000 t $year^{-1}$); previously submitted data or independently available data indicate hazards; the substance is put to new uses. Irrespective of these conditions a regular submittal of data on marketed quantities is foreseen (e.g. every 3 years). In individual cases the competent authorities can require informations beyond the Stufenplan.
5. A list of existing (old) substances will be established.
6. Good laboratory practices, including the maintenance of records, can be regulated.
7. There will be one national notification office and several agencies responsible for the hazard evaluation of notified substances with respect

to: environmental hazards, labour safety, health hazards. (The Umweltbundesamt will be charged to evaluate the environmental hazards and make recommendations as to the regulation of substances).

8. The new law will have provisions for restricting and/or eliminating specific uses of chemicals and their marketing.

In 1975 the OECD recognized that considerable non-tariff barriers to trade may arise among its member countries unless existing and future testing and evaluation procedures were harmonized.[2] A series of expert groups was therefore agreed upon in 1977 with the mandate to report by the end of 1979 on harmonized methods to be applied by exporting and importing countries alike. It seems reasonable to expect that these reports will contain test guidelines for a considerable number of environmentally significant properties of substances in the areas of: physical chemistry, persistence, accumulation potential, ecotoxicity and human toxicity, as well as general guidelines to be observed when evaluating[3] the environmental hazards of (new) substances.

It is interesting to note that in spite of a very large number of existing standards (e.g. DIN, ISO, ASTM) very few testing procedures have as yet been designed and internationally agreed upon which can be applied to elucidate the properties of all *new* that is entirely unknown, substances.

The Federal Republic of Germany is significantly engaged in all on-going harmonization activities and intends to issue internationally harmonized testing and evaluation procedures as regulations in conjunction with the 'Chemikaliengesetz'.

The Umweltbundesamt in West Berlin will be charged to evaluate the environmental hazards of new and existing substances and make recommendations for regulatory actions. In this work it will be supported by external experts from science, industry, the unions and other competent sectors of society.

REFERENCES

1. Schmidt-Bleek, F., Wagenknecht, P., Boehringer, U., Gnevther, K.O., Haberland, W., Kayser, D., Poremski, H.J., Schultze-Wolfhagen, S., and Storm, P.C. (1979). Problems of environmental chemical law in the Federal Republic of Germany. *Chemosphere*, **8**(9), 583–721.
2. Schmidt-Bleek, F., and Bonberg, W. Proceedings of the International Conference on Control of Chemicals, Swedish Product Control Board, April 1978; Schmidt-Bleek, F., and Bonberg, W. (1978). *Umschau*, **17**, 536.
3. Schmidt-Bleek, F., and Haberland, W. (1979). *Z. Umweltpolitik*, **2**, 127.

Pollutants and Their Ecotoxicological Significance
Edited by H. W. Nürnberg

CHAPTER 29

New Challenges to Ecotoxicology

J. SMEETS AND R. AMAVIS

SUMMARY

The current international development legislation on chemical substances leads to new dimensions in operational ecotoxicology and reasearch. Mandatory rules are established or envisaged to present the marketing of chemical substances or mixture known to be dangerous or to present an unreasonable risk, in one way or another, to man and his environment. These legislative measures require an important effort in order to test chemical products to be manufactured, imported or marketed and which have to be notified to the competent authorities.

Reference will be made to

1. The proposed directive of the Commission of the European Communities with respect to the 6th Amendment to the 1967 directive on dangerous substances which is now under discussion at political level.
2. The Toxic Substances Control Act (TSCA) of the USA, covering all existing, as well as new, chemical substances.

In this presentation, the notification mechanism and procedures will be discussed with respect to the required testing, testing methods and data.

INTRODUCTION

The current international development legislation on chemical substances leads to new dimensions in operational ecotoxicology and research. Several countries, such as Japan, the USA, Canada, France, Sweden and Switzerland, have already adopted legislative measurements in order to protect man and his environment against the risks caused by the use of hazardous chemical substances and mixtures. Others, like the United Kingdom, the European Community and the

Federal Republic of Germany, are in advanced stage of preparation. Regarding the European Community, I must say that we are expecting a final decision at the next Council of Ministers on 19 June 1979.

It is not the aim of this presentation to discuss the many regulatory aspects involved, not to compare in detail the contents, the field of application, the administrative procedures or the implications of these regulations on the world market of chemical substances and mixtures. I simply wish to draw your attention to some points of major interest in order to orientate and stimulate work to be accomplished in a few years. As will be shown, there is a great need!

It can be said that mandatory rules are established or envisaged to prevent the marketing of chemical substances or mixtures known to be dangerous, or to present an unreasonable risk, in one way or another, to man or his environment. The general and wide ranging coverage of these rules must be stressed: the regulatory concepts are no longer limited to certain products or groups of chemicals, such as pesticides, pharmaceuticals, food additives. Some countries limit the field of application only to new chemical substances which have not been marketed before. Others, like the Toxic Substances Control Act (TSCA) of the USA cover all existing, as well as new, chemical substances.

Control is often reflected through an obligatory notification system to a competent authority of the central government by the producer or importer of the substances in question. This notification requires information on the results of research and tests which characterize and identify the properties and the possible risks of the substances. The marketing or production of an incriminated product can be forbidden on the basis of this information. This characterization and identification concerns physical and chemical parameters and a study of the possible toxicological and ecotoxicological risks.

Some legislations, such as the Japanese law and the proposed European Community directive, indicate, in the notification scheme, the tests to be performed. Others, leave the judgment for the tests to the producer or importer of chemical substances: TSCA foresees a guidance for testing.

It is evident that these legislative measures require an important effort from man and will place important financial charges on the budget of the manufacturer of chemicals. The cost of testing might be considerable. On the other hand, this development is not surprising, considering some very doubtful experiences with chemicals in the past. A major characteristic of this legislation is that it covers risks for man as well as for the total environment.

In this chapter, reference will be made, among others, to the proposed directive of the Commission of the European Communities which was submitted to the Council of Ministers in 1967 and after long discussions at policy level in expert groups should be adopted in the future meeting of Council of Ministers on the 19th of June 1979. This directive envisages the harmonization of legislation with respect to chemical substances in the nine Common Market countries, to avoid unacceptable trade barriers between the member states.

PROPOSED COMMUNITY DIRECTIVE

The proposed directive of the European Communities is the so-called '6th Amendment of the Council Directive of the 27th of June 1967 relating to the classification, packaging and labelling of dangerous substances'. The purpose of this directive is to approximate the laws, regulations and administrative provisions of the member states on the notification of substances and the classification, packaging and labelling of dangerous substances which are placed on the market in the member states of the Community.

The procedures of notification are well defined in a special article of the directive:

'Any manufacturer or importer into the Community of a substance shall be required to submit to the competent authority of the Member State in which the substance is produced or into which it has been imported from a third country, at the latest 30 days before the substance is placed on the market a notification including:

(a) a technical dossier supplying the information necessary for evaluating the foreseeable risks, whether immediate or delayed, which the substance may entail for man and the environment and containing at least the information of the studies referred to in Annex VII of the directive (called 'mandatory basic dossier'), together with a detailed and full description of the studies conducted and of the methods used or a bibliographical reference to them;
(b) a declaration concerning the unfavourable effects of the substance in terms of the various uses envisaged;
(c) the proposed classification and labelling of the new substance in accordance with the directive;
(d) proposals for any recommended precautions relating to the safe use of the substance.

In addition, and if it is justified, the notifier shall also be required to inform the competent authority of the studies carried out in accordance with Annex VIII of the directive, which is called 'Additional information and tests required'.

In fact, in these Annexes VII and VIII, the directive proposes *notification* by a hierarchical or step-sequence test system composed of three levels or stages; first, a mandatory *base-set or level 0*, later a *level 1* and a *level 2*.

When giving such a pre-marketing notification, the technical dossier requires, among others, the *results of tests*. These data are required in order to assess the real or potential hazard of the new substance: the characteristics of these tests are based on existing internationally recognized parameters.

First, some information regarding exceptions for notification. Exceptions for notification are foreseen, such as for substances placed on the market before the entry into force of this regulation and for substances in limited quantities intended solely for laboratory research and analysis. Furthermore, for substances placed on the market for commercial purposes in limited quantities, the following parameters should be notified.

Level 0

1. Identity of the substance
   Name in the IUPAC nomenclature.
   Empirical and structural formula.
   Composition of the chemical product (purity or nature of impurities).
   Methods of detection and determination.
2. Information on the substance
   Proposed uses.
   Fields of application: closed systems; open systems.
   Estimated production and/or imports.
   Recommended methods and precautions concerning in particular handling, storage, transport, fire.
3. Physico-chemical properties of the substance
   Melting point, boiling point, relative density, vapour pressure, surface tension, water solubility, fat solubility, partition coefficient between *n*-octanol and water, flash point, flammability, explosiveness, auto-flammability, oxidation.
4. Toxicological studies
   Acute toxicity, in particular: LD_{50} on rat, by oral administration, by inhalation, by cutaneous administration; skin irritation on rabbit; eye irritation on rabbit.
   Sub-acute toxicity: 28-day periods.
   Other effects, in particular: mutagenicity (including carcinogenic prescreening test).
5. Ecotoxicological studies
   Effects on organisms: acute lethal concentration for fish LC_{50} one species or more; acute lethal concentration for daphnia LC_{50}.
   Degradation: biotic; abiotic.
6. Possibilities for rendering the substance harmless
   For industry (neutralization, destruction possibilities).
   For the public at large.

Level 1

The competent authority may require additional tests to those defined in the mandatory base set. Taking into account: current knowledge of the substance; known and planned uses; the results of the tests carried out in the context of the base set; the competent authority *may require* the results of the following additional tests where the quantity of a substance placed on the market by a notifier reaches a level of 10 t year^{-1} or a total of 50 t and if the conditions specified below after each test are fulfilled in the case of that substance.

Toxicological studies

1. Fertility studies (one species, one generation, male and female, most appropriate route of administration). If there are equivalent findings in the first generation, then study of a second generation is required. It is also possible in this study to obtain evidence on teratogenicity. If there are indications of teratogenicity, full evaluation of teratogenic potential may require a study in a second species.
2. Teratology study (one species, most appropriate route of administration). This study is required, if teratogenicity has not been examined or evaluated in the preceding fertility study.
3. Sub-chronic and/or chronic toxicity study, including special studies (one species, male and female, most appropriate route of administration). If the results of the subacute study in level 0, or other relevant information, demonstrate the need for further investigation, this may take the form of a more detailed examination of effects, on a more prolonged exposure, e.g. 90 days or longer (even up to 2 years). The effects which should indicate the need for such a study would include, for example: serious or irreversible lesions; a very low or no 'no effect' level; a clear relationship in chemical structure with other substances which have been proved dangerous.
4. Additional mutagenesis testing (including screening for carcinogenesis) following defined conditions.

Ecotoxicological study

1. An algal test. One-species, growth-inhibition test.
2. Prolonged toxicity study with *Daphnia magna*. Twenty-one days, this study should also include the determination of the 'no effect level' for reproduction and the 'no effect level' for lethality.
3. Test on a higher plant.
4. Test on an earthworm.
5. Prolonged toxicity study with fish. At least a period of 14 days. This study should also include the determination of the 'threshold level'.
6. Test for species accumulation. One species, preferably fish.
7. Prolonged biodegradation study. If sufficient biodegradation in level 0 has not been proved, another test (dynamic) shall be chosen with lower concentrations and with another inoculum (e.g. flow-through system).

Remarks

In any case, the notifier shall inform the competent authority if the quantity of a substance placed on the market reaches a level of 100 t $year^{-1}$ or a total of 500 t. On receipt of such notification and if the requisite conditions are fulfilled the competent authority, with a time limit it will specify, *shall require* that the above test be carried out unless, in particular cases, an alternative preferred scientific study is indicated.

Level 2

If the quantity of a substance placed on the market by a notifier reaches 1000 t $year^{-1}$ or a total of 5000 t, *the notifer shall inform* the competent authority. The latter shall then draw up a programme of tests to be carried out by the notifier in order to enable the competent authority to evaluate the risks of the substance for man and the environment. The test programme shall cover the following aspects unless there are strong reasons to the contrary, supported by evidence.

Toxicology

1. Chronic toxicity study.
2. Carcinogenicity study.
3. Fertility study (e.g. three-generation study). Only if an effect on fertility has been established on Level 1.
4. Teratology study (non-rodent species). Verification study to level 1. Teratology study and additional experiment to level 1 study, if effects on embryos/foetuses have been established.
5. Acute and sub-acute toxicity study on second species. Only if results of level 1 studies indicate a need for this. Also results or biotransformation studies and studies on pharmacokinetics may lead to such studies.
6. Additional toxicokinetic studies.

Ecotoxicology

1. Additional testing for accumulation, degradation and mobility. The purpose of this study should be to determine any accumulation in the food chain. For further bioaccumulation studies special attention should be paid to the solubility of the substance in water and to its *n*-octanol/water partition coefficient. Results of the level 1 accumulation study and the physicochemical properties may lead to a large-scale, flow through test.
2. Prolonged toxicity testing with fish (including reproduction).

3. Additional toxicity study (acute and sub-acute) with birds (e.g. quails), if accumulation factor is greater than 100.
4. Additional toxicity study with other organism, if this proves necessary.
5. Adsorption-desorption study where the chemical product is not particularly degradable.

Remarks

As you can see in this type of scheme, *the dialogue* and *flexibility* are major characteristic of the procedure foreseen in the European Communities regulations. The diversity of chemical substances and products manufactured and the wide range of their potential effects on human health and environment make it impossible to require universally applicable testing for all chemical substances. A certain flexibility should be considered. But how flexible should the system be and what are the criteria for controlling flexibility without allowing the system to become arbitrary?

Manpower and cost are particularly important factors to be taken into consideration on the one hand, in relation to the quantity of the chemical in question to be produced or imported and, on the other hand, considering the possible harmfulness of the product. As you can see in the proposed directive of the European Communities, a step-sequence testing, or to use the common German expression, 'Stufen-plan', has been proposed.

Following a defined trigger mechanism a certain degree of automaticity for step-sequential tests has been developed. The first step is the information to be considered in the so-called base-set dossier; this mechanism can be expressed in yearly production quantities or in accumulated import quantities, and if preceeding tests demonstrate the presence of risks, the next steps should be performed.

OTHER REGLEMENTATIONS

The Japanese Law

Another good example of this procedure is the Japanese Law.

1. The first test step to be performed is biodegradability testing.
2. The second step test is bioaccumulation, but this step is only required, if the chemical substance has proved with the first test to have a low degradability.
3. The third test step, fish toxicity testing, is required only for a substance which has proved during the second test to be highly bioaccumulative. In fact, the Japanese only foresee information on these three tests in their notification scheme. The factual data of PCB for degradability, accumulation and toxicity are used as quantitative references.

Toxic Substances Control Act (TSCA)

The recently published guidance for premanufacture testing in the framework of the USA TSCA calls for manufacturers' notifications including acute and clinic effects on terrestrial and on aquatic ecosystems and their components. The initial responsibility for selecting and conducting a testing scheme adequate to support such an evaluation must be borne by the manufacturer. In this draft guidance the following ecological effects and tests organisms have been selected: microbial effects; plant effect tests; animal effect tests.

As I already said, manufacturers *themselves* are responsible for identifying the risks associated with the new substance and for limiting exposure. They must *themselves* select appropriate test programmes and estimate the potential risk presented by new substances. The authority, the Environmental Protection Agency, will assess, *after notification*, the adequacy of the data submitted, the risks which new substances may present and the reasonableness of those risks which are significant.

However, EPA is now involved in constructing testing guidelines in order to facilitate the effective and efficient implementation of TSCA. They informed us that is their *long-term goal* to develop a tiered scheme of testing, that will provide data needed for risk assessment.

Remarks

It is not the aim of this chapter to discuss in depth the value of both systems. I shall limit myself to state that the European industry, as well as the European governmental authorities, *prefer* to indicate in their regulations the type and extent of the testing required; the testing decisions *are not*, in the first instance—as in TSCA—the responsibility of the notifier. We feel a substantial uncertainty as to the nature of the information required in the TSCA system, particularly with respect to the assessment of whether or not a substance represents an unreasonable risk; the absence of clear and specific risk criteria may place a supplementary burden of *responsibility and costs* on the manufacturer. A great deal of discretion is left to the authorities to decide whether or not a satisfactory evaluation has been made. The US authorities have been informed about our feelings in this respect.

The EPA publication on testing policy of 16 March 1979, discussed in depth the policy to be followed in the USA together with the advantages and disadvantages of the different approaches and alternatives. This publication also contains a series of tests and methods which the EPA has identified as appropriate for testing certain properties and effects of chemical substances. However, they have been selected to function as 'reference tests'—that is 'recommended tests to which a person may refer *once* a decision has been made to test for a particular effect or property'—EPA feels that 'more difficult than selecting tests and methods is the problem of providing guidance concerning their use'. That is guidance concerning 'whether to test' and 'type and extent of testing'. They also feel that 'it is very difficult with recommended tests to develop specific guidance to apply them to *all* such *substances* and *situations*'. In the EPA discussion document, different approaches to testing are analysed, distinguishing between :

1. Comprehensive base test, defined as a set of tests to produce data presumed to be necessary for every new chemical substance.
2. Categorial base set, defined as a base set of tests for *categories of chemicals*.
3. Flexible base set, in this base set of tests, more flexibility is built than in the base set mentioned under 1 and 2.
4. Tiered base set, which is testing arranged in hierarchical structure.

From this summarizing analysis of the two proposed systems, we can notice an important difference in policy and strategy.

The European Community believes that a minimal base set of tests should be coupled to a *tiered* testing scheme based on production volume and risk of exposure as well as on toxicological and ecotoxicological data. Flexibility could still be retained in the context of a minimum list of clearly mandatory data. Certain tests *would not* have to be performed, provided justification is given. An authority could still retain the right of requiring further testing after examination of the first test data, exposure data and marketing provisions.

In *our opinion, the tiered testing scheme offers many advantages*:

1. With clear decision criteria, it eliminates one element of uncertainty as far as the supply of required data is concerned.
2. One of the major difficulties for foreign manufacturers and suppliers, i.e. the necessity of a dialogue, will be considerably reduced.
3. If quantities represent one type of trigger, the disproportionate burden on small manufacturers and importers would be partially alleviated. Notification costs would also be kept more in proportion with the volume of trade and with the exposure of man and the environment.
4. With a clearly defined hierarchical system of tests, appropriate international harmonization, including mutual recognition of common test data is likely to be facilitated and encouraged.

TESTING METHODS AND DATA

Many activities are currently underway, nationally as well as internationally, to consider the appropriate testing methods and techniques to be used in the implementation of this legislation. Reference can be made to the work going on in

ISO, the work of the OECD Working Groups and the EEC interlaboratory studies. This work is aimed at developing test methods to produce data which are reliable and also appropriate for the evaluation of risks for man and his environment. It is evident, data and results must be reliable and comparable within themselves, whatever their origin. *Systematic and random errors* must be reduced as much as possible.

Standardisation of tests has undoubtedly certain advantages, however, it leads to a certain inflexibility when experience and techniques progress. Standard tests are appropriate as reference methods, particularly in case of a dispute concerning the validity of test results. It might be more desirable to put the technical effort into the *harmonization* of testing *data*, thus envisaging the *intercomparability of test results*. The European Community feels that, for the time being, it is too early to envisage a standardization of *all details* for *all test techniques*; in its mandatory testing scheme it only indicates the required tests characteristics, e.g. LC_{50} fish-toxicity tests should be performed during *x* hours or *y* days and the same for the other tests. More technical and experimental details will be elaborated at a later stage.

In order to assess, (a) *the performance* of each laboratory, (b) *the reliability* of test results, and (c) *the applicability* of the testing methods currently in use, the CEC organizes interlaboratory intercomparison programmes or ring-test studies. Based on the results of these ring tests we hope to propose reference methods.

Programmes have been undertaken concerning (human) LD_{50} toxicity tests, degradability and fish and daphnia acute toxicity tests and physicochemical parameters. Other programmes are foreseen. Quality control and assurance programmes will now, and in the future, be indispensable to check the quality of the testing performed by different laboratories, it might even be included in a procedure for the accreditation of laboratories.

In this respect the establishment of so-called *good laboratory practice* (GLP) must be considered. Standards for all non-clinical experimental studies in relation to submission of data on safety and with respect to health and the environment, are currently being developed in the framework of the European Communities' legislation as well as in the US TSCA. It must be emphasized that the principle of applying standards for GLP is not new; however, this new comprehensive legislation on chemical substances gives a new dimension to this problem. Generally GLP procedures will refer to the facilities, equipment, personnel, test systems, testing practices, procedures and data records.

The use of *reference substances and reference data* will be indispensable to evaluate the test data produced for a *new* chemical. Relevant properties, of *well-known* existing chemical substances are necessary in order to evaluate, by intercomparison, the same properties of *unknown* new substances. The research and development of such vectors is now underway in the CEC.

A major problem, for which a satisfactory solution has not yet been found for all parties, is industrial and commercial *secrecy* and also the *confidentiality* of test data. It is evident that one reason for the confidential treatment of data, especially of the chemical identity of a substance, is to allow a manufacturer of a new substance to establish a firm position on the market before his competitors. Therefore, it is essential that the chemical identity of a new substance be treated confidentially and with the necessary precautions before it is put on the market. But generally, legislators *stress* that industrial and commercial *secrecy* shall not apply to any data contained in the notification dossier *relating to safety advice and hazard classification*. Furthermore, it is recognized by most legislators that supporting experimental observations, such as testing, should be regarded as *industrial property*. A second manufacturer of a previously notified substance must either purchase the data from the original notifier or arrange to carry out the necessary tests.

This *cost factor* of testing for notification is indeed a very important one. TSCA states clearly—and this statement should be applicable to all similar legislations—that 'authority over chemical substances and mixtures should be exercised in such a manner as not to impede unduly or create unnecessary barriers to technological innovation while fulfilling the primary purpose of the Act to assure that such innovation and commerce in such chemical substances and mixtures do not present an unreasonable risk of injury to health or the environment'. In fact, the costs involved in this context are similar to the expenditures so far only spent by industry for drugs, pesticides and food additives. It is clear that the cost of testing will vary according to the tests to be

Table 29.1 Estimated cost of tests in $US (Annexe VII and VIII EEC proposal)

Base set	
(a) Physicochemical properties	5000
(b) Toxicological studies	30000
(c) Ecotoxicological studies	3000
Total	38000
Level 1	
(a) Toxicological studies	215000–385000
(b) Ecotoxicological studies	16500
Total a + b 230000–400000 including chronic toxicity 100000 excluding chronic toxicity	
Level 2	
(a) Toxicological studies	320000–890000
(b) Ecotoxicological studies	18500–93000
Total a + b 340000–1000000	
Total base set + Level 1 + Level 2	480000–1138000

performed and that undertaking the tests in accordance with good laboratory practice may affect the cost considerably.

Many data on cost evaluation are available but they vary widely. European experts estimate the cost for biodegradation testing, for example, in the neighbourhood of $1000–2000. The costs for this test, according to EPA estimates, are between $5000 and $12 000. Certainly the cost will be dependent on the choice of the type of experiments (see Table 29.1).

CONCLUSIONS

It seems appropriate to consider as a conclusion the practical impact of these legislations and more particularly the number of technical dossiers to be notified yearly. It is estimated that about 1000 new chemicals are introduced every year on the market of which about 500–600 are in the Community.

Many new chemicals currently in commercial production are low volume speciality chemicals. Estimates from EPA show that approximately 75 per cent of all chemicals in production in the USA have an annual production volume of less than 45 t. 'Presumably, many new chemical substances will share some of these features.' In the European Community it is estimated that only about 2–3 per cent of all new chemicals reach a production level of 1000 t $year^{-1}$, which means ten to twelve substances yearly (see Table 29.2). Following the established quantitative trigger mechanism, those chemicals should be tested as enumerate in the level 2 scheme of the hierarchical system. However, it might be expected that only very few substances must undergo the *complete* level 2 testing, considering the knowledge of the substance, the results of the testing performed in the base set and level 1, the known and planned uses, and products expected in the application of this legislation like phytopharmaceuticals.

The same estimates indicate that about 20 per cent of all new chemicals are produced in quantities more than 100 t $year^{-1}$, which means mandatory testing following the level 1 scheme. The total number amounts to about 100–120 compounds, so even in level 1 the total number of compounds to be tested will be limited. The number of compounds to undergo *total* level 1 testing is even less.

The competent authorities have taken the decision that an appropriate control is necessary in order to prevent new serious ecotoxicological threats by chemical

Table 29.2 Percentage of production quantities (estimates)

t $year^{-1}$	%
1	95
10	75–80
100	20
1000	2–3

substances. This is the objective of the new comprehensive legislation. On the other hand, the legislator acknowledges the role of chemicals in improving our standard of living. In this dilemma of risk benefit, the legislators in the European Communities have very seriously considered the question of the limitation of the cost for the manufacture of new substances. We must recognize a lack of experience in the implementation of those regulations, for the time being, for both parties. However, the current presentation and proposals seem to be the most appropriate and reasonable approach to solve the problems raised on the level of the European Community in order to prevent a too detrimental influence on innovation of chemicals.

ADDENDUM

From the date of my presentation, on 6 June 1979, the proposed directive of the 6th Amendment of the 1967 Directive on dangerous substances was adopted by the Council of Ministers on 19 June, 1979. The definitive text will be published in the *Official Journal of the European Communities.*

Pollutants and Their Ecotoxicological Significance
Edited by H. W. Nürnberg

CHAPTER 30

Industry and the Environmental Challenge

Paolo Schmidt di Friedberg

The environmental debate over the last few years has often focused on the industrial enterprise considered, rightly or wrongly, as one of the fundamental factors in the degradation of the modern world. Faced by such criticisms, the industrial enterprise at first adopted an uncertain attitude, either denying that the problem really existed, or debating whether or not it was its concern to face it. Subsequently some empirical solutions were adopted based, in general, on the principle of solving problems as they arose and according to the specific demands of each individual case (pollution and purification, energy crisis and nuclear reactors, etc.).

Therefore, with some exceptions, no coherent strategic framework was devised to avoid this piecemeal approach because it was thought to be of little practical use and problems, as a rule, were faced separately without a wider overall solution in mind. As a consequence, several solutions were subsequently found to be inadequate, because the complexity of the subject was not fully grasped (a typical example is that of nuclear energy projects which met with unexpected difficulties of a not purely technological kind); whilst the whole debate gradually became more chaotic because of the specious wisdom of those who confused prophesy with prediction, and groundless ambitions with far-sightedness. This prevented the development of a more practical discussion and did not allow a generic identification of aims (about which there is a widespread agreement: clean air and water, countryside free from pollution, well-preserved historical centres etc.) to be translated into a precise definition of the measures that needed to be taken to achieve them. Recently, however, the aggravation of certain factors like the petrol crisis, giving more weight to the economic values, greater pragmatism has become imperative. Accordingly there can no longer be any illusion that a serious commitment of industry to the environment is not necessary.

The aim of this chapter is, therefore, to help such evolution, delineating some elements that the industrial enterprise needs to consider for a strategic approach to the problem. Particular emphasis will be placed on, (a) the different aspects of the environmental problem, (b) how to face the same, and (c) the relationship between the environmental problem and the general main evolution lines of the modern world.

THE DIFFERENT ASPECTS OF THE ENVIRONMENTAL PROBLEM

From the industrial enterprise point of view the environmental problem can be seen at least under five different headings, economic, social, political, legal, technological.

The Economic Aspects

It is worth remembering that the industrialized countries spend between 0.4 and 2.6 per cent of their national incomes on the environment (Table 30.1), so that

Table 30.1 Total expenditure of the new pollution control programme as a percentage of GNP

		Investments	Running costs	Total
A	1971–75			
	USA	0.5	0.2	0.8
	West Germany	0.6	0.2	0.8
	Sweden	0.5	nd	0.6–1.0
	Italy	0.3	0.1	0.4
	Japan	2.6	nd	3.5–5.0
	Holland	0.7	0.1	0.8
B	1976–80			
	USA	0.7	0.9	1.7
	Italy	0.6	0.6	1.2
	Holland	1.2	0.2	1.4
C	1971–80			
	USA	0.6	0.7	1.4
	Holland	1.0	0.2	1.2
	United Kingdom	0.2	0.9	1.1
	Italy	0.5	0.4	0.9

Source: OECD, *Economic Implications of Pollution Control: A General Assessment*, 1973.
Some adjustments have been made by OECD to United Kingdom, Japan and Sweden data to make them comparable. (From E. Gerelli, *Economia e tutela dell'ambiente*, Universal Paperbacks il Mulino, Bologna 1974.)

Table 30.2 Pollution control investment by industry in the USA (million $)

	Total reported expenditures 1976	Pollution control as a percentage of total plant and equipment investment				
		Actual 1973	Actual 1974	Actual 1975	Actual 1976	Planned 1977
All industries	6762	4.9	5.0	5.8	5.6	5.6
Manufacturing	4382	8.3	8.0	9.3	8.3	7.9
Durable goods	1560	8.1	7.3	8.1	6.6	7.0
Primary metals	923	23.4	16.6	17.2	15.7	17.3
Blast furnaces, steel works	446	16.3	12.1	13.5	15.1	19.1
Non-ferrous metals	405	31.1	21.8	24.1	18.9	17.1
Electrical machinery	148	4.5	6.8	5.8	5.6	4.1
Machinery, except electrical	80	2.3	1.8	1.8	1.6	2.4
Transportation equipment	125	5.6	3.7	3.4	3.4	4.1
Motor vehicles	90	6.4	4.1	3.9	3.6	4.3
Aircraft	32	3.8	2.9	2.8	3.3	2.8
Stone, clay and glass	103	9.6	12.9	14.3	6.1	6.7
Other durables	181	4.9	4.5	5.3	3.9	3.9
Non-durable goods	2821	8.5	8.7	10.3	9.6	8.
Food including beverages	175	5.0	4.7	5.2	4.5	4.
Textiles	37	3.7	3.3	4.6	4.4	3.
Paper	511	18.8	19.3	16.8	14.7	14.
Chemicals	765	9.6	8.3	10.9	11.4	10.
Petroleum	1275	10.3	10.1	11.8	10.9	9.
Rubber	37	3.1	3.2	4.0	3.4	2.
Other non-durables	23	1.2	1.8	2.8	1.4	1.
Non-manufacturing	2381	2.9	3.0	3.2	3.5	3.
Mining	86	3.3	1.8	1.9	2.2	2.
Railroad	27	8	1.2	1.4	1.1	0.
Air transport	16	0.6	0.0	0.6	1.2	0.
Other transport	38	0.7	2.3	1.4	1.1	1.
Public utilities	2032	7.6	7.9	8.4	9.1	9.
Electric	1990	8.7	8.9	9.7	10.5	10.
Gas and other	42	1.5	1.5	1.5	1.2	0.
Communications, commercials and others	182	0.6	0.6	0.6	0.5	0.

Source: Eighth Annual Report of the CEQ (1977), p. 325.

Table 30.3 Capital expenditure on pollution control in Japanese industry[a]

	As a percentage of total capital expenditure			
	1973	1974	1975	1976
All industries	10.6	16.2	19.8	
Iron and steel	17.3	18.5	18.7	22.
Oil	18.5	33.8	40.5	34.
Thermal power stations	28.4	44.7	40.4	
Pulp and paper	22.1	24.3	30.9	22.
Non-ferrous metals	8.4	11.8	17.0	
Chemicals (except for petrochemicals)	17.1	27.8	34.0	24.
Machinery	4.0	5.5	6.7	
Petrochemicals	15.7	22.0	21.5	17
Mining (except for coal)	24.4	31.0	36.2	
Textiles	10.1	15.4	27.7	
Cement	11.2	16.7	14.6	
Pottery (except for cement)	9.9	10.3	10.7	
Gas	2.3	3.7	4.2	
Coal	4.0	3.6	9.5	
Miscellaneous	8.6	10.3	10.1	
Construction equipment	5.9	6.2	9.2	
Electricity (except for thermal power stations)	1.1	1.0	0.8	

Source: Ministry of International Trade and Industry, Japan.
[a] Estimates based on engineering studies.

Table 30.4 Environmental expenditure of the German chemical industry (million DM)

	1974		1975		1976		1977	
	Investment	Running cost	Investment	Running cost	Investment	Running cost	Investment	Running cost
Water	474	608	343	670	301	796	236	928
Air	176	353	214	345	257	428	255	654
Waste	59	225	27	287	66	307	69	358
Noise	9	21	10	24	27	22	25	35
Sub total	718	1207	594	1326	651	1553	585	1975
Total	1925		1920		2204		2560	

Source: Ehreburg des Verbandes der Chemischen Industrie e.V.

Table 30.5 Expected increasing of prices due to water treatment (Italy)

	%
Food industry	1.1
Non-ferrous metals	1.0
Chemical and allied industries	1.9
Rubber processing	1.8
Paper and board	3.6
Mining and quarrying	0.7
Textile, wool	0.7

Source: G. Scaiola: 'Situazione costi e problemi attuali della depurazione delle acque'.

this constitutes a substantial proportion of public spending (about 6 per cent in France). The figures for the expenditure of private enterprise are less certain; however Tables 30.2 and 30.3 set out the available information for USA and Japan, showing that the investment in the environment varies, according to different branches of industry, between 0.5 and 40 per cent of total investments. Table 30.4 on the other hand shows a subdivision of environmental spending in the German chemical industry, between sectors (air, water, solid wastes, noise) and between investments and running costs.

It should be noted that, whereas the first of the last two is in decline, both absolutely and relatively to total investments (from 14 per cent in 1974 to 12 per cent in 1976), the second is steadfastly increasing. Also worth noting is the increase in investments relating to noise control, which is the most advanced phase in the fight against environmental pollution. For increasing of prices related to such expenditures, Table 30.5 sets out some information by sectors of industry. Finally it is interesting to consider some examples showing the size of some specific environmental expenditures.

1 The SYNCRUDE project (oil from the bituminous sands of Alberta) involves an expenditure of $300 million for environmental control (impact analysis and pollution control).
2 The cost of a complete ecological and toxicological evaluation of a new chemical is ranging between 700 and 1000 million Lira. To stand such a burden is possible only for chemicals with large markets; therefore, this will undoubtedly have a negative effect on the innovation of fine chemicals that often do not satisfy such a condition.
3 Pollution control in the titanium dioxide industry by the sulphate process, means an increase in the cost of the production of about 20 per cent.

The inflationary effect of all such environmental expenditures has been the subject of much discussion: experts now tend to regard this as negligible (for instance see the paper by Poitier). Among the economic overall aspects of the environmental problem one of the most interesting is the effect on trade. In fact the differences between pollution control costs in the various countries, due to differing regulations or to differences in their environmental situation, tend to create a request for protection and/or to cause the relocation of certain industrial plants (the so-called export of pollution).

Likewise product-control regulations can disguise hindrances to free competition (see, for example, the debate concerning the American Toxic Substances Control Act).

Finally it should be added that discrepancies in the governmental supports for fighting pollution (financing at reduced rate, advanced depreciations, capital contributions, etc.), may also create protectionist pressures. This is the real reason for the EEC introducing the 'pollutioner pays' principle since, beyond its public relations value, in fact it is always the consumer, in the end, who pays.

Social Aspects

The shifting of resources towards the environment has obvious effects on employment; the following examples can be recalled.

1 Between 1971 and 1973 Sweden enjoyed an expansion of 0.7 per cent in the employment in the industrial and municipal sectors due to increased government subsidies aimed at reducing pollution levels.
2 In the USA jobs lost from 1971 to 1976 as a result of plant closures for environmental reasons, rose to 19 500 and those threatened to 31 007. These figures must be compared with the 678 000 jobs created by the pollution control industry.
3 Econometric models of several countries (Norway, Japan, Italy etc.) show a general, more or less marked increase in employment corresponding to an increase in anti-pollution activities.

Coming to the problem more specifically concerning the industrial enterprise, it should be noted that its use of the natural resources of a community (air, water, land, countryside etc.) often leads to conflicts with the population (or its representatives) which can have heavy effects on the industrial activity.

In addition to the well-known cases of plants in various parts of the world which were never opened or whose opening was delayed (nuclear reactors for example) or which made heavy losses for environmental reasons (the titanium dioxide production plants of Scarlino for example), a classification of 102 other conflicts of this type in the chemical industry is shown in Table 30.6. Considering this table, the important role of peripheral administrative and political structures in these conflicts should be particularly noted.

Health plays a dominant role in the social aspect of the environmental question, both in terms of the tragic consequences of the pollution of the

Table 30.6 102 Conflicts in the chemical industry

Production processes	
Pollution	27
Work environment	21
Accidents	12
Waste disposal	4
Transports	6
New plants	6
Total chemical industry	76
Chemical products	
Work environment	3
As above (not chemical work)	10
Diffused pollution	5
Consumer goods	8
Total products	26
Grand total	102

Actors		Areas of conflict		Results	
International organizations	6	Water	27	Administrative inquiries	14
Central administrations	12	Air	28	Judicial inquiries	18
Regions	16	Land	5	Judicial procedures	20
Provinces	24	Noise	1	Administrative measure	13
Municipalities	42	Environmental hygiene	44	Judicial conviction	4
Magistructure	45	Safety	16	Stop or renunciation	13
Trade unions	53	Public health	14	Delay	4
Citizens groups	21			Transfer or reconversion	3

environment (it is enough to remember the terrible case of the 110 deaths resulting from mercury pollution in the Bay of Minamata, or the more recent case of Seveso) and in the not less important area of the pollution of the work site, with all the complexities relating to its control, in terms of expenditures, work organization and labour relations. As far as the last item is concerned, it should be particularly noticed how important the role played by environmental considerations in the negotiations of labour contracts is becoming (see, for example, the introduction in some countries of the principle that risk can not be exchanged with money).

Some information on the attitude of people with respect to the environmental problem is given by a research summed up in Table 30.7 showing that there is a tendency to shift from a moralist approach to an egocentric one when we pass from generalities to personal choice. The importance of such trend for the industrial enterprise needs no further comments.

Table 30.7 Inquiry on relationships between public and environment (Archivio di Psicologia, Neurologia, Psichiatria, Anno XXXVI, fasc. III)

	No (%)	Intermediate (%)	Yes (%)
Generally speaking the environment is important	1.2	8.6	90.2
The benefits of protecting the environment are higher than the cost	2.3	39.6	58.1
Personal availability to habits change as a consequence of environmental protection	9.2	73.5	17.3

Note: A discrepancy, having a direct consequence on the industrial enterprise, exists between practice and theory. In fact the industrial enterprise is blamed on a broad moral level as supplier of goods required on a self-centred level.

Political Aspects

The environmental question, in so far as some of its aspects are of fundamental concern for man (survival, health etc.) logically also involves the political world.

Among the institutional aspects of the problem we may remember: the rise of the 'green parties' (9 per cent of votes in French local government elections in 1977, 3.7 per cent in those of West Berlin in 1978 etc.); the numerous referenda on environmental issues, especially for nuclear plants; the creation of specific political authorities at different levels (from ministers to town councillors) which take care of environmental problems; the numerous electoral campaigns based on environmental issues; and so on.

Among the less formal aspects, beyond the fact that all politicians sooner or later have discussed the quality of life, several examples can be referred.

1 The many movements for the defence of the environment (from temporary underground movements to the regular criticisms of the official opposition).
2 The stance adopted by scientists, either alone or in groups, on many environmental questions (nuclear energy, cancer, etc.).
3 The intentional politicization of some specific environmental situations which could have been better approached from a less broad angle.
4 The birth of slogans such as 'ecology is politics' etc.

Finally the environmental problem also has a bearing on international politics. Apart from the foundation of specific bodies (UNEP, WHO, etc.) which, among their other activities, have organized international debates on different subjects having world wide repercussions (Stockholm, Bucharest, Vancouver, Rio de la Plata, etc.), the following items should be remembered:

1 the various international conventions relating to the use of the sea;
2 the transfrontier pollution (for example, the 'acid rains' in Scandinavia originated from England, France and Germany);
3 the efforts being made to control on a world wide basis some products (fluorocarbons for instance).

Particularly the transfrontier pollution is slowly leading to such innovations as:

1 the compulsory consultations between bordering countries,
2 the identification of a Common Court of Justice for all the countries interested in the same pollution,
3 the concession of the same rights to all the inhabitants on international borders irrespective of their nationality.

Legal dimensions

As a consequence of what said above, the number of laws and regulations concerning the environment is increasing rapidly all over the world.

Some of them are very general: see, for instance, the dictates of some constitutions (the Italian one states that 'the Republic defends the countryside') whereas others are much more specific: see for instance Figure 30.1 regarding the increasing number of them in West Germany.

Laws and regulations express the formal sensibility reached by a community about the environment; given that this is in continuous evolution, an industrial enterprise that bases its activity only on existing laws and regulations and overlooks the general direction of their developments is likely to be continually behind and consequently to meet with enormous difficulties (consider, for instance, the influence of the trend of Figure 30.1 on the planning of the enterprise).

The case of the production plant of titanium dioxide at Scarlino serves as the best example: the system of dumping at sea the 'red muds' was universally accepted at the time of the feasibility studies, but, when going into production only a few years after, this was no longer the case.

One of the characteristics of the evolution of laws and regulations all across the world is the emphasis placed on the prevention of pollution; this is the case, for instance, of the procedures concerned with the environmental impact assessment whose aim is to compel to act in advance before damage is done to the environment and when project modifications are still possible.

Another interesting aspect is the increasing extension of the right to bring actions by associations or by private citizens, as a recognition of the bearing that environmental damage has on their legitimate interests (either physical, moral or aesthetic). Regarding this the decision of the US Supreme Court, in 1970, to introduce the concept of 'aesthetic damage' was sensational. Such evolution can also be seen as a progressive abandoning of the concept of 'free goods', in view of the opposing idea that water, air and countryside are not *res nullius*, available in unlimited supply, but are also consumable and belong to the whole community.

Another development to be emphasized is the tendency of the penalties that can be incurred by neglecting environmental issues to become more and more severe (i.e. no longer only fines but also imprisonment). This development goes hand in hand with an extension of the concept of responsibility, so that damage to the environment is less and less frequently considered unpremeditated (for

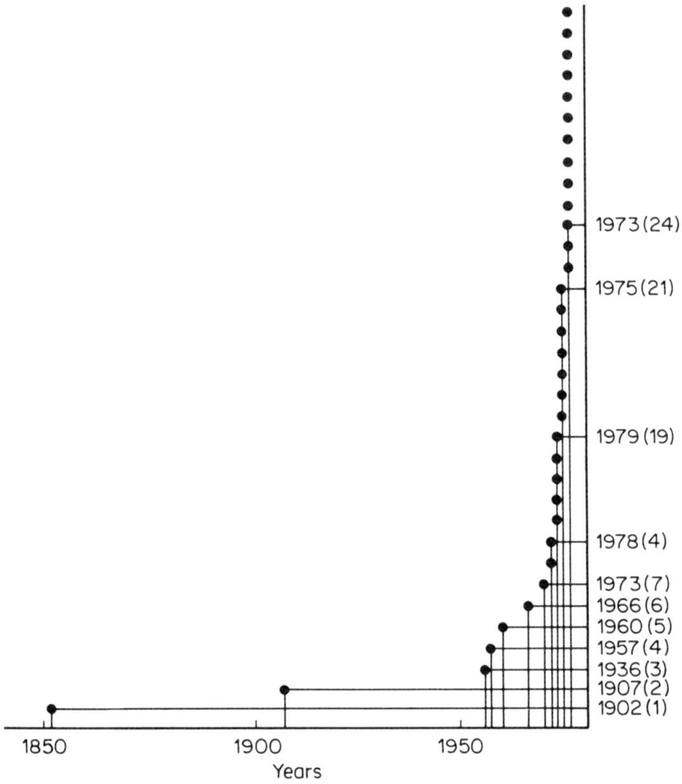

Fig. 30.1 Legislation (national and international) on the control of the water pollution in Germany (*Source:* Peine, 1976, BASF, F.R. Germany

example, see the international convention establishing the absolute responsibility of the ownership in the case of an oil-tanker causing pollution, except *force majeure*, war, fraud by third parties, governmental negligence or negligence on the part of the injured parties).

Another principle under revision is that of *nemo reo sine judicio*, in that the onus of proof of being not guilty has tended to pass to the accused party. This revision, which is due to the difficulties of defining the often distant consequences of environmental changes, is obviously of the greatest importance to industry.

Furthermore, the following points should also be noted.

1 The limitations of the concept of 'ownership' when confronted with the so-called social duties.
2 The tendency to codify the management of resources belonging to the whole of man-kind (see the above mentioned international conventions relating to the use of the sea, the EEC directives concerning migrating birds, the 1972 convention regarding the protection of cultural and natural heritage etc.).

3 The introduction by many countries of the principle that regulations should be applied on the basis of the 'best technological means' available.
4 The introduction of the principle of the integrated management of the environment; and many others.

TECHNOLOGICAL ASPECTS

As far as technology is concerned, there is no need to extend comments on the fact that the environmental problem is a major stimulus to innovation. This is true both for what concern the development of new technologies needed to counteract the negative aspects of industrial progress (pollution, etc.) and for what concerns the development of new products and processes that can improve the quality of life without negative side effects.

To the first of these belongs the technology for purifying water and the atmosphere, for disposing of solid wastes (incineration, conversion into compost, pyrolysis, etc.) for protecting the landscape, for recovering wastes etc. To the second belong products which do not create pollution, the so-called clean technologies (i.e. those which optimize the consumption of raw materials and energy, quantity and quality of production, cost and social expectations), closed-cycle processes etc. The importance of these problems for industrial enterprise is well known and is further demonstrated by the fact that the proportion of research investments in these areas is continually increasing.

HOW TO APPROACH THE ENVIRONMENTAL PROBLEM

The preceding pages show that the relationships between industry and the environment concerns all the aspects of industrial activity. There are three basic ways by which industry may approach this situation.

1 Passive ecology: the environment is considered only as a constraint and industry limits itself to correcting the negative aspects of its activity as individual cases arise. It is the approach that opposes every production plant with a purification plant and puts the emphasis on relative costs.
2 Active ecology: the environment is also considered an area of ample opportunity for industrial activity. It is the approach which leads to stopping of pollution, rather than to the fighting pollution, to recycling wastes, to the so-called clean products etc. Basically it is concerned with extracting some advantages from the situation.
3 Integrated ecology: environmental problems are seen in the broadest context possible, trying to spread deeply the environment in the enterprise behaviour. It is the approach which considers the solution of the environmental problems not as something special but just as the result of healthy management.

In a general sense, modern industry is on the verge of the third position, researching the ways for translating it into practice. To approach this last topic it is convenient to divide it into two steps.

1 Analysis of the relationships between industry and the environment (the ecology of enterprise).
2 Means of managing effectively this relationship (the environmental management).

The Ecology of Enterprise

The ecology of enterprise is primarily concerned with the global relationships between industry and its environment; as a consequence it is entirely different from the industrial ecology which is concerned essentially with the problem of purification and/or with certain specific productions. The ecology of enterprise—as ecology *tout court*—is based on the concept of the ecosystem: a functional unity comprising the environment and the population within it and corresponding to a higher level of organization; almost at the top of the scale from that of protoplasm, through cells, tissues, organs, groups of organs, populations, communities and ecosystems, to the biosphere.

A consequence of the concept of the ecosystem, the full understanding of one of its two components (population or environment), may be reached only if any one of them is considered in relation to the other; otherwise, as they say: 'You can't see the wood for the trees'. Conversely the knowledge of the wood is needed for a full understanding of the single tree.

In the case of the human ecosystem (of which the industrial ecosystem is a part), the environment includes not only physical but also political, social and economic aspects. The relationship between man and the environment is summarized well in two declarations, one made at the ONU conference in Stockholm in 1972, and the other at the UNESCO conference on education at Tbilisi in 1977:

> Man is both creature and moulder of his environment, which gives him the opportunity for intellectual, moral, social and spiritual growth.... Both aspects of man's environment, the natural and the man-made are essential to his well-being.

> Environment is the sum of the resources available in a given place and in a given moment to satisfy yearnings and needs of man.

In this framework there are some noteworthy analogies between industrial enterprise and living organisms: each type of industrial activity, like every organism, is born, develops and declines, occupies a certain ecological niche and has a determined function within the ecosystem; similarly for each of them the struggle for survival is the norm. Within an ecosystem—whatever its nature—certain equilibriums are established and maintained basically unchanged by various self-regulating mechanisms: these are the so-called homeostatic (self-regulating) mechanisms which allow the ecosystem to preseve its identity. By definition those equilibriums are dynamics, i.e. they are variations around a mean point rather than rigidly fixed values, in as much as stillness within an ecosystem means death.

The self-regulating mechanisms originate in a chain of interdependences binding together all the ecosystem's components, interdependence is the fundamental principle of the universe, and the chain, input–transformation–output, is the norm. Man, whose principal biological characteristic is his ability to modify his environment, generally tries to remove himself from this chain of interdependence by simplifying the ecosystem; but this way of ignoring the problem only creates others. The examples are endless in the most diverse field. If we pump too much water from the ground beneath cities we will deplete the aquifer and as a result slacken the process of urbanization; if we pollute the atmosphere and the water, the cost of purifying both slows down industrial growth which is the basis of our development; if we neglect agriculture in favour of industry, the importing of foodstuffs consume the foreign currency needed for industrial imports; if industry takes more services and infrastructures from urban areas than it can give back, they become impoverished and will hamper the industrial activities; if we do not provide enough funds for educational purposes, our communities will lack political cohesion and, insufficiently informed, make unrealistic demands which strain the fabric of society still further and endanger the model of life; and so on.

Industrial enterprise too often cooperates in this refusal of a systemic approach to reality, with grave consequences for its activity, as it is shown by the above examples and the many others that could be quoted.

From what we have already seen it is obvious that neither component of the ecosystem—whether it is the organism or the environment—can be favoured without eventually damaging the other, because of the self-regulating mechanisms within the ecosystem. This suggests the concept of an integrated managing of the ecosystem: in other words it is necessary to emphasize the overall functioning of the ecosystem rather than of any single component within it, because it is only in this way that these components can really thrive (Figure 30.2). Along this line of thought there are three important courses of action to be taken into account.

The first of these is the identification of the fundamental needs of each ecosystem component without which the whole ecosystem would sooner or later lose its identity. For example in the case of the environment the following needs should be taken into consideration: the maximization of the use of renewable resources; the minimization of the use of non-renewable resources; the minimization of irrevocable choices (i.e. the reduction of the risk of irrevocable errors); the respect due to the specific role of each component of the environment (territory, subsystem etc.); a gradual change of equilibria; a rational utilization

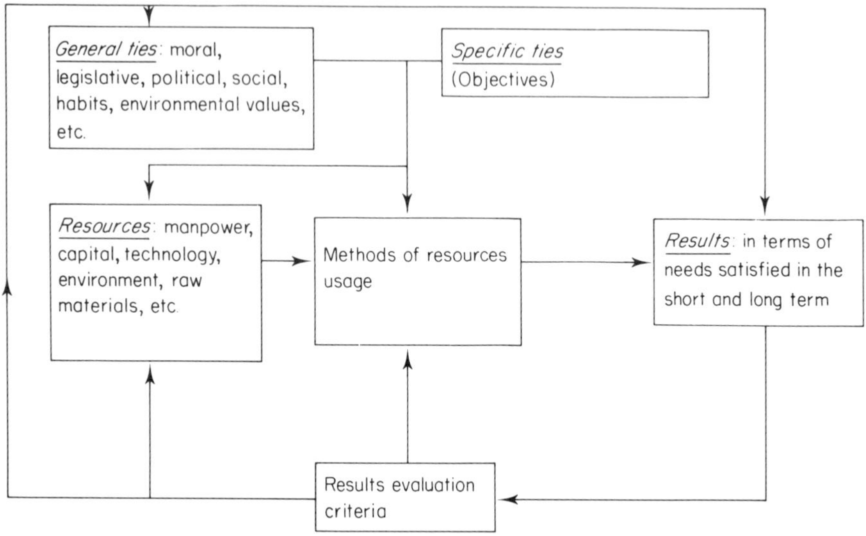

Fig. 30.2 Integrated management of urban ecosystem.

of energy levels (for example: the production of a vegetable diet requires ten times less solar energy than the production of a diet of meat).

As far as man is concerned, on the other hand, we may quote: the rising above the 'poverty level' which too often limits the choices of men; the respect due to innate characteristics produced by the evolution of the species; the respect due to the diversity (technical, cultural, social, etc.) which should be considered as a source of innovations indispensable to an evolving world; the respect due to traditions, patient accumulation of expertise over the centuries; a consideration of man as a species and not only as an individual. As an explanation of this last point it may be remembered that natural selection tends to promote qualities which are useful to the species (such as solidarity), whilst social selection tends to promote characteristics useful to the individual (such as aggression). As a consequence the even fiercer social selection that we are fostering at the moment tends toward a weak collectivity composed of efficient individuals.

The second of the three lines of action concerns the need to foresee not only the immediate consequences of any act that may effect the ecosystem but also the long-term ones. This is the field of application of all the methodologies to forecast the environmental impact of a project, the only way to foster the long-term objectives that are too often sacrificed for immediate gains.

The third line of action concerns the necessity, if the management of the ecosystem is to prove advantageous to all its components and therefore also to man, to take into account the desires and expectations of the population. This leads directly to the heart of one of the fundamental problems of the modern world: to the relationships between science and democracy. On one side there are specialists whose knowledge suggests decisions which sooner or later have repercussions on people; on the other hand there are people who have experienced the fallibility of these specialists and who know, moreover, that they must pay for the failure of their experiments.

Following this line of thought it should be noted, that often it is said that the failures of the modern world are a result of its complexity, a consequence of the fact that technological advances have out-stripped social, psychological and moral advances. Such delay, however, can also be expressed in terms of the desocialization of information, that is in terms of an increase of the amount and the complexity of information that is not accompanied by an equal drive towards its socialization. The solution of this problem lies in the democratic process and in the decentralization of the centres of power, as to make them more responsive to the mood and expectations of the population: a decentralized democratic power, in fact, falls better within the ecosystem logic, given that it better preserves the equilibria pursued by the homeostatic mechanisms.

If we examine our civilization in the light of the above considerations the reasons for the contemporary world crisis can be seen clearly: our society is based on the wasting of non-renewable resources, on social selection, on the depreciation of traditions, on the cancelling out of all variety, on brutal changes that are not adequately prepared, on the prevalence of short-term gains over long-term planning, on the autonomous proliferation of technology beyond the interests of the population, etc. In other words exactly the opposite of what is suggested by a rational approach to the problem. It is not surprising, therefore, that all political and social structures in the modern world are under pressure.

Coming to more concrete grounds from the point of view of industrial enterprise, all this means the need of respecting the specific inclinations of the different areas, not wasting natural resources, adopting integrated technologies, recycling and recovering products, using technologies appropriate to the specific social, economic and environmental situation etc. Furthermore it means respecting public opinion as the only way of making possible freedom of production to continue. Failure in this respect would lead, in fact, to public reaction which would eventually give rise to excessive limitations on the activity of the enterprise even in its legitimate areas.

Summing up, the modern management of enterprise requires a sensitivity to which we are not accustomed; it is the traditional concept of development—based only on efficiency—which is in a state of crisis, in that man refuses both the role of the arrogant conqueror of creation and that of the slave to machinery or to poverty, and longs instead for a unification of the natural and social sciences in an attempt to allow social and economic well-being and respect for the environment.

In other words the environment must turn into a component of the development, to be brought into all decision-making processes. The possibility of achieving this depends on man's capacity to build scenarios of the future and to choose and follow in a systemic way the best of them, abandoning the traditional approach by sector of Cartesian type.

Moreover it is based on accepting technology as an evolutionary force pertaining to nature; i.e. capable of contributing to the successful performance of the phenomenon 'life' on planet Earth, as an integration of the capacities of the biological tool that has prevailed up to now (for a fully exhaustive treatment of this subject see Meyer's book quoted in the bibliography).

In this framework it should be noted that the environment, for the system specialist, is all that can influence the behaviour of a system, despite the fact that it is outside it. As a consequence, in the case of the human ecosystem, an effective environmental policy will eliminate the same concept of environment, or better still will cause its getting interior within the ecosystem.

This apparent paradox offers the key for choosing between the different available policies, by means of a comparison of the social and ecological impacts of the alternatives from the point of view of the degree to which the getting interior can be achieved.

Along this road there are numerous different types of difficulty to overcome; as an example it is worth outlining a few.

Hysteresis of the socio-economical system

Many of the problems with which the modern world is presented develop exponentially, so that they are often already too serious when they first attract our attention, and there is too little time available for resolving them. This tendency is clearly illustrated in Figure 30.3 which shows how three-quarters of oil reserves will be consumed in the last quarter of time.

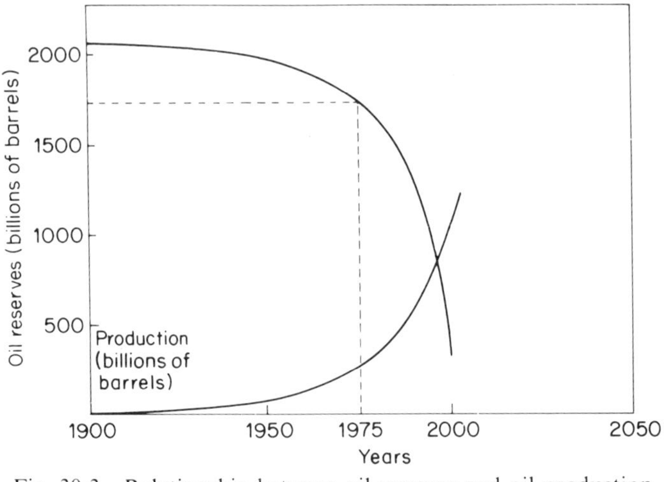

Fig. 30.3 Relationship between oil reserves and oil production

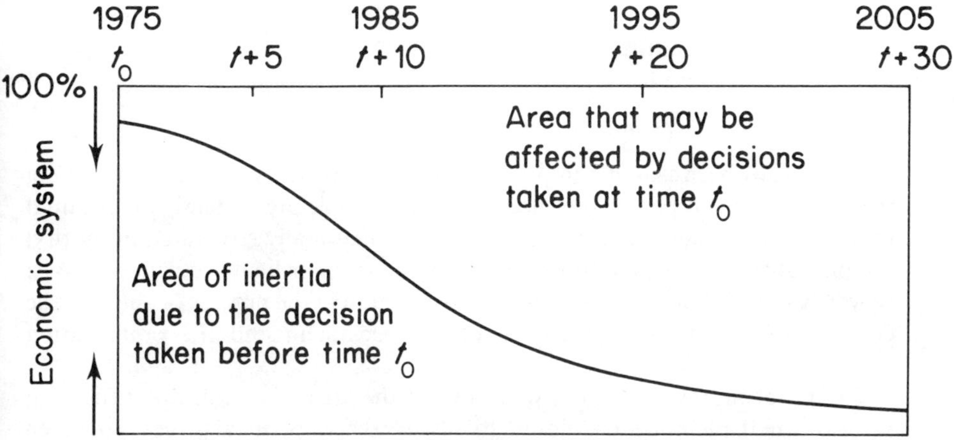

Fig. 30.4

The graph in Figure 30.4, on the other hand, illustrates the margin for free decision left to man as a consequence of the increasing complexity of our social system: only for the next 25 years do we have a completely free choice, the dictates of the past having already shaped the decision to be taken in the more immediate future.

The opposition of the two phenomena—the urgency of taking decisions and the impossibility of taking them—is at the root of the frustration felt nowadays, and the cause of the consequent escapism, often taking the form of irresponsible and useless projects.

This situation is further complicated by the expanding gap between our potential capacity to cause social and ecological catastrophes and our ability to foresee them (Figure 30.5).

The lack of a common language

The various elements involved in our social framework often do not consider problems from the same point of view. A typical case is that of the so-called 'free goods' (air, water, landscape, etc.) which do not even enter into the accounts of industry, whilst people regard them as irreplacable wealth to be tapped with great circumspection.

As a practical example, it is worth remembering the many millions of tonnes of chemical gypsum dumped into the sea each year, whilst at the same time a similar amount of gypsum is quarried nearby. The cost of the natural gypsum is certainly lower than that of the chemical one once it has been purified, but this would not be the case if the cost of relevant environmental damage was added to the first (damage to the landscape, use of the sea, etc.). Lessons could certainly be learnt from the fact that authorities who authorize the mining operations do not realize that their narrow perspectives create the conditions in which gypsum wastes should be dumped into the sea.

The problem of power in a systemic world

As has already been said, it is impossible to manage the ecosystem without taking the expectations of people into account, since people are certainly an essential part of the ecosystem. This would require a horizontally structured authority, not only able to override sectional interests and to co-ordinate activities across a very broad field, but also to involve the population in the process of defining the community's needs, the potentials of the ecosystem and the organization necessary to develop it.

The problem obviously goes way beyond the area of a centralized (or even decentralized) planning procedure to involve basic political issues, so much discussed throughout the world but still far from a practical solution (see, for instance, the so-called 'Tocqueville effect' i.e. the increasing of resentment as inequality decreases). Naturally the problem is present also within industry, where it is taking the shape of a search for new organizational structures and new decision-making procedures.

The complexity of the environmental problem

The environmental problem, by definition, considers 'the total interests of the total man in a total world' (Vincent Ostrom); at the same time it is clear that such an omniscience cannot be achieved and therefore it is obvious that man must be satisfied with tackling the limited sector of reality that the intrinsic capabilities of his species allow him.

In this respect the environmental problem seems—both theoretically and practically—particularly difficult to face because of its tendency to complicate and give rise to unexpected reversals that throw out the most carefully prepared plans and theories.

The attitude of mind trending toward a single and definitive solution seems particularly open to mistakes, in that it overlooks the complexity of the situation and fosters the illusion that a world may exist where difficulties have been suppressed.

As far as industrial enterprise is concerned the complexity of the environmental problem is reflected in the ever increasing number of regulations to observe (Figure 30.1) which is the result of a compromise between an increasing pressure for the defence of the environment and ever rising prices of such defence (Figure 30.5). One practical example of the complexity of the environmental problem is to be found in the case of fluorocarbons, used mainly for refrigerators and sprays.

The theory of Rowlands and Molina formulated in 1974 suggests that there is a

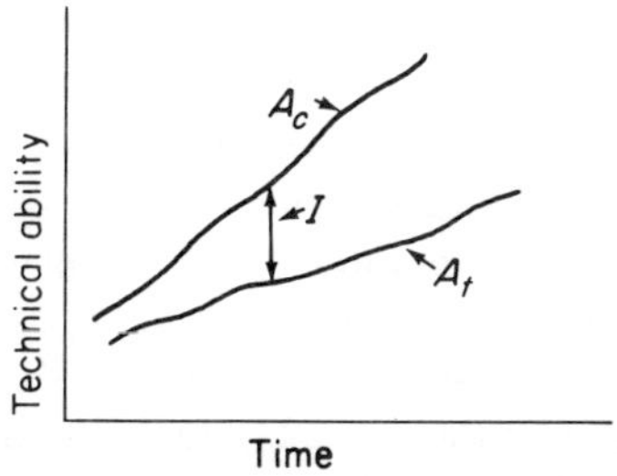

Fig. 30.5 Expanding gap between our ability to change A_c, and our ability to foresee the consequences of such change, A_t.

concentration of fluorocarbons in the upper atmosphere, where they would interfere with the ozone layer which limits the penetration of ultra-violet rays to the biosphere. In other words their low reactivity, which was one of the reasons they were chosen in the manufacture of certain products, would prove a double-edged weapon in that, through their persistence in the environment, it could increase the amount of skin cancer due to the ultra-violet rays.

This hypothesis naturally made a great sensation, particularly because of the triviality of the cause (sprays) and the gravity of the consequences (skin cancer); as a consequence a ban on fluorocarbons was requested by many people and different organizations. At the same time researches were conducted, the only clear results of which being to reveal how little was known about the physical and chemical properties of the upper atmosphere: substances were found there which we did not previously expect to find, and more and more sophisticated mathematical models were worked out, without getting a much better picture of things than before. These complexities even led some experts to suggest that the presence of fluorocarbons was beneficial for the equilibrium of the ozone layer, in that it counteracted the effect of carbon dioxide, which is increasing because of combustion and which lowers the temperature of the upper atmosphere and therefore increases the concentration of ozone within it.

In such a complex state of affairs the idea that scientific research could quickly resolve the dilemma—to ban or not to ban the fluorocarbons—rather than further complicate the matter, was sheer madness. The choice obviously had to be made according to political criteria. That is, either on the hypothesis that the play was not worth the risk and therefore no risk whatsoever was to be taken (this being the case of the US ban) or on the hypothesis that the problem was not sufficiently urgent to justify the negative effects of a ban on the economy (this being the case of the European position).

In both cases scientific research, too limited with respect to size of the problem, does not become an element of clarity but support of preconceived theses.

From all that has been said until now, it is clear that a healthy ecosystem is able to maintain its identity through a proper equilibrium between its components. In other words it must be sufficiently complex to allow its homeostatic mechanisms to operate efficiently, without excessive fluctuations, whilst materials used must be retained within the ecosystem rather than expelled permanently from its cycles. Seen from this point of view the industrial ecosystem seems to fluctuate too much as it is shown: by structural crisis of overproduction with their consequent social tensions (synthetic fibres); by waste of precious raw materials (methane used as combustible fuel); by industrial accidents of greater or lesser magnitude (Seveso, Flixborough, etc.); by pollution effects more and more difficult to foresee and increasingly complex (from toxic to cancerous, to mutagenic, to interfering with the natural cycles); by real struggles against certain types of industries (nuclear reactors); by diffuse sense of antipathy for the whole of industry (ecological movements) etc.

In general it can be said that the industrial ecosystem has hardly been able to confront its problems rapidly enough and/or with the necessary understanding of their dimensions, given that they arise from what is essentially an alien source, i.e. from cultural considerations. In other words man has found it hard to adapt himself to the fundamental revolution of the values that are at the same basis of our society, that is taking place before our eyes.

The Environmental Management

Industry's answer to the problems mentioned above can be collected under three main headings:

1 A change of attitude: i.e. to abandon the schizophrenia by which the environment has been traditionally viewed, which has led to the abuse of natural resources as producers and the deploring of this abuse as citizens.
2 Structural changes: i.e. to shape the enterprise's organization according to the above new attitude.
3 Functional changes: i.e. to shape the production process according to the preceding points (construction of purification plants, modification of other plants, etc.).

The first two points belong to the field of the so-called 'environmental management', a methodology which sets out to compare the aims of the organization with the expectations of the population, to identify potential areas of conflict and all the relevant actors, to evolve a strategy of avoiding conflicts or resolving them. Such strategy, where necessary, should include also the change of the aims of the organization when they cannot be reconciled to the population's wishes (see Figure 30.6).

The environmental management is made necessary by the fact that traditional industry has numerous and valid management tools at its disposal in the

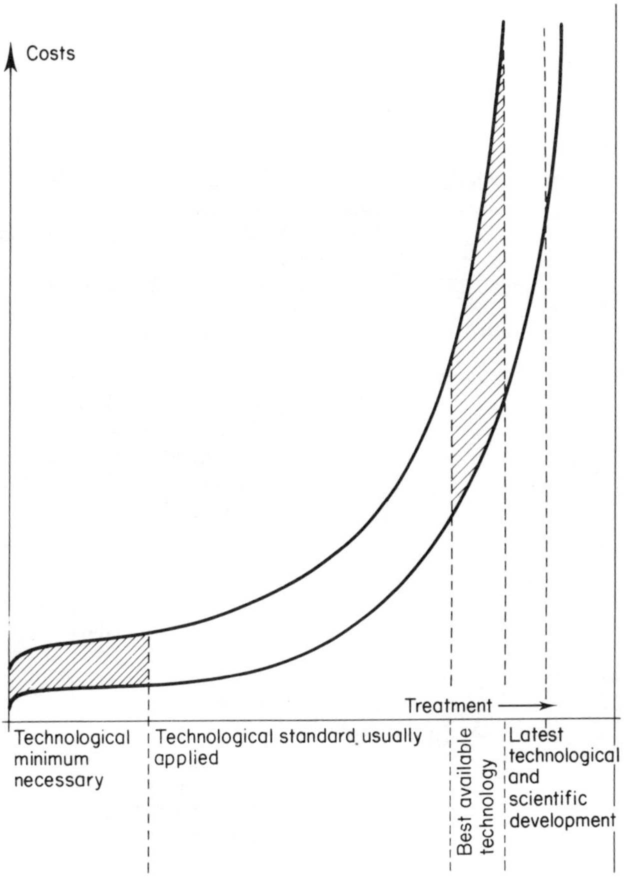

Fig. 30.6 Specific costs of treatment of waste waters

economic field; while it has no tools at all in the environmental and social fields where, as a consequence, it has been forced to adopt a passive and negative attitude which has been reflected in its economic performance.

The examples, even with reference only to Italy, are numerous: beyond the classic case of nuclear reactors, which continue to be delayed by the population's opposition despite receiving an official and lawful go-ahead; there are the bioprotein plants that have already been built and still await official approval to go into operation which is continually postponed; or the threatened ban of fluorocarbons because they are considered a threat to the ozone layer in the upper atmosphere; or still further there are numerous industrial plants, big and small, spread throughout the country, which have gone into production late or have been modified because of environmental demands which were not foreseen at the planning stage (see Table 30.8).

Table 30.8 Outcomes of 504 environmental conflicts relating to industrial projects

Final outcomes of conflicts		
Plants shut down or reduced in capacity		36 cases
of these	2 cases after block	
	2 cases after delay	
	2 cases after modification	
	1 case after delay and modification	
Programmes or projects modified (changed in physical structure or design)		99 cases
of these	2 cases after relocation	
	16 cases after shutting down or reduction in capacity	
	27 cases after delay	
	2 cases after delay and shutting down	
Programmes delayed (either in starting construction or operations)		209 cases
Projects cancelled or abandoned		50 cases
of these	20 cases after delay	
	11 cases after block	
	1 case after modification	
	11 cases after delay and block	
	1 case after delay and postponement	
Projects posponed or deferred		16 cases
of these	12 cases after delay	
	1 case after delay and block	
	1 case after delay and modification	
Projects relocated		47 cases
of these	11 cases after block	
	7 cases after delay	
	16 cases after shutting down or reduction in capacity	
	1 case after block and modification	
	4 cases after delay and blocks	
	3 cases after delay and cancelling	
	2 cases after delay, blocks and cancelling	
	1 case after delay, blocks and modification	
Projects or programmes blocked		47 cases
of these	15 cases after delay	
	1 case after delay and postponement	
No outcomes on projects or programmes		0 cases
Of these 504 conflicts (and allowing for multiple issues)		
	48 involved shutting down or reduction in capacity	
	332 involved delay	
	108 involved modification	
	93 involved block	
	17 involved postponement	
	42 involved relocation	
Consequently		
All cases involved economic effects		
A good half of cases involved substantial modifications of projects or activities		
A fifth part of the cases involved the abandonment of projects of activities		

The point of departure for environmental management is that the problems of the environment (in so far as they involve man's fundamental concerns) should be faced openly; only in this way can short-term interests, which often gain way because of immediate necessities, be reconciled with the long-term interests that are often, and in a sense understandably, overlooked, because the future always seems to be further away than it actually is.

There is nothing mystical about 'environmental management' in that its single ideas have been known to experts and often to non-specialists for some time. What is new and interesting, however, is their overall fusion into a coherent methodology which can be studied through the discussion of real cases. Basically, what is involved is the addition of a new dimension to the usual mix of management tools; with the view of facing social environmental problems in a rational quantitative manner. At its best it will request a continuous dialogue between industry and society, with the purpose of comparing requirement and proposal of the first with those of the latter considered both as a whole and as relevant components.

What must be stressed above all is that this new approach should not be used as a means of realizing more easily the will of an industrial organization in regard to certain projects, but as a way for industry to become more responsive to the demands made upon it by society. If this were not the case 'environmental management' instead of fostering the equilibrium in the relationships between man and its environment would itself be a source of mystification and of further confusion.

It is similarly essential that industry does not overstep its limits, since enterprise's sociality should not clash with that of other organizations and/or institutions.

Introducing environmental management techniques into traditional industry is neither easy nor painless, and generally cannot be done unless the fourth stage of the customary reaction to the discovery of the environmental problem has been reached. The first of these is to deny the existence of the problem; the second to emphasize that all negative situations, even those that are not directly relevant, are an unavoidable price to be paid in order to enjoy the advantages of the product and of the increased employment. The third stage is to manoeuvre in such a way as to quieten the opposition; while the final stage is to accept all the implications involved, and to give the problem the same care and attention that are given to traditional constraints (economic, legislative, etc.).

Once the decision has been taken to introduce environmental management procedures (usually accompanied by the pronouncement of an official political line), it is quite common that two contradictory processes are experienced; on the one hand the desire to respond to the environmental challenge in the most correct fashion can interfere with the efficiency of the industry as a supplier of goods and services; on the other hand, the tendency to improve and modernize production can hinder the way the industry responds to the environmental challenge. This leads, especially in large decentralized industries, to a bureaucratic agreement on new environmental principles, which are then not put into full practice by the managers of single profit centres who know full well that their careers will continue to be judged in the traditional way, i.e. on sales and/or on productivity (see Ackerman's article cited in the bibliography).

The solution of these problems requires a revision of the way individual careers are judged, an up-dating of the training of personnel, changes in the organization, new procedures etc., none of which are easy to achieve and all of which depend on the type and structure of the industry concerned.

In general these problems can only be overcome by a strong commitment on the part of the company directors, accompanied by the strongest possible determination to transform the environmental issue from a problem into an opportunity; in other words, as has already been said, from a 'passive ecology' which involves high costs, the symbol of which is the purification plant, into an 'active ecology' which leads to profits, the symbol of which is recycling and the revision of production cycles to decrease pollution levels and, at the same time, boost income.

The importance of the role directors can play is demonstrated by the fact that investigations conducted in the USA and Canada show that 9 out of 10 of the 516 companies involved determined their environmental policy at board or president level. Further confirmation of this can be seen from the fact that in 96 per cent of cases environmental decisions are taken at head office and in only 3 per cent at individual factories. Moreover about 40 per cent of the above companies based their decisions on environmental policies clearly defined in relevant documents, ranging from declarations of principle to detailed procedures relevant to the whole range of the companies' activities.

The organizational structure in the environment area is generally more influenced by the type of industry than by the size or operational complexity of the plants: it is obvious in fact that a company with many factories and a great variety of products needs a more systematized political outlook than a company which, however important, has only one factory and produces only one product.

There are two basic organizational solutions possible: one in which the direction and definition of the environmental policy of the company is given to a specific office, and one in which this task is given to a committee or a task force. The essential difference between these two solutions consists in the way in which the policy relating to the environment is coordinated with the rest of the company's activities. This is a basic problem particularly for companies with some importance, whose environmental policy will necessarily involve many activities far beyond those of the simple struggle against pollution. In the first case, therefore, the office in charge of the environment is also in charge of the contacts with all other offices involved in the problem and with the world outside the company.

This explains why this structure is frequently headed by someone near to the top level of the company, 78.5 per cent of cases (at vice-president level in American companies).

In the second case, on the contrary, a committee becomes the centre for co-ordinating all the elements involved in the formulation of an environmental policy and for ensuring that it is applied, whilst the most important decisions are reserved for the highest levels of the company, for which the committee acts as a consultative body.

In general it can be said that, in recent years, with a greater understanding and acceptance of the environment on the part of industry, this issue has become less and less an isolated problem to be considered alone, and more and more a part of the normal management. This process has requested a spreading of the cultural aspects of the environmental problem within the industrial bureaucracy and has led to the emergence of what is called the environmental co-ordinator (see Figure 30.7).

The teaching of the environmental management is based on the technique of cases: having outlined thoroughly the general field of its application and the knowledge required for its proper utilization, a series of concrete problems are examined; in order to research their solutions *in vitro*, in terms of identification of the areas of potential conflict and relevant actors and of formulating an overall strategic answer by the enterprise. In this way problems, which at first sight seem to escape to any rational approach, can be dealt with rationally and brought into line with the customary decision-making process of the enterprise. This method of rationalizing the problem is based on numerous techniques, some of which can

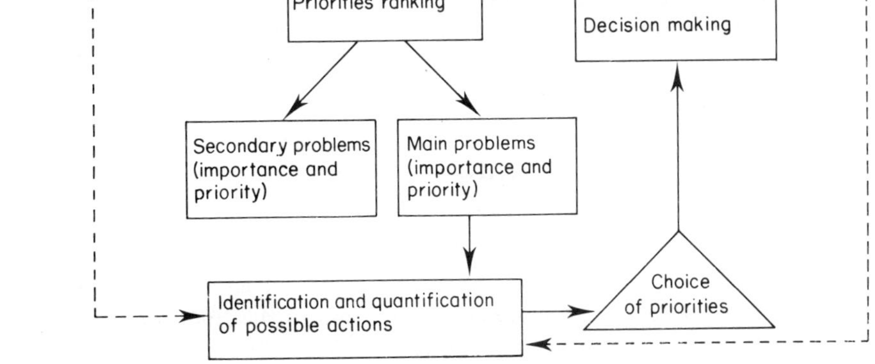

Fig. 30.7 Environmental management

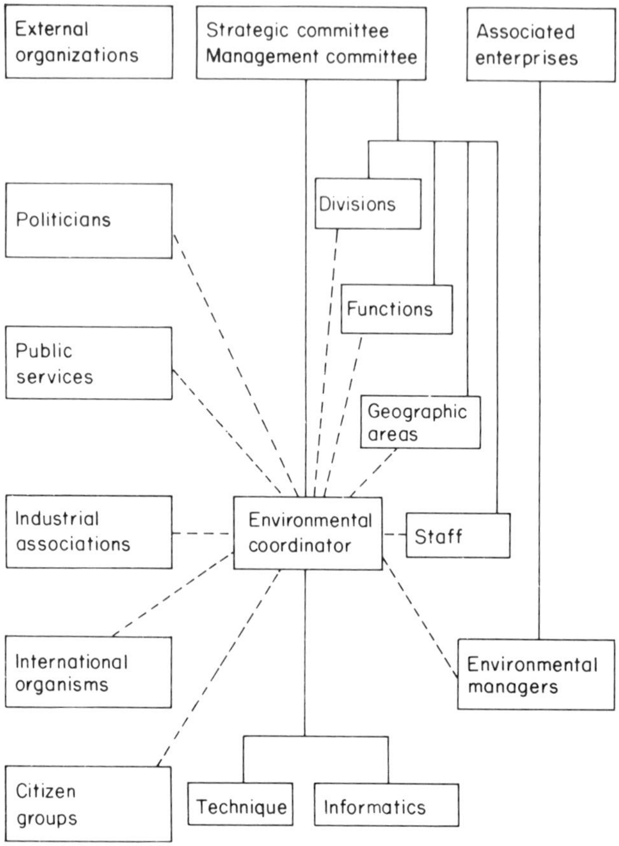

Fig. 30.8 The environmental coordinator role

be considered, even if only conventionally, as pertaining to the planning stage (environmental impact statement, risk analysis etc.) whilst others are used in the final balance stage (social–environmental balance etc.).

THE ENVIRONMENTAL PROBLEM AND THE MAIN EVOLUTION OF THE MODERN WORLD

As often happens in the case of innovations of a methodological or a behavioural kind, the theoretical definition of environmental management was preceded by a whole series of sectorial hints (no waste technology, impact analysis etc.) which were foretelling a better and more coherent vision of the enterprise.

From the conceptual point of view the most important of these starting points was without doubt the distinction between, (a) industrial activities orientated towards production and the maintenance of sales levels at all costs (i.e. activities closed around their own problems), and (b) industrial activities geared to the market (i.e. those open to the requirement of the external world). Such distinction is as much important as the one between crop gatherers and farmers, since both are based on the idea of sowing for harvesting, i.e. on the management of the future.

A market orientation, seen in terms of the environment, is certainly the best means of using to advantage and coordinating the different methodologies already used in the various sectors of the enterprise and relating either directly or indirectly to the world outside it. In such framework the three fundamental tasks of marketing (identifying, stimulating and satisfying the demand) must be understood not only in the traditional term of consumer demand but also in terms of environmental demand which, until recently, has not been much in evidence. In Table 30.9 some of the above-mentioned methods are listed by sectors of activity of the enterprise; it should be noted that all of them pay close attention to the external projections of the company and to the needs of those with whom it has contact. Some of these methods have so far only been considered in a traditional light, without identifying specifically their

Table 30.9 The tools of environmental management

Research and development	appropriate technologies clean technologies value analysis energy cost analysis cost/benefit analysis key factor analysis participatory planning biological analysis of products
Building	impact audit risk analysis industrial landscaping
Purchase	environmental product analysis
Production	clean technologies appropriate technologies environmental monitoring (inside and outside works) emergency plans environmental control plans environmental product analysis
Logistic	risk analysis
Marketing	image audit social budget environmental oriented marketing
Organization	environmental planning of careers environmental training

connections with the environmental problem; it is enough to think here of value analysis. But even in this case the change of emphasis away from limiting the costs of production and sales towards defining the minimum costs necessary to satisfy a function requested by market demand (i.e. from an introvert point of view to an extrovert one) is such that this methodology automatically falls within the area of environmental management as defined above.

If we accept, therefore, the principle of reacting with efficiency to the stimulation of the outside world as an essential characteristic of a healthy company—a company that is capable of producing added value—it follows that the greater or the lesser attention it pays to environmental management and to its methodology is a good indication of how well the company is in fulfilling its role. It is not a question, therefore, of having to suffer ecology for its economic, political and social effects on the industrial enterprise, but of making the qualitative shift of accepting a wider management philosophy that will involve, among other things, a greater attention to ecology. It should be noted that, while this new approach is growing within industry, the outside world's acceptance of industry's aims is also improving (often to a much greater extent than is true the other way round), with the result of putting to a lot of trouble managers used to traditional industrial procedures. And this is not so much because they are acting as scape-goats for a power considered to be intrinsically evil—as they would like people to believe—than because they are professionally inefficient.

In other words, traditional-styled managers can no longer hide themselves behind the claim that the population is making 'absurd emotive demands' and 'is not prepared to pay the necessary price for progress'. Instead it has to respond for its obvious inability to satisfy these demands, within a world that accepts, to a far greater extent than is sometimes suggested, the vital contribution that industry makes to the quality of life. In short the crisis is not one of technology *per se*, but of its management.

This situation has widespread consequences also in the public relations field, because the population accepts more and more reluctantly the connection between the advantages of a product and the disadvantages involved in producing it or using it (if you like this, you must have that as well!) and looks to technology and those who manage it for a positive solution to the dilemma. ('What kind of an expert are you if you can't solve the problem?') In this situation not even the coming era of underfed cows and the consequent crisis of the development myth and the strengthening of the traditional, severe view of life is of any use to a manager who is environmentally incompetent; given that external difficulties do not act in his favour but underline the need for a new and more efficient approach to the problem, better suited to the circumstances.

Another consequence of an environmentally minded market orientation is the tendency of ecology to be used as a competitive weapon: by this it is meant not the generally unsophisticated kind of propaganda which aims to highlight the real or imaginary environmental advantages of a product over those with which it is in competition, but the much more important strategy of identifying the 'environmentally weak' products of other producers and of replacing them with others which do not have this drawback. To the watchful observer these situations are revealed by the unexpected appearance, from various sources, of information emphasizing the negative aspects of a certain product and which had hitherto circulated only in closed circles and in specialized publications. This subject varies widely according to the imagination of different business operators, and according to specific cases, but in any case should be borne in mind at the planning stage of business activities.

Many of the points made above are often more relevant to large companies than to small ones, which can sometimes operate on a day-to-day basis having a shorter lifespan which can allow them to avoid some of the long-term consequences of their actions. On the other hand their greater flexibility, which makes them less prone to the myth of the called 'scale-economy' (which sometimes means a waste of resources on a bigger scale) allows them to react better to circumstance as they arise, so as to find an 'ecological niche' more easily. In other words they are particularly suited to a complex world such as the one we are living in.

This has been expressed by the successful formula 'small is beautiful' which, apart from its public relations value, has a grain of truth and is worth underlining also from this point of view. It is worth remembering here that a systemic management of the enterprise, aimed at complying with the relevant homeostatic mechanisms, takes out the sting of the accusation of stasis and of the abandoning of the pursuit of innovation without which industrial enterprise cannot survive.

The answer to this criticism comes in inserting all that has so far been said about the subject into a still broader context: that is, the context of the characteristic intellectual development of modern times which tends away from reductivism, an analytic method of understanding the world in terms of cause and effect, and towards a vision of the world, compatible with it but much broader, of an expansionistic kind, which focuses not so much on the components of reality as on the complex whole. The starting point of such an approach is the consideration that a set is always displaying some peculiarity not shown by its components, so that if we consider the first as a mere addition of the latter we loose a dimension of the reality. The transition to the new interpretation took place in gradual stages: to an universal and hierarchical model of scientific logic, with no casual interactions (nature as order, i.e. the behaviour of the soldiers) was added a probabilistic model (nature as homogeneity, i.e. the behaviour of the trade unionist) and a homeostatic one (nature as equilibrium, i.e. the behaviour of the politician). Only recently, and particularly as a result of work done in the biological and social sciences, has a fourth model been added. This tends to emphasize heterogeneity as an effect of positive feed back: it is the so-called morphogenic logic of reciprocal not deterministic causality (nature as diversity). Among its many implications, this model of logic suggests that it is necessary to take account of non-hierarchical interactions within the eco-system. It also emphasizes the importance of keeping in mind that, due to the presence of positive feed-back, different experimental results do not necessarily correspond to different starting conditions; furthermore it underlines that diversity is not to be interpreted in terms of superiority and inferiority, nor necessarily in terms of conflict. In short it is a view which connects rather than separates things.

The consequences of this logic are enormous (it is not accidental that it has been considered the greatest intellectual revolution for 2500 years), but far from having been explored. Especially because too often it is difficult to get free from the traditional ways of thinking. For example, the ecological movement, which is in many respects one of the most cherished offspring of the new logic, has in many cases transformed the evolutionary harmony of natural system into pure conservationism. Or still worse, it has criticized the biblical exhortation to cultivate and preserve the Garden of Eden (interpreting it as an invitation to exploit nature) rather than accepted it as an invitation to cooperate to the world's creation.

One of the main consequences of the new line of thinking has certainly been the revaluation of the teleological approach with respect to the mechanistic one in other words of the tendency to explain certain phenomena by concentrating on what they have produced (or tried to produce) rather than on what provoked them. This puts the accent on the future and re-evaluates our freedom of choosing between the different answers available to us, according to the logic of diversity, i.e. recovers our power to plan ahead.

As a consequence we may now choose, in a rational way, between the four attitudes available, according to Ackoff, in the realm of planning.

1 Inactive: unsatisfied with the present and its evolution, we follow the current passively in the hope of not aggravating the situation (the conservative position).
2 Reactive: preferring the past to the present and the future, all efforts are made to go back to it swimming against the current (the reactionary position).
3 Preactive: satisfied with the way things are progressing, but not of the present situation, we prepare for the future, exploiting the current as much as possible; i.e. seeking modifications within the system rather than of the system (the liberal position).
4 Interactive: unsatisfied both by the present and the future we try, irrespective of where the current is moving, not so much to prepare for but to change the future (the radical position).

The logic of nature as diversity tends, of course, to encourage the last of these approaches: as a consequence ideals are considered not as utopian abstractions but as an essential phase of the process of planning, to be constantly revised. Furthermore the field of planning encompass the whole reality and not only some restricted parts of it: while humanities and science are two sides of the same culture (rather than being two different cultures) both of which are indispensable. Accordingly, technology is seen as neither intrinsically good nor bad, but good and bad according to how it is used.

All these factors give the interactive approach a particular interest and importance, both because it enables a better understanding of the problems faced by the modern world (since it shares with them the same logical roots) and because it represents a positive approach to the future, i.e. a re-evaluation of man's role within nature. It is important to note that in such a picture technology is the main tool given to man by evolutionary forces to allow him to assert himself on planet earth; and this is because it is the most flexible variable in the complex process of integrating social, economic and environmental objectives by which this assertion is reached.

The concept of 'ecodevelopment' follows logically from this; that is the utilization of the specific natural resources of an area to satisfy needs of its population, established in realistic terms and independently of other regions.

Ecodevelopment is the practical application of the technological flexibility underlined above: i.e. technology should depend on circumstances in such a way that for each specific situation there is an 'appropriate technology'. This approach supersedes the classical concept of economic growth through industrialization at any cost, according to which the produced income should miraculously spread itself amongst the population, and underlines instead the priority of satisfying the fundamental needs of the population. This emphasis on the indispensable needs of a community gives to the ecodevelopment a general significance that goes much farther than the case of developing regions, because the option between needs (not only of the many) and surplus (not only of the few) is present at every level of development.

Starting out from the economic, political, technological and social effects of the environmental problem we have arrived at theoretical considerations which may seem to be far removed from the practical world of industrial enterprise; this has been necessary to show that the 'ecology' is only the tip of an iceberg, and that the great changes taking place in the world today must be fully appreciated if they are not to have a disastrous effect on the future of the industrial enterprise. On the other hand, fully grasping them now would greatly smooth the road ahead.

It is not a fashion then, or a temporary phenomenon or an emotional excess to be exorcized in the hope that it will disappear as quickly as possible, but a fundamental necessity of our civilization to be faced by good managers utilizing all the instruments at their disposal in a comprehensive and fearless way.

Particularly because, it should be stressed again, the essential role of technology discussed above gives industry, as a place where technology and reality come together, a central position with respect to the problems of the modern world, so that there is no alternative to managers then to confront themselves with the task.

Subject Index